Biofoundry Techniques for Biotechnology Applications

Biofoundry Techniques for Biotechnology Applications

Edited by

Anuj Kumar Chandel
Renewable Carbon and Biology System Laboratory-ReCABS
Department of Biotechnology
Engineering School of Lorena, University of São Paulo (EEL-USP)
Lorena, São Paulo, Brazil

WILEY

This edition first published 2026

Registered Offices
John Wiley & Sons, Inc., 111 River Street, Hoboken, NJ 07030, USA
John Wiley & Sons Ltd, New Era House, 8 Oldlands Way, Bognor Regis, West Sussex, PO22 9NQ, UK

For details of our global editorial offices, customer services, and more information about Wiley products visit us at www.wiley.com.

The manufacturer's authorized representative according to the EU General Product Safety Regulation is Wiley-VCH GmbH, Boschstr. 12, 69469 Weinheim, Germany, e-mail: Product_Safety@wiley.com.

Wiley also publishes its books in a variety of electronic formats and by print-on-demand. Some content that appears in standard print versions of this book may not be available in other formats.

Library of Congress Cataloging-in-Publication Data Applied for:
Hardback ISBN: 9781394309924

Cover Design: Wiley
Cover Images: © sevector/stock.adobe.com, Picture Office/stock.adobe.com, Stockr/stock.adobe.com, evegenesis/stock.adobe.com

Set in 9.5/12.5pt STIXTwoText by Straive, Pondicherry, India

Contents

List of Contributors

Leila Maria Aguilera Campos
Postgraduate Program in Chemical Engineering
Federal University of Bahia (UFBA)
Federação
Salvador de Bahia, Brazil

Shruti Ahlawat
Department of Microbiology
Faculty of Allied Health Sciences
SGT University
Gurugram
Haryana, India

Jesús Jiménez Ascencio
Department of Biotechnology
Engineering School of Lorena
University of São Paulo (EEL-USP)
Lorena
São Paulo, Brazil

Gabriel Baioni
Graduate Program of Chemical Engineering
Federal University of São Carlos
São Carlos
São Paulo, Brazil

Nagamani Balagurusamy
Laboratorio de Biorremediación
Facultad de Ciencias Biológicas
Universidad Autónoma de Coahuila
Torreón, Coahuila, México

Henrique Macedo Baudel
SENAI Bom Retiro Biotecnologia
São Paulo, São Paulo
Brazil; Bioenergy Department
America Biomass Technologies Division
Brix Brasil, São Paulo
Brazil; Department of Chemical Engineering
Federal University of Pernambuco (UFPE)
Recife, Pernambuco
Brazil

Sreelatha Beemagani
Department of Microbiology
Chaitanya (Deemed to be University)
Hyderabad
Telangana, India

Anuj Kumar Chandel
Renewable Carbon and Biology Systems Laboratory-ReCABS
Department of Biotechnology
Engineering School of Lorena
University of São Paulo (EEL-USP)
Lorena
São Paulo, Brazil

Mónica Ma. Cruz-Santos
Department of Biotechnology
Engineering School of Lorena
University of São Paulo (EEL-USP)
Lorena
São Paulo, Brazil

Adilson José da Silva
Department of Chemical Engineering
Federal University of São Carlos, São Carlos
São Paulo, Brazil;
Graduate Program of Chemical Engineering
Federal University of São Carlos
São Carlos, São Paulo, Brazil

Alisson Dias da Silva Ruy
Postgraduate Program in Chemical Engineering
Federal University of Bahia (UFBA)
Federação, Salvador de Bahia
Brazil

Paula Zaghetto de Almeida
SENAI Bom Retiro Biotecnologia
São Paulo, São Paulo, Brazil

Juliana Velasco de Castro Oliveira
SENAI Bom Retiro Biotecnologia
São Paulo
São Paulo, Brazil;
Postgraduate Program in Genetics and Molecular Biology
Institute of Biology
University of Campinas (UNICAMP)
Campinas, São Paulo, Brazil

Christian de Oliveira Martins
Graduate Program of Chemical Engineering
Federal University of São Carlos
São Carlos
São Paulo, Brazil

Allan H. Félix de Mélo
SENAI Bom Retiro Biotecnologia
São Paulo
São Paulo, Brazil; Postgraduate Program in Food Engineering
Faculty of Food Engineering
University of Campinas (UNICAMP)
Campinas, São Paulo, Brazil

Yaimé Delgado-Arcaño
Center of Studies for Sustainable Development
ECOTEC Technological University 10 (CEDS/ECOTEC) Samborondón
Guayas, Ecuador;
Center for Renewable Energy Technology Study (CETER)
Technological University of Havana (CUJAE)
Faculty of Mechanical Engineering
Marianao, Havana
Cuba

Felipe Fernando Furlan
Department of Chemical Engineering
Federal University of São Carlos
São Carlos
São Paulo, Brazil;
Graduate Program of Chemical Engineering
Federal University of São Carlos
São Carlos, São Paulo
Brazil

Fabricio M. Gomes
Department of Chemical Engineering
Department of Biotechnology
Engineering School of Lorena
University of São Paulo (EEL-USP)
Lorena, São Paulo
Brazil

Somya Gupta
Department of Life Sciences
Faculty of Allied Health Sciences
SGT University
Gurugram, Haryana
India

Diego Hernandez
Departamento de Ciencias Biológicas
Universidad de los Andes
Bogotá
Colombia

Nevena Ilić
Innovation Centre of the Faculty of Technology and Metallurgy
Belgrade, Serbia

Prisha Inani
Department of Biotechnology
College of Engineering and Technology
Faculty of Engineering and Technology
School of Bioengineering
SRM Institute of Science and Technology
Chengalpattu
Tamil Nadu, India

Samuel Jacob
Department of Biotechnology
College of Engineering and Technology
Faculty of Engineering and Technology
School of Bioengineering
SRM Institute of Science and Technology
Chengalpattu
Tamil Nadu, India

Ana Paula Jacobus
SENAI Bom Retiro Biotecnologia
São Paulo
São Paulo
Brazil; São Paulo State University (Unesp)
Institute for Research in Bioenergy
Rio Claro
São Paulo, Brazil

Galina Jevđenović
Faculty of Technology and Metallurgy
University of Belgrade
Belgrade, Serbia

Simab Kanwal
High-Value Food from Mushrooms and Bioactive Plants in the Green Economy Value Chain Research Group
The Institute of Biotechnology and Genetic Engineering
Chulalongkorn University
Pathumwan, Bangkok, Thailand

Aphichart Karnchanatat
Center of Excellence in Bioconversion and Bioseparation for Platform Chemical Production
The Institute of Biotechnology and Genetic Engineering
Chulalongkorn University
Pathumwan
Bangkok
Thailand

Andreza A. Longati
Department of Chemical Engineering
Federal University of São Carlos
São Carlos
São Paulo 13565-905
Brazil; Institute of Science and Technology
Federal University of Alfenas
Poços de Caldas
Minas Gerais
Brazil

Carlos Lopez-Ortiz
Gus R. Douglass Institute and Department of Biology
West Virginia State University
Institute, WV, USA

Sounak Maitra
Department of Biotechnology
College of Engineering and Technology
Faculty of Engineering and Technology
School of Bioengineering
SRM Institute of Science and Technology
Chengalpattu
Tamil Nadu, India

Giovani Maltempi-Mendes
Renewable Carbon and Biology Systems Laboratory-ReCABS
Department of Biotechnology
Engineering School of Lorena
University of São Paulo (EEL-USP)
Lorena, São Paulo
Brazil

Katarina Mihajlovski
Faculty of Technology and Metallurgy
University of Belgrade, Belgrade, Serbia

Thais Suzane Milessi
Department of Chemical Engineering
Federal University of São Carlos
São Carlos
São Paulo, Brazil;
Graduate Program of Chemical Engineering
Federal University of São Carlos
São Carlos, São Paulo, Brazil

Marija Milić
Faculty of Technology and Metallurgy
University of Belgrade
Belgrade, Serbia

José Daniel Cano Montoya
Departamento de Ciencias Biológicas
Universidad de los Andes
Bogotá, Colombia;
Departamento de Química
Universidad de los Andes
Bogotá, Colombia

Samatha Paladugu
Prameela Life Sciences Private Limited
Hyderabad, Telangana, India;
Department of Microbiology
Chaitanya (Deemed to be University)
Hyderabad, Telangana, India

Mamatha Potu
9MSS UK Ltd.
Aberystwyth, Ceredigion, UK

Carina Prado
Renewable Carbon and Biology Systems Laboratory-ReCABS
Department of Biotechnology
Engineering School of Lorena
University of São Paulo (EEL-USP)
Lorena, São Paulo, Brazil

Lucas Ramos
Renewable Carbon and Biology Systems Laboratory-ReCABS
Department of Biotechnology
Engineering School of Lorena
University of São Paulo (EEL-USP)
Lorena
São Paulo, Brazil

Ritika Yadav
Department of Microbiology
Faculty of Allied Health Sciences
SGT University
Gurugram, Haryana, India

Danielle Matias Rodrigues
Postgraduate Program in Bioenergy
Faculty of Food Engineering
University of Campinas (UNICAMP)
Campinas
São Paulo, Brazil;
SENAI Bom Retiro Biotecnologia
São Paulo
São Paulo, Brazil;
Bioenergy Department
America Biomass Technologies Division
Brix Brasil
São Paulo, Brazil

Uirajá C. M. Ruschoni
Department of Biotechnology
Engineering School of Lorena
University of São Paulo (EEL-USP)
Lorena
São Paulo
Brazil

Sher Zaman Safi
Department of Biochemistry
Faculty of Medicine
MAHSA University
Jenjarom
Selangor
Malaysia

Sarepalli Sai Sathwik
Prameela Life Sciences Private Limited
Hyderabad
Telangana
India; Department of Microbiology
Chaitanya (Deemed to be University)
Hyderabad
Telangana, India

Muskaan Sethi
Department of Biotechnology
College of Engineering and Technology
Faculty of Engineering and Technology
School of Bioengineering
SRM Institute of Science and Technology
Chengalpattu
Tamil Nadu, India

Krishna Kant Sharma
Laboratory of Enzymology and Gut Microbiology
Department of Microbiology
Maharshi Dayanand University
Rohtak
Haryana
India

Palak Shrivastava
Department of Biotechnology
College of Engineering and Technology
Faculty of Engineering and Technology
School of Bioengineering
SRM Institute of Science and Technology
Chengalpattu
Tamil Nadu, India

C. Shriya
Department of Biotechnology
College of Engineering and Technology
Faculty of Engineering and Technology
School of Bioengineering
SRM Institute of Science and Technology
Chengalpattu
Tamil Nadu
India

Messias Borges Silva
Department of Chemical Engineering
Department of Biotechnology
Engineering School of Lorena
University of São Paulo (EEL-USP)
Lorena, São Paulo
Brazil

Aldo Sosa
Laboratorio de Biorremediación
Facultad de Ciencias Biológicas
Universidad Autónoma de Coahuila
Torreón
Coahuila
México

Piroonporn Srimongkol
High-Value Food from Mushrooms and Bioactive Plants in the Green Economy Value Chain Research Group
The Institute of Biotechnology and Genetic Engineering
Chulalongkorn University
Pathumwan, Bangkok
Thailand

Maria Teresa Ferreira Ramos Raimundo
Department of Biotechnology
Engineering School of Lorena
University of São Paulo (EEL-USP)
Lorena
São Paulo
Brazil

Oscar Daniel Valmaña-García
Center of Studies for Sustainable Development
ECOTEC Technological University 10 (CEDS/ECOTEC), Samborondón
Guayas, Ecuador

Josman Velasco
Departamento de Ciencias Biológicas
Universidad de los Andes
Bogotá, Colombia

James Villar
Department of Biotechnology
Engineering School of Lorena
University of São Paulo (EEL-USP)
Lorena, São Paulo, Brazil

Vinícius Pereira Shibukawa
Department of Biotechnology
Engineering School of Lorena
University of São Paulo (EEL-USP)
Lorena, São Paulo, Brazil

Alan Chavez-Hita Wong
Laboratorio de Biorremediación
Facultad de Ciencias Biológicas
Universidad Autónoma de Coahuila
Torreón, Coahuila, México

About the Editor

Dr. Anuj Kumar Chandel is a professor in the Department of Biotechnology, Engineering School of Lorena, University of São Paulo, Brazil. He has over 20 years' research experience working on various aspects of industrial biotechnology (production of industrial enzymes, biofuels, and bio-based products), and membrane-based separation of fats, proteins, and viruses. Dr. Chandel did his Ph.D. in India in 2009 and then did three postdoctoral fellowships at Stellenbosch University, South Africa; Engineering School of Lorena, University of São Paulo, Brazil; and University of Arkansas, Fayetteville, USA, for almost 6 years. He has also worked at Dalas Biotech Ltd, Bhiwadi, India, for penicillin acylase production for 3 years and for cellulosic ethanol production at Centro de Tecnologia Canaviera (CTC)-Piracicaba, São Paulo, for 2 years. He has published 140 articles in peer-reviewed journals and 65 book chapters and has edited 19 books on various aspects of industrial biotechnology. He has recorded one Brazilian patent also. He has traveled to many countries like USA, Canada, Mexico, Japan, Turkey, and Ireland. Currently, Anuj is coordinating two research projects on lignin valorization into PHA, biolipids, biogas, and catalytic link funded by FAPESP-São Paulo State Funding Agency, Brazil.

Anuj Kumar Chandel, Ph.D., M.B.A.
Renewable Carbon and Biology Systems Laboratory-ReCABS
Department of Biotechnology
Engineering School of Lorena (EEL),
University of São Paulo (USP)
Lorena, São Paulo, Brazil

Preface

Lignocellulosic biomass is a widely available renewable feedstock for the production of biofuels, bioenergy, food/feed, commodity, and specialty biochemicals under the biorefinery concept. A continuous demand for food, feed, and fuels together with climate change and energy security concerns has necessitated the sustainable supply of energy and food/feed security. Taking this into consideration, this is the dire need of the hour to harness the potential of lignocellulosic biomass into these products sustainably at competitive prices. Now, the major industrial biochemicals and fuels producers like DSM, DuPont, BASF, Raizen, Novonesis, and several others are focusing on development of biofuels, biochemicals, and biomaterials

Recent biotechnological advancements have paved the way for lignocellulosic biomass conversion into bio-renewables at competitive prices, catalyzing the circular economy growth with lower carbon and water footprints. However, this success has been limited up to the laboratory, pilot, and demonstration levels. There are only a few large-scale plants producing cellulosic ethanol, organic acids, polyhydroxyalkanoates, and other biochemicals. Biomass collection, round the year availability, biomass handling at shop floor, and production cost of fuels and renewables are important issues in large-scale production. Biofoundry-based developments have shown tremendous growth in the last two to three decades, which enabled the development of novel designer biocatalysts for the production of biorenewables with desired yields and productivities growing on a variety of sugars even in the presence of various inhibitors. Though, development in biofoundry led to contributions in the economic production of food, feed, fuels, and chemicals at laboratory scale. However, the use of these biocatalysts at commercial scale and getting a consistency in renewables production is still a big challenge. Use of each fraction of lignocellulosic biomass (carbohydrates and lignin) assessing designer biocatalysts manufactured following the principles of biofoundry is the cornerstone of the success of lignocellulose biorefinery, which in turn minimizes the carbon and water footprints.

This book entitled *Biofoundry Techniques for Biotechnology Applications* is a research reference work presenting the cutting-edge scientific advancements and valuable insights in the biofoundry techniques aligning with lignocellulose conversion via biological transformation. The biofoundry area, in conjunction with bioprocess engineering, is getting significant attention worldwide as the conventional petroleum-based fuels and chemicals are not renewable and cause environmental pollution.

This book has 14 chapters focusing on various aspects of lignocellulose conversion into fuels, chemicals, materials, etc. **Chapter 1: Biomass Refining and Biofoundry: Key Products, Process Limitations, and Future Aspects,** by **Ramos and co-authors,** made a comprehensive appraisal on biomass refining approaches toward the production of the top 12 biochemicals selected by the US Department of Energy (USDOE) that have the potential to drive the bioeconomy and achieve net-zero carbon emissions. This chapter presents the developments in genetic engineering, mathematical modeling, and process simulation to harness renewable carbon sources in the top 12 biochemicals, fostering a circular economy aligned with sustainability pillars. **Chapter 2: Structural Carbohydrates Conversion into Sugars, Fuels, Chemicals, and Sustainable Materials** by **Mihajlovski et al.** fill in the gap of literature made in structural carbohydrates conversion into value-added products through biotechnological interventions. **Chapter 3: Integrating Lignocellulosic Biomass Processing, Biomanufacturing, and Biofoundries: Innovations and Challenges in the Bioeconomy** by **Delgado-Arcaño et al.** concludes the technological advancements through interdisciplinary collaboration and policy support in order to harness the full potential of biofoundry and biorefinery and innovation and strategic efforts to achieve sustainable development. **Chapter 4: Lignin Valorization Is the Key for a Win–Win Situation in a Biomass Refinery** by **Ramos et al.** elucidated the lignin conversion into value-based products through biological and chemical routes. Lignin is the second most available renewable carbon entity on Earth, having the potential of producing renewable chemicals (bulk and specialty chemicals), fuels, and materials. Furthermore, the integration of lignin conversion with carbohydrate is a key factor in the development of consolidating biorefinery with lower environmental impact.

Chapter 5: Sustainable Production of Advanced Alcohol-Based Biofuels in Biorefinery: From Alcohols to Sustainable Aviation Fuels, written by **Rodrigues et al.,** summarizes the valuable insights on biofoundry for the production of advanced alcohol-based biofuels, particularly ethanol, butanol, and 2,3-BDO, mostly used for SAF production, thus decarbonizing the bioenergy-based industry.

In **Chapter 6: Biomanufacturing of Smart Packaging Materials, Cosmetics, Therapeutics, and Nanomaterials Through Lignocellulosic Biorefinery Framework, Maitra and co-authors** elaborated the developments made in the production of biopolymers (PLA, PHA, and PHB) and synthesis of bioactive compounds for cosmetics, antibacterial agents for therapeutic purposes, and cellulose-derived nanomaterials for potential applications. In particular, this chapter appraised the innovative bioprocesses aligning biorefinery framework, together with environmental and economic benefits. **Chapter 7: White Biotechnology for Skincare: Unveiling the Power of Bioactives for the Cosmetic Industry** by **Paladugu et al.** critically assesses the market gap of fermentation-based cosmetic formulations including scale up, downstream processing, market of cosmetic bioproducts, safety and efficacy studies, regulatory documentation, licensing, and marketing. Metabolism profiles of lactic acid bacteria enable access to a wide variety of green bioactive compounds directly having applications in the food, pharmaceutical, textile, cosmetics, and other allied industries. **Paladugu et al.,** in **Chapter 8: Biotechnological Advancements in Lactic Acid Bacteria Fermentation: Metabolic Pathways and Metabolite Profiles,** provides an in-depth analysis of recent advancements in the metabolic pathways and profiles of lactic acid bacteria. Specifically, the chapter presents an overview on production of D-amino acids and polyols and the utilization of malic acid and citric acids. Furthermore, the chapter reviews the fermentation processes employing lactic acid bacteria, summarizing the production of organic acids, bacteriocins, vitamins, short-chain fatty acids, exopolysaccharides, and antioxidants. **Chapter 9: Biofoundry in Microbial Protein Production: Process Challenges and Future Scenario**, by **Kanwal et al.**, critically made an appraisal on the production processes and future advancements in genetic engineering and highlights the challenges in optimizing conditions for large-scale production and scaling up fermentation processes for microbial protein production. Technological advancements in improvement of biocatalysts for the production of microbial production employing the basic ingredients of synthetic biology and bioengineering at large scale are summarized in this chapter. **Chapter 10: Nanotechnological Interventions in the Advancement of Lignocellulose Biofoundry: Current Status and Future Prospects** by **Lopez-Ortiz et al.** reviews the role of engineered nanomaterials, catalytic nanostructures, enzyme immobilization techniques, and adsorptive nanoparticles in improving biomass conversion efficiency and reducing environmental impact, among others. Furthermore, the integration of nanosensors, synthetic biology, and machine learning with nanotechnology is analyzed in this chapter. Nanotechnology has the potential to achieve sustainable, high-yield biomass valorization within the framework of circular bioeconomy.

Synthetic biology-based methods within the larger scope of biofoundry have shown tremendous progress toward designer energy crops, designer cellulase producers, and microorganisms for desired metabolites production with satisfactory yields, productivities, while withstanding inhibitors. In this line, **Chapter 11: Synthetic Biology in the Realm of Genome Engineering for Improved Biocatalysts and Production** by **Montoya et al.** presents the methods to modify cellular genomes for the construction of cell factories, revolutionizing the production systems, enabling biologically based industries for the global economy of bio-renewables at competitive prices. **Chapter 12: Multi-omics Technologies Paving the Way for the Success of Biorefinery** by **Ahlawat and co-authors** summarizes competitively the latest advancements in the field of -omics approaches along with systems biology, synthetic biology, and genome editing for re-constructing and re-designing selective biological pathways. Particularly, the chapter is focused on the role of -omics approaches in decoding major genes and their manipulation through genetic engineering tools to develop potential microbial cell factories to produce value-added biochemicals and biofuels. Life cycle analysis and techno-economic analysis of biofuels and biochemicals is an important area for the acceptability of the technology at pilot or commercial scale. **Chapter 13: Sustainability Assessment of Genetically Engineered Biocatalysts Producing Biofuels and Biochemicals** by **Longati et al.** rolls out the potential of genetically modified organisms (GMOs) for the effective conversion of biomass into biofuels and value-added biochemicals. This chapter comprehensively assesses the economic, environmental, and social parameters directly addressing sustainability and security concerns and ensuring industrial viability.

Lean manufacturing, derived from the Toyota Production System, can be pivotal for the success of biorenewables industry. **Chapter 14: Lean Manufacturing Toward Minimum Waste Discharge and Potential Gains in the Biorefinery and Biotechnology Industries** by **Gomes et al.** summarizes the lean principles, including value definition, value stream mapping, continuous flow, pull production, and optimizing biotechnology processes. Lean principles can promote continuous improvement, operational excellence, and sustainable practices in biofoundry-based biorefineries.

This book is a unique book covering a broad aspect of biofoundry in the advancement of lignocellulose biorefinery. This book is the need of the students, academicians, researchers, bio-business analysts, and policy makers who are working on sustainable fuels, chemicals, and materials. The major readership includes undergraduate, graduate, post-graduate, and research students from academic institutions all over the world. Additionally, engineers, scientists, and technology analysts working in industries, institutes, and universities will also find this book interesting. In addition, policy makers, business analysts, and entrepreneurs will also find this book quite interesting. Investors and policy makers from the governmental bodies may find it useful for making policies, rules/regulations, and investments.

Anuj Kumar Chandel gratefully acknowledges the financial support from FAPESP (Project numbers: 2020/12559-6; 2022/13184-1) and CNPq for scientific productivity program (Process number: 304451/2025-8). Editor also thanks his research group—Renewable Carbon and Biological Systems Laboratory (ReCABS) at Engineering School of Lorena, University of São Paulo, Brazil, for their cooperation during the editing of the book.

All the chapters included in the book have been written by leading international experts from biorefinery research, synthetic biology, bioprocess engineering, and lean manufacturing. This book aims to disseminate the landmark information and recent advances made in the lignocellulose bioconversion in the realm of biofoundry. The book essentially appraises critical insights into microorganism-centric biorefinery, reviews current research, and presents future perspectives in this area. Editor (Anuj Kumar Chandel) is confident that the book will serve as a very useful source of information for researchers, academicians, and students of graduate and post-graduate level in the field of microbial biotechnology, applied microbiology, biochemical engineering, environmental science/engineering, and other related disciplines.

1

Biomass Refining and Biofoundry: Key Products, Process Limitations, and Future Aspects

Lucas Ramos[1], Jesús Jiménez Ascencio[2], James Villar[2], Mónica Ma. Cruz-Santos[2], and Anuj Kumar Chandel[1]

[1] *Renewable Carbon and Biology Systems Laboratory-ReCABS, Department of Biotechnology, Engineering School of Lorena, University of São Paulo (EEL-USP), Lorena, São Paulo, Brazil*
[2] *Department of Biotechnology, Engineering School of Lorena, University of São Paulo (EEL-USP), Lorena, São Paulo, Brazil*

1.1 Introduction

The vast variety of biomass types available on the planet presents opportunities for consolidating sustainable processes with low environmental impact. The use of biomass presents opportunities that support the circular economy and foster sustainable processes, generating social, environmental, and economic benefits. This biomass waste can be fully utilized, as long as the part of the biomass that does not compete with the food supply chain is utilized. In this regard, enhancing methods that use biomass is essential for decreasing carbon emissions and making these processes financially sustainable (Brienza et al., 2024; Wang et al., 2025).

Biomass refining is becoming increasingly relevant, with the entire biomass fraction being utilized for bioproduct production. The cellulosic fraction is used for sugar extraction through enzymatic processes, which minimize the formation of fermentation inhibitors. The extracted sugars are fermentable and can be converted into a wide range of bioproducts, mainly biopolymers and biofuels (Wang et al., 2025).

Another component of plant biomass that has been gaining attention is lignin. This polyphenolic macromolecule can be purified and fractionated into different molecular sizes depending on its intended application. After refinement, the aromatic compounds derived from lignin can be converted through chemical or biological pathways into various chemical intermediates (Lou et al., 2024). The method of extraction and stabilization of the lignin will directly impact the quantity and quality of the lignin-derived monomers. Research efforts will be necessary to make the process viable, from lignin extraction to its conversion into a bioproduct of interest.

In this context, technologies like pretreatment, enzymatic hydrolysis, fermentation, and gasification are improving, and a set of challenges must be overcome to maximize conversion rates and minimize energy spent. Additionally, process viability is a crucial

Biofoundry Techniques for Biotechnology Applications, First Edition. Edited by Anuj Kumar Chandel.

consideration; economic factors such as raw material availability, cost-effectiveness, and scalability must be addressed. The ability to integrate renewable biomass feedstocks and reduce dependency on fossil fuels makes these processes attractive, but significant improvements are needed to make them fully competitive with conventional chemical production (Chandel et al., 2020). Given this, the identification of the top 12 biochemicals by the U.S. Department of Energy (USDOE) represents a promising pathway for the sustainable production of key chemicals and will play a crucial role in the current bioeconomy, driving the production of clean and efficient energy to meet market demand. The integration of biofoundries into the lignocellulosic bioeconomy represents a paradigm shift in biomass processing, leveraging automation, computational design, and synthetic biology to engineer microbial systems for industrial applications (Chandel et al., 2021). Biofoundries facilitate rapid prototyping and optimization of microbial strains, and this approach significantly improves the conversion of lignocellulosic hydrolysates into high-value products, such as biofuels, platform chemicals, and specialty biomaterials (Asghar et al., 2022; Wang et al., 2025).

Given these challenges and proposed solutions, the way to achieve the decarbonization of the planet is through collaborative research efforts, government involvement, and the participation of large companies. Cooperation between sectors is essential for achieving net-zero carbon emissions and biofoundry development (Liang et al., 2023; Wang et al., 2025). This chapter comprehensively discusses the biomass diversity and biomass refining methods (both lignin-first and sugars-first approaches), as well as the basic principles and applied aspects of biofoundry toward sustainable biorefinery development.

1.1.1 Biomass Diversity and Major Principle Feedstock in the World

Biomass is defined as organic matter derived from plants, algae, or animal manure that can be used as a source of energy, including wood, agricultural residues, forestry residues, and urban waste (Dhamodharan et al., 2019). Biomass can be further categorized based on its origin into microbial, plant-based (agricultural and forest), and animal-derived biowaste. The diversity of biomass and major raw materials in the world is strongly influenced by the availability and type of lignocellulosic resources, which include agricultural and forest residues. The main lignocellulosic materials include corn, sugarcane, wood, and agricultural residues, which are key sources for the production of biofuels (Velidandi et al., 2023). A significant increase in lignocellulosic resources is expected, according to the Billion-Ton Study. The current availability is estimated at approximately 365 million dry tons, and it is projected to increase to 826 million dry tons by 2040 (Langholtz et al., 2016).

The treatment of large quantities of solid waste and wastewater from urban activities (domestic, municipal, and industrial) poses significant challenges. These wastes, often dumped or sent to treatment plants, cause pollution but are rich in nutrients and could be used to produce energy and chemicals. These biomass feedstocks can be transformed into bioenergy through a variety of processes and conversion technologies. Biomass undergoes several processes to produce a variety of products, which are classified into two categories: energetic products, as primarily fuels, and non-energetic products, which include biochemicals (Kumar & Verma, 2020). Energetic products are alternatives to conventional fuels like firewood, petrol, and diesel. They are categorized into solid, liquid, and gaseous fuels. Non-energetics from the decomposition of waste biomass can result in the generation of various value-added chemical products (Kumar et al., 2017). Biomass could be categorized according to its origin, summarized in the following types:

Agricultural Biomass: This includes dedicated crops such as sugarcane and sugar crops; starch-based crops such as wheat and corn; lignocellulosic crops such as switchgrass; and oil-based crops such as rapeseed and jatropha, as well as post-harvest crop residues such as straw, husks, or cobs, which are used in the production of biofuels (Aich & Datta, 2020).

Forestry Biomass: Comprises waste generated from forests like wood residues, sawdust, and bark, as well as forest residues like short twigs and leaves that fall during timber harvests (Kumari & Jain, 2020).

Aquatic Biomass: Derived from marine sources like algae (microalgae and macroalgae), which can be used for producing good biofuels like biodiesel (Kumari & Jain, 2020).

Industrial Biomass: By-products from industries such as pulp and paper mills, including process residues like sawdust or bagasse (Aich & Datta, 2020).

Municipal Solid Waste: Includes organic waste from households and food industries that can be converted into biofuels like biogas, which is a virtuous raw material for biofuel production (Kumari & Jain, 2020).

Animal-Derived Biomass: Includes waste generated during animal farming, such as manure, feathers, or bedding materials (Kumar & Verma, 2020).

1.1.1.1 Major Feedstocks

Globally, the primary biomass feedstocks vary depending on geographic region and local resources. The primary feedstocks for biomass energy globally include:

Corn (Maize): Widely used in the production of bioethanol, especially in countries like the United States. Corn is a key feedstock due to its high starch content, which is easily converted into ethanol (Kazemi Shariat Panahi et al., 2020).

Sugarcane: Predominantly used in Brazil for bioethanol production (Amorim & Lopes, 2005). Sugarcane has a high sugar content, making it an efficient feedstock for fermentation into ethanol (Silveira et al., 2015).

Soybean and Palm Oil: Major feedstocks for biodiesel production. Soybeans are primarily used in the United States, while palm oil is extensively used in Southeast Asia (especially in Indonesia and Malaysia) due to its high oil yield (Sut & Kataki, 2020).

Lignocellulosic Biomass: Includes agricultural residues like wheat straw and rice husks, as well as forestry residues such as wood chips and sawdust. This is used for second-generation biofuels globally, particularly in advanced biofuel production (Antizar-Ladislao & Turrion-Gomez, 2008).

Algae: A growing feedstock for biodiesel production due to its fast growth rate and high oil content. While not yet widespread, algae hold significant potential for future biofuel production (Singh et al., 2020).

Municipal Solid Waste: Organic components such as food and yard waste are increasingly being used for biogas and waste-to-energy projects around the world (Gurjar & Behera, 2020).

Palm Oil: Extensively used for biodiesel production in countries like Malaysia and Indonesia due to its large-scale cultivation and high oil yield (Sut & Kataki, 2020).

These biomass feedstocks represent a diverse range of materials used in renewable energy, biofuels, and bioproducts. The listed references provide in-depth information on their production, conversion technologies, and global significance.

1.1.2 Biomass Refining Methods

1.1.2.1 Sugars-First Approach

Lignocellulosic biomass components, such as cellulose, hemicellulose, and lignin, serve as an alternative carbon source for the sustainable production of liquid fuels, chemicals, and polymeric materials (Zhu & Pan, 2022). In the context of biorefinery, the recovery of sugars from biomass is essential, since they are used as raw materials; however, there are limitations that do not allow the direct use of sugars (Shibukawa et al., 2023). Therefore, pretreatment processes that facilitate the deconstruction and extraction of various lignocellulosic fractions are essential. These pretreatment methods can generally be classified into four categories: physical, chemical, physicochemical, and biological (Hernández-Beltrán et al., 2019).

Each method has distinct advantages and disadvantages. For example, physical-mechanical pretreatments reduce biomass particle size without generating inhibitory byproducts, but often involve high operating costs (Diyanilla et al., 2020). Acidic chemical pretreatments effectively break down hemicellulose, enhancing enzymatic hydrolysis; however, they are often corrosive and can produce toxic byproducts. In contrast, alkaline pretreatments promote delignification and reduce biomass crystallinity but are more expensive and generate more waste. Finally, biological pretreatments are economically viable but suffer from lower efficiency and longer processing times compared to conventional methods (Awogbemi & Von Kallon, 2022).

Pretreatment processes are typically applied sequentially. For example, a physical process may be followed by a chemical process, or a chemical process may precede a biological process. This sequential approach is necessary because each method targets different fractions of the biomass, such as cellulose (sugar C6) and hemicellulose (sugars C5 and C6) (Chaturvedi et al., 2022; Sheldon, 2021).

The success of sugar recovery processes, as well as fermentation processes, is reflected in the need to produce different sustainable products. For example, ethanol is a common product produced by different biomass. Twenty-nine billion tons of ethanol are currently produced worldwide (Rajeswari et al., 2022). On the other hand, the production of biopolymers is also of great interest, including polyhydroxyalkanoates (PHAs), where their current production is based on the use of low-cost raw materials (Sen & Baidurah, 2021). Another product example is xylitol, where biotechnological routes allow the bioconversion of C5 and C6, taking advantage of lignocellulosic biomass (Jain & Ghosh, 2023). As well as these products mentioned above, there is a wide range of second-generation biofuels and other compounds of biological origin generated, involving processes for the use of different lignocellulosic biomasses in a biorefinery context (Ashokkumar et al., 2022; Lu et al., 2022).

Sugarcane bagasse, one of the most widely used lignocellulosic by-products, consists of approximately 40% cellulose and 25% hemicellulose. Various studies have been conducted to obtain C5 and C6 sugars. For instance, Antunes et al. (2021) reported that acid pretreatment with 100 mg H_2SO_4 per gram of biomass is optimal for recovering the hemicellulosic fraction (C5 and C6), achieving up to 66% recovery in the liquid fraction after pretreatment. Moreover, in the same study, sequential alkaline pretreatment of the cellulignin fraction led to an 80% reduction in lignin, yielding a solid fraction rich in cellulose (C6).

In another study, various residual parts of corn (including the flower, stem, cob, husk, and leaf) were analyzed and subjected to acid pretreatment followed by enzymatic hydrolysis. After 72 hours of hydrolysis, glucose yields of 94.2%, 86.6%, 77.8%, 90.5%, and 86.0% were obtained for the cob, leaf, husk, stem, and flower, respectively. The study concluded that the structural differences among these residues influence sugar recovery, with the cob being the byproduct with the highest yield for ethanol production (Li et al., 2016).

Zhu et al. (2021) pretreated corn stalks for sugar recovery using conventional and microwave (MW)-assisted thermochemical methods. MW-assisted pretreatment led to greater biomass decomposition. Specifically, MW-assisted H_2SO_4 pretreatment resulted in significantly higher sugar availability—up to 4.5 times more than the other tested methods—while also reducing reaction time and energy consumption.

In other studies, the use of more environmentally friendly catalysts has been promoted, aiming to reduce the production of inhibitory compounds. Citric acid, for example, exhibits these characteristics. Bittencourt et al. (2022) conducted a study on the production of fermentable sugars from soybean husk through a sequential pretreatment process. First, the husk underwent citric acid pretreatment at a liquid-to-solid ratio of 5:1 for 105 minutes at 120 °C. This was followed by enzymatic hydrolysis (with 20% solids and an enzyme concentration of 15 FPU/g), yielding 72.03 g/L of glucose and 31.13 g/L of xylose. Additionally, the authors reported a mass yield of 80% for the citric acid pretreatment, demonstrating its potential as an effective method for sugar recovery and the production of new bio-based molecules (Bittencourt et al., 2022). In this context, Table 1.1 presents the major technologies discussed in the production of first-generation sugars and the products obtained through this processing.

1.1.2.2 Lignin-First Approach

The lignin-first biorefining strategy represents an innovative and complementary approach to traditional biomass valorization, shifting the focus from the sugars-first approach, which prioritizes carbohydrate valorization, to the selective extraction and conversion of lignin (Abu-Omar et al., 2021; Brienza et al., 2024). While conventional biorefining methods often degrade lignin into a low-value byproduct primarily used for heat and power generation, lignin-first biorefining prioritizes its preservation and controlled depolymerization, thereby unlocking its potential for high-value applications (Goswami et al., 2022; Wong et al., 2020). Consequently, this approach aligns with the principles of sustainability and circular bioeconomy by promoting atom-efficient biomass utilization and addressing a key limitation of sugars-first strategies—the underutilization of lignin's complex aromatic structure (Wang et al., 2025).

More specifically, lignin-first biorefining is based on the selective removal and stabilization of lignin from native lignocellulosic biomass before significant carbohydrate degradation occurs (Abu-Omar et al., 2021; Brienza et al., 2024; Renders et al., 2017). This strategy not only prevents the formation of highly recalcitrant condensed lignin but also enables the production of valuable aromatic compounds (Xu et al., 2022). Furthermore, stabilization is achieved through techniques such as reductive catalysis, chemical protection of reactive

Table 1.1 Major Technologies Based on the Sugar-First Approach and the Related Bioproducts

Technology	Product File	Key Products	Viability	References
Enzymatic Hydrolysis	Conversion of lignocellulosic biomass into fermentable sugars	Bioethanol, bio-butanol, bio-hydrogen	High, scalable, but dependent on biomass feedstock availability	Zhao et al. (2012)
Gasification to Syngas	Biomass gasified to syngas ($CO+H_2$), converted to sugar intermediates	Biofuels, chemicals (methanol, ethanol)	High potential, but technology is capital-intensive and requires optimization	(Bridgwater, 2012)
Fermentation of Sugars	Fermentation of sugars (glucose, xylose) from biomass to biofuels and chemicals	Ethanol, lactic acid, succinic acid	Commercially viable for bioethanol, R&D stage for advanced chemicals	(Balan, 2014)
Chemical Catalysis of Sugars	Catalytic conversion of sugars into platform chemicals	Furan-based chemicals (hydroxymethylfurfural [HMF], Furan di-carboxylic acid [FDCA]), lactic acid	Promising, but catalyst efficiency and costs need improvement	(Bond et al., 2013)
Microbial Electrolysis Cells (MECs)	Using microorganisms to produce electricity and chemicals from biomass sugars	Acetate, hydrogen, methane	Emerging technology, low efficiency, and cost-intensive	Escapa et al. (2016)
Aqueous Phase Reforming (APR)	Reforming of biomass-derived sugars into hydrogen and hydrocarbons	Hydrogen, alkanes, methanol	Limited commercial success due to catalyst costs and stability issues	Cortright et al. (2010)

groups, and selective hydrogenation of intermediates (Zhang et al., 2023). The success of lignin-first processes relies on three fundamental principles: (a) selective solvolysis to prevent unwanted condensation reactions and preserve lignin's structure, (b) catalytic stabilization through hydrogenation or chemical modification, and (c) controlled fractionation and efficient recovery of lignin to obtain homogeneous and valuable product streams (Abu-Omar et al., 2021; Schutyser et al., 2018). Together, these principles enhance both the efficiency and sustainability of lignocellulose utilization.

Several key technologies have emerged within the lignin-first framework, each offering distinct advantages in terms of product selectivity and yield, as shown in Table 1.2. Among these, one of the most extensively studied approaches is reductive catalytic fractionation (RCF), which employs catalytic hydrogenation under a hydrogen atmosphere to generate

Table 1.2 Major Technologies Based on the Lignin-First Approach and Their Product Profiles

Technology	Mechanism	Notes	Key Products	References
Reductive Catalytic Fractionation (RCF)	Catalytic hydrogenation under H_2.	High yields demand high H_2 and catalyst consumption	Aromatic monomers (guaiacols, syringols), deoxygenated aromatic hydrocarbons for fuels.	Kumar et al. (2023)
Catalytic Lignin Arylation Fractionation (CLAF)	Catalytic fractionation with selective arylation.	Controls lignin condensation to produce specific high-value products.	Lignin-derived bisphenols and polymer precursors.	Li et al. (2024)
Aldehyde-Assisted Fractionation (AAF)	Chemical protection of intermediates using aldehydes.	High selectivity uses formaldehyde for stabilization.	Stabilized intermediate compounds and aromatic monomers.	Wu et al. (2024)
Diol-Assisted Fractionation (DAF)	Capture of reactive lignin groups using diols.	Environmentally friendly, lower monomer yields.	Acetals, stabilized phenolic compounds.	De Santi et al. (2021)
Catalytic Organosolv	Solubilization using organic solvents.	Catalysts are a challenge in solvent recovery and catalyst efficiency.	Phenolic monomers, low molecular weight oligomers.	Liu et al. (2019)
Tandem Flow-Through Reactor Systems	Two sequential flow-through reactors, solvolytic release and reductive processing.	Optimized for continuous flow operation, improved process control.	Lignin oil with controlled molecular diversity.	Jang et al. (2022)

phenolic monomers, bio-oils, and aromatic compounds (Kenny et al., 2024). Specifically, RCF integrates lignin extraction, depolymerization, and stabilization using transition metal catalysts (e.g., Ru/C, Pd/C) in the presence of hydrogen or hydrogen-donor solvents like methanol (Yu et al., 2024). Typically operating at 180–250 °C, RCF achieves high delignification yields (~90% for hardwoods) and near-theoretical monomer yields (10–35%) (Renders et al., 2019; Schutyser et al., 2015). Additionally, it preserves cellulose in a digestible form suitable for bioethanol production. However, challenges such as high operating pressures, catalyst sintering, and metal leaching must be addressed to enhance process scalability and cost-effectiveness (Mondal et al., 2024).

Similarly, another promising approach is catalytic lignin arylation fractionation (CLAF), which utilizes selective arylation reactions to convert lignin into bisphenol derivatives with applications in polymer manufacturing (Li et al., 2024; Kang et al., 2025). By transforming lignin condensation into controlled arylation, CLAF achieves high product selectivity (~93%) and significant yields of S-lignin-derived bisphenols (34.2 wt%) (Li et al., 2024). Notably, these bio-based bisphenols exhibit significantly lower biotoxicity than conventional bisphenol A, making them attractive for sustainable plastic production (Li et al., 2024). Nevertheless, optimizing reaction conditions and reducing production costs remain key challenges for broader implementation (Wu et al., 2024).

In addition, aldehyde-assisted fractionation (AAF) employs aldehydes like formaldehyde to chemically protect lignin intermediates, preventing undesired condensation and facilitating further depolymerization (Brienza et al., 2024; König-Mattern et al., 2024). Conducted at moderate temperatures (80–120 °C), AAF achieves high lignin extraction yields (~80%) and monomer selectivity (>90% for certain phenolics) (Behaghel et al., 2020; Dick et al., 2022; Vendamme et al., 2020). Despite these advantages, the use of toxic aldehydes and the complexity of solvent recovery pose environmental and economic challenges that require further refinement (Vendamme et al., 2020).

Alternatively, diol-assisted fractionation (DAF) offers a different stabilization strategy, utilizing diols such as ethylene glycol under acid-catalyzed conditions to capture reactive lignin groups and generate stabilized phenolic compounds (Brienza et al., 2024; Wu et al., 2024). Operating at ~140 °C, DAF eliminates the need for metal catalysts, thereby improving its environmental footprint (Abu-Omar et al., 2021; De Santi et al., 2021). However, moderate monomer yields (~9% for softwoods) and the additional cost to recovery remain significant barriers to industrial adoption (Witthayolankowit et al., 2023).

Moreover, catalytic organosolv processes leverage organic solvents such as ethanol or methanol, often in combination with acid or base catalysts, to solubilize lignin and produce oligomeric fractions suitable for further upgrading into functional polymers (Agwu et al., 2024). Besides, the incorporation of heterogeneous catalysts (e.g., Ni/hydrogenated ZSM-5 [HZSM-5]) enhances lignin depolymerization, yielding phenolic monomers and low-molecular-weight oligomers (Liu et al., 2019). While operating at approximately 180 °C, this method improves lignin extraction and stabilization but faces challenges in solvent recovery and catalyst efficiency (Vaidya et al., 2022).

Additionally, tandem flow-through reactor systems represent a cutting-edge development in lignin-first biorefining, decoupling solvolytic lignin release (160–180 °C) from subsequent reductive processing (200–240 °C) (Anderson et al., 2018; Dixon et al., 2024; Jang et al., 2022). This dual-stage approach enhances process control and efficiency, yielding high-quality cellulose pulps with improved xylan retention and lignin oils with controlled molecular diversity (Dixon et al., 2024; Jang et al., 2022). Nevertheless, catalyst deactivation and separation issues remain critical obstacles to widespread industrial application (Dixon et al., 2024).

Recent advances in lignin-first biorefining have significantly expanded the potential for sustainable biomass utilization. The phenolic monomers generated through these processes serve as precursors for bioplastics, adhesives, and resins, while bio-oils provide a renewable source for sustainable fuels (Jeffri et al., 2024). Furthermore, integrating lignin-first technologies with carbohydrate valorization offers a more holistic approach to biomass processing, maximizing overall feedstock efficiency. However, several challenges persist, including the need for more selective and economically viable processes, as well as the seamless integration of these technologies within existing industrial infrastructures (Argyropoulos et al., 2023).

Indeed, process efficiency, cost, and catalyst management represent key limitations to the large-scale implementation of lignin-first biorefining (Sun et al., 2020). Since the complexity of reaction pathways necessitates advanced reactor designs to optimize energy and mass balances, overcoming these limitations is crucial (Shapiro et al., 2023). Additionally, high costs associated with reagents such as formaldehyde and hydrogen, along with catalyst recovery, require pilot-scale validation to assess economic feasibility (Jeffri et al., 2024). At the same time, catalyst deactivation due to polymerization and condensation reaction and solvent degradation introduces technical hurdles, prompting the development of innovative strategies like magnetic separation and porous catalyst supports (Tian et al., 2023; Bugli et al., 2024). Looking ahead, future research efforts will focus on industrial scale-up, integration with existing biorefineries, and genetic engineering of biomass to enhance lignin extractability. Specifically, in planta modifications, such as increasing the S/G ratio, could improve lignin depolymerization efficiency and overall process economics, further advancing the feasibility of lignin-first biorefining as a cornerstone of sustainable biomass utilization (Guo et al., 2023).

Ultimately, lignin-first biorefining represents a transformative paradigm in biomass processing, facilitating the efficient valorization of lignin while contributing to the broader objectives of the bioeconomy. With continued advancements in catalysis, process engineering, and industrial integration, this approach has the potential to play a crucial role in the transition toward a renewable, carbon-based economy. Thus, overcoming existing limitations through innovative technologies and process optimization will be key to fully realizing the economic and environmental benefits of lignin-first strategies, positioning them as a fundamental component of next-generation biorefining.

1.1.3 Key Products from Biorefinery Based from Listed Top 12 Biochemicals from U.S. Department of Energy

Biorefineries, which convert biomass into fuels, chemicals, and materials, offer a promising pathway to a sustainable bioeconomy. The USDOE has identified 12 biochemicals with high commercial potential, as shown in Figure 1.1 and Table 1.3 (Biddy et al., 2016; Liang et al., 2023; Markus and Zwicky, 2020). This study examines key products derived from these biochemicals, emphasizing their technical feasibility and process viability. These bio-based alternatives can significantly reduce greenhouse gas emissions, enhance sustainability, and support a circular bio-based economy (Bozell and Petersen, 2010; Liang et al., 2023), presenting critical opportunities to replace petrochemicals with renewable solutions.

The USDOE evaluates major biochemicals based on commercial viability, environmental impact, and technological readiness. To qualify, these biochemicals must exhibit strong market potential, including high demand and growth, and seamlessly integrate into biorefineries utilizing renewable feedstocks (Bozell & Petersen, 2010; Liang et al., 2023). They should also function as versatile platforms for derivatives. Environmental criteria focus on minimizing greenhouse gas emissions and fossil energy use, while production technologies must be mature or scalable. Additionally, products should target cost competitiveness with petrochemical equivalents (Liang et al., 2023). This strategic approach maximizes sustainability and economic impact.

1,4-Butanediol: Biodegradable plastics, such as polybutylene succinate, elastomers, and polyurethane products, are produced through a biotechnological process that uses

Figure 1.1 U.S. Department of Energy's top listed 12 biochemicals and their molecular structure.

microorganisms like *Escherichia coli* and *Corynebacterium glutamicum* to ferment renewable sugars and obtain 1,4-butanediol. Although 1,4-butanediol production from renewable sources shows potential due to advances in metabolic engineering, scaling up to commercial levels remains a challenge because of high fermentation costs and the complexity of downstream processing, which requires optimization for commercial viability (Radoš et al., 2015).

1,3-Butadiene: Key products such as synthetic rubber, plastics, and elastomers used in tires and automotive components are made from renewable feedstocks like ethanol, which are converted into 1,3-butadiene through catalytic processes such as oxidative dehydrogenation. Although the production of 1,3-butadiene from biobased sources is emerging, it faces challenges in catalyst development, reactor design, and product purification (Moncada et al., 2018).

Table 1.3 U.S. Department of Energy's Top Listed 12 Biochemicals and Their Assessment of Their Commercial Applications, Key Products and Viability

Biochemical	Commercial Applications	Key Products	Viability	Bibliographic References
1,4-Butanediol	Solvents, fibers, and plastics	Polyurethane, biodegradable plastics	High market demand due to applications in spandex	(Biddy et al., 2016; Liang et al., 2023)
1,3-Butadiene	Synthetic rubber production	Tires, adhesives	Growing demand due to constrained petroleum supply	(Biddy et al., 2016; Liang et al., 2023)
Ethyl Lactate	Green solvents in coatings and pharmaceuticals	Biodegradable solvents	High potential as a bio-based alternative	(Biddy et al., 2016; Liang et al., 2023)
Fatty Alcohols	Personal care products and detergents	Surfactants, cosmetics	Established market, mainly from plant-based sources	(Biddy et al., 2016; Liang et al., 2023)
Furfural	Solvent for refining lubricants and resin production	Adhesives, plastics	Viable, supported by lignocellulosic biomass	(Biddy et al., 2016; Liang et al., 2023)
Glycerin	Pharmaceuticals, cosmetics, and food additives	Soaps, detergents, lubricants	Market saturated with crude glycerin from biodiesel	(Biddy et al., 2016; Liang et al., 2023)
Isoprene	Synthetic rubber production	Tires, elastomers	High viability due to limited petroleum sources	(Biddy et al., 2016; Liang et al., 2023)
Lactic Acid	Bioplastics and food preservation	PLA (polylactic acid), food preservatives	Growing market for biodegradable plastics	(Biddy et al., 2016; Liang et al., 2023)
Propanediol (1,3-)	Anti-freeze, resins, and food preservatives	Solvents, unsaturated polyester resins	Increasing demand due to versatile applications	(Biddy et al., 2016; Liang et al., 2023)
Propylene Glycol	Antifreeze, de-icing agents, and food additives	Engine coolant, personal care products	Economically viable in the United States and Europe	(Biddy et al., 2016; Liang et al., 2023)
Succinic Acid	Bioplastics, solvents, food, and beverage additives	Polyurethanes, coatings	High growth potential due to green chemistry push	(Biddy et al., 2016; Liang et al., 2023)
Xylene (para)	Plastics, textiles, and synthetic fibers	Polyethylene terephthalate (PET)	Strong demand in plastics and textiles	(Biddy et al., 2016; Liang et al., 2023)

Ethyl Lactate: Key products, such as solvents for coatings, cleaners, and polymer production, include ethyl lactate, which is obtained through the esterification of lactic acid with ethanol, both derived from fermentation processes using renewable feedstocks. Ethyl lactate is a well-established green solvent, and its production is highly viable, with competitive costs and environmental benefits compared to petrochemical alternatives (Pereira et al., 2011).

Fatty Alcohols: Key products, such as detergents, cosmetics, lubricants, and surfactants, are made from fatty alcohols produced through the hydrogenation of fatty acids, typically derived from vegetable oils like palm oil or algal oil. The sustainable production of fatty alcohols from biomass is economically viable due to the established supply chains for vegetable oils, though competition with food crops remains a concern (Ye & Ren, 2014).

Furfural: Key products such as furfuryl alcohol, resins, solvents, and fuel additives are produced from furfural, which is obtained through the degradation of hemicellulose followed by dehydration of agricultural residues like corn cobs and sugarcane bagasse. The production of furfural is economically viable due to the abundance of lignocellulosic feedstocks, although the process remains challenging due to its high energy consumption (Zeitsch, 2000).

Glycerol: Key products such as propylene glycol, acrolein, and biofuels are obtained from glycerin, a by-product of biodiesel production that can be transformed into value-added chemicals through dehydration, hydrogenation, and oxidation processes. Glycerin is abundant due to biodiesel production, making its conversion economically attractive; however, an oversupply could reduce market prices for glycerin-derived products (Gu and Jérôme, 2010).

Isoprene: Key products such as synthetic rubber, elastomers, adhesives, and plastics can be obtained from biobased isoprene, which is produced through the fermentation of sugars using engineered strains of *E. coli* or *Bacillus subtilis*. Although biobased isoprene production is feasible, current commercial production is dominated by petrochemical routes, making it less competitive unless further process optimization is achieved (Beller et al., 2015).

Lactic Acid: Key products include polylactic acid (PLA), which is used in biodegradable plastics, food additives, and pharmaceuticals. Lactic acid is produced through microbial fermentation of sugars and can be polymerized into PLA or used directly as a platform chemical. The production of lactic acid from renewable feedstocks is commercially viable, with well-established industrial fermentation processes, and its use in biodegradable plastics has increased demand (Andres et al., 2013).

Propanediol: Key products include biodegradable plastics, cosmetics, and antifreeze, which are obtained from 1,3-propanediol produced through the fermentation of glycerol or glucose by microbial strains such as Clostridium. The biological production of 1,3-propanediol is considered efficient and competitive, especially when glycerol is used as the feedstock due to its low cost as a biodiesel by-product (Rajagopalan et al., 2024).

Propylene Glycol: Key products include antifreeze, de-icing fluids, and unsaturated polyester resins, produced through the catalytic hydrogenation of glycerol, which is abundant due to biodiesel production. The production of propylene glycol from glycerol is considered commercially viable because of the excess glycerol supply from biodiesel, providing a cost advantage (Johnson & Taconi, 2007).

Succinic Acid: Key products include biodegradable plastics, solvents, and resins, which are obtained through the fermentation of glucose or other sugars by engineered bacterial strains such as *Corynebacterium* or *E. coli*. Biobased succinic acid is economically viable, and several companies are scaling up production due to its importance in bioplastics and solvents, positioning it as a leading candidate to replace petrochemical-derived counterparts (Mckinlay et al., 2007).

p-Xylene: Key products include polyethylene terephthalate (PET), used in bottles, packaging, and fibers, which can be synthesized from renewable feedstocks such as biomass-derived sugars through catalytic reforming. Although biobased production methods are under development, p-xylene is still primarily produced from petrochemical sources, and scaling up biobased routes remains a challenge (Dutta, 2023).

The technical aspects of bioproduct production include biochemical pathways where most products are obtained through microbial fermentation using engineered microorganisms that convert sugars and biomass into valuable chemicals. In products like isoprene and xylene, catalytic conversion of fermentation intermediates is used to obtain the final products. Biorefineries combine fermentation, hydrolysis, and catalysis processes to transform biomass into biocompounds and biofuels. Established technologies, such as those for 1,4-butanediol and lactic acid, are in commercial production, while emerging technologies like biobased isoprene show potential. Economic viability is driven by the demand for biodegradable and sustainable products, as well as the reduction of greenhouse gas emissions, and these processes are moving toward greater commercialization supported by technological advancements and favorable trends (Biddy et al., 2016).

1.1.4 Process Limitations and Net-Zero Environment

Lignin is a complex macromolecule of aromatic compounds (a planar ring of atoms stabilized by the interaction of bonds in the ring) that is difficult to disentangle. In terms of bonds, lignin is made up of easy-to-break and weak carbon–oxygen bonds and much stronger carbon–carbon bonds.

Catalytic methods break the weak carbon–oxygen bonds with ease, but leave the carbon–carbon bonds untouched, thus limiting the yield of useful products from lignin (Luo et al., 2024). The main challenge in applying lignin to produce bioproducts is the heterogeneity and instability of this molecule. Its complex structure, after undergoing depolymerization, results in different monomers, dimers, trimers, and oligomers with aromatic rings present in the chemical structure (Kent et al., 2018; Menezes et al., 2023).

The diversity of the obtained monomers needs to be properly elucidated for their use as substrates in the production of fermentation products, or to use as raw material for chemical products (Luo et al., 2024).

Given the diversity of depolymerization methods reported in the literature, the choice of method, process severity, reaction time, and different types of catalysts will strongly influence the cleavage of bonds and, consequently, the size of the oligomers. It is important to standardize these processes according to the product of interest (Luo et al., 2024; Ravi et al., 2017).

The structure of lignin varies greatly depending on the source of the plant and the extraction method, among other factors. The diversity in size and structure results in an unstable molecule presenting poor solubility. The solubility of lignin in common solvents is low, requiring specific solvents for the solubilization of this molecule. However, solubility is crucial for the conversion of this molecule into a bioproduct (Ajao et al., 2019; Lou et al., 2024).

Conventional lignin extraction processes (kraft, soda, and sulfite) are known for their use of high temperatures; these are severe processes that require significant energy expenditure and can be costly. The need for sustainable and lower-cost processes is a key point for the development of the sector and the reduction of greenhouse gas emissions. In addition to these technical aspects of the extraction process, political incentives in the sector are essential for consolidating cleaner processes such as cold pulping, the use of organic solvents, supercritical fluids, and biological pulping (enzymes and microorganisms), among other extraction methods (Ajao et al., 2019).

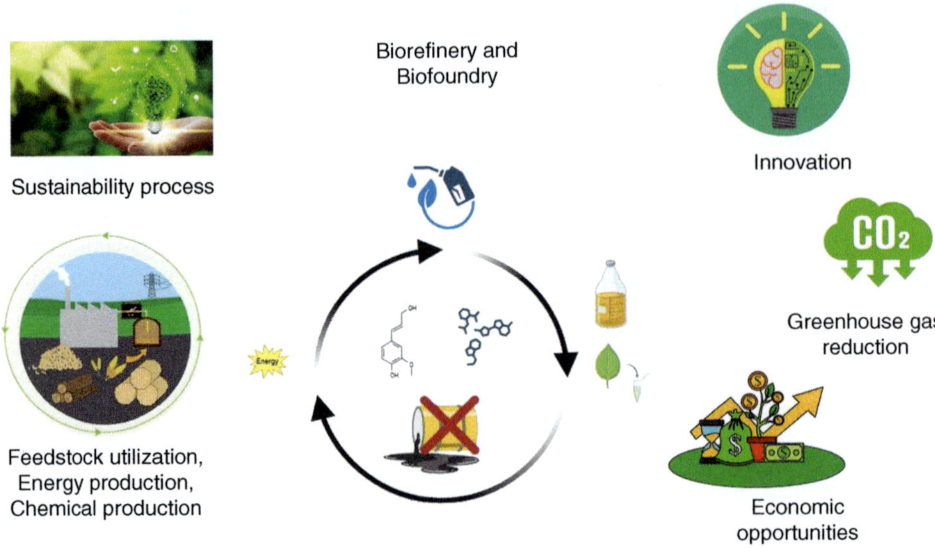

Figure 1.2 Key domains of biorefinery and biofoundry: interlinking and their impact toward a net-zero environment. *Source:* peampath/Adobe Stock Photos.

The regulation of the sector and tax incentives are essential for replacing production routes with milder and more environmentally friendly processes. Lignin is a biomass with enormous potential to substitute for fossil fuels; however, the need for greater development and support in the sector is crucial for the economic viability of a lignin-based chemical platform (Garlapati et al., 2020). However, for the upgrading the lignin stream, catalytic depolymerization of lignin via selective approaches while maintaining the structural preservation is essential (Ramos et al., 2025a, b).

A lignin-based chemical platform offers numerous environmental advantages and promotes the development of new materials, biodegradable polymers, nanomaterials, epoxy resin, phenolic resins, adhesives, and nanocomposites, among other products (Garlapati et al., 2020). A schematic sketch of the lignin production chain as an opportunity for the generation of new products, coupled with the reduction of greenhouse gas emissions, is presented in Figure 1.2.

In addition to the development of new materials and production routes, the valorization of lignin is rooted in the concept of biorefineries and promotes a diversity of opportunities within the pillars of sustainability (environmental, social, and economic). It also offers the opportunity to integrate sugarcane biorefineries with those of eucalyptus (Garlapati et al., 2020).

1.1.5 Biofoundry and Advanced Bioeconomy

The integration of biofoundries into the lignocellulosic bioeconomy marks a paradigm shift in biomass processing, leveraging automation, computational design, and synthetic biology to engineer microbial systems for industrial applications (Philp 2023; Aggarwal et al., 2024). Unlike traditional biorefineries that rely on thermochemical and enzymatic processes (Asghar et al., 2022), biofoundries enable rapid prototyping and optimization of microbial strains through design-build-test-learn (DBTL) cycles (Tellechea-Luzardo et al., 2022). The biofoundry approach enhances the conversion of lignocellulosic hydrolysates into high-value products such as biofuels, platform chemicals, and specialty biomaterials. For instance, engineered *Saccharomyces cerevisiae* strains have demonstrated enhanced co-fermentation of glucose and xylose from corn cob hydrolysates, increasing yields of second-generation bioethanol and xylitol (Baptista et al., 2018). Similarly, advances in *Klebsiella oxytoca* engineering have enabled 2,3-butanediol production from mixed sugars in lignocellulosic hydrolysates (Cha et al., 2020). Furthermore, engineered *Aspergillus* species, optimized through biofoundry platforms, have shown promising efficiency in producing 3-hydroxypropionic acid—a key precursor for biodegradable plastics—from corn stover hydrolysate (Dai et al., 2023), while engineered *Yarrowia lipolytica* produces β-farnesene, a precursor to biofuels and cosmetics, from low-cost glucose in lignocellulosic hydrolysates (Bi et al., 2022).

Despite their advantages, biofoundries face challenges in lignocellulose valorization, including substrate complexity, scalability issues, lignin underutilization, and techno-economic constraints (Devi et al., 2024). The recalcitrant nature of lignocellulose necessitates pretreatment, which generates microbial inhibitors such as furfural and 5-hydroxymethylfurfural (5-HMF), requiring the engineering of robust strains with enhanced tolerance (Jilani & Olson, 2023). Additionally, while biofoundries excel at laboratory-scale optimization, transitioning to industrial-scale production remains difficult due to high capital costs and inconsistent biomass supply chains (Ahlawat et al., 2024). Lignin, an abundant but underutilized component, poses another challenge, as its structural complexity limits microbial conversion into valuable aromatic compounds (Dixon et al., 2024). Addressing these limitations demands innovations in metabolic engineering, automation, and artificial intelligence (AI)-driven strain design to improve efficiency and cost-effectiveness.

Recent advancements highlight the growing role of biofoundries in driving the advanced bioeconomy (Figure 1.3). Clustered regularly interspaced short palindromic repeats (CRISPR)-based metabolic engineering has enabled microbial co-utilization of glucose and xylose, boosting product titers above 100 g/L for some products (Gao et al., 2023). High-throughput screening platforms have accelerated strain selection, as demonstrated by *C. glutamicum*'s 30% improvement in itaconic acid yields through iterative DBTL cycles (Hanh et al., 2023). Integrated biorefinery strategies now explore lignin valorization by engineering microbes to convert lignin-derived intermediates into biofuels, alongside bioethanol production, thereby enhancing overall feedstock utilization (Beckham et al., 2016; Dixon et al., 2024). Additionally, biofoundries complement emerging green pretreatment methods, like hydrothermal and deep eutectic solvent, by tailoring microbial performance to hydrolysates with minimal inhibitors, boosting sugar yields and fermentation efficiency (Otoupal et al., 2022).

Compared to traditional biorefineries, biofoundries offer a more flexible and targeted approach to biomass conversion. Whereas syngas biorefineries convert lignocellulose into fuels via gasification—an energy-intensive process requiring large-scale investment—biofoundries enable microbial engineering for diverse product outputs (Devi et al., 2024).

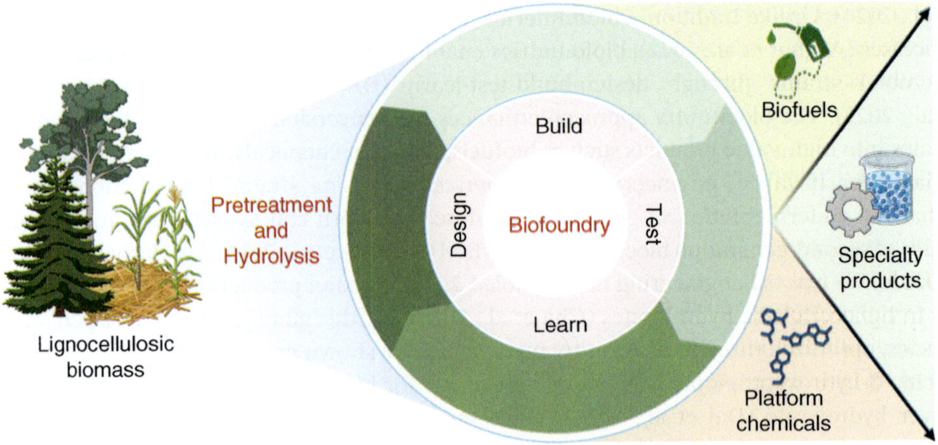

Figure 1.3 Implicit role of biofoundry in advancing bioeconomy. This schematic illustrates how biofoundries integrate lignocellulosic biomass into a sustainable bioeconomy through automated strain engineering and the DBTL cycle, yielding diverse products from biofuels to specialty biomaterials.

This adaptability positions biofoundries as a key player in the bioeconomy, complementing existing technologies rather than replacing them.

Looking ahead, biofoundries must overcome current limitations to fully integrate into industrial biorefineries. Advances in synthetic biology could unlock lignin's potential through microbial consortia capable of degrading and converting lignin into high-value aromatics and bioplastics (Zhao et al., 2024). AI-driven predictive modeling could further optimize DBTL cycles, reducing development time and costs (Liu et al., 2022). Aligning biofoundries with circular bioeconomy principles will also be critical, fostering zero-waste biorefineries where all lignocellulose fractions contribute to sustainable production by engineering microbes (Philp, 2023). Moreover, policy support and strategic investments, as seen in European bioeconomy initiatives (Liobikienė & Miceikienė, 2023), will be crucial in accelerating biofoundry adoption and commercialization.

1.1.6 Conclusion and Future Directions

The vast amount and variety of biomass waste have the potential to provide the necessary carbon to produce bioproducts, chemical intermediates, and energy through environmentally sustainable production pathways. Although it is not yet an industrial reality worldwide, biomass refining presents challenges that will create both economic and environmental opportunities. It is essential to adopt and enhance biotechnological pathways to improve the quantity and quality of bioproducts and energy. These processes are still expensive and consume a lot of energy when compared to conventional processes. Achieving zero carbon emissions will depend not only on legislation that benefits companies through tax reductions, but also on the advancement of innovative technologies. In this context, process control tools, genetic engineering, mathematical modeling, and process simulation will play a key role in guiding new production pathways and integrating biorefineries, thereby supporting decision-making and bringing these routes closer to economic viability. Thus, it is possible to harness all renewable carbon sources, fostering a circular economy aligned with sustainability pillars.

Acknowledgments

L. Ramos and A.K. Chandel gratefully acknowledge São Paulo State Funding Agency-FAPESP, Brazil (grant numbers: 2020/12559-6, 2022/13184-1, and 2023/08226-0) for the financial project. A.K. Chandel also gratefully acknowledges the Brazilian National Council for Scientific and Technological Development (CNPq), Brazil, for the scientific productivity program (process number: 304451/2025-8).

References

Abu-Omar, M. M., Barta, K., Beckham, G. T., Luterbacher, J. S., Ralph, J., Rinaldi, R., ... Wang, F. (2021). Guidelines for performing lignin-first biorefining. *Energy & Environmental Science*, *14*(1), 262–292.

Aggarwal, N., Pham, H. L., Ranjan, B., Saini, M., Liang, Y., Hossain, G. S., ... Chang, M. W. (2024). Microbial engineering strategies to utilize waste feedstock for sustainable bioproduction. *Nature Reviews Bioengineering*, *2*(2), 155–174.

Agwu, K. A., Ekeoma, B. C., Enguita, J. M., Daymude, J. W., Bara, J. E., & Sheehan, J. D. (2024). Glycerol-Derived Solvents for Tractable Organosolv Extraction of Softwood Lignins. *ACS Sustainable Chemistry & Engineering*, *12*(41), 15089–15102.

Ahlawat, Y. K., Srivastava, V., Samani, M., Harun, S., Rajuloo, V., & Chaudhary, D. (2024). Computational Strategies for Maximizing Biomass and Metabolite Yields for Bioproduction. In *Medicinal and Aromatic Plants: Current Research Status, Value-Addition to Their Waste, and Agro-Industrial Potential (Vol I)* (pp. 143–161). Springer Nature: Cham, Switzerland.

Aich, S., & Datta, S. (2020). In P. Verma (Ed.), *Efficient Utilization of Lignocellulosic Biomass: Hydrolysis Methods for Biorefineries BT - Biorefineries: A Step Towards Renewable and Clean Energy* (pp. 273–295). Singapore: Springer Singapore. https://doi.org/10.1007/978-981-15-9593-6_11

Ajao, O., Jeaidi, J., Benali, M., Abdelaziz, O. Y., & Hulteberg, P. C. (2019,121799, ISSN 0960-8524,). Green solvents-based fractionation process for kraft lignin with controlled dispersity and molecular weight. *Bioresource Technology*, *291*. https://doi.org/10.1016/j.biortech.2019.121799

Amorim, H. V, Lopes, M.L., 2005. Ethanol production in a petroleum dependent world: the Brazilian experience.

Anderson, E. M., Stone, M. L., Hulsey, M. J., Beckham, G. T., & Román-Leshkov, Y. (2018). Kinetic studies of lignin solvolysis and reduction by reductive catalytic fractionation decoupled in flow-through reactors. *ACS Sustainable Chemistry & Engineering*, *6*(6), 7951–7959.

Andres, F., Martinez, C., Marcos, E., Dom, M., & De Souza, R. P. (2013). Lactic acid properties, applications and production. *A review*, *30*, 70–83. https://doi.org/10.1016/j.tifs.2012.11.007

Antizar-Ladislao, B., & Turrion-Gomez, J. L. (2008). Second-generation biofuels and local bioenergy systems. Biofuels. *Bioprod. Biorefining*, *2*, 455–469. https://doi.org/10.1002/bbb.97

Antunes, F. A., Thomé, L. C., Santos, J. C., Ingle, A. P., Costa, C. B., Dos Anjos, V., ... Da Silva, S. S. (2021). Multi-scale study of the integrated use of the carbohydrate fractions of sugarcane bagasse for ethanol and xylitol production. *Renewable Energy*, *163*, 1343–1355.

Argyropoulos, D. D., Crestini, C., Dahlstrand, C., Furusjö, E., Gioia, C., Jedvert, K., ... Wimby, M. (2023). Kraft lignin: a valuable, sustainable resource, opportunities and challenges. *ChemSusChem*, *16*(23), e202300492.

Asghar, A., Sairash, S., Hussain, N., Baqar, Z., Sumrin, A., & Bilal, M. (2022). Current challenges of biomass refinery and prospects of emerging technologies for sustainable bioproducts and bioeconomy. *Biofuels, Bioproducts and Biorefining*, *16*(6), 1478–1494.

Ashokkumar, V., Venkatkarthick, R., Jayashree, S., Chuetor, S., Dharmaraj, S., Kumar, G., ... Ngamcharussrivichai, C. (2022). Recent advances in lignocellulosic biomass for biofuels and value-added bioproducts-A critical review. *Bioresource Technology*, *344*, 126195.

Awogbemi, O., & Von Kallon, D. V. (2022). Pretreatment techniques for agricultural waste. *Case Studies in Chemical and Environmental Engineering*, *6*, 100229.

Balan, V. (2014). Current Challenges in Commercially Producing Biofuels from Lignocellulosic Biomass. *ISRN Biotechnol.*, *2014*, 1–31. https://doi.org/10.1155/2014/463074

Baptista, S. L., Cunha, J. T., Romaní, A., & Domingues, L. (2018). Xylitol production from lignocellulosic whole slurry corn cob by engineered industrial Saccharomyces cerevisiae PE-2. *Bioresource Technology*, *267*, 481–491.

Beckham, G. T., Johnson, C. W., Karp, E. M., Salvachúa, D., & Vardon, D. R. (2016). Opportunities and challenges in biological lignin valorization. *Current Opinion in Biotechnology*, *42*, 40–53.

Behaghel de Bueren, J., Heroguel, F., Wegmann, C., Dick, G. R., Buser, R., & Luterbacher, J. S. (2020). Aldehyde-assisted fractionation enhances lignin valorization in endocarp waste biomass. *ACS Sustainable Chemistry & Engineering*, *8*(45), 16737–16745.

Beller, H. R., Lee, S., & Katz, L. (2015). Natural Product Reports Natural products as biofuels and bio-based chemicals : fatty acids and isoprenoids. *Natural Product Reports*, *32*, 1508–1526. 10.1039/C5NP00068H

Bi, H., Xv, C., Su, C., Feng, P., Zhang, C., Wang, M., ... Tan, T. (2022). β-Farnesene production from low-cost glucose in lignocellulosic hydrolysate by engineered Yarrowia lipolytica. *Fermentation*, *8*(10), 532.

Biddy, M. J., Scarlata, C., & Kinchin, C. (2016). Chemicals from Biomass: A Market Assessment of Bioproducts with Near-Term Potential. *United States.* https://doi.org/10.2172/1244312

Bittencourt, G. A., de Souza Vandenberghe, L. P., Valladares-Diestra, K. K., & Soccol, C. R. (2022). Soybean hull valorization for sugar production through the optimization of citric acid pretreatment and enzymatic hydrolysis. *Industrial Crops and Products*, *186*, 115178.

Bond, J. Q., Alonso, D. M., & Dumesic, J. A. (2013). Catalytic strategies for converting lignocellulosic carbohydrates to fuels and chemicals. *Aqueous Pretreat. plant biomass Biol. Chem. Convers. to fuels Chem.*, 61–102.

Bozell, J. J., & Petersen, G. R. (2010). Technology development for the production of biobased products from biorefinery carbohydrates—the US Department of Energy's "top 10" revisited. *Green Chemistry*, *12*, 539–555. https://doi.org/10.1039/b922014c

Bridgwater, A. V. (2012). Review of fast pyrolysis of biomass and product upgrading. *Biomass and Bioenergy*, *38*, 68–94. https://doi.org/10.1016/j.biombioe.2011.01.048

Brienza, F., Cannella, D., Montesdeoca, D., Cybulska, I., & Debecker, D. P. (2024). A guide to lignin valorization in biorefineries: traditional, recent, and forthcoming approaches to convert raw lignocellulose into valuable materials and chemicals. *RSC Sustainability*, *2*(1), 37–90.

Bugli, F., Baldelli, A., Thomas, S., Sgarzi, M., Gigli, M., Crestini, C., ... Tabanelli, T. (2024). Improved Reductive Catalytic Fractionation of Lignocellulosic Biomass through the Application of a Recyclable Magnetic Catalyst. *ACS Sustainable Chemistry & Engineering*, *12*(45), 16638–16651.

Cha, J. W., Jang, S. H., Kim, Y. J., Chang, Y. K., & Jeong, K. J. (2020). Engineering of Klebsiella oxytoca for production of 2, 3-butanediol using mixed sugars derived from lignocellulosic hydrolysates. *GCB Bioenergy*, *12*(4), 275–286.

Chandel, A. K., Garlapati, V. K., Kumar, S. P. J., Hans, M., Singh, A. K., & Kumar, S. (2020). The role of renewable chemicals and biofuels in building a bioeconomy. *Biofuels, Bioproducts and Biorefining.*, *14*, 830–844.

Chandel, A. K., Forte, M. B. S., Gonçalves, I. S., Milessi, T. S., Arruda, P. V., Carvalho, W., & Mussatto, S. I. (2021). Brazilian biorefineries from second generation biomass: critical insights from industry and future perspectives. *Biofuels, Bioproducts and Biorefining.* https://doi.org/10.1002/bbb.2234

Chaturvedi, T., Hulkko, L. S. S., Fredsgaard, M., & Thomsen, M. H. (2022). Extraction, isolation, and purification of value-added chemicals from lignocellulosic biomass. *PRO*, *10*(9), 1752.

Cortright, R. D., Davda, R. R., & Dumesic, J. A. (2010). Hydrogen from catalytic reforming of biomass-derived hydrocarbons in liquid water. Mater. Sustain. Energy A Collect. Peer-Reviewed Res. *Rev. Artic. from Nat. Publ. Gr.*, *418*, 289–292. https://doi.org/10.1142/9789814317665_0043

Dai, Z., Pomraning, K. R., Deng, S., Kim, J., Campbell, K. B., Robles, A. L., Hofstad, B. A., Munoz, N., Gao, Y., Lemmon, T., Swita, M. S., Zucker, J. D., Kim, Y. M., Burnum-Johnson, K. E., & Magnuson, J. K. (2023). Metabolic engineering to improve production of 3-hydroxypropionic acid from corn-stover hydrolysate in Aspergillus species. *Biotechnology for Biofuels and Bioproducts* 2023, *16*(1), 53. https://doi.org/10.1186/s13068-023-02288-1

De Santi, A., Monti, S., Barcaro, G., Zhang, Z., Barta, K., & Deuss, P. J. (2021). New mechanistic insights into the lignin β-O-4 linkage acidolysis with ethylene glycol stabilization aided by multilevel computational chemistry. *ACS Sustainable Chemistry & Engineering*, *9*(5), 2388–2399.

Devi, N. B., Pugazhenthi, G., & Pakshirajan, K. (2024). Synthetic biology approaches and bioseparations in syngas fermentation. *Trends in.* Biotechnology

Dhamodharan, K., Ahlawat, S., Kaushal, M., & Rajendran, K. (2019). *Economics and cost analysis of waste biorefineries.* Refining Biomass Residues for Sustainable Energy and Bioproducts: Technology, Advances, Life Cycle Assessment, and Economics. Elsevier Inc. https://doi.org/10.1016/B978-0-12-818996-2.00025-9

Dick, G. R., Komarova, A. O., & Luterbacher, J. S. (2022). Controlling lignin solubility and hydrogenolysis selectivity by acetal-mediated functionalization. *Green Chemistry*, *24*(3), 1285–1293.

Dixon, R. A., Puente-Urbina, A., Beckham, G. T., & Román-Leshkov, Y. (2024). Enabling lignin valorization through integrated advances in plant biology and biorefining. *Annual Review of Plant Biology*, *75*(1), 239–263.

Diyanilla, R., Hamidon, T. S., Suryanegara, L., & Hussin, M. H. (2020). Overview of pretreatment methods employed on oil palm biomass in producing value-added products: a review. *BioResources*, *15*(4), 9935.

Dutta, S., 2023. Catalytic synthesis of renewable p -xylene from biomass-derived 2, 5-dimethylfuran : a mini review 541–554.

Escapa, A., Mateos, R., Martínez, E. J., & Blanes, J. (2016). Microbial electrolysis cells: An emerging technology for wastewater treatment and energy recovery. from laboratory to pilot plant and beyond. *Renewable and Sustainable Energy Reviews*, *55*, 942–956. https://doi.org/10.1016/j.rser.2015.11.029

Gao, J., Yu, W., Li, Y., Jin, M., Yao, L., & Zhou, Y. J. (2023). Engineering co-utilization of glucose and xylose for chemical overproduction from lignocellulose. *Nature Chemical Biology*, *19*(12), 1524–1531.

Garlapati, V. K., Chandel, A. K., Jeevan-Kumar, S. P., Sharma, S., Sevda, S., Ingle, A. P., & Pant, D. (2020). *Renewable and Sustainable Energy Resources*, *130*, 109977.

Goswami, L., Kayalvizhi, R., Dikshit, P. K., Sherpa, K. C., Roy, S., Kushwaha, A., ... Rajak, R. C. (2022). A critical review on prospects of bio-refinery products from second and third generation biomasses. *Chemical Engineering Journal*, *448*, 137677.

Gu, Y., & Jérôme, F. (2010). Glycerol as a sustainable solvent for green chemistry. *Green Chemistry*, *12*, 1127–1138. 10.1039/C001628D

Guo, Y., Wang, S., Yu, K., Wang, H. L., Xu, H., Song, C., ... Lin, J. (2023). Manipulating microRNA miR408 enhances both biomass yield and saccharification efficiency in poplar. *Nature Communications*, *14*(1), 4285.

Gurjar, R., & Behera, M. (2020). In P. Verma (Ed.), *Biopolymer: Production from Biomass BT - Biorefineries: A Step Towards Renewable and Clean Energy* (pp. 371–390). Singapore: Springer Singapore. https://doi.org/10.1007/978-981-15-9593-6_14

Hanh, D. D., Elkasaby, T., Kawaguchi, H., Tsuge, Y., Ogino, C., & Kondo, A. (2023). Enhanced production of itaconic acid from enzymatic hydrolysate of lignocellulosic biomass by recombinant Corynebacterium glutamicum. *Journal of Bioscience and Bioengineering*, *136*(1), 7–12.

Hernández-Beltrán, J. U., Hernández-De Lira, I. O., Cruz-Santos, M. M., Saucedo-Luevanos, A., Hernández-Terán, F., & Balagurusamy, N. (2019). Insight into pretreatment methods of lignocellulosic biomass to increase biogas yield: current state, challenges, and opportunities. *Applied Sciences*, *9*(18), 3721.

Jain, V., & Ghosh, S. (2023). Biotransformation of lignocellulosic biomass to xylitol: an overview. *Biomass Conversion and Biorefinery*, *13*(11), 9643–9661.

Jang, J. H., Brandner, D. G., Dreiling, R. J., Ringsby, A. J., Bussard, J. R., Stanley, L. M., ... Beckham, G. T. (2022). Multi-pass flow-through reductive catalytic fractionation. *Joule*, *6*(8), 1859–1875.

Jeffri, N. I., Rawi, N. F. M., Kassim, M. H. B. M., & Abdullah, C. K. (2024). Unlocking the potential: Evolving role of technical lignin in diverse applications and overcoming challenges. *International Journal of Biological Macromolecules, 133506.*

Jilani, S. B., & Olson, D. G. (2023). Mechanism of furfural toxicity and metabolic strategies to engineer tolerance in microbial strains. *Microbial Cell Factories*, *22*(1), 221.

Johnson, D. T., & Taconi, K. A. (2007). The glycerin glut: Options for the value-added conversion of crude glycerol resulting from biodiesel production. *Environmental Progress*, *26*, 338–348.

Kang, F., Qi, H., Huang, Z., & Hu, J. (2025). *New horizons in lignin-first strategies for upgrading lignocellulose*. Green Carbon.

Kazemi Shariat Panahi, H., Dehhaghi, M., Aghbashlo, M., Karimi, K., & Tabatabaei, M. (2020). Conversion of residues from agro-food industry into bioethanol in Iran: An under-valued biofuel additive to phase out MTBE in gasoline. *Renewable Energy*, *145*, 699–710. https://doi.org/10.1016/j.renene.2019.06.081

Kenny, J. K., Neefe, S. R., Brandner, D. G., Stone, M. L., Happs, R. M., Kumaniaev, I., ... Beckham, G. T. (2024). Design and Validation of a High-Throughput Reductive Catalytic Fractionation Method. *JACS Au*, *4*(6), 2173–2187.

Kent, M. S., Zeng, J., Rader, N., Avina, I. C., Simoes, C. T., Brenden, C. K., Busse, M. L., Watt, J., Giron, N. H., Alam, T. M., Allendorf, M. D., Simmons, B. A., Bell, N. S., & Sale, K. L. (2018). *Green Chemistry*, *20*, 3024.

König-Mattern, L., Medina, E. I. S., Komarova, A. O., Linke, S., Rihko-Struckmann, L., Luterbacher, J. S., & Sundmacher, K. (2024). Machine learning-supported solvent design for lignin-first biorefineries and lignin upgrading. *Chemical Engineering Journal*, *495*, 153524.

Kumar, B., & Verma, P. (2020). In P. Verma (Ed.), *Techno-Economic Assessment of Biomass-Based Integrated Biorefinery for Energy and Value-Added Product BT - Biorefineries: A Step Towards Renewable and Clean Energy* (pp. 581–616). Singapore: Springer Singapore. https://doi.org/10.1007/978-981-15-9593-6_23

Kumar, K., Yadav, A. N., Kumar, V., Vyas, P., & Dhaliwal, H. S. (2017). Food waste: a potential bioresource for extraction of nutraceuticals and bioactive compounds. *Bioresources and Bioprocessing*, *4*. https://doi.org/10.1186/s40643-017-0148-6

Kumar, A., Jindal, M., Rawat, S., Kumar, J., Sripadi, P., Yang, B., & Thallada, B. (2023). Upgradation of sugarcane bagasse lignin: Fractionation to cyclic alcohols production. *Catalysis Today*, *408*, 182–193.

Kumari, D., & Jain, Y. (2020). In P. Verma (Ed.), *Enhanced Biofuel Production from Lignocellulosic Biomass: An Overview of Advanced Physico-Chemical and Biological Technologies BT - Biorefineries: A Step Towards Renewable and Clean Energy* (pp. 151–172). Singapore: Springer Singapore. https://doi.org/10.1007/978-981-15-9593-6_7

Langholtz, M. H., Stokes, B. J., & Eaton, L. M. (2016). *2016 billion-ton report: advancing domestic resources for a thriving bioeconomy*. Washington, DC (United States): EERE Publication and Product Library.

Li, P., Cai, D., Luo, Z., Qin, P., Chen, C., Wang, Y., ... Tan, T. (2016). Effect of acid pretreatment on different parts of corn stalk for second generation ethanol production. *Bioresource Technology*, *206*, 86–92.

Li, N., Yan, K., Rukkijakan, T., Liang, J., Liu, Y., Wang, Z., ... Wang, F. (2024). Selective lignin arylation for biomass fractionation and benign bisphenols. *Nature*, *630*(8016), 381–386.

Liang, C., Gracida-alvarez, U. R., Hawkins, T. R., & Dunn, J. B. (2023). *Life-Cycle Assessment of Biochemicals with Clear Near-Term Market Potential.* https://doi.org/10.1021/acssuschemeng.2c05764

Liobikienė, G., & Miceikienė, A. (2023). Contribution of the European Bioeconomy Strategy to the Green Deal Policy: Challenges and opportunities in implementing these policies. *Sustainability*, *15*(9), 7139.

Liu, X., Jiang, Z., Feng, S., Zhang, H., Li, J., & Hu, C. (2019). Catalytic depolymerization of organosolv lignin to phenolic monomers and low molecular weight oligomers. *Fuel*, *244*, 247–257.

Liu, J., Wang, X., Dai, G., Zhang, Y., & Bian, X. (2022). Microbial chassis engineering drives heterologous production of complex secondary metabolites. *Biotechnology Advances*, *59*, 107966.

Lu, H., Yadav, V., Bilal, M., & Iqbal, H. M. (2022). Bioprospecting microbial hosts to valorize lignocellulose biomass–Environmental perspectives and value-added bioproducts. *Chemosphere*, *288*, 132574.

Luo, Z., Liu, C., Radu, A., et al. (2024). Carbon–carbon bond cleavage for a lignin refinery. *Nature Chemical Engineering*, *1*, 61–72. https://doi.org/10.1038/s44286-023-00006-0

Markus, J., & Zwicky, M. (2020). *Addressing environmental sustainability of biochemicals.* https://doi.org/10.1038/s41893-019-0442-8

Mckinlay, J. B., Vieille, C., & Zeikus, J. G. (2007). *Prospects for a bio-based succinate industry 727–740.* https://doi.org/10.1007/s00253-007-1057-y

Menezes, F. F., Nascimento, V. M., Gomes, G. R., Rocha, G. J. M., Strauss, M., Junqueira, T. L., & Driemeier, C. (2023). Depolymerization of enzymatic hydrolysis lignin: Review of technologies and opportunities for research. *Fuel*, *342*, 127796, ISSN 0016-2361,. https://doi.org/10.1016/j.fuel.2023.127796

Moncada, J., Gursel, I. V., Worrell, E., & Ramírez, A. (2018). Production of 1, 3-butadiene and ε-caprolactam from C6 sugars: Techno-economic analysis. Biofuels. *Bioprod. Biorefining*, *12*, 600–623.

Mondal, S., Ruidas, S., Chongdar, S., Saha, B., & Bhaumik, A. (2024). Sustainable Porous Heterogeneous Catalysts for the Conversion of Biomass into Renewable Energy Products. *ACS Sustainable Resource Management*, *1*(8), 1672–1704.

Otoupal, P. B., Geiselman, G. M., Oka, A. M., Barcelos, C. A., Choudhary, H., Dinh, D., ... Gladden, J. M. (2022). Advanced one-pot deconstruction and valorization of lignocellulosic biomass into triacetic acid lactone using Rhodosporidium toruloides. *Microbial Cell Factories*, *21*(1), 254.

Pereira, C. S. M., Silva, V. M. T. M., & Rodrigues, E. (2011). *Green Chemistry Ethyl lactate as a solvent : Properties, applications and production processes – a review 2658–2671.* https://doi.org/10.1039/c1gc15523g

Philp, J. (2023). Bioeconomy and net-zero carbon: lessons from Trends in Biotechnology, volume 1, issue 1. *Trends in Biotechnology*, *41*(3), 307–322.

Radoš, D., Carvalho, A. L., Wieschalka, S., Neves, A. R., Blombach, B., Eikmanns, B. J., & Santos, H. (2015). glutamicum for the production of 2, 3- butanediol. *Microbial Cell Factories, 1–14.* https://doi.org/10.1186/s12934-015-0362-x

Rajagopalan, G., Jain, A., Poosarla, V. G., Krishnan, C., & Yang, K. L. (2024). High level of 1,3-propanediol production from crude glycerol by Clostridium strain BOH3: critical roles of inoculum and substrate concentration. *Journal of Chemical Technology & Biotechnology, 99*(6), 1445–1458.

Rajeswari, S., Baskaran, D., Saravanan, P., Rajasimman, M., Rajamohan, N., & Vasseghian, Y. (2022). Production of ethanol from biomass–Recent research, scientometric review and future perspectives. *Fuel, 317*, 123448.

Ramos, L., Maltempi-Mendes, G., Siqueira, A. F., Napoleão, D. A. S., & Chandel, A. K. (2025a). Kinetic analysis of kraft lignin conversion via the Fenton process: process optimization and stochastic modelling. *Reaction Chemistry & Engineering, 10*(1), 119–129. 10.1039/D4RE00401A

Ramos, L., Maltempi-Mendes, G., Santos, J. C., & Chandel, A. K. (2025b). Kraft lignin depolymerization by the Fenton process assisted by hydrodynamic cavitation. *Sustainable Energy & Fuels.* 2025 10.1039/D5SE00436E

Ravi, K., García-Hidalgo, J., Gorwa-Grauslund, M. F., & Lidén, G. (2017). Conversion of lignin model compounds by Pseudomonas putida KT2440 and isolates from compost. *Applied Microbiology and Biotechnology, 101*(12), 5059–5070.

Renders, T., Van den Bosch, S., Koelewijn, S. F., Schutyser, W., & Sels, B. F. (2017). Lignin-first biomass fractionation: the advent of active stabilisation strategies. *Energy & Environmental Science, 10*(7), 1551–1557.

Renders, T., Van den Bossche, G., Vangeel, T., Van Aelst, K., & Sels, B. (2019). Reductive catalytic fractionation: state of the art of the lignin-first biorefinery. *Current Opinion in Biotechnology, 56*, 193–201.

Schutyser, W., Van den Bosch, S., Renders, T., De Boe, T., Koelewijn, S. F., Dewaele, A., ... Sels, B. F. (2015). Influence of bio-based solvents on the catalytic reductive fractionation of birch wood. *Green Chemistry, 17*(11), 5035–5045.

Schutyser, W., Renders, A. T., Van den Bosch, S., Koelewijn, S. F., Beckham, G. T., & Sels, B. F. (2018). Chemicals from lignin: an interplay of lignocellulose fractionation, depolymerisation, and upgrading. *Chemical Society Reviews, 47*(3), 852–908.

Sen, K. Y., & Baidurah, S. (2021). Renewable biomass feedstocks for production of sustainable biodegradable polymer. *Current Opinion in Green and Sustainable Chemistry, 27*, 100412.

Shapiro, A. J., O'Dea, R. M., Li, S. C., Ajah, J. C., Bass, G. F., & Epps, T. H., III (2023). Engineering innovations, challenges, and opportunities for lignocellulosic biorefineries: leveraging biobased polymer production. *Annual Review of Chemical and Biomolecular Engineering, 14*(1), 109–140.

Sheldon, R. A. (2021). *Biomass processing with biocatalysis* (pp. 113–146). Biomass Valorization: Sustainable Methods for the Production of Chemicals.

Shibukawa, V. P., Ramos, L., Cruz-Santos, M. M., Prado, C. A., Jofre, F. M., de Arruda, G. L., ... dos Santos, J. C. (2023). Impact of product diversification on the economic sustainability of second-generation ethanol biorefineries: a critical review. *Energies, 16*(17), 6384.

Silveira, M. H. L., Morais, A. R. C., Da Costa Lopes, A. M., Olekszyszen, D. N., Bogel-Łukasik, R., Andreaus, J., & Pereira Ramos, L. (2015). Current Pretreatment Technologies for the Development of Cellulosic Ethanol and Biorefineries. *ChemSusChem, 8*, 3366–3390. https://doi.org/10.1002/cssc.201500282

Singh, M., Chavali, M., Enamala, M. K., Obulisamy, P. K., Dixit, R., & Kuppam, C. (2020). In P. Verma (Ed.), *Algal Bioeconomy: A Platform for Clean Energy and Fuel BT - Biorefineries: A*

Step Towards Renewable and Clean Energy (pp. 335–370). Singapore: Springer Singapore. https://doi.org/10.1007/978-981-15-9593-6_13

Sun, Z., Cheng, J., Wang, D., Yuan, T. Q., Song, G., & Barta, K. (2020). Downstream processing strategies for lignin-first biorefinery. *ChemSusChem*, *13*(19), 5199–5212.

Sut, D., & Kataki, R. (2020). In P. Verma (Ed.), *A Biorefinery Based Zero-Waste Utilization of Non-edible Oilseeds for Biodiesel and Biofuel Production Along with Chemicals and Biomaterials* (pp. 21–55). Singapore: Springer Singapore. https://doi.org/10.1007/978-981-15-9593-6_2

Tellechea-Luzardo, J., Otero-Muras, I., Goñi-Moreno, A., & Carbonell, P. (2022). Fast biofoundries: coping with the challenges of biomanufacturing. *Trends in Biotechnology*, *40*(7), 831–842.

Tian, Q., Xu, P., Huang, D., Wang, H., Wang, Z., Qin, H., ... Zhao, Y. (2023). The driving force of biomass value-addition: Selective catalytic depolymerization of lignin to high-value chemicals. *Journal of Environmental Chemical Engineering*, *11*(3), 109719.

Vaidya, A. A., Murton, K. D., Smith, D. A., & Dedual, G. (2022). A review on organosolv pretreatment of softwood with a focus on enzymatic hydrolysis of cellulose. *Biomass Conversion and Biorefinery*, *12*(11), 5427–5442.

Velidandi, A., Kumar Gandam, P., Latha Chinta, M., Konakanchi, S., Reddy Bhavanam, A., Raju Baadhe, R., Sharma, M., Gaffey, J., Nguyen, Q. D., & Gupta, V. K. (2023). State-of-the-art and future directions of machine learning for biomass characterization and for sustainable biorefinery. *Journal of Energy Chemistry*, *81*, 42–63. https://doi.org/10.1016/j.jechem.2023.02.020

Vendamme, R., Behaghel de Bueren, J., Gracia-Vitoria, J., Isnard, F., Mulunda, M. M., Ortiz, P., ... Eevers, W. (2020). Aldehyde-assisted lignocellulose fractionation provides unique lignin oligomers for the design of tunable polyurethane bioresins. *Biomacromolecules*, *21*(10), 4135–4148.

Wang, P., Wang, S., Su, S., Zhang, D., Liao, Y., Song, G., & Wang, L. (2025). A techno-economically feasible and sustainable C-lignin biorefinery. *Green Chemistry*.

Witthayolankowit, K., Ramazanova, L., Baddigam, K. R., Marson, A., Apostolopoulou-Kalkavoura, V., Lebedeva, D., ... Samec, J. S. (2023). Valorization of tops and branches to textile fibers and biofuel: value chain explored experimentally; environmental sustainability evaluated by life cycle assessment. *ACS Sustainable Chemistry & Engineering*, *12*(1), 526–533.

Wong, S. S., Shu, R., Zhang, J., Liu, H., & Yan, N. (2020). Downstream processing of lignin derived feedstock into end products. *Chemical Society Reviews*, *49*(15), 5510–5560.

Wu, X., Smet, E., Brandi, F., Raikwar, D., Zhang, Z., Maes, B. U., & Sels, B. F. (2024). Advancements and perspectives toward lignin valorization via O-Demethylation. *Angewandte Chemie International Edition*, *63*(10), e202317257.

Xu, J., Zhou, P., Zhang, C., Yuan, L., Xiao, X., Dai, L., & Huo, K. (2022). Striding the threshold of photocatalytic lignin-first biorefining via a bottom-up approach: From model compounds to realistic lignin. *Green Chemistry*, *24*(14), 5351–5378.

Ye, X. P., & Ren, S. (2014). Value-added chemicals from glycerol. *Soy-based Chem. Mater.*, 43–80. https://doi.org/10.1021/bk-2014-1178.ch003

Yu, Z., Kong, W., Liang, W., Guo, Y., Cui, J., Yang, H., Sun, Z., Elangovan, S., & Fuqing, X. (2024). Heterogeneously catalyzed reductive depolymerization of lignin to value-added chemicals. *ChemSusChem*, e202401399.

Zeitsch, K. J. (2000). *The chemistry and technology of furfural and its many by-products.* Elsevier.

Zhang, B., Meng, Q., Liu, H., & Han, B. (2023). Catalytic conversion of lignin into valuable chemicals: Full utilization of aromatic nuclei and side chains. *Accounts of Chemical Research, 56*(24), 3558–3571.

Zhao, X., Zhang, L., & Liu, D. (2012). Biomass recalcitrance. Part I: the chemical compositions and physical structures affecting the enzymatic hydrolysis of lignocellulose. *Biofuels, Bioprod. Biorefining, 6*, 465–482. https://doi.org/10.1002/bbb.1331

Zhao, Y., Xue, L., Huang, Z., Lei, Z., Xie, S., Cai, Z., ... Xie, S. (2024). Lignin valorization to bioplastics with an aromatic hub metabolite-based autoregulation system. *Nature Communications, 15*(1), 9288.

Zhu, J. Y., & Pan, X. (2022). Efficient sugar production from plant biomass: Current status, challenges, and future directions. *Renewable and Sustainable Energy Reviews, 164*, 112583.

Zhu, Z., Liu, Y., Gómez, L. D., Wei, T., Yang, X., Simister, R., ... Macquarrie, D. J. (2021). Thermochemical pretreatments of maize stem for sugar recovery: Comparative evaluation of microwave and conventional heating. *Industrial Crops and Products, 160*, 113106.

2

Structural Carbohydrates Conversion into Sugars, Fuels, Chemicals, and Sustainable Materials

Katarina Mihajlovski[1], Nevena Ilić[2], Galina Jevđenović[1], and Marija Milić[1]

[1] *Faculty of Technology and Metallurgy, University of Belgrade, Belgrade, Serbia*
[2] *Innovation Centre of the Faculty of Technology and Metallurgy, Belgrade, Serbia*

2.1 Introduction

The widespread reliance on non-renewable resources such as coal, petroleum, and natural gas for producing industrially valuable chemicals and fuels presents significant environmental and sustainability challenges. As a result, there is a growing interest in exploring alternative sources of raw and renewable materials. Biomass-derived carbohydrates have emerged as a promising solution due to their abundance and cost-effectiveness. Also, replacing petroleum-based energy and chemicals is vital for a sustainable and effective circular economy in the future. The main constituents of plant biomass, cellulose (30–45%) and hemicellulose (15–35%) with pectin, are predominantly used for chemicals and biofuels production. This is very challenging due to the recalcitrant nature of some carbohydrates (e.g., cellulose) in biomass. Converting structural carbohydrates into industrially important materials holds great potential for reducing the dependence on non-renewable resources, addressing environmental concerns, and aggregating economical value to different biotechnological processes. Accordingly, covered by the latest knowledge in literature, this chapter describes the main structural carbohydrate (originated from biomass) conversions for obtaining biotechnological value-added products. Thus, cellulose, hemicellulose, and pectin, together with oligosaccharide conversions are discussed for the production of sugars and relevant chemicals, followed by biofuels and sustainable materials, highlighted through the viability potential of each of the selected process.

2.1.1 What are Structural Carbohydrates?

Carbohydrates are the main constituents of biomass and could be structural or non-structural (Figure 2.1).

Structural carbohydrates are long-chained molecules that build and solidify biomass structures (cell walls, stalks, and stems). They are part of the plant cell wall and bound in

Biofoundry Techniques for Biotechnology Applications, First Edition. Edited by Anuj Kumar Chandel.

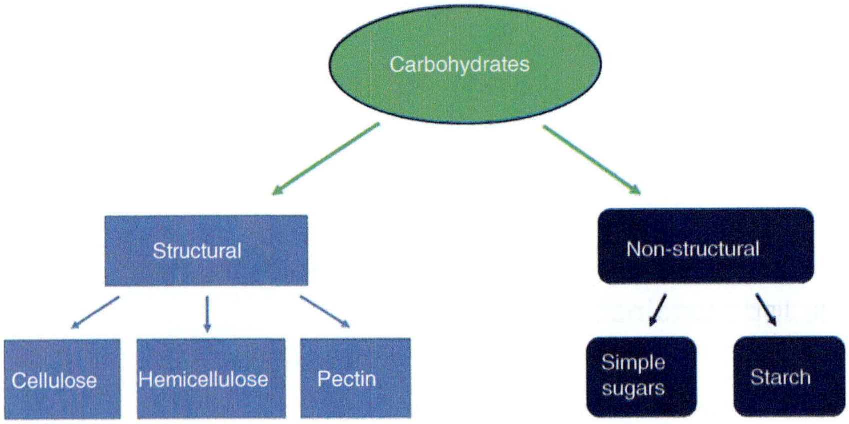

Figure 2.1 Carbohydrates division into structural and non-structural.

the matrix of the biomass, while non-structural carbohydrates are found within the plant cell and serve for the primary and secondary plant metabolism, and can be easily removed from biomass. The main structural carbohydrates in biomass are cellulose, hemicellulose, and pectin. Non-structural carbohydrates are sugars and starch (Hartmann & Trumbore, 2016; Sluiter et al., 2008).

The structural carbohydrates can be converted to simple sugars, oligosaccharides, chemicals, fuels, and sustainable chemicals by applying various processes and methods.

2.1.2 Cellulose

Cellulose, a primary polysaccharide, is a linear homopolymer of D-glucose units linked by stable β-1,4-glycosidic bonds. Its structure is highly crystalline, ranging from 50 to 80%, due to an extensive network of hydrogen bonds formed by the three hydroxyl groups on each glucose unit (Dutta, 2024). Despite being the most abundant biopolymer on Earth, with approximately 180 billion metric tons produced annually, only a small fraction is used in the production of organic chemicals. While cellulose presents challenges as a substrate for catalytic processes, its high abundance from non-food sources, low cost, and structural uniformity (consisting exclusively of glucose units) make it a highly promising feedstock for carbohydrate-centric biorefineries.

A thorough understanding of the molecular structure of these carbohydrates is crucial for developing efficient catalytic processes to convert them into valuable products. Cellulose conversion is a multifaceted and synergistic process (Katarina et al., 2024; Mihajlovski and Milić, 2022). It can occur through either chemical or biochemical means. Biochemical conversion offers a significant advantage over thermochemical methods, as it preserves the carbohydrate structure in the form of monomeric sugars. In contrast, thermochemical processes often lead to the destruction of these carbohydrates. Biochemical conversion is widely regarded as the most sustainable technology for saccharification. However, its efficiency is often hindered by the inherent recalcitrance and limited accessibility of cellulose, which poses a major challenge in biorefinery processes (Datta, 2024; Mihajlovski & Milić, 2022).

2.1.2.1 Cellulose Conversions to Sugars

2.1.2.1.1 Oligosaccharides

Cello-oligosaccharides are small carbohydrate molecules derived from cellulose through partial hydrolysis, sharing the same chemical structure but with a significantly smaller polymerization degree. Unlike cellulose, which is insoluble in water, cello-oligosaccharides with 2–6 degree of polymerization are fairly soluble at room temperature, while larger molecules can dissolve in hot water. The solubility of cello-oligosaccharides is crucial for their separation from unreacted cellulose (Chen et al., 2021).

During the depolymerization of cellulose, glycosylation can lead to the formation of various glycosidic linkages, including α-1,4, α/β-1,3, and α/β-1,6, particularly under water-deficient conditions. These secondary linkages can introduce branching, potentially enhancing solubility. Enzymatic synthesis of cello-oligosaccharides typically involves modifying cellulase enzyme cocktails to minimize the conversion of soluble oligosaccharides to glucose, as cellulase comprises endoglucanases, exoglucanases, and β-glucosidases (Chen et al., 2021; Mihajlovski & Milić, 2022). Endoglucanases cleave internal β-1,4 bonds in cellulose, creating chain ends, while exoglucanases produce cellobiose from these ends. β-Glucosidases further hydrolyze cellobiose and cello-oligosaccharides into glucose, which is undesirable for cello-oligosaccharide synthesis. Therefore, strategies focus on removing or inhibiting β-glucosidases, which can be separated by filtration after cellulase adsorption onto cellulose due to their lack of a cellulose-binding domain. A multi-stage reaction system effectively enhances product yield by removing the product from the reaction media, thereby mitigating inhibition through washing and hydrolyzing the substrate with fresh enzymes (Chen et al., 2021). Chu et al. (2014) demonstrated that their three-stage process achieved a cello-oligosaccharides yield of 52%, in contrast to only 25% yield in a single-stage process under identical conditions (Chu et al., 2014). Development of an efficient process for xylo- and cello-oligosaccharides production could be strategically important for sugarcane biomass-based biorefinery (Chandel et al., 2019). Silveira et al. (2018) studied the extraction and recovery of higher xylo- and cello-oligosaccharides (60 g/L) production from steam-exploded sugarcane bagasse (0.37) than sulfuric acid-catalyzed bagasse (14 g/L).

Alternative approaches for cello-oligosaccharide synthesis include homogeneous and heterogeneous acidic hydrolysis using acid catalysts, which also offer viable methods for obtaining cello-oligosaccharides. Historically, sulfuric acid and hydrochloric acid have been employed for the homogeneous acid hydrolysis of cellulose to produce cello-oligosaccharides. However, these strong acids tend to catalyze further hydrolysis into glucose, making them unsuitable for cello-oligosaccharide synthesis (Chen et al., 2021). In contrast, Liebert et al. (2008) developed an alternative method by initially treating cellulose with 85% phosphoric acid at room temperature for 30 minutes, followed by an extended hydrolysis period of 20 hours at 55 °C. This approach resulted in a cello-oligosaccharides yield of 53%, with an average degree of polymerization of 7.5 and a polydispersity index of 1.7 (Liebert et al., 2008). Additionally, impregnating cellulose with concentrated phosphoric acid under helium pressure and subsequently heating the substrate to 100 °C induced oligosaccharide formation with a 40% yield (Hirayama et al., 2007).

The increasing interest in cello-oligosaccharides is linked to their potential applications as prebiotics and elicitors in healthcare and agriculture. Enzymatic methods and heterogeneous catalysts are particularly promising due to their ability to produce processable cello-oligosaccharide solutions while allowing control over the reaction conditions. Semi-flow processes utilizing heterogeneous catalysts help prevent secondary hydrolysis of cello-oligosaccharides, maintaining their integrity (Chen et al., 2021). Additionally, carbon-based heterogeneous catalysts uniquely adsorb cellulose for hydrolysis, enabling the production of linear cello-oligosaccharides through aqueous phase reactions. For example, CMK-3 (mesoporous carbon material), a mesoporous carbon material, achieved a cello-oligosaccharides yield of 24% during the hydrolysis of cellulose (Datta, 2024; Kobayashi et al., 2010). In contrast, nonthermal technologies and mechano-catalysis in the absence of water can generate oligosaccharides featuring α-1,6-glycosidic branching. The versatility and potential of cello-oligosaccharides highlight their significance in various industrial applications (Chen et al., 2021).

2.1.2.1.2 Simple Sugars

Enzymatic degradation of cellulose is a synergistic process that involves specialized enzyme systems (endoglucanases, exoglucanases, and β-glucosidases) designed to break down this complex polysaccharide. Together, these enzymes act in concert to break down the cellulose structure at the interfaces, making the individual fibers more accessible for hydrolysis (Mihajlovski & Milić, 2022).

Commercial cellulase mixtures are primarily formulated from a diverse cellulolytic enzyme cocktail (Katarina et al., 2024; Mihajlovski & Milić, 2022). In a study by Obama et al. (2012), the enzymatic hydrolysis of raw and pre-processed miscanthus biomass was conducted using Celluclast 1.5 L (mainly cellobiohydrolase with endoglucanase) derived from *Trichoderma reesei*, along with a β-glucosidase preparation (Novozym 188) from *Aspergillus niger*, at loadings of 40 IFPU (international filter paper unit) and 80 IU per gram of cellulose, respectively. Both raw miscanthus and pre-hydrolyzed materials exhibited low to very low cellulose-to-glucose conversion rates, remaining below 20%. However, following organosolv treatment, the cellulose became more amenable to enzymatic hydrolysis, resulting in significantly improved glucose yields (approx. 80% at 170 °C, 1% sulfuric acid, 60 minutes) (Obama et al., 2012).

The limitations of enzyme systems are evident due to their low efficiency and high costs. Much research has focused on using mineral acids, particularly sulfuric acid, for the hydrolysis of cellulose to glucose. However, large-scale acid use presents issues like reactor corrosion, challenges in catalyst recovery, and the need to manage acid residues, resulting in significant waste. These factors must be considered in industrial process design, along with considerations of cost, simplicity, efficiency, and environmental impact. Heterogeneous acids may offer a solution to some of these challenges, as they facilitate easier catalyst separation. Recent advancements have explored various catalysts with large pore sizes and strong acidity to improve this process (Huang & Fu, 2013). Developing highly acidic, thermally stable nanometer-sized catalysts presents a challenge, but it is a promising area of research. Focusing on solid catalysts with unique characteristics, such as paramagnetic properties, could lead to the creation of practical systems for biomass hydrolysis (Guo et al., 2012). Utilizing nano Zn–Ca–Fe oxide as a catalyst in the hydrolysis of crystalline cellulose yielded impressive results. The catalyst demonstrated strong catalytic activity, achieving a cellulose conversion of 42.6% and a glucose selectivity of 69.2% (Zhang et al., 2011). In addition, other innovations such as ionic liquids, microwave irradiation, and supercritical water can facilitate the hydrolysis of cellulose under milder conditions (Guo et al., 2012). For example, Wang et al. (2011) demonstrated that cellulose extraction rates of up to 62% could be achieved, with microwave irradiation significantly reducing the reaction time from 15 hours to just 2 hours, in contrast to traditional oil bath heating (Wang et al., 2011).

2.1.2.2 Cellulose Conversions to Fuels

Lignocellulosic biomass, particularly cellulose, plays a crucial role in the advancement of second-generation biofuels, offering a sustainable alternative to first-generation biofuels derived from starch and sucrose. Ethanol, which currently dominates the global biofuel market, is primarily produced from these sources. In contrast, cellulose provides a more abundant feedstock for ethanol production through two main processes: hydrolysis and fermentation (Katarina et al., 2024; Mihajlovski & Milić, 2022).

Ethanol production typically employs either separate hydrolysis and fermentation (SHF) or simultaneous saccharification and fermentation (SSF). SHF allows for optimized conditions that speed up hydrolysis, while SSF improves overall productivity by combining hydrolysis and fermentation, though it may face challenges related to enzyme activity under varying conditions. A hybrid approach, semi-simultaneous saccharification and fermentation (SSSF), balances the benefits of both methods to enhance yield (Katarina et al., 2024; Mihajlovski & Milić, 2022). In a study focused on microcrystalline cellulose (Avicel PH101), researchers explored three different processes for ethanol production: 24 hours of hydrolysis followed by 48 hours of SSF (SSSF), 72 hours of SSF, and 24 hours of hydrolysis followed by 48 hours of fermentation (SHF). Hydrolysis was conducted at 50 °C and pH 4.8 using 2 g of Novozymes enzyme, while fermentation occurred at 36 °C, pH 4.8, and 300 rpm with *Saccharomyces cerevisiae*. The results indicated that the SSSF process yielded the highest ethanol productivity at 0.222 g/Lh, surpassing 0.194 g/Lh for SSF and 0.176 g/Lh for SHF. Additionally, the maximum ethanol concentrations reached were 16 g/L for SSSF, 14 g/L for SSF, and 12.6 g/L for SHF, translating to theoretical yields of 70.5%, 61.8%, and 56.1%, respectively (Shen & Agblevor, 2010).

For butanol production, cellulose must first be converted to glucose, as most butanol-fermenting microorganisms primarily metabolize glucose. Historically, the process involves enzymatic hydrolysis of polysaccharides to monosaccharides, followed by fermentation to produce biosolvents (Fatehi, 2013). However, this method is energy-intensive and costly. *Clostridia* species, such as *Clostridium acetobutylicum, Clostridium beijerinckii,* and *Clostridium pasteurianum* are commonly used for this purpose, though they require careful management of cell growth and sporulation to optimize butanol yield (Bramono et al., 2011).

Cellulose can also serve as a substrate for hydrogen production. Biological methods, employing thermophilic bacteria like *Clostridium thermocellum* and *Caldicellulosiruptor saccharolyticus,* are operating at lower temperatures compared to chemical methods, thus reducing energy input. These bacteria enhance enzymatic reactions and biomass pretreatment, making them effective for large-scale hydrogen production. For example, using mesophilic *Clostridium* strain BOH3, both glucose and xylose at concentrations of 30 g/L each resulted in butanol production of 4.67 and 4.63 g/L, respectively. However, increasing the inoculated cell concentration enhanced butanol yields to 7.05 and 7.41 g/L. Additionally, when cellulose and xylan were used at 10 g/L each, hydrogen production of 2.47 and

1.93 mmol was observed, indicating that strain BOH3 exhibits both xylanolytic and cellulolytic capabilities (Bramono et al., 2011).

Additionally, a new avenue in biofuel production involves the chemical conversion of cellulose into hydroxymethyl furfural (HMF), which can be further transformed into 2,5-dimethylfuran (DMF). This process can utilize hydrothermal conversion or reactions with concentrated acids, yielding high-purity products (Fatehi, 2013). The hydrothermal conversion of cellulose to HMF at temperatures between 275 and 300 °C for less than 30 minutes in homogeneous systems with sulfuric acid yielded approximately 20% of HMF. In contrast, the same process under neutral conditions resulted in a lower yield of less than 16%, although the purity of the HMF produced was higher, ranging from 50 to 60% (Yin et al., 2011). Conversely, treating cellulose with hot compressed water at 523 K for 5 minutes (at a pressure of 2.5 MPa under nitrogen) produced 15% HMF. Notably, the addition of 10% (wt) TiO_2 to this system enhanced the yield to 28% (Binder & Raines, 2009).

2.1.2.3 Cellulose Conversions to Sustainable Materials

Cellulose, a naturally occurring biopolymer from lignocellulosic biomass, is increasingly recognized for its environmental benefits, including biodegradability and recyclability. Its high mechanical strength, purity, and non-toxic properties make it suitable for diverse applications, including cellulose nanocrystals (CNCs), nanofibrils (CNFs), hydrogels, and aerogels (Kamalesh et al., 2023). CNCs and CNFs, collectively known as cellulose nanomaterials (CNs), are produced from sources like wood and bacteria. CNCs are obtained via acid hydrolysis, which disintegrates amorphous regions and yields stable crystalline domains. In contrast, CNFs are created through mechanical disintegration methods. Both types exhibit unique properties, including high surface area and mechanical resistance, making them ideal for various applications, including wastewater treatment (Mohammed et al., 2018).

Recent developments in cellulose-based hydrogels have attracted interest due to their ability to store moisture and biodegradability. These hydrogels can be synthesized from biomass, such as cotton linter pulp, and possess high electrostatic stability, which enhances their structural integrity. Cellulose aerogels, recognized for their low density and high porosity, are another promising application. They are derived from cellulose, which is inexpensive and sustainable. The production process for cellulose aerogels involves dispersing nanocellulose particles in water and can include techniques like crash-freezing and freeze-casting. Despite the higher energy costs associated with their production, cellulose aerogels offer a viable alternative as support materials due to their renewability, biocompatibility, and large specific surface area (Kamalesh et al., 2023).

Cellulose acetate (CA) is primarily produced by esterifying renewable cellulose sources such as wool, oak, rice husks, and bagasse. One method involves treating cellulose with acetic acid and using sulfuric acid as a catalyst, leading to the formation of cellulose triacetate, which is then hydrolyzed to produce cellulose acetate. CA exhibits desirable properties such as good mechanical strength, biodegradability, non-toxicity, and moisture management, making it suitable for applications in membranes, films, fibers, biodegradable plastics, and filters. However, it also has limitations, including high water consumption during purification and potential environmental impacts from acetic acid contamination, which can contribute to water eutrophication. Additionally, CA's sensitivity to water and fragility restrict its use in composite films (Kamalesh et al., 2023).

2.1.3 Hemicellulose

Hemicelluloses, the second most common polysaccharides in nature, represent about 20–35% of lignocellulosic biomass. Xylan is a main component of hemicellulose. Hemicelluloses are heterogeneous polymers that consist of pentoses (xylose and arabinose), hexoses (mannose, glucose, and galactose), and a small number of other saccharides, mainly rhamnose and fructose. Also, hemicellulose contains some uronic acids and acetyl groups. Pentose and hexose sugars are linked by β-1,4 and β-1,3 glycosidic bonds. The backbone of xylan is composed of xylose residues linked by β-1,4 linkages. Besides xylose, xylans may contain arabinose, glucuronic acid or its 4-O-methyl ether, and acetic, ferulic, and p-coumaric acids (Katarina et al., 2024; Saha, 2003).

2.1.3.1 Hemicellulose Conversion to Sugars

2.1.3.1.1 Oligosaccharides

In the last decades, the production of xylooligosaccharides (XOS) from hemicellulose (xylan) has gained more attention. Namely, XOS have many benefits to human and animal health, from potent prebiotics to prevention of colon cancer, antioxidant, and inflammatory role.

XOS are oligomers of 2–10 xylose units linked by β-1,4 bonds and are produced from hemicellulose using chemical, hydrothermal, and enzymatic hydrolysis. These oligomers could contain various side groups like acetlyl groups, thus forming a branched structure. Producing XOS from xylan presents key challenges that must be addressed for effective outcomes. First, reaching the optimal degree of polymerization (DP) is vital; the prebiotic benefits of XOS hinge on this, making molecules with a DP of 2–6 essential for success. Furthermore, the release of xylose or other degradation byproducts must be minimal as this directly impacts the purity of the final product (Poletto et al., 2020). There have been many studies published on the production of XOS, including auto-hydrolysis, enzymatic hydrolysis, and acid hydrolysis. Auto-hydrolysis, which is also known as hydrothermal hydrolysis, is a method that only uses water as a medium; however, it should be conducted at high temperatures (170–230 °C) and high pressures (5 Mpa). Due to the high requirements on equipment, auto-hydrolysis is not suitable or feasible for large-scale industrialization. Although enzymatic hydrolysis can produce XOS with a relatively high purity, the high cost, lower output, and non-recyclability of enzymes made it unfeasible for scaled production. Comparatively, acidic hydrolysis of xylan or xylan-rich biomass for producing XOS, with the advantages of rapid reaction, low cost, and high efficiency, has become increasingly recognized and accepted. Currently various acids, such as sulfuric, acetic, gluconic, and furoic, have been applied to assist xylan hydrolysis into XOS. Organic acids with a higher pKa were more effective in the production of XOS, while furfural was almost nonexistent. Among the tested organic acids (sulfuric, maleic, lactic, and acetic), lactic acid was the best catalyst for the hydrolysis of xylan to XOS (Zhang et al., 2023).

On the other hand, some authors stated that XOS production using acid hydrolysis is not suitable for industrial application due to the high corrosivity, high production costs, and the use of strong acids. Enzymatic hydrolysis is one of the most preferable methods for industrial XOS production due to the use of mild conditions, releasing less monosaccharides, less unwanted byproducts and reducing the amount of waste. Enzymes involved in xylan hydrolysis are β-xylosidase and endo xylanase, with minor activities of β-xylosidase.

β-xylosidase degrades xylobiose and XOS from the non-reducing end, releasing xylose. However, released xylose decreased the prebiotic effect of XOS. Conversions of xylan into XOS accessory enzymes such as α-l-arabinofuranosidase, α-d-glucuronidase, acetylxylan esterase, and ferulic acid esterases that are responsible for degradation of the side chain of the xylan backbone (Kaur et al., 2024).

A recent trend in enzymatic conversions of xylan is immobilized xylanase due to its enhanced stability, activity, and reusability. In line with that, a group of researchers in an environmentally friendly manner, immobilized endo-1,4-β-xylanase (recombinant endoxylanase from *Bacillus subtilis* KCX006) onto ordered mesoporous silica. A maximum yield of XOS reached 797.61 mg/g from Beechwood xylan. Their performance in terms of operational stability and ability to be reused was assessed through 10 cycles of XOS production. Obtained XOS was with a low degree of polymerization from 2 to 6 units (Kartik et al., 2024). A high yield of XOS could be obtained by using an integrated process based on autohydrolysis of xylan and using endo-xylanase. Alkali-extracted xylan (xylan fraction after alkaline treatment of corn straw) was subjected to autohydrolysis, where 48.37% of xylan was degraded into oligomers of which 57.83% was XOS in autohydrolysate. Byproducts and xylose were removed by nanofiltration. Adding endo-xylanase to autohydrolysates, and after the hydrolysis, an XOS yield was 96–98%. A total of 31.58 g XOS could be obtained from 100 g raw alkali-extracted xylan (Lian et al., 2020).

Mannans are long-chain carbohydrates present in hemicellulose found in the cell wall of most plants. They consist of β-(1→4)-linked D-mannose units forming linear chains. Four different types of mannan comprise linear mannan, galactomannan, glucomannan, and galactoglucomannan. Manno-oligosaccharides (MOS) are widely used in the animal feed industry, namely as prebiotics, with the aim of promoting the growth of beneficial gut bacteria in humans as well as in animals by suppressing enteric pathogens (*Salmonella* sp., *Escherichia coli*, *Campylobacte*r sp., and others) (Hlalukana et al., 2021). MOS could be obtained by cleaving mannan polysaccharides through chemical, physical, or enzymatic processes. MOS with 2–4 monosaccharide units has better health-promoting properties compared to oligosaccharides that consist of higher monosaccharide units. For example, end products derived from the enzymatic hydrolysis of mannan from copra meal by endo-β-mannanase (previously produced by *Aspergillus quadrilineatus* RSNK-1) revealed production of MOS with 2–4° of polymerization. These MOS promoted *Lactobacilli* growth and biofilm formation, promoting probiotic growth in the gut and helping to resolve gut-related problems. Furthermore, the obtained MOS in a concentration of 1000 μg/Ml showed cytotoxicity on Caco-2 cells (the Caco-2 cell line is from a colon carcinoma), reducing the viability of Caco-2 cells up to 40% (Suryawanshi & Kango, 2021).

Recently, a new method for producing MOS from guar gum was developed. Guar gum has a rich galactomannan content and is increasingly being explored for prebiotic production via MOS. To degrade mannan successfully, a combination of citric acid and β-mannanase was used. From 10 g of guar gum, after citric acid treatment (300 Mm, 130 °C, 1 h) and after enzymatic hydrolysis, 4.84 g of MOS was obtained. The obtained yield of MOS, 61.8%, was the highest reported in literature so far. It is interesting to note that after ion exchange chromatography for desalination, and spray drying, 4.57 g of MOS with a high purity of 90% was obtained (Xin et al., 2024).

2.1.3.1.2 *Simple Sugars*

The main route of converting hemicellulose is its hydrolysis to monosaccharides. This process is maintained by cleaving the glycosidic bonds in the polysaccharides by the addition of water in the presence of mineral, organic, or solid acids. Common acids for the hydrolysis are mineral acids, e.g., sulfuric acid, phosphoric acid, nitric acid, and hydrochloric acid, as well as organic acids such as trifluoroacetic acid, oxalic acid, and acetic acid.

An efficient hydrolysis of xylan for obtaining 100% of xylose yield was conducted during very acidic conditions at pH 0.5 using HCl and 0.5 wt.% of initial xylan at 90 °C for 20 hours of hydrolysis. Furthermore, for this conversion, temperature is an important factor due to the formation of furfural from monosaccharides at temperatures higher than 90 °C. When the pH of the reaction mixture is higher than 0.5 (pH near to 3) the reaction rate of xylan degradation is slow, and hardly any xylose was detected. On the other hand, if pH is less than 0.5, the undesired products of xylan hydrolysis occur (Hilpmann et al., 2016).

In the one-pot method, Brönsted acidic ionic liquid (BAIL)-catalyzed hydrolysis of hardwood hemicellulose (xylan from birchwood) at 160 °C in water media within 1 hour to produce C5 sugars, xylose, and arabinose, yielding 87% (Matsagar & Dhepe, 2015).

2.1.3.2 Hemicellulose Conversion to Chemicals

2.1.3.2.1 *Carboxylic Acids*

After hydrolysis, the hemicellulose fraction obtained from lignocellulosic material becomes rich in pentoses and hexoses. Therefore, it can be utilized for the production of C1–C3 carboxylic acids. In this single-step process, homogeneous catalysts are used. This process yields oxalic, levulinic, and lactic acids, which find applications in the food, pharmaceutical, and cosmetic industries. The highest yield of lactic acid of 64.57% was only obtained using a low catalyst concentration (0.1 M H_2SO_4), a temperature of 100 °C, and a short process time (1 hour). At higher concentrations of sulfuric acid (1 M H_2SO_4), oxalic acid is mainly formed with a yield above 99%, while a prolonged time of 5 hours and higher temperatures of 200 °C lead to only 29% of levulinic acid yield (Sobuś & Czekaj, 2023).

2.1.3.2.2 *Xylitol*

Xylitol, a sugar alcohol, belongs to a platform chemical obtained from the hemicellulose part xylan. As platform chemical xylitol is further converted into propylene glycol (PG), ethylene glycol (EG), glycerol, xylaric acid, lactic acid, and a mixture of hydroxyfurans. As a sugar alcohol, xylitol is added to foods and beverages as a substitute for sucrose. Xylitol is obtained from hemicellulose by either chemical or biotechnological methods. These methods include pretreatment of biomass, hydrolysis or saccharification of hemicellulose, detoxification of the obtained mixture rich in xylose, followed by fermentation of xylose in biotechnological methods, i.e., hydrogenation of xylose in chemical methods (Subroto & Hayati, 2020).

The xylitol yield in the fermentation medium is determined by selected lignocellulosic biomass, the conditions (pH, temperature, and duration) during pretreatment and hydrolysis, the detoxification strategy, and the strain of microorganism applied (Table 2.1).

Table 2.1 Xylitol Yield Obtained Through Fermentation of Hemicellulose from Different Feedstocks

Microorganism strains	Raw material	Pretreatment method + hydrolysis method	Fermentation conditions	Detoxification method	Xylitol,(g/L)	References
Kluyveromyces marxianus ATCC 36907	Oil palm fond	Acid pretreatment and hydrolysis (4.0% HNO_3, 130 °C, 20 min, HNO_3 to dry matter ratio 8 : 1)	Initial yeast extract concentration 3 g/L, 30 °C, 200 rpm, 96 h	$Ca(OH)_2$ + activated carbon	5.5	(Manaf et al. 2018)
Scheffersomyces amazonensis UFMG-HMD-26.3	Sugarcane bagasse and straw	Diluted acid pretreatment and hydrolysis (100 mg H_2SO_4 per g dry biomass, 121 °C, 20 min, 2 atm)	Initial inoculum concentrates on 1.0 g/L, 30 °C, 450 rpm, pH 5.5, 96 h	CaO + activated charcoal	21.56	(Silva et al. 2020)
Escherichia coli BL21(DE3)	Freshly-harvested kenaf stems	Diluted acid pretreatment and hydrolysis (2% [*v/v*] HNO_3, 130 °C, 60 min, HNO_3 to dry matter ratio 10 : 1)	30 °C, pH 7.0 ± 0.05, 96 h	Coconut shell granular activated carbon	6.8	(Shah et al. 2020)
Candida guilliermondii FTI 20037	Poplar tree	Diluted acid pretreatment and hydrolysis (1.75% [*w/w*] H_2SO_4, 120 °C, 120 min)	Inoculum (concentration 1×10^7 cells/mL) was added to a level of 10% (v/v), 32 °C, 150 rpm, pH 5.5, 120 h	Ion exchange resin Vacuum evaporation + solvent extraction	3.46 28.34	(Dalli et al. 2017)

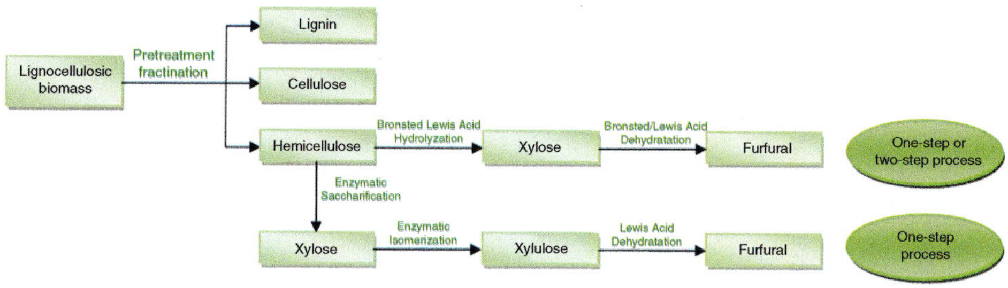

Figure 2.2 Block diagram of the hemicellulose conversion process from biomass to furfural.

2.1.3.2.3 Furfural

Obtaining furfural from renewable feedstocks is a crucial process due to its inherent sustainability. Furfural, derived from lignocellulosic biomass, is of great significance as it acts as a precursor for the production of numerous vital biochemicals, biofuels, and sustainable materials. Furfural and its derivatives are primarily used in the pharmaceutical, agrochemical, and plastics industries (Kabbour and Luque 2020).

To obtain furfural from hemicellulose, pretreatment is required. The purpose of pretreatment is to break the structure of lignocellulosic biomass so that the carbohydrate constituents become more accessible for further transformation. The efficiency of utilizing hemicellulose from biomass for furfural production is highly dependent on the pretreatment process, which can significantly enhance both yield and selectivity (Luo et al. 2019). The four main types of biomass pretreatment are physical, chemical, biological, and physico-chemical (Beluhan et al. 2023).

The process of converting hemicellulose to furfural may occur via a one-step or two-step process, utilizing either homogeneous or heterogeneous catalysts (Figure 2.2). The one-step process involves the direct production of furfural from hemicellulose. The two-step process includes a mild pretreatment in the first step, resulting in pentosan-rich hydrolysates. The second step involves filtration, resulting in a filtrate enriched with xylose and xylooligosaccharides. This filtrate is then subjected to harsh reaction conditions to yield furfural (Bao et al. 2024).

The yield of furfural obtained from hemicellulose depends on the raw material used, the conditions applied during the reaction, and the selected catalyst. Pretreatment conditions and catalysts that have provided a high furfural yield from hemicellulose obtained from specific lignocellulosic biomass are shown in Table 2.2.

2.1.3.3 Hemicellulose Conversion to Fuels

The production process of bioethanol and biobutanol from hemicellulose biomass occurs through several key steps, which include pretreatment, hydrolysis, fermentation, and distillation (purification) of the obtained bioethanol. After the "release" of hemicellulose by the action of an appropriate pretreatment, hydrolysis into fermentable sugars is performed. Enzymatic hydrolysis is frequently used, as it is fast, occurs under mild conditions, and eliminates the need for harmful chemicals (Yuan et al. 2021). After hydrolysis, certain toxic substances, such as formic acid, acetic acid, furfural, phenolic compounds, and

Table 2.2 Pretreatment Conditions and Catalysts that Resulted in a High Yield of Furfural from Hemicellulose Obtained from Specific Biomass

Type of process	Characteristics of the process	Raw Material	Pretreatment	Type of catalyst	Catalyst + organic solvent	Optimal furfural yield	References
One-step	Harsh pretreatment conditions or the addition of a large number of chemicals	Maple wood	170 °C, 30 min, 0.1 M	Homogeneous	$FeCl_3$ + tetrahydrofuran	95.04%	(Cai et al. 2014)
		Corncob	195 °C, 120 min, 0.1 g	Heterogeneous	Phosphotungstic acid-functionalized biochar + γ-valerolactone/subcritical water	96.06%	(Li et al. 2022)
Two-step	Separation of pre-hydroisolates formed in the pretreatment under mild conditions and harsh treatment conditions	Corn straw	120 °C, 60 min, 15 g/L $MnCl_2$ + 5 mL deep eutectic solvent	Homogeneous	$MgCl_2$ and ChCl/ HCl + 2-methyltetrahydrofuran	86.94%	(Hu et al. 2023)
		Pubescens	200 °C, 120 min, 5%	Heterogeneous	NaCl + tetrahydrofuran and γ-valerolactone	76.9 mol%	(Luo et al. 2017)

others, may remain in the reaction mixture. Consequently, detoxification can be applied after hydrolysis through different techniques, including evaporation or using activated carbon, ion-exchange resins, ammonium hydroxide, and more. While detoxification is optional, it may significantly improve bio-alcohol yields during the fermentation process (López-Linares et al. 2020). Hemicellulose hydrolysis primarily produces fermentable pentose sugars, with xylose being the most abundant. Microorganisms capable of fermenting pentoses into bioethanol are mostly yeasts, such as *Scheffersomyces stipitis*, *Pachysolen tannophilus*, *Candida shehatae*, *Pichia stipitis*, and others. Besides yeasts, certain intestinal bacteria are also capable of fermenting pentoses into bioethanol (Broda et al. 2022). Bioethanol yield that can be achieved from the fermentation of hemicellulose hydrolysates obtained from different feedstocks previously treated with appropriate pretreatment is shown in Table 2.3.

The fermentation of carbohydrates for biobutanol production is typically performed by strains of the *Clostridium sp.* (Chacón et al. 2020; Wu et al. 2020). Biobutanol and bioethanol yield can be achieved from the fermentation of hemicellulose hydrolysates obtained from various types of lignocellulosic biomass previously treated with appropriate pretreatment. For example, when fermenting previously pretreated sugarcane bagasse with diluted acid, *Clostridium saccharoperbutylacetonicum* DSM 14923 produced 10 g/L of both biobutanol and bioethanol (Chacón et al. 2020). When growing *C. beijerinckii F-6* on alkali-pretreated rice straw, 4.22 g/L of biobutanol and 3.0 g/L of bioethanol could be obtained (Wu et al. 2020).

2.1.3.4 Hemicellulose Conversion to Sustainable Materials

Xylan, as the main polysaccharide of hemicellulose, could be further used as a packaging material after the extraction from wood. However, due to their weak mechanical qualities, excessive moisture absorption, and poor film-forming ability, xylan-based packaging materials have limited practical applicability. Meanwhile, various approaches, such as blending and molecular engineering, have been put forth to create xylan packing materials with improved properties. Recently, xylan plastic was fabricated by a strategy of double cross-linking through etherification combined with hot-pressing. Some mechanical properties, like the toughness of xylan plastic, were significantly improved using the chemical and physical cross-linking domains. For instance, the tensile strength, toughness, and modulus of xylan plastic can reach up to 55 MPa, 2.2 MJ/m^3, and 1.7 GPa, respectively, which are superior to most traditional plastics. This xylan plastic is a thermoplastic, sustainable, biodegradable, biocompatible, and non-cytotoxic material with certain possible industrial applications (Jia et al. 2023).

2.1.4 Pectin

Pectin is a complex structural polysaccharide that is a major component of the primary cell walls of higher plants. Structurally, pectin is formed by a main chain, which is mainly composed of α-1,4-D-galacturonic acid (GalA), and numerous types of side chains containing almost 17 different monosaccharides (L-rhamnose (Rha), L-arabinose (Ara), and D-galactose (Gal), etc.), and over 20 different linkages. The linear regions are often called "smooth regions" or homogalacturonan (HG). Regions with side chains are known as

Table 2.3 Bioethanol Yield Achieved from the Fermentation of Hemicellulose Hydrolysate Obtained from Different Feedstock Previously Treated with Appropriate Pretreatment

Microorganism strains	Raw material	Pretreatment method + Hydrolysis method	Fermentation conditions	Detoxification method	Bioethanol(g/L)	References
Escherichia coli SL100	Exhausted olive pomace	Water extraction (100 °C, 30 min) + Acid hydrolysis (2% H_2SO_4, 170 °C, 0 min, 20% dry matter)	Initial cell concentration 0.2 g/L, 37 °C, 200 rpm, pH 6.5, 142 h (overlimed detox. hydrolysate) or 31 h (activated charcoal detox. hydrolysate)	Overliming	13.6	(López-Linares et al. 2020)
Scheffersomyces stipitis CECT 1922	Paulownia elongata × fortunei	Hydrothermal pretreatment (210 °C, 150 rpm, liquid to solid ratio 6 g/g) + Acid hydrolysis (1.5% H_2SO_4, 115 °C, 75 min)	Final cell concentration 10 g/L (quantified by dry cell weight), 30 °C, 100 rpm, 72 h	Overliming+ Activated charcoal+ Ion exchange resins	14.2	(Domínguez et al. 2021)
Scheffersomyces stipitis CBS6054	Brewers' spent grain	Acid pretreatment (1% H_2SO_4, 30 °C, 26 min, 12.5% dry matter) + Enzymatic hydrolysis (Novozymes A/S substrate concentration 5% [w/v], pH 4,8, 50 °C, 150 rpm, 24 h)	Initial cell concentration 1 g/L, 30 °C, 150 rpm, pH 5, 60 h	Non detoxification	11.4	(Rojas-Chamorro et al. 2020)
Scheffersomyces stipitis ATCC 58376	Sweet sorghum bagasse	Acid pretreatment and hydrolysis (1.75% H_2SO_4, 121 °C, 40 min, 5% dry matter)	Initial cell concentration 0.5 g/L, pH 5,5, 30 °C, 200 rpm, 78 h	pH change + Activated charcoal	22.0	(Camargo et al. 2019)

"hairy regions" and are represented by rhamnogalacturonans (RG-I and RG-II) and xylogalacturonan (XGA).

HG is the primary pectin domain, making up around 65% of total pectin content. It consists of GalA residues linked by α-1,4 glycosidic bonds, with methyl-esterified C-6 carboxyl groups or acetylated groups at C-2 and C-3. RG-I represents 20–35% of pectin and consists of GalA and rhamnose units (α-1,2 rhamnopyranose residues) with side chains of D-galactose and L-arabinose. The RG-II domain consists of 10% of pectin and consists of GaLA residues or HG, which is branched by different sugars such as rhamnose, D-galactose, L-fucose, D-xylose, D-glucose, D-apiose, D-galacturonic acid, borate, D-3-deoxy-lyxo-2-heptulosaric acid, and D-3-deoxy-manno-2-octulosonic acid. Methyl ester is also present in the side chains of RG-II. The XG domain presents HG with a single unit of D-xylose (Li et al. 2024).

The galacturonic acid residues in pectin are partly esterified and present as the methyl ester. The degree of esterification is defined as the percentage of carboxyl groups esterified. Pectin with a degree of esterification (DE) above 50% is named high methyl ester (HM) pectin or high ester pectin, and one with a DE lower than 50% is referred to as low methyl ester (LM) pectin or low ester pectin (Christensen et al. 2012).

Since the pectin has a very complex structure, a wide range of conversions exists. Structural conversions of pectin include many enzymatic and non-enzymatic reactions. One of the important issues related to pectin conversions is understanding the reasons behind its degradation. Various physicochemical conditions can disrupt the pectin structure, leading to degradation. Although the degradation of pectin is crucial for understanding its structure, the products resulting from pectin degradation or modification havc significant biotechnological value.

2.1.4.1 Pectin Conversions to Sugars

2.1.4.1.1 Oligosaccharides

Smaller fractions of pectin fragments are beneficial for juice quality by adding nutritional value and stabilizing polyphenols. For the degradation of sugar beet pulp pectin (1%), an ultrasound treatment (UST) is widely spread in the food processing industry. Larsen et al. (2021) combined UST together with the commercial enzymatic mix Rohapect® Classic into the continuous system for obtaining higher amounts of oligomers with medium molecular weight (Mw) around 150 kDa, with an enriched fraction of smaller oligomers (less than 30 kDa). These smaller oligomers are HG-derived, RG I-derived (RG-I), and rhamnogalacturonan (RG) II-derived residues. Oligomers of low Mw form soluble complexes with polyphenols like anthocyanins, leading to a stabilization of these value-adding compounds. It was assumed that HG-derived residues stabilize anthocyanins rich in hydroxy groups, while RG-derived residues probably play a role in the complex formation of anthocyanins rich in methoxy groups (Larsen et al. 2021).

2.1.4.1.2 Simple Sugars

The pectin network holds the cellulose and hemicellulose together, blocking their exposure to degradative enzymes (Xiao and Anderson 2013). Involving pectinase in cell wall degradation, the yield of galacturonic acid and neutral sugars improved the accessibility of cellulose to degradative enzymes (cellulase and xylanase) and the production of reducing sugars (Pakarinen et al. 2012). Arabinose (Ara) is the starting material to produce arabinitol, one of the 12 top value-added chemicals from biomass, which is used as a building block for unsaturated polyester resins. Arabinose in sugar beet pectin arabinans is predominantly present as terminal (1 → 5)-linked, (1 → 3,5)-linked, and (1 → 2,3,5)-linked residues. When hydrolyzing sugar beet pulp with α-l-arabinofuranosidase, previously immobilized on an epoxy carrier, arabinose is released. Cardenas-Fernandez et al. (2018) performed continuous hydrolysis of sugar beet (SB)-pectin using a packed bed reactor with immobilized arabinofuranosidase. During this process, continuous separation of Ara from the galacturonic acid-rich backbone was achieved in 7 days with 92% arabinose recovery (Cárdenas-Fernández et al. 2018).

To release various monosaccharides from pectin saccharides, trifluoroacetic acid (TFA) was used by Wu and coworkers (Wu et al. 2024). When the pectin (rhamnogalacturonan-I [RG-I] enriched pectin, Mw = 12.1 kDa, 56.29% RG-I content) concentration was 0.5 mg/mL, and 2 M TFA was employed at 110 °C for 6 hours, the highest level of galacturonic acid was observed, along with the preservation of galactose and arabinose. After 12 hours of pectin hydrolysis, xylose, rhamnose, arabinose, and glucose were fully liberated. The galacturonic acid content of different pectic polysaccharides was in the range between 68 and 80 mol% (mandarin peel, pear peel, pomelo peel and membrane, lemon peel, and red pomelo membrane) (Wu et al. 2024).

2.1.4.2 Pectin Conversions to Fuels

During the hydrolysis of lignocellulose biomass, microbial enzymatic cocktails showed a synergistic effect in the degradation of pectin, cellulose, and hemicellulose, liberating reducing sugars, which are then converted into ethanol using yeast fermentation. Table 2.4 listed some recent findings regarding bioethanol production from pectin-rich lignocellulose biomass using enzyme cocktails.

Apple pomace, citrus waste, and sugar beet pulp are pectin-rich biomass and thus are promising substrates for bioethanol production. However, during the fermentation of sugar beet pulp both ethanol and acetic acid is produced. In some cases, acetic acid, as a side product, decreases the amount of obtained ethanol. During fermentation of sugar beet pulp (50 g/L) *Clostridium thermocellum* produced 3.2 g/L of ethanol and 7.8 g/L of acetic acid (Spinnler et al., 1986).

2.1.4.3 Pectin Conversions to Chemicals

2.1.4.3.1 Uronic Acids

Uronic acids (L-arabinuronic acid, D-xyluronic acid, D-galacturonic acid, D-glucuronic acid, and D-mannuronic acid) are a group of monosaccharides, aldohexoses, in which terminal group hydroxyl groups have been oxidized to carboxylic acids. Uronic acids, particularly D-galacturonic and D-glucuronic acid, are highly valuable chemicals used in the pharmaceutical, cosmetic, and food industries as, among other applications, antioxidants, detoxifying and inactivating agents of various substances in the human body, precursors of ascorbic acid, and chelating reagents with anti-cancer properties.

Recently, subcritical water technology was used for the hydrolysis of high methyl ester citrus-apple pectin for obtaining uronic acids. During this treatment, 0.7 g of uronic acids

Table 2.4 Some Recent Findings in Bioethanol Production from Lignocellulose Biomass Using Enzyme Cocktails (Microbial and Commercial)

Enzymes	Lignocellulose biomass for hydrolysis	Reducing sugars	Bioethanol	References
Crude enzyme cocktail – pectinase, cellulase (CMCase and Avicelase), amylase, and xylanase produced by *Streptomyces fulvissimus* CKS 7	Horsetail (*Equisetum arvense*)	2.55 g/L	4.6 g/L	(Mihajlovski et al. 2021a)
	Yellow gentian (*Gentiana lutea*)	1.81 g/L	–	
Crude enzyme cocktail—pectinase and cellulase (CMCase and Avicelase) produced by *Hymenobacter* sp. CKS3	Corn stover	3.85 g/L	3.7 g/L	(Mihajlovski et al. 2021b)
Crude enzyme cocktail—pectinase, cellulase (CMCase and Avicelase), amylase, xylanase and laccase produced by *Bjerkandera adusta* TMF1	Brewer spent grain	5.16 g/L	0.94 g/L	(Ilić et al. 2022)
	Corn stover	4.94 g/L	0.86 g/L	
Pectinex® Ultra SP-L (a blend of pectinases, hemicellulases and beta-glucanases)	Banana biomass (rich in pectin)	268.08 g/L	28.02 g/L	(Favaretto et al. 2023)
	Apple biomass (rich in pectin)	150.22 g/L	20.11 g/L	
Celluclast® 1.5 L (cellulase- exoglucanase with endoclucanase) and Pectinex Ultra SP-L	Mixed agricultural waste (fruit and vegetable waste) (rich in pectin)	91% of sugar conversion	84.9% of ethanol yield	(Chang et al. 2024)

at 155 °C and for 36 minutes could be obtained from 1 kg of used pectin (degree of esterification more than 50%) (Pińkowska et al., 2019).

D-galacturonic acid is a uronic acid and could be obtained from pectin using the enzyme polygalacturonase. Polygalacturonase belonged to hydrolases and split the glycosidic linkage in the presence of water molecules across the oxygen bridge, forming a monomer of D-galacturonic acid (Khattab, 2022). The highest reported D-galacturonic acid concentration, 247 mmol/L for pectin hydrolysis, was reported when fermented solid was added directly to a 10% w/v pectin solution. Fermented solid with pectinolytic activity was previously produced using *Aspergillus oryzae* (Leh et al., 2017).

In the food industry, pectin conversions under high-pressure technology are applied over the past two decades for processing fruits and vegetables. When applying high-pressure technology to orange juice, the pressure 30–150 MPa increased the galacturonic acid (GalA) content and the linearity of pectin, while decreasing the molecular weight, pectin branching, and RG contribution, and cracks and pores of different sizes formed on the surface of pectin, implying depolymerization. Modification of pectin structure improved the stability of orange juice and turbidity significantly positively correlated with GalA and pectin linearity (Yu et al., 2021).

2.1.4.3.2 Methanol

It has been widely recognized that methanol is mainly formed from demethylation of pectin during fermentation and aging. However, although methanol is toxic to humans, nearly all fruits and alcoholic beverages contain methanol in safe concentrations, and thus it is essential to understand the production of methanol in apple wine fermentation for further controlling methanol content. Pectin methyl esterase and pectin lyase play important roles in controlling methanol production during apple wine fermentation or aging. Generally, pectin methyl-esterase breaks down the methyl ester group of pectin, releasing methanol and converting pectin into pectic acid or pectate via a single-chain mechanism. Lyases, as transeliminases, break down pectinate polymers by catalyzing the polygalacturonate depolymerization and pectin esterification by splitting the C-4 of the glycosidic linkage, followed by hydrogen removal from the C-5, releasing an unsaturated product. Han and Du (2022) reported that the key points for low-methanol-content apple wine making were to avoid adding Ca^{2+} and inhibit the pectin methyl esterase and pectin lyase activities effectively during fermentation and aging. Also, inhibiting the pectinases' activities, or shortening the aging period, would play an important role in decreasing the methanol content in apple wine (Han & Du, 2022).

2.1.4.3.3 Furfural

Furfural can be obtained through pectin hydrolysis using a catalyst such as sulfuric acid or some alkali solution. A maximum concentration of 54.7 g/L of furfural was obtained during the hydrolysis of tangerine pectin with $Ca(OH)_2$ for 120 minutes. Mango and tamarind pectins showed lower furfural conversions than tangerine pectin (citrus pectin) due to the lower DE (Guízar-Amezcua et al., 2022). Higher DE, as in tangerine pectin, favors the rupture of glycosidic bonds and subsequently GalA decomposition and furfural production. When 0.05 M sulfuric acid was used for mango peel pectin hydrolysis, 72.88 g/L of furfural was obtained for 90 minutes (López-Mercado et al., 2018). In both studies, a higher yield of furfural was obtained when using hydrolysis with acid/base and the Maillard reaction (77.5 g/L for the Maillard reaction with H_2SO_4 and 71.8 g/L for the alkaline Maillard reaction) (Guízar-Amezcua et al., 2022; López-Mercado et al., 2018).

2.1.4.3.4 Organic Acids

Pectin and pectin oligosaccharides, consisting of galacturonic acid (GalA), rhamnose (Rha), arabinose (Ara), and galactose (Gal), have been used as prebiotic substrates, as these compounds stimulate the activity and growth of beneficial gut bacteria. Pectin-derived oligosaccharides have shown properties of stimulation as these enhanced the growth of beneficial bacteria in the colon, apoptosis of colon cancer cells, and protection against various pathogens (Chandel et al., 2022). Pectic substrates are slowly but completely fermented. By the activity of gut microbiota, short-chain fatty acids (SCFAs) may be obtained. As a result of intestinal fermentation of pectin, acetate, propionate, and butyrate are produced (Roman-Benn et al., 2023). For degradation of pectin and the production of SCFAs, the following gut microbiota species are responsible: *Bifidobacterium*, *Faecalibacterium*,

Anaerostipes, Roseburia, Clostridium, Lachnospira, Faecalibacterium, etc. (Roman-Benn et al., 2023). The structural characteristics of pectic substrates have been shown to influence their utilization by the gut microbiota, and thus the formation of SCFAs. For example, low-methoxy pectin (LM; they have less than 50% of their carboxyl groups esterified with methanol) is fermented faster than high-methoxy pectins (high methoxyl [HM]; have over 50% of their carboxyl groups esterified with methanol), while HM pectins can induce a higher SCFA production, especially propionate (Pascale et al., 2022).

2.1.4.4 Pectin Conversions to Sustainable Materials

Hydrothermal conversion of biomass is a useful technique for biomass conversion into valuable carbon compounds. These carbon compounds are biotechnological value-added products and may play a main role in reducing pollution and environmental protection. Pectin as a carbon precursor was used for synthesizing nanoporous carbon materials by hydrothermal carbonization and subsequent chemical activation. The obtained carbon materials showed a good surface area and nanoporosity with a complex particle morphology. Fourier transform infrared spectroscopy (FTIR) analysis confirmed the very low content of oxygen-containing functional groups and the efficiency of the activation procedure using KOH (potassium hydroxide) as the activating agent. This material presented a remarkable CO_2 capture capacity (13–7 mmol CO_2/gADS) at room temperature and 1 atm (Lagazzo et al., 2016). To avoid the use of hazardous chemicals, researchers developed carbon microspheres based on pectin for removing the synthetic organic dyes from wastewater. Citrus pectin-derived carbon microspheres were synthesized by a facile hydrothermal method without any hazardous chemicals, showing superior adsorption capacity, a high adsorption rate, and good reusability for methylene blue adsorption, thus holding great potential for treating wastewater (Zhang & Zhou, 2017).

Recently, pectin-based plastic films were developed as food packaging materials. These films have the potential to replace petroleum-based plastics as packaging materials. The development of applied technology is crucial, but still challenging to create food packaging film with both protective capability and biodegradability. Scientists developed a multifunctional pectin-based material by grafting vanillin to pectin chains and introducing Fe^{3+} ions. Vanillin-grafted pectin was synthesized via an acid-catalyzed acetal reaction between the pectin hydroxyl groups and the vanillin aldehyde groups. Then Fe^{3+} ions are introduced to form metal–ligand bonding between pectin chains and vanillin. This film possesses high mechanical strength, processability, gas barrier properties, moisture resistance, and ultraviolet (UV)-light barrier. Since the pectin is biodegradable, these plastic films can be completely degraded in soil within 7 weeks, and fragmented into small hydrogel pieces in seawater. Furthermore, they exhibited much better freshness preservation performance than polyethylene and also presented superior processability (Qiang et al., 2024).

To expand the use of pectin, new pectin derivatives with potential antimicrobial activity were developed. For this purpose, carboxylic and hydroxyl functional groups of pectin were used to obtain improved pectin products. For example, the carboxylic acid groups of pectin have been utilized to form zirconium (IV) selenotungstophosphate nanocomposite for the removal of methylene blue and malachite green dyes with antimicrobial activity against Gram-negative *Escherichia coli* (Gupta et al., 2015). Recently, researchers developed pectin derivatives using copper and iron terpyridine (Tpy-Cu and Tpy-Fe) complexes. First, pectin was extracted from the sugar beet pulp and then derivatives with terpyridine-metal-pectin were made. The terpyridine-metal-pectin derivatives retained the known emulsification activity of sugar beet pectin plus antibacterial properties. Cytotoxicity tests on HepG2, MCF7, and HCT116 cell lines showed good safety for the terpyridine-Fe-pectin and moderate cytotoxicity for the terpyridine-Cu-pectin derivative. The results indicate the potential use of the terpyridine pectin derivatives as antibacterial materials and also in preparing emulsions with antibacterial properties (Hassan et al., 2021).

2.1.5 Oligosaccharides

The development of biorefinery processes to platform chemicals for most lignocellulosic substrates results in side processes to intermediates such as oligosaccharides. Agrofood wastes are most amenable to produce such intermediates, as many common plant food contain oligosaccharides including fruits, wheat, soybean, rye, barley, mustard, and vegetables, like artichoke, chicory, leek, garlic, asparagus, sugar beet, onion, tomato, and bamboo shoots (Cano et al., 2020). These compounds show a remarkable bioactivity as prebiotics, elicitors in plants, food complements, healthy coadjutants in certain therapies, and more. They are medium to high added-value compounds with an increasing impact in the pharmaceutical, nutraceutical, cosmetic, and food industries (Cano et al. 2020). Thus are regarded as functional oligosaccharides, and may be classified into several groups, like, sucrose-related oligosaccharides (fructooligosaccharides [FOS]), starch-related oligosaccharides (maltooligosaccharides (MAOS), isomaltooligosaccharides (IMAOS), cyclodextrines (CD), trehalose), and others-oligosaccharides (xylooligosaccharides (XOS), soy-oligosaccharides (raffinose, stachiose, verbascose), arabinooligosaccharides (AOS)) (Belorkar & Gupta, 2016; Ibrahim, 2018; Kherade et al., 2021).

2.1.5.1 Oligosaccharides Conversions to Sugars

The hydrolysis of oligosaccharides into monomeric sugars is crucial in an integrated biorefinery framework. Industrially important simple sugars may be obtained by the mineral acid or enzymatic hydrolysis of naturally derived oligosaccharides. For example, xylose may be obtained by the activity of the β-xylosidases on the XOS (Rohman et al., 2019). β-xylosidases are crucial hydrolytic enzymes since they hydrolyze more glycosidic bonds than any of the other xylanolytic enzymes. Lagaert et al. (2011) used two recombinant β-xylosidases, XylB and XylC, with different substrate specificities from *Bifidobacterium adolescentis* to degrade XOS. XylB shows only weak activity on xylobiose and prefers xylooligosaccharides with a degree of polymerization above two. The remaining xylobiose is efficiently hydrolyzed by the second *B. adolescentis* β-xylosidase, XylC, a glycoside hydrolase (GH) of family 43. Furthermore, XylB releases more xylose from arabinose-substituted xylooligosaccharides than XylC (30% and 20%, respectively) (Lagaert et al., 2011).

There is another valuable investigation of Zhou et al. (2018) who reported about novel amylolytic enzyme CoMA from *Corallococcus* sp. EGB, with the unique ability to catalyze the conversion of MAOS to maltose as the sole hydrolysate. CoMA showed both hydrolysis and transglycosylation activities toward α-1,4-glycosidic bonds but not to α-1,6-linkages. Moreover, glucosyl transfer was postulated to be the major transglycosidation reaction for producing a high level of maltose without the attendant production of glucose. These results indicated that CoMA possesses some unusual properties that distinguish it from maltogenic amylases and typical α-amylases (Zhou et al., 2018). Feller et al. (2021) reported about α-amylase paralogue Amyrel from *Drosophila* that possesses both the hydrolytic α-amylase activity (EC 3.2.1.1) and a 4-α-glucanotransferase (EC 2.4.1.25) transglycosylation activity. Amyrel reacts specifically on α-(1–4) glycosidic bonds of MAOS, releasing glucose as the major end-product, followed by maltose.

A thorough hydrolysis of pectin-derived acidic oligosaccharides by the commercial enzymatic preparations allows the recovery of monomers (Bonnin et al., 2014). Galacturonic acid, arabinose, and galactose are among the main building units, whereas rhamnose and xylose are present in minor amounts. Galacturonic acid can be transformed into surface-active agents by esterification with various fatty acids or into L-ascorbic acid by a yeast conversion. Arabinose has some interesting health applications, such as its derivatization to 5-deoxy-L-arabinose for anti-Parkinson treatment. Rhamnose can be chemically transformed into aromas such as furaneol used in caramel, roasted, and fruit flavor applications (Bonnin et al., 2014). L-arabinose may be obtained, as well, by the use of a thermostable recombinant α-L-arabinofuranosidase from *Caldicellulosiruptor saccharolyticus* for the complete hydrolysis of arabinooligosaccharides (within 13 hours in the presence of 30 U/mL of enzyme, at 75 °C), demonstrating the potential in the commercial production of L-arabinose in concert with endoarabinanase and/or xylanase (Lim et al., 2010). Another group of authors characterized three arabinofuranosidases, AbfA, AbfB, and AXH-d3 from *B. adolescentis*, belonging to GH family, which released only arabinose and no xylose from arabinoxylanoligosaccharides (Lagaert et al., 2010).

Soy-oligosaccharides may be a good source of fermentable sugars, by the activity of the α-galactosidase (Ibrahim, 2018). Industrial application of many studied α-galactosidases has been hindered for a long time due to the low enzymatic activity, high concentration of raffinose family oligosaccharides (RFOs), and inhibition of galactose, as well as the slow and incomplete hydrolysis toward RFOs (Cao et al., 2007). However, hydrolysis of 99% of stachyose was achieved with 10 U/mL RmgalB from *Rhizomucor miehei* in 8 hours (Chen et al., 2015), followed by a novel study by Huang et al. (2018) who revealed that raffinose may be almost completely hydrolyzed by AgaB in just 30 minutes, while 52% of stachyose was hydrolyzed. Furthermore, after *in vitro* fermentation of raffinose by the human gut bacteria, in the supernatant of *Enterococcus avium* E5, the authors detected 3.1 ± 0.2 and 0.4 ± 0.0 g/L of glucose and fructose, respectively (Mao et al., 2018). Later, the authors used a purified α-galactosidase from the dried fruit bodies of *Tremella aurantialba*, which completely hydrolyzed stachyose and partially hydrolyzed raffinose to galactose (at 50 °C, within 6 hours); afterward, the content of reducing sugar reached 4.36 mg/mL (Geng et al., 2020).

β-fructofuranosidase (EC 3.2.1.26) may liberate glucose and fructose from FOS, such as β-fructofuranosidase from *Bifidobacterium lactis* DSM 10140T, expressing a high affinity to cleave a variety of FOS (Janer et al., 2004).

A recent study by Lee et al. (2023) reported about oligosaccharide hydrolysis by the rat intestinal extract, affected by the action of both mammalian α-glucosidases and the intestinal microbiota. Thereby, XOS were partially degraded to xylose monomers, while IMAOS were almost entirely broken down into glucose, as compared to the action of fungal amyloglucosidase, which was enabled to fully hydrolyze IMAOS because of its very low hydrolytic activity on α-1,6 linkages (Song et al., 2022).

2.1.5.2 Oligosaccharides Conversions to Fuels

Early studies reported about ethanol obtaining after β-glucooligosaccharides, derived from oat bran, were fermented by the use of lactic acid bacteria (LAB), whereby 5.0–5.7 nmol/mL were produced (Kontula et al., 1998). Those authors found that the ethanol production was more inducible on the XOS compared to other oligosaccharides, and counted 14–15 nmol/mL (Kontula et al., 1998). Arenahalli Ningegowda and Siddalingaiya Gurudutt (2012) also used LAB, *Lactobacillus plantarum* CFR 2194, an isolate from kanjika, for obtaining ethanol in a much higher concentration of 80 mmol/L, after 18 hours of FOS fermentation. Lappa et al. (2015b) performed acidogenesis of raffinose and a mixture of raffinose and sucrose by bacterial anaerobic culture immobilized on γ-alumina and kissiris for ethanol production. This approach was addressed as a promising for new-generation ester-based biofuel production from industrial wastes. Depending on the reaction performance, different outputs were derived, whereby ethanol volume production varied up to seven times (from 0.20 to 1.40 mL/L). Previously, Atiyeh and Duvnjak (2003) investigated the production of ethanol and high fructose syrups using raw materials based on raffinose. They used the mutant *Saccharomyces cerevisiae* ATCC 36858 in the production of ethanol and/or fructose from synthetic media in batch processes with raffinose. The mutant was able to hydrolyze the raffinose and to selectively ferment glucose and galactose to ethanol while fructose accumulated in the fermentation medium. The fructose fraction in the carbohydrate content of the produced syrup was more than 96%, while the ethanol yield was 72% of the theoretical value in the media with raffinose (Atiyeh & Duvnjak, 2003).

The work of Gong et al. (2014) provided inspiring information for oligosaccharides utilization, which should facilitate more efficient production of biofuels. It has been demonstrated that *Cryptococcus curvatus* can utilize oligoxyloses as the sole carbon source for microbial lipid production. After 72 hours, lipid content and lipid coefficient were 30.3% and 0.17 g/g consumed sugar, respectively. The lower lipid coefficient on oligoxyloses suggests that *C. curvatus* probably metabolizes xylose through the pentose phosphate pathway rather than through the phosphoketolase pathway. Relative fatty acid content (%) of oligoxyloses fermentation by *C. curvatus* was as follows: 0.8 ± 0.0, 42.1 ± 0.8, 13.8 ± 0.3, 41.0 ± 1.5, and 0.9 ± 0.1 for myristic, palmitic, stearic, oleic, and linoleic acid, respectively (Gong et al., 2014). In an early study, oligoxyloses recovered from steam-exploded wheat straw were consumed by *Microsphaeropsis* sp. for lipid production, since the oleaginous fungus was able to secrete xylanase to degrade oligoxyloses (Xiaowei & Hongzhang, 2012). Because biomass hydrolysates usually contain monosaccharides as well as various amounts of oligosaccharides, direct utilization of oligosaccharides for lipid production should further promote full utilization of biomass (Gong et al., 2014).

2.1.5.3 Oligosaccharides Conversions to Chemicals

2.1.5.3.1 Organic Acids

Organic acids are the final products of oligosaccharide fermentation by anaerobic intestinal microorganisms (Wichienchot et al., 2017). Oligosaccharides are more challenging substrates than monosaccharides for microorganisms, because assimilation of oligosaccharides may require additional hydrolytic enzymes and transportation systems. To enable the consumption of oligosaccharides, microbes should secrete or surface display glycosidase to enable extracellular hydrolysis or harbor a dedicated transport system to take up oligosaccharides for intracellular utilization (Gong et al., 2014). Microorganisms that can directly assimilate biomass-derived oligosaccharides for the production of biofuels or biochemicals would be much more advantageous. Also, the rate at which oligosaccharides are fermented depends on the degree of polymerization, sugar and glycosidic linkage and degree of branching, synergy between bacteria during fermentation, relationship between substrate bacteria and fermentation products, nature of the fermentations, and saccharolytic capacity (Mussatto & Mancilha, 2007).

A team of researchers from China explored the effect of different oligosaccharides on the production of small organic acids, SCFAs and others, during *in vitro* fermentation by human fecal microbiota (Yao et al., 2022). FOS fermentation resulted in the highest production of acetate, followed by lactate, propionate, and butyrate, respectively, whereby the amounts of monosaccharides glucose and fructose were also found (Yao et al., 2022). A considerable amount was previously obtained by Stewart et al. (2008), 45.8 mM of total SCFA, after only 4 hours, with the acetic acid as the main product (39.7 mM), followed by propionic (4.1 mM) and butyric acids (12.8 mM). FOS was degraded into small molecules through a series of enzymes. The intestinal microbes convert oligosaccharides to monosaccharides through different metabolic pathways and further to pyruvate, which is the key intermediate for subsequent SCFAs and lactate. Thus, lactate and acetate were mainly produced via the pentose phosphate pathway and pyruvate pathway, while propionate was produced via the succinate pathway (Yao et al., 2022). According to the heatmap results of taxa correlation with SCFA production during oligosaccharide fermentation by fecal microbiota, strains *Olsenella, Veillonella, Clostridium, Bifidobacterium, Weissella, Megamonas, Collinsella*, and *Oenococcus* were found to be significantly positively correlated with the production of lactate, acetate, and propionate (Yao et al., 2022), and notably, *Lactococcus, Megamonas, Erysipelotrichaceae, Dorea, Caenorhabditis*, and *Pseudomonas* showed a highly significant positive correlation with butyrate production (Yao et al., 2022).

Similar to anaerobic intestinal microorganisms, many LAB were also found to ferment oligosaccharides and produce organic acids, which vary in their type and concentration. Arenahalli Ningegowda and Siddalingaiya Gurudutt (2012) distinguished *L. plantarum* CFR 2194, an isolate from kanjika, for obtaining of lactic acid as the main metabolic end product, in concentration of near 200 mM, after 18 hours of FOS fermentation. The following metabolites were acetic acid, butyric acid, and formic acid, in approx. Concentrations of 20, 40, and 30 mM, respectively (Arenahalli Ningegowda & Siddalingaiya Gurudutt, 2012). A more recent study reported about lower organic acids concentrations after FOS fermentation by *Lactiplantibacillus* and *Levilactobacillus* strains (up to 20.42, 16.17, 14.21, and 69.45 mM, for lactic, acetic, citric, and oxalic acid, respectively). However, this study cares an importance for obtaining a high concentrations of oxalic acid, an industrially important chemical, used mainly in cleaning and washing of surfaces, equipment and textile, and in drugs manufacturing and wood preservation (Abouloifa et al., 2020). A recent study revealed the possible combination of probiotic *Lactiplantibacillus* and *Levilactobacillus* strains with XOS that may be used in production of mixture of organic acids dominated by malic acid. Malic acid is used in food industry, as favoring and acidulant agent, and in pharmaceuticals and cosmetics development. The lactic, acetic, citric, oxalic, and malic acids values ranges obtained were 41.27–51.13, 13.20–15.94, 28.34–39.20, 8.10–12.87, and 57.87–61.53 mmol/L, respectively (Abouloifa et al., 2020).

Raffinose-type oligosaccharides were also found to be industrially important for the purpose of organic acid production by bioconversion processes. For example, the intestinal microbiota showed preference for raffinose as a substrate, showing the highest consumption value at 48 hours, while increasing the total production of lactate and short-chain fatty acids (up to 179.6 ± 0.6 mmol/L) (Amorim et al., 2020). Hernandez-Hernandez et al. (2011) found that acetic acid was the main short-chain fatty acid produced during the fermentation process of raffinose-based oligosaccharides, particularly pentasaccharides, at 10 hours (63.11 mM). Another group of authors declared that lactic acid was the major product after raffinose fermentation by the inoculum containing lactate-producing bacteria (*Streptococcus* and *Enterococcus*), while the acetic and butyric acids were prevalent at fermentation with the enrichment of *Catenibacterium* (Adamberg et al., 2018). A group of authors from Greece performed acidogenesis of raffinose and a mixture of raffinose and sucrose by bacterial anaerobic culture immobilized on γ-alumina (Lappa et al., 2015a) and kissiris (Lappa et al., 2015b), as promoter, for organic acid production, including acetic, propionic, butyric, lactic, succinic, malic, valeric, and isovaleric acids.

Furthermore, a recent study by Hu et al. (2024) reported about effects of pectin oligosaccharides during the fermentation by intestinal flora on the SCFA production. By the mass spectrum, the number of differential fermentation products was classified, whereby three major categories of substances were distinguished as acids and their derivatives and organic heterocyclic compounds. The above 68 significantly enriched pathways consisted of 38 differential fermentation products. Acetic acid was the highest content of all SCFAs, followed by butyric acid and propionic acid, and the lowest content was valeric acid (Hu et al., 2024). In the work of Bonnin et al., (2014) feruloyl esterase was used for the release of ferulic acid from pectin rich agro-industrial by-products. Ferulic acid and ferulic acid derivatives have antioxidant properties and could therefore play a role in a wide spectrum of therapeutic effects including the prevention of cardiovascular diseases. Ferulic acid can also be used as a feedstock for its biocatalytic conversion into other valuable molecules such as styrenes, polymers, epoxides alkylbenzenes, and catechol. Attention has been paid to its bioconversion in aromas and particularly in vanillic acid and vanillin (Bonnin et al., 2014).

Some authors performed in vitro fermentation by the human feces of arabinoxylan oligosaccharides from wheat bran and psyllium seed husk. Acetic, propionic, and butyric acids were the main, followed by the isobutyric and isovaleric acids, within 24–48 hours of incubation (Pollet et al., 2012). Similar were obtained after fermentation of AOS from sugar beet, where acetate, propionate, and butyrate derived were 27.73, 2.54, and 2.97 nM, respectively (Vigsnæs et al. 2011). Later, the same group of authors marked off phenylalanine, xanthine, linoleic acid, or its derivatives, produced by AOS fermentation by human intestinal microbiota (Vigsnæs et al., 2011). Also, a certain fermentation fraction had a higher content of metabolites corresponding to 3-oxoalanine, tyramine, and homoveratric acid, N′-acetylspermidine, cysteine, and aminobenzoic acid. Xanthine and allantoic acid are products of bacterial purine metabolism. 3-oxooctadecanoic acid and (R)-3-hydroxy-octadecanoic acid are building blocks of the unsaturated fatty acids. The co-occurrence of the clostridial taxa (Cl and Cc) and fatty acid derivative no. 14 indicates involvement of clostridial butyrate producers in the microbial interaction leading to production of this particular derivative (Sulek et al., 2014).

2.1.5.3.2 Vitamins

There is a valuable investigation of Wichienchot et al. (2017) who reported about vitamin B1 and folic acid production in a concentration of 19.19 and 20.00 μg/mL, respectively, after fecal fermentation of FOS, produced specifically by FTase isolated from Jerusalem artichoke.

2.2 Conclusions

This chapter provides comprehensive insights into ongoing developments in structural carbohydrate conversions, deriving outcomes applicable in diverse industrial sectors. These sectors primarily encompass biofuel production, being seen as a growth spurt necessary alternative to fossil fuels, but also many non-energy applications. For sustainable use, efficient technologies in biomass utilization for deriving industrially important products are necessary. Thus, new research trends in conversion processes are dedicated to the integration of green solvents, advanced nanoparticle technologies, and functional catalysts, poised to enable the transition from chemical processes to biomass-derived alternatives. In addition, as regards to bioconversions, microorganisms that can directly assimilate biomass-derived structural carbohydrates for the production of biofuels or biochemicals would be much more advantageous.

References

Abouloifa, H., Khodaei, N., Rokni, Y., et al. (2020). The prebiotics (Fructo-oligosaccharides and Xylo-oligosaccharides) modulate the probiotic properties of *Lactiplantibacillus* and *Levilactobacillus* strains isolated from traditional fermented olive. *World Journal of Microbiology and Biotechnology*, *36*, 1–12.

Adamberg, K., Adamberg, S., Ernits, K., et al. (2018). Composition and metabolism of fecal microbiota from normal and overweight children are differentially affected by melibiose, raffinose and raffinose-derived fructans. *Anaerobe*, *52*, 100–110.

Amorim, C., Silvério, S. C., Cardoso, B. B., et al. (2020). In vitro fermentation of raffinose to unravel its potential as prebiotic ingredient. *LWT*, *126*, 109322.

Arenahalli Ningegowda, M., & Siddalingaiya Gurudutt, P. (2012). In vitro fermentation of prebiotics by Lactobacillus plantarum CFR 2194: selectivity, viability and effect of metabolites on β-glucuronidase activity. *World Journal of Microbiology and Biotechnology*, *28*, 901–908.

Atiyeh, H., & Duvnjak, Z. (2003). Utilization of raffinose and melibiose by a mutant of Saccharomyces cerevisiae. *Journal of Chemical Technology & Biotechnology*, *78*(10), 1068–1074.

Bao, Y., Du, Z., Liu, X., et al. (2024). Furfural production from lignocellulosic biomass: one-step and two-step strategies and techno-economic evaluation. *Green Chemistry*.

Belorkar, S. A., & Gupta, A. (2016). Oligosaccharides: a boon from nature's desk. *Amb Express*, *6*, 1–11.

Beluhan, S., Mihajlovski, K., Šantek, B., et al. (2023). The production of bioethanol from lignocellulosic biomass: Pretreatment methods, fermentation, and downstream processing. *Energies*, *16*(19), 7003.

Binder, J. B., & Raines, R. T. (2009). Simple chemical transformation of lignocellulosic biomass into furans for fuels and chemicals. *Journal of the American Chemical Society*, *131*(5), 1979–1985.

Bonnin, E., Garnier, C., & Ralet, M.-C. (2014). Pectin-modifying enzymes and pectin-derived materials: applications and impacts. *Applied microbiology and biotechnology*, *98*, 519–532.

Bramono, S. E., Lam, Y. S., Ong, S. L., et al. (2011). A mesophilic Clostridium species that produces butanol from monosaccharides and hydrogen from polysaccharides. *Bioresource technology*, *102*(20), 9558–9563.

Broda, M., Yelle, D. J., & Serwańska, K. (2022). Bioethanol production from lignocellulosic biomass—challenges and solutions. *Molecules*, *27*(24), 8717.

Cai, C. M., Nagane, N., Kumar, R., et al. (2014). Coupling metal halides with a co-solvent to produce furfural and 5-HMF at high yields directly from lignocellulosic biomass as an integrated biofuels strategy. *Green Chemistry*, *16*(8), 3819–3829.

Camargo, D., Sydney, E. B., Leonel, L. V., et al. (2019). Dilute acid hydrolysis of sweet sorghum bagasse and fermentability of the hemicellulosic hydrolysate. *Brazilian Journal of Chemical Engineering*, *36*, 143–156.

Cano, M. E., García-Martin, A., Comendador Morales, P., et al. (2020). Production of oligosaccharides from agrofood wastes. *Fermentation*, *6*(1), 31.

Cao, Y., Yang, P., Shi, P., et al. (2007). Purification and characterization of a novel protease-resistant α-galactosidase from Rhizopus sp. F78 ACCC 30795. *Enzyme and microbial technology*, *41*(6–7), 835–841.

Cárdenas-Fernández, M., Hamley-Bennett, C., Leak, D. J., et al. (2018). Continuous enzymatic hydrolysis of sugar beet pectin and L-arabinose recovery within an integrated biorefinery. *Bioresource technology*, *269*, 195–202.

Chacón, S. J., Matias, G., dos Santos Vieira, C. F., et al. (2020). Enabling butanol production from crude sugarcane bagasse hemicellulose hydrolysate by batch-feeding it into molasses fermentation. *Industrial Crops and Products*, *155*, 112837.

Chandel, A. K., Albarelli, J. Q., dos Santos, D. T., Chundawat, S. P. S., Puri, M., & Meireles, M. A. A. (2019). Comparative Analysis of Key Technologies for Cellulosic Ethanol Production from Brazilian Sugarcane Bagasse at the Commercial-Scale. *Biofuels, Bioproducts and Biorefining*, *13*, 994–1014.

Chandel, V., Biswas, D., Roy, S., et al. (2022). Current advancements in pectin: extraction, properties and multifunctional applications. *Foods*, *11*(17), 2683.

Chang, J., Song, Y., Lee, D.-S., et al. (2024). Efficient enzymatic saccharification of agricultural wastes for the production of bioethanol, D-allulose and lactic acid. *Process Biochemistry*, *144*, 54–63.

Chen, Z., Yan, Q., Jiang, Z., et al. (2015). High-level expression of a novel α-galactosidase gene from Rhizomucor miehei in Pichia pastoris and characterization of the recombinant enyzme. *Protein expression and purification*, *110*, 107–114.

Chen, P., Shrotri, A., & Fukuoka, A. (2021). Synthesis of cello-oligosaccharides by depolymerization of cellulose: A review. *Applied Catalysis A: General*, *621*, 118177.

Christensen, S. H., Hansen, K. M., & Trudso, J. E. (2012). *Process for making de-esterified pectins their composition and uses thereof*. Google Patents.

Chu, Q., Li, X., Xu, Y., et al. (2014). Functional cello-oligosaccharides production from the corncob residues of xylo-oligosaccharides manufacture. *Process Biochemistry*, *49*(8), 1217–1222.

Dalli, S. S., Patel, M., & Rakshit, S. K. (2017). Development and evaluation of poplar hemicellulose prehydrolysate upstream processes for the enhanced fermentative production of xylitol. *Biomass and bioenergy*, *105*, 402–410.

Datta, R. (2024). *Enzymatic degradation of cellulose in soil: A review*. Heliyon.

Domínguez, E., PGd, R., Romaní, A., Garrote, G., & Domingues, L. (2021). Hemicellulosic bioethanol production from fast-growing Paulownia biomass. *Processes*, *9*(1), 173.

Dutta, S. (2024). Catalytic Transformation of Carbohydrates into Renewable Organic Chemicals by Revering the Principles of Green Chemistry. *ACS omega*, *9*(25), 26805–26825.

Fatehi, P. (2013). Production of biofuels from cellulose of woody biomass. *Cellulose-Biomass Conversion*, 46–74.

Favaretto, D. P. C., Rempel, A., Lanzini, J. R., et al. (2023). Fruit residues as biomass for bioethanol production using enzymatic hydrolysis as pretreatment. *World Journal of Microbiology and Biotechnology*, *39*(6), 144.

Feller, G., Bonneau, M. D., & Lage, J.-L. (2021). Amyrel, a novel glucose-forming α-amylase from Drosophila with 4-α-glucanotransferase activity by disproportionation and hydrolysis of maltooligosaccharides. *Glycobiology*, *31*(9), 1134–1144.

Geng, X., Yang, D., Zhang, Q., et al. (2020). Good hydrolysis activity on raffinose family oligosaccharides by a novel α-galactosidase from Tremella aurantialba. *International journal of biological macromolecules*, *150*, 1249–1257.

Gong, Z., Wang, Q., Shen, H., et al. (2014). Conversion of biomass-derived oligosaccharides into lipids. *Biotechnology for biofuels*, *7*, 1–11.

Guízar-Amezcua, M.-G., Pineda-Santana, A., González-Domínguez, M.-I., et al. (2022). Evaluation of pectin extractions and their application in the alkaline Maillard reaction. *Scientific Reports*, *12*(1), 19834.

Guo, F., Fang, Z., Xu, C. C., et al. (2012). Solid acid mediated hydrolysis of biomass for producing biofuels. *Progress in Energy and Combustion Science*, *38*(5), 672–690.

Gupta, V. K., Sharma, G., Pathania, D., et al. (2015). Nanocomposite pectin Zr (IV) selenotungstophosphate for adsorptional/photocatalytic remediation of methylene blue and malachite green dyes from aqueous system. *Journal of Industrial and Engineering Chemistry*, *21*, 957–964.

Han, Y., & Du, J. (2022). Relationship of the methanol production, pectin and pectinase activity during apple wine fermentation and aging. *Food Research International*, *159*, 111645.

Hartmann, H., & Trumbore, S. (2016). Understanding the roles of nonstructural carbohydrates in forest trees–from what we can measure to what we want to know. *New phytologist*, *211*(2), 386–403.

Hassan, E. A., Abou Elseoud, W. S., Abo-Elfadl, M. T., et al. (2021). New pectin derivatives with antimicrobial and emulsification properties via complexation with metal-terpyridines. *Carbohydrate Polymers*, *268*, 118230.

Hernandez-Hernandez, O., Côté, G. L., Kolida, S., et al. (2011). In vitro fermentation of alternansucrase raffinose-derived oligosaccharides by human gut bacteria. *Journal of agricultural and food chemistry*, *59*(20), 10901–10906.

Hilpmann, G., Becher, N., Pahner, F.-A., et al. (2016). Acid hydrolysis of xylan. *Catalysis Today*, *259*, 376–380.

Hirayama, J., Sahar, S., Grimaldi, B., et al. (2007). CLOCK-mediated acetylation of BMAL1 controls circadian function. *Nature*, *450*(7172), 1086–1090.

Hlalukana, N., Magengelele, M., Malgas, S., et al. (2021). Enzymatic conversion of mannan-rich plant waste biomass into prebiotic mannooligosaccharides. *Foods*, *10*(9), 2010.

Hu, M., Yu, Y., & Liu, Y. (2023). Integrated strategies for furfural production and lignocellulose fractionation in aqueous deep eutectic solvents. *Industrial Crops and Products*, *204*, 117334.

Hu, H., Zhang, P., Liu, F., et al. (2024). Regulations of Citrus Pectin Oligosaccharide on Cholesterol Metabolism: Insights from Integrative Analysis of Gut Microbiota and Metabolites. *Nutrients*, *16*(13), 2002.

Huang, Y.-B., & Fu, Y. (2013). Hydrolysis of cellulose to glucose by solid acid catalysts. *Green chemistry*, *15*(5), 1095–1111.

Huang, Y., Zhang, H., Ben, P., et al. (2018). Characterization of a novel GH36 α-galactosidase from Bacillus megaterium and its application in degradation of raffinose family oligosaccharides. *International journal of biological macromolecules*, *108*, 98–104.

Ibrahim, O. O. (2018). Functional oligosaccharides: Chemicals structure, manufacturing, health benefits, applications and regulations. *J Food Chem Nanotechnol*, *4*(4), 65–76.

Ilić, N., Davidović, S., Milić, M., et al. (2022). Valorization of lignocellulosic wastes for extracellular enzyme production by novel Basidiomycetes: screening, hydrolysis, and bioethanol production. *Biomass conversion and biorefinery*, 1–12.

Janer, C., Rohr, L. M., Peláez, C., et al. (2004). Hydrolysis of oligofructoses by the recombinant β-fructofuranosidase from Bifidobacterium lactis. *Systematic and applied microbiology*, *27*(3), 279–285.

Jia, S., Lv, Z., Rao, J., et al. (2023). *Xylan plastic. ACS nano*, *17*(14), 13627–13637.

Kabbour, M., & Luque, R. (2020). Furfural as a platform chemical: From production to applications. *Biomass, biofuels, biochemicals*, 283–297.

Kamalesh, T., Kumar, P. S., Hemavathy, R., et al. (2023). A critical review on sustainable cellulose materials and its multifaceted applications. *Industrial Crops and Products*, *203*, 117221.

Kartik, J. P. M., Dutta, P., Chandraraj, K., et al. (2024). Xylooligosaccharides as emerging prebiotics and their sustainable generation from xylan catalysed by endoxylanase immobilized ordered mesoporous silica. *Molecular Catalysis*, *564*, 114287.

Katarina, M., Ilić, N., & Milić, M. (2024). *The Realm of Smart Biomass Degrading Enzymes in Low-Carbon Fuels and Chemicals Production, Biorefinery and Industry 4.0: Empowering Sustainability* (pp. 113–179). Springer.

Kaur, G., Kaur, P., Kaur, J., et al. (2024). Xylanase, xylooligosaccharide and xylitol production from lignocellulosic biomass: Exploring biovalorization of xylan from a sustainable biorefinery perspective. *Industrial Crops and Products*, *215*, 118610.

Khattab, A. M. (2022). *The microbial degradation for pectin.* IntechOpen: Pectins-The New-Old Polysaccharides.

Kherade, M., Solanke, S., Tawar, M., et al. (2021). Fructooligosaccharides: A comprehensive review. *J. Ayurvedic Herb. Med, 7*, 193–200.

Kobayashi, H., Komanoya, T., Hara, K., et al. (2010). Water-tolerant mesoporous-carbon-supported ruthenium catalysts for the hydrolysis of cellulose to glucose. *ChemSusChem: Chemistry & Sustainability Energy & Materials, 3*(4), 440–443.

Kontula, P., von Wright, A., & Mattila-Sandholm, T. (1998). Oat bran β-gluco-and xylo-oligosaccharides as fermentative substrates for lactic acid bacteria. *International journal of food microbiology, 45*(2), 163–169.

Lagaert, S., Pollet, A., Delcour, J. A., et al. (2010). Substrate specificity of three recombinant α-L-arabinofuranosidases from Bifidobacterium adolescentis and their divergent action on arabinoxylan and arabinoxylan oligosaccharides. *Biochemical and biophysical research communications, 402*(4), 644–650.

Lagaert, S., Pollet, A., Delcour, J. A., et al. (2011). Characterization of two β-xylosidases from Bifidobacterium adolescentis and their contribution to the hydrolysis of prebiotic xylooligosaccharides. *Applied microbiology and biotechnology, 92*, 1179–1185.

Lagazzo, A., Finocchio, E., Petrini, P., et al. (2016). Hydrothermal synthesis of pectin derived nanoporous carbon material. *Materials Letters, 171*, 212–215.

Lappa, K., Kandylis, P., Bastas, N., et al. (2015a). New generation biofuel: continuous acidogenesis of sucrose-raffinose mixture simulating vinasse is promoted by γ-alumina pellets. *Biotechnology for biofuels, 8*, 1–10.

Lappa, K., Kandylis, P., Bekatorou, A., et al. (2015b). Continuous acidogenesis of sucrose, raffinose and vinasse using mineral kissiris as promoter. *Bioresource Technology, 188*, 43–48.

Larsen, L. R., van der Weem, J., Caspers-Weiffenbach, R., et al. (2021). Effects of ultrasound on the enzymatic degradation of pectin. *Ultrasonics Sonochemistry, 72*, 105465.

Lee, W.-M., Song, Y.-B., Han, K.-S., et al. (2023). Hydrolysis of oligosaccharides in the gastrointestinal tract alters their prebiotic effects on probiotic strains. *Food Science and Biotechnology*, 1–6.

Leh, D. S., Biz, A., De Paula, D., et al. (2017). Conversion of citric pectin into D-galacturonic acid with high substrate loading using a fermented solid with pectinolytic activity. *Biocatalysis and Agricultural Biotechnology, 11*, 214–219.

Li, X., Lu, X., Hu, W., et al. (2022). Phosphotungstic acid functionalized biochar for furfural production from corncob. *Fuel Processing Technology, 229*, 107178.

Li, J., Peng, C., Mao, A., et al. (2024). An overview of microbial enzymatic approaches for pectin degradation. *International Journal of Biological Macromolecules, 254*, 127804.

Lian, Z., Wang, Y., Luo, J., et al. (2020). An integrated process to produce prebiotic xylooligosaccharides by autohydrolysis, nanofiltration and endo-xylanase from alkali-extracted xylan. *Bioresource Technology, 314*, 123685.

Liebert, T., Seifert, M., & Heinze, T. (2008). Efficient Method for the preparation of pure, water-soluble cellodextrines. *Macromolecular symposia. Wiley Online Library*, 140–149.

Lim, Y. R., Yoon, R. Y., Seo, E. S., et al. (2010). Hydrolytic properties of a thermostable α-l-arabinofuranosidase from Caldicellulosiruptor saccharolyticus. *Journal of applied microbiology, 109*(4), 1188–1197.

López-Linares, J. C., Gómez-Cruz, I., Ruiz, E., et al. (2020). Production of ethanol from hemicellulosic sugars of exhausted olive pomace by *Escherichia coli*. *Processes*, *8*(5), 533.

López-Mercado, J., Nambo, A., Toribio-Nava, M.-E., et al. (2018). High and low esterification degree pectins decomposition by hydrolysis and modified Maillard reactions for furfural production. *Clean Technologies and Environmental Policy*, *20*, 1413–1422.

Luo, Y., Li, Z., Zuo, Y., et al. (2017). A simple two-step method for the selective conversion of hemicellulose in pubescens to furfural. *ACS Sustainable Chemistry & Engineering*, *5*(9), 8137–8147.

Luo, Y., Li, Z., Li, X., et al. (2019). The production of furfural directly from hemicellulose in lignocellulosic biomass: A review. *Catalysis Today*, *319*, 14–24.

Manaf, S. F. A., Jahim, J. M., Harun, S., et al. (2018). Fractionation of oil palm fronds (OPF) hemicellulose using dilute nitric acid for fermentative production of xylitol. *Industrial Crops and Products*, *115*, 6–15.

Mao, B., Tang, H., Gu, J., et al. (2018). In vitro fermentation of raffinose by the human gut bacteria. *Food & function*, *9*(11), 5824–5831.

Matsagar, B. M., & Dhepe, P. L. (2015). Brönsted acidic ionic liquid-catalyzed conversion of hemicellulose into sugars. *Catalysis Science & Technology*, *5*(1), 531–539.

Mihajlovski, K. R., & Milić, M. D. (2022). The role of plant cell wall degrading enzymes in biorefinery development. *Lignocellulose bioconversion through white biotechnology*, 99–135.

Mihajlovski, K., Buntić, A., Milić, M., et al. (2021a). From agricultural waste to biofuel: enzymatic potential of a bacterial isolate *Streptomyces fulvissimus* CKS7 for bioethanol production. *Waste and Biomass Valorization*, *12*, 165–174.

Mihajlovski, K., Pecarski, D., Rajilić-Stojanović, M., et al. (2021b). Valorization of corn stover and molasses for enzyme synthesis, lignocellulosic hydrolysis and bioethanol production by *Hymenobacter* sp. *CKS3. Environmental Technology & Innovation*, *23*, 101627.

Mohammed, N., Grishkewich, N., & Tam, K. C. (2018). Cellulose nanomaterials: promising sustainable nanomaterials for application in water/wastewater treatment processes. *Environmental Science: Nano*, *5*(3), 623–658.

Mussatto, S. I., & Mancilha, I. M. (2007). *Non-digestible oligosaccharides: A review. Carbohydrate polymers*, *68*(3), 587–597.

Obama, P., Ricochon, G., Muniglia, L., et al. (2012). Combination of enzymatic hydrolysis and ethanol organosolv pretreatments: Effect on lignin structures, delignification yields and cellulose-to-glucose conversion. *Bioresource technology*, *112*, 156–163.

Pakarinen, A., Zhang, J., Brock, T., et al. (2012). Enzymatic accessibility of fiber hemp is enhanced by enzymatic or chemical removal of pectin. *Bioresource technology*, *107*, 275–281.

Pascale, N., Gu, F., Larsen, N., et al. (2022). The potential of pectins to modulate the human gut microbiota evaluated by in vitro fermentation: A systematic review. *Nutrients*, *14*(17), 3629.

Pińkowska, H., Krzywonos, M., Wolak, P., et al. (2019). Production of uronic acids by hydrothermolysis of pectin as a model substance for plant biomass waste. *Green Processing and Synthesis*, *8*(1), 683–690.

Poletto, P., Pereira, G. N., Monteiro, C. R., et al. (2020). Xylooligosaccharides: transforming the lignocellulosic biomasses into valuable 5-carbon sugar prebiotics. *Process Biochemistry*, *91*, 352–363.

Pollet, A., Van Craeyveld, V., Van de Wiele, T., et al. (2012). In vitro fermentation of arabinoxylan oligosaccharides and low molecular mass arabinoxylans with different structural properties from wheat (Triticum aestivum L.) bran and psyllium (Plantago ovata Forsk) seed husk. *Journal of agricultural and food chemistry*, *60*(4), 946–954.

Qiang, T., Ren, W., & Chen, L. (2024). Biodegradable, high mechanical strength, and eco-friendly pectin-based plastic film. *Food Hydrocolloids*, *149*, 109539.

Rohman, A., Dijkstra, B. W., & Puspaningsih, N. N. T. (2019). β-Xylosidases: Structural diversity, catalytic mechanism, and inhibition by monosaccharides. *International Journal of Molecular Sciences*, *20*(22), 5524.

Rojas-Chamorro, J. A., Romero, I., López-Linares, J. C., et al. (2020). Brewer's spent grain as a source of renewable fuel through optimized dilute acid pretreatment. *Renewable Energy*, *148*, 81–90.

Roman-Benn, A., Contador, C. A., Li, M.-W., et al. (2023). Pectin: An overview of sources, extraction and applications in food products, biomedical, pharmaceutical and environmental issues. *Food Chemistry Advances*, *2*, 100192.

Saha, B. C. (2003). Hemicellulose bioconversion. *Journal of industrial microbiology and biotechnology*, *30*(5), 279–291.

Shah, S. S. M., Luthfi, A. A. I., Jahim, J. M., et al. (2020). An improvement in fermentability of acid-hydrolysed hemicellulose from kenaf stem for xylitol production. *International Journal of Food Engineering*, *16*(10), 20190230.

Shen, J., & Agblevor, F. A. (2010). Modeling semi-simultaneous saccharification and fermentation of ethanol production from cellulose. *Biomass and Bioenergy*, *34*(8), 1098–1107.

Silva, D. D., Dussán, K. J., Idarraga, A., et al. (2020). Production and purification of xylitol by *Scheffersomyces amazonenses* via sugarcane hemicellulosic hydrolysate. *Biofuels, Bioproducts and Biorefining*, *14*(2), 344–356.

Silveira, M. H. L., Chandel, A. K., Vanelli, B. A., Sacilotto, K. S., & Cardoso, E. B. (2018). Production of hemicellulosic sugars from sugarcane bagasse via steam explosion employing industrially feasible conditions: Pilot scale study. *Bioresource Technology Reports*, *3*, 138–146.

Sluiter, A., Hames, B., Ruiz, R., et al. (2008). Determination of structural carbohydrates and lignin in biomass. *Laboratory analytical procedure*, *1617*(1), 1–16.

Sobuś, N., & Czekaj, I. (2023). Catalytic transformation of biomass-derived hemicellulose sugars by the one-pot method into oxalic, lactic, and levulinic acids using a homogeneous H2SO4 catalyst. *Catalysts*, *13*(2), 349.

Song, Y.-B., Lamothe, L. M., Rodriguez, N. E. N., et al. (2022). New insights suggest isomaltooligosaccharides are slowly digestible carbohydrates, rather than dietary fibers, at constitutive mammalian α-glucosidase levels. *Food Chemistry*, *383*, 132456.

Spinnler, H.-E., Lavigne, B., & Blachere, H. (1986). Pectinolytic activity of Clostridium thermocellum: its use for anaerobic fermentation of sugar beet pulp. *Applied microbiology and biotechnology*, *23*, 434–437.

Stewart, M. L., Timm, D. A., & Slavin, J. L. (2008). Fructooligosaccharides exhibit more rapid fermentation than long-chain inulin in an in vitro fermentation system. *Nutrition Research*, *28*(5), 329–334.

Subroto, E., & Hayati, F. (2020). Chemical and biotechnological methods for the production of xylitol: a review. *Int J*, *8*(6).

Sulek, K., Vigsnaes, L. K., Schmidt, L. R., et al. (2014). A combined metabolomic and phylogenetic study reveals putatively prebiotic effects of high molecular weight arabino-oligosaccharides when assessed by in vitro fermentation in bacterial communities derived from humans. *Anaerobe*, *28*, 68–77.

Suryawanshi, R. K., & Kango, N. (2021). Production of mannooligosaccharides from various mannans and evaluation of their prebiotic potential. *Food Chemistry, 334*, 127428.

Vigsnæs, L. K., Holck, J., Meyer, A. S., et al. (2011). In vitro fermentation of sugar beet arabino-oligosaccharides by fecal microbiota obtained from patients with ulcerative colitis to selectively stimulate the growth of Bifidobacterium spp. and Lactobacillus spp. *Applied and environmental microbiology, 77*(23), 8336–8344.

Wang, X., Li, H., Cao, Y., et al. (2011). Cellulose extraction from wood chip in an ionic liquid 1-allyl-3-methylimidazolium chloride (AmimCl). *Bioresource technology, 102*(17), 7959–7965.

Wichienchot, S., Prakobpran, P., Ngampanya, B., et al. (2017). Production, purification and fecal fermentation of fructooligosaccharide by FTase from Jerusalem artichoke. *International Food Research Journal, 24*(1).

Wu, J., Dong, L., Liu, B., et al. (2020). A novel integrated process to convert cellulose and hemicellulose in rice straw to biobutanol. *Environmental Research, 186*, 109580.

Wu, J., Zhu, K., Li, J., et al. (2024). An optimize adaptable method for determining the monosaccharide composition of pectic polysaccharides. *International Journal of Biological Macromolecules, 277*, 133591.

Xiao, C., & Anderson, C. T. (2013). Roles of pectin in biomass yield and processing for biofuels. *Frontiers in plant science, 4*, 67.

Xiaowei, P., & Hongzhang, C. (2012). Hemicellulose sugar recovery from steam-exploded wheat straw for microbial oil production. *Process biochemistry, 47*(2), 209–215.

Xin, D., Yin, H., & Ran, G. (2024). Efficient production of High-Purity manno-oligosaccharides from guar gum by citric acid and enzymatic hydrolysis. *Bioresource Technology, 401*, 130719.

Yao, D., Wu, M., Dong, Y., et al. (2022). In vitro fermentation of fructooligosaccharide and galactooligosaccharide and their effects on gut microbiota and SCFAs in infants. *Journal of Functional Foods, 99*, 105329.

Yin, S., Pan, Y., & Tan, Z. (2011). Hydrothermal conversion of cellulose to 5-hydroxymethyl furfural. *International Journal of Green Energy, 8*(2), 234–247.

Yu, W., Cui, J., Zhao, S., et al. (2021). Effects of high-pressure homogenization on pectin structure and cloud stability of not-from-concentrate orange juice. *Frontiers in Nutrition, 8*, 647748.

Yuan, Q., Liu, S., Ma, M.-G., et al. (2021). The kinetics studies on hydrolysis of hemicellulose. *Frontiers in Chemistry, 9*, 781291.

Zhang, W., & Zhou, Z. (2017). Citrus pectin-derived carbon microspheres with superior adsorption ability for methylene blue. *Nanomaterials, 7*(7), 161.

Zhang, F., Deng, X., Fang, Z., et al. (2011). Hydrolysis of microcrystalline cellulose over Zn-Ca-Fe oxide catalyst. *Petrochemical Technology, 40*(1), 43–48.

Zhang, Q., Lian, Z., Zhou, X., et al. (2023). Kinetic studies and predictions for xylo-oligosaccharide production by acidic hydrolysis of xylan with various acids. *Fuel, 333*, 126325.

Zhou, J., Li, Z., Zhang, H., et al. (2018). Novel maltogenic amylase CoMA from Corallococcus sp. strain EGB catalyzes the conversion of maltooligosaccharides and soluble starch to maltose. *Applied and Environmental Microbiology, 84*(14), e00152–e00118.

3

Integrating Lignocellulosic Biomass Processing, Biomanufacturing, and Biofoundries: Innovations and Challenges in the Bioeconomy

Yaimé Delgado-Arcaño[1,2], Alisson Dias da Silva Ruy[3], Leila Maria Aguilera Campos[3], and Oscar Daniel Valmaña-García[1]

[1] *Center of Studies for Sustainable Development, ECOTEC Technological University 10 (CEDS/ECOTEC), Samborondón, Guayas, Ecuador*
[2] *Center for Renewable Energy Technology Study (CETER), Technological University of Havana (CUJAE), Faculty of Mechanical Engineering, Marianao, Havana, Cuba*
[3] *Postgraduate Program in Chemical Engineering, Federal University of Bahia (UFBA), Federação, Salvador de Bahia, Brazil*

3.1 Introduction

Pursuing a sustainable future, underpinned by global commitments such as the United Nations' Sustainable Development Goals and the Paris Agreement, has ignited widespread interest in bioproduct development derived from renewable biomass. Such bio-based alternatives present an opportunity to lower the dependence on fossil-based fuels, chemicals, and materials, thereby reducing greenhouse gas emissions and dependency on finite resources. At the center of this transition are biorefineries, which integrate various technologies to convert biomass into a suite of valuable products, including bioethanol, bioplastics, and platform chemicals. However, such a changeover is far from simple, necessitating a focus on technological, economic, and policy challenges that can hinder bioproducts' widespread adoption and commercialization.

Lignocellulosic biomass (LCB), called as second-generation (2G) feedstock, is an abundant residue rich in carbohydrates that plays crucial structural roles, providing strength, impermeability, and protection against microbiological attacks. LCB is mainly composed of cellulose (40–50%), hemicellulose (25–30%), lignin (10–25%), and smaller fractions of extractives, proteins, pectin, and ash. This composition gives LCB a highly resistant polymeric structure, making it difficult to degrade. This same structural strength offers significant potential for valorization, positioning LCB as a promising raw material for various conversion processes and bioproduct manufacture (Guimarães et al., 2023). On the other hand, due to its complex chemical structure, pretreatment is necessary to facilitate the efficient recovery of valuable products, whether in a bioproduct-based economy or the production of biofuels.

Biofoundry Techniques for Biotechnology Applications, First Edition. Edited by Anuj Kumar Chandel.

Based on the above, the present chapter considers innovative methods and technologies for LCB processing, 2G sugar purification, and bioproduct development within the framework of modern biorefineries, aiming for key challenges in the bioeconomy. Key topics include pretreatment strategies to enhance sugar production and advanced separation and purification techniques for 2G sugars. The economic competitiveness of some bioproducts of commercial interest is questioned, alongside the transformative role of biofoundries and biomanufacturing powered by the Design-Build-Test-Learn (DBTL) framework in synthetic biology. Life cycle assessment (LCA) and technical-economic analysis (TEA) of bioproducts, specifically bioethanol, lactic acid, succinic acid, and xylitol, are analyzed. Finally, the role of automation and synthetic biology in enhancing biorefinery operations, and the economic challenges and opportunities associated with producing commercially viable bioproducts are investigated.

3.2 Advances in Biomass Processing: Pretreatment and Purification Strategies

Pretreatment technique is essential to the success of bioconversion processes. To make the biomass more accessible for subsequent hydrolytic processes, it is necessary to break down the recalcitrant lignocellulose matrix, remove lignin, hydrolyze hemicellulose, and reduce cellulose crystallinity. Consisting of critical steps for refining biomass efficiently, which is essential to optimize this resource's conversion and valorization processes, pretreatment methods have received considerable attention in the last two decades. Besides, this stage has a major impact on the profitability of biorefineries, as was claimed by Patel and Shah (2021). Then, by combining one or more isolation and purification processes, high value biochemicals must then be separated. Some of the most studied biomass-derived products are xylo-oligosaccharides, xylitol, 2,5-furandicarboxylic acid, 2,3-butanediol, lactic acid, succinic acid, levulinic acid, chitosan, 3-hydroxypropionic acid, 5-hydroxymethyl furfural, and lignin-based derivatives, like aromatic diacids, quinones, vanillin, vanillic acid, and syringaldehyde (Chandel et al., 2018; Chaturvedi et al., 2022).

3.2.1 Pretreatment Methods of Lignocellulosic Biomass

LCB can be pretreated via physical, chemical, physicochemical, and biological techniques (see Fig. 3.1).

Joshi and Manjare (2024) published a thorough review of the most recent pretreatment techniques and how they affect biomass valorization. Furthermore, Antunes et al. (2019) discussed promising LCB pretreatment methods, their challenges, and the strategic role of nano- and biotechnological approaches concerning the viability and sustainability of biorefineries. Generally, a single pretreatment is not used but rather a mixture of some of them, depending on the raw material and the target product. There is extensive literature on the different pretreatment methods, their advantages, and their disadvantages. For this reason, we will briefly describe and discuss the methods most currently used, as illustrated in Figure 3.1.

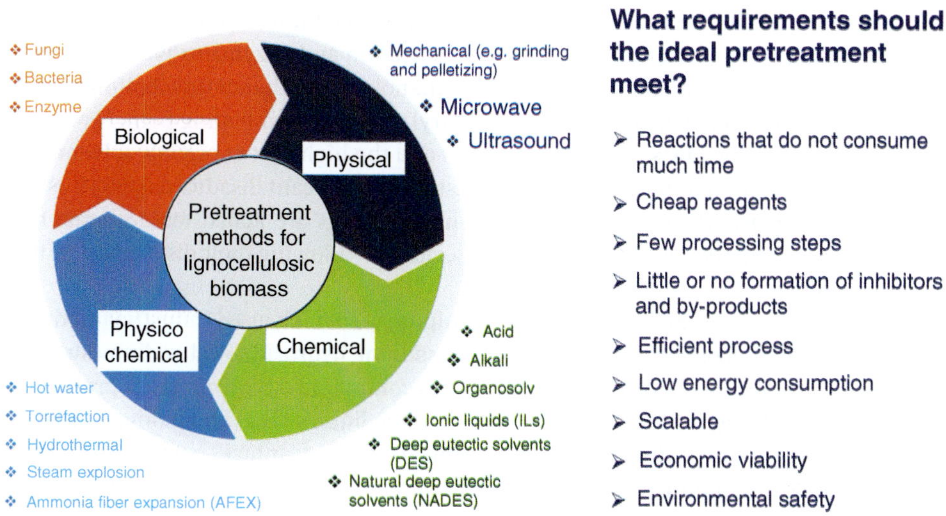

Figure 3.1 Some lignocellulosic biomass pretreatments and requirements for the ideal one.

Among physical pretreatments, mechanicals, such as grinding and pelletizing, are essential to increase bulk density and standardize biomass, facilitating its handling and processing. Chemical pretreatments, including leaching with water or acids, effectively remove metallic species that can cause problems during gasification at high temperatures. In addition, physicochemical pretreatments have stood out for their ability to transform raw biomass into higher value-added products, such as torrefied biomass and hydrochar through thermal torrefaction and hydrothermal carbonization respectively, improving the efficiency and performance of conversion processes. Biological pretreatment, like anaerobic digestion, alters biomass composition and structure, improving the efficiency of subsequent processes (Shen, 2024).

According to Zhang et al. (2023), each pretreatment acts in a different manner on biomass. Baral et al. (2022) stated that acid pretreatments tend to remove approximately 85% of hemicellulose and up to 40% of lignin, while alkaline processes can remove about 75% of lignin and retain 70–85% of holocellulosic sugars, along with the loose fibril structure with high porosity suitable for mass transfer, increasing cellulose digestibility. However, chemical hydrolysis also generates inhibitory compounds, such as acetic acid, furfural, and 5-(hydroxymethyl)furfural (HMF), which impair the growth of microorganisms and consequently reduce the overall fermentation efficiency. Alkaline pretreatment is known for its lower formation of inhibitors than acid pretreatment (Woiciechowski et al., 2020). Thus, the choice of method depends on the chemical composition of biomass, the hydrolysis capacity of cellulose, the generation of inhibitors, water consumption, lignin recovery, economic viability, and environmental safety (Delgado-Arcaño et al., 2024; Lobato-Rodríguez et al., 2023).

According to Ferreira and Taherzadeh (2019), organosolv pretreatment is more effective in dissolving lignin than hydrothermal, acid, and alkaline methods, efficiently removing it from the biomass and dissolving it in the solvent. This process also breaks the internal chemical bonds of cellulose and hemicellulose, increasing cellulose's pore volume and surface area, and improving its availability for saccharification and enzymatic hydrolysis. Due to the high polarity of the biomass components, ethanol and methanol (low boiling point) and glycerol, ethylene glycol, and tetrahydrofurfuryl alcohol (high boiling point) are typically used. However, the low solubility of biomass and the high cost of organic solvents, in addition to their flammability and volatility, constitute significant disadvantages. Thus, to increase pretreatment efficiency, acidic or alkaline catalysts are added to accelerate the hydrolysis of aryl ether bonds in lignin and hemicellulose, allowing their dissolution in the organic solvent.

Sewsynker-Sukai et al. (2020) reported Kraft pulping is a widely used industrial method, which employs non-ecological solvents such as sodium hydroxide and sodium sulfide, resulting in a liquor with degraded lignin and other compounds. To avoid these problems, solvents that prevent lignin condensation, such as ionic liquids (ILs) and biomass organic solvents, are being studied. However, these solvents have disadvantages, such as high cost, synthesis complexity, high toxicity, and low recoverability, which can reduce their economic and practical viability (Lobato-Rodríguez et al., 2023).

Pretreatments with ILs and deep eutectic solvents (DES) are promising alternatives to toxic industrial solvents, considered environmentally friendly solvents, according to the principles of green chemistry. ILs, especially those based on imidazole, dissolve cellulose, hemicellulose, and lignin, facilitating the conversion of biomass into biofuels; however, they are expensive, toxic to cellulases, and increase viscosity by lignocellulosic residues, limiting their application. DES, in turn, are non-toxic, economical, and compatible with enzymes, promoting the breakdown of lignin and hemicellulose and improving biomass valorization, although they also face residual viscosity challenges (Achinivu et al., 2024; Ziaei-Rad & Pazouki, 2024). To overcome these limitations, DES has received great attention in the last decade as a substitute for conventional organic solvents in biomass pretreatment (Contreras-Gámez et al., 2023). In addition to DES, there are natural deep eutectic solvents (NADES), which are composed of mixtures of two or three components derived from natural sources, including non-toxic quaternary ammonium salts, such as choline chloride and betaine, and uncharged hydrogen bond donors, such as amines, sugars, polyalcohols, carboxylic acids, and amino acids, which are capable of associating through intramolecular hydrogen bonds. One of the most notable characteristics of NADES is their ability to be fine-tuned for specific purposes; that is, by modifying the components and their respective molar proportions, it is possible to adjust their physicochemical properties to meet the specific needs of each process. Thus, NADES have proven to be excellent solvents for carrying out various chemical or enzymatic reactions, extracting plant-based components, and pre-treating LCB (Panić et al., 2021).

Liquid hot water (LHW) pretreatment, or hydrothermal, is an economical and sustainable method for processing LCB (Singh et al., 2023). It separates biomass into solid fractions, containing cellulose and lignin, and liquid fractions, with dissolved hemicellulose, increasing cellulose accessibility for enzymatic hydrolysis and improving the profitability of biorefining. Operating between 160 °C and 270 °C and pressures above 5 MPa, it uses autohydrolysis to facilitate the depolymerization of polysaccharides and lignin, improving the efficiency of enzymatic hydrolysis and the economic viability of biorefinery processes (Unyay et al., 2024). Steam explosion pretreatment is widely used in LCB processing, applying steam at high temperature and pressure to modify its structure and increase the efficiency of enzymatic hydrolysis (Hoang et al., 2023). Performed between 160 °C and 240 °C and pressures ranging from 0.7 to 4.8 MPa, the process involves rapid depressurization, resulting in explosive decompression. This procedure primarily hydrolyzes hemicellulose, improving cellulose accessibility and increasing the hydrolysis rate. Hydrolysis of the acetyl groups in hemicellulose releases acetic acid, which catalyzes further hydrolysis of hemicellulose, converting it into fermentable glucose and xylose monomers (Silveira et al., 2018). Advantages include high efficiency in sugar recovery, cost-effectiveness, and environmental benefits. Despite stringent pretreatment requirements and high energy consumption, this method continues to be widely adopted in industrial applications requiring high sugar recovery (Yu et al., 2022).

In agreement with Zabochnicka et al. (2022), biological pre-treatment with fungi or bacteria is environmentally benign and energy efficient. For example, fungi such as *Phanerochaete chrysosporium* effectively degrades lignin and hemicellulose, although the process is slower than employing chemical or physical methods. This pre-treatment can be adjusted to achieve specific degradation, increasing biomass fermentability without adding aggressive chemicals or high-energy inputs.

Recently, magnetic nanoparticle use has increased the effectiveness of pretreatment methods, promoting cost-effective renewable sugar obtention, essential for biofuel and biochemical production. Several nanomaterials, such as nanoparticles, magnetic nanocatalysts, carbon nanotubes, polymers, and hybrid alkaline nano-shearing procedures, have been used for LCB pretreatment. These nanomaterials have high penetration power, facilitating the exposure of the matrix content for subsequent enzymatic hydrolysis and the production of biofuels, like bioethanol, biohydrogen, biogas, and biodiesel. Nanotechnology-based pretreatment methods offer significant advantages, such as high reusability, cost efficiency, and reduced use of harmful chemicals. The use of acid-functionalized magnetic nanoparticles is a promising technique for the pretreatment of LCB, altering its structure at the molecular level and facilitating recovery and reuse. Furthermore, nanomaterials are environmentally friendly compounds used in lignolytic and saccharolytic enzyme immobilization for the disintegration of LCB, increasing the production of biofuels with minimal release of inhibitory compounds. The hybrid alkaline nanoscale shear method uses volatile solvents, chemical catalysts, and high shear force to remove lignin from biomass efficiently. Carbon nanotubes have been employed in the pretreatment of LCB due to their high tensile strength, durability, and thermal and electrical stability. Summarizing, the use of nanomaterials in the pretreatment of LCB for the economical production of second-generation sugar, followed by its biotransformation into renewable fuels and chemicals, strengthens the sustainable economy (Verma et al., 2023).

Finally, several conditions must be met to achieve efficient pretreatment of lignocellulose. First, particle size reduction is desirable to release sugars, but grinding is highly energy-consuming. The chemical reagents used must be low-cost, both in the pretreatment and in the subsequent neutralization and conditioning steps. In addition, it is crucial to use non-corrosive reagents. Ideally, fast reactions and the process should have low energy demand or allow energy to be reused in other steps. Minimizing the sugar and lignin degradation products is equally important, as these can act as fermentation inhibitors. Pretreatment should result in high concentrations of sugar production in the liquid fraction, facilitating the recovery of lignin and other constituents. Lastly, efficient pretreatment must have low operational and investment costs, contributing to the economic viability of the process (Hu et al., 2024).

Definitively, each pretreatment method has its advantages and limitations, influencing the choice based on the type of biomass, desired efficiency, cost, and environmental impact. Continuous optimization of these processes is essential to produce more economically viable and environmentally sustainable platform molecules, bio-compounds, and biofuels.

3.2.2 Separation and Purification of the Interest Compounds

Separation and purification processes are of paramount importance, especially because many applications, such as in the medical and pharmaceutical fields, require high-purity products. In addition, according to Alves de Oliveira et al. (2020), there are also growing environmental concerns, with increasingly stringent legislation on the use of solvents and waste generation, aiming at a more sustainable production chain.

The pretreatment of LCB mainly aims to increase the accessibility of sugars present in cellulose and hemicellulose. However, with the growing interest in value-added chemical production, compounds previously considered inhibitors, such as HMF and furfural, have come to be seen as molecules of great interest. Furfural is the precursor to produce tetrahydrofuran, 2-methyltetrahydrofuran, tetrahydrofurfuryl, furoic acid, fumaric acid, and tetrahydrofuran. HMF is a precursor to produce levulinic acid, 2,5-dimethylfuran, adipic acid, caprolactone, polyamide 6, 2,5-furandicarboxylic acid, and others (Chaturvedi et al., 2022).

Significant advances in the processing of LCB have allowed researchers to develop methods for the extraction, isolation, and purification of these compounds efficiently and cost-effectively. These methods include techniques such as solvent extraction, ultrasound extraction, enzymatic extraction, membrane filtration, and adsorption processes, optimized to maximize the recovery of HMF and furfural, minimizing the formation of unwanted by-products. Examples of those compounds include polymers and resin chitosan (Chaturvedi et al., 2022).

Despite its great importance for scalability, the separation of final products in a biorefinery still requires a lot of research. While ethanol separation can be performed by simple distillation, this method is not suitable for furanic compounds due to their low thermal stability. Therefore, it is necessary to evaluate different separation and purification techniques, like liquid-liquid extraction, precipitation, adsorption, and membrane technologies, to determine the most suitable for each specific product. Molecular distillation is promising for thermally sensitive compounds, operating at low temperatures and with short residence times. Liquid-liquid extraction relies on the affinity of solutes between immiscible liquid phases, while precipitation uses the solubility of the compound for recovery. Membrane technologies, such as microfiltration and electrodialysis, are environmentally friendly and scalable. Salting-out extraction, based on the partition coefficient, offers low cost and easy recovery of components. Adsorption methods use adsorbents with an affinity for the product, followed by recovery with a suitable solvent, requiring a combination with another separation process (Alves De Oliveira et al., 2020; Kruschitz & Nidetzky, 2020; Lecona-Vargas et al., 2022).

In summary, the pretreatment of LCB now includes the efficient recovery of valuable compounds. Continuing research and innovating new technologies are essential to improve biomass conversion processes.

3.3 Bioeconomy and Biofoundries: How Automation and Synthetic Biology can Enhance Biorefineries

The development of engineering biological systems—powered by experimental technologies, artificial intelligence (AI), cloud computing, bioinformatics, and laboratory robotics—has accelerated the creation of new treatments, materials, energy solutions, and agricultural advancements (Densmore et al., 2023). Additionally, bioprocessing applications like bioremediation and biofuel production demonstrate the potential of sustainable solutions to address global challenges (Pandey et al., 2024). These technological developments have gained momentum in trade endeavors. This is why the term bioeconomy arises to refer to economic activities driven by life sciences research, including products like environmental biosensors, biosynthetic spider silk fabrics, and chemicals produced from yeast or bacteria (Frisvold et al., 2021). The facilities or platforms that leverage advanced technologies, such as synthetic biology, automation, robotics, AI, and data science, to design, build, and test biological systems or genetically modified organisms are called biofoundries (Gupta et al., 2024; Holowko et al., 2020).

One of the biotechnological process's advantages is that it generally operates at lower pressures and temperatures than conventional petrochemicals. That is why biotechnology performs a crucial role in net-zero carbon expectations and the United Nations Sustainable Development Goals (SDGs), specifically the number 12: Responsible Consumption and Production and number 13: Climate Action. According to Chui et al. (2023), with the bio-route applications in different production sectors, it is expected that by 2040–2050 the emissions will be reduced by 7–9% compared to 2018. With these expectations, the bioeconomy aims to substitute oil and gas gradually, yet the cost of oil frequently plays a crucial role in the viability of any alternative technology. On the other hand, organic chemistry products contain carbon, so many industrial sectors cannot be decarbonized although they may be "defossilized" (Pavan, 2022). To meet these objectives and promote the circular industry economy, biomass use is essential. Biomass is the planet's most abundant solid renewable feedstock and the primary source material for biotechnological processes. For better valorization and processing of LCB, bioengineering could be efficient in the creation of new microbial strains for the sustainable and renewable production of chemicals, fuels, and pharmaceuticals. The design, construction, and testing of recombinant, industrially relevant microbial chassis are crucial for the success of this emerging field (Banner et al., 2021).

Increasing efforts have been made to boost biomass transformation into high-value products and their commercialization. Programs such as the Biomass Crop Assistance, and the BioRefinery Assistance were launched to increase initiatives to facilitate the purchase of bio-based goods. Furthermore, the Billion-Ton Biomass Report, published in 2016 by the

U.S. Department of Energy (USDOE), evaluated the possibility of generating and using one billion tons of renewable biomass in the country. According to the paper, 50 billion pounds of bio-based chemicals and goods and 50 billion gallons of biofuels may be produced each year. By 2030, this output may provide 1.1 million direct U.S. jobs, contribute to a 450-million-ton reduction in CO_2 emissions, and supply 25% of the country's liquid transportation fuels (Frisvold et al., 2021).

Another case in point is the U.S. National Science Foundation (NSF) BioFoundrie, integrated facilities that enable researchers to rapidly design, create, test, and streamline the development of tools and products to accelerate research to advance the bioeconomy. These facilities seek to increase biotechnology capabilities, democratize access to vital research infrastructure, and foster science, technology, engineering, and mathematics (STEM) education and workforce development by collaborating with marginalized groups and educational institutions (NSF, 2024).

One of the beneficiaries of the NSF's BioFoundries program is the University of Georgia's Complex Carbohydrate Research Center (CCRC). The target is to launch the BioFoundry: Glycoscience Resources, Education, and Training (GREAT). The initiative aims to increase awareness, interest, knowledge, and participation in carbohydrate science. GREAT seeks to democratize glycobiology research and expertise and expand its applications beyond medical life sciences to biofuel production and plant-based new materials. The CCRC's Analytical Services & Training (AST-CCRC) unit provides instrumentation services and hands-on training for clients worldwide and will be part of the BioFoundry user facility (Terrazas, 2024).

3.3.1 Bibliometric Analysis

The term biofoundry is so recent that there is few research related to this novel topic. This was confirmed after a bibliometric analysis, using sugar purification, biorefinery, biotechnology and biofoundry as keywords. A search and identification of documents were carried out in the Scopus, ScienceDirect and Web of Science databases, considering the period from 2014 to 2024. The same search algorithms were used in each one. This process was done on November 20, 2024. Table 3.1 shows the publication number for each keyword in each database in the period analyzed.

As expected, research on sugar purification, 2G sugar purification, biorefinery, and biotechnology yields an extensive number of publications, as these topics have been widely studied for decades. However, when we looked for biofoundry, the number of published works decreased significantly, demonstrating the freshness of the area and the necessity to continue working in this direction. Even when connectors such as AND, and OR were used to link to the previous keywords, the number of publications about biofoundries is so small.

Of the three databases consulted, Scopus is the one that provides detailed document information by year, journals, author, affiliation, country, subject area, type, and funding sponsor. Therefore, we chose it to find out which countries have published the most on this topic (a) and the trend of publications in the period analyzed (b) as is shown in Figure 3.2. To create a keyword co-occurrence map (Figure 3.3), VOSviewer software was employed. The different colors represent the various clusters, the size is according to the frequency of the term, and the links between nodes show the co-word association.

Table 3.1 Publications About Sugar Purification, Biorefinery, Biotechnology, and Biofoundry in the Scopus, ScienceDirect and Web of Science Databases from 2014 to 2024

	Database		
Keywords	**Scopus**	**ScienceDirect**	**Web of Science**
Sugar purification	6,795	72,647	3,583
2G sugar purification	19	1,268	15
Biorefinery	14,145	26,214	21,878
Biotechnology	80,213	513,537	290,181
Biofoundry	71	285	268
Biofoundry AND Sugar purification	0	16	3
Biofoundry AND Biorefinery	2	26	5
Biofoundry AND Biotechnology OR Biorefinery	12	26,404	21,981

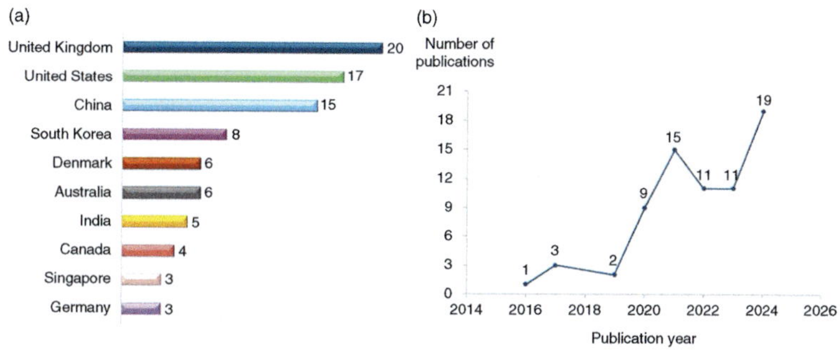

Figure 3.2 Trends in publications about biofoundries according to Scopus from 2014–2024. (a) Top 10 countries with publications about biofoundry. (b) Publications about biofoundry in the analyzed period.

The lines show the interconnection between the different branches of knowledge, emphasizing how crucial a global and multidisciplinary approach is to research advancements.

North America, Europe, and the Asia-Pacific region are the ones that lead in the number of publications on biofoundry, led by the United Kingdom and the United States. The first publication was in 2016 when the USDOE launched Agile BioFoundry (ABF). Since then, interest in the topic has been increasing, and this trend is expected to continue to rise. Regarding the co-occurrence network map, by concentrating just on keywords from authors' papers, we made sure that focus areas and trends as determined by subject matter experts were accurately represented. In correspondence with the larger size of the clusters, the most recurrent keywords were biofoundry, synthetic biology, automation, genetic engineering, and metabolic engineering.

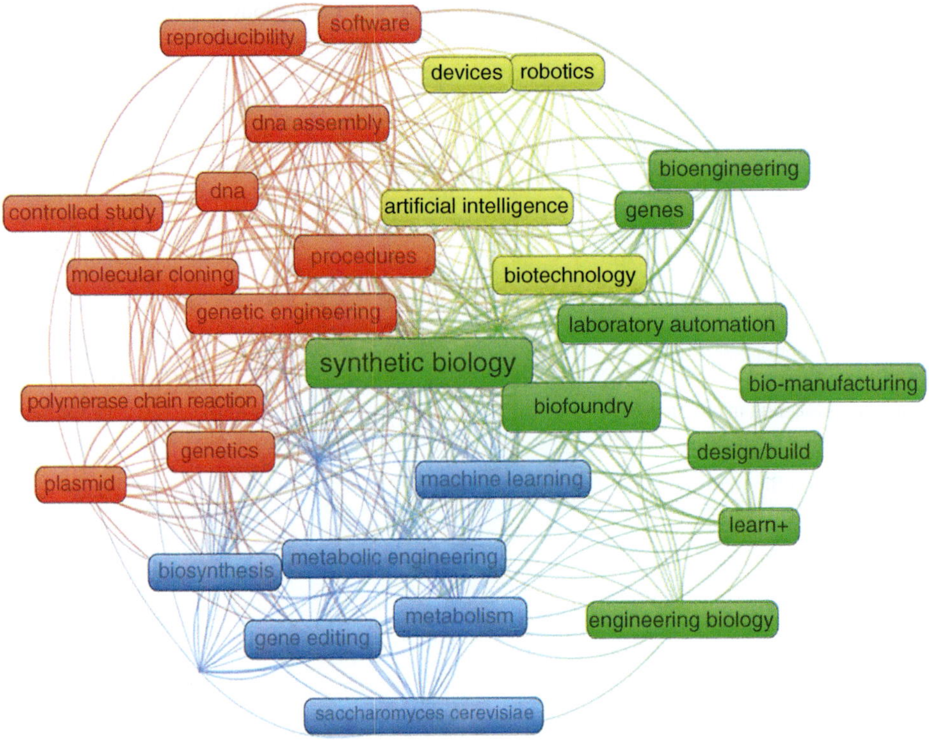

Figure 3.3 Keyword co-occurrence network map for biofoundry from 2014 to 2024.

3.3.2 Design-Build-Test-Learn (DBTL) in Biofabrication and Synthetic Biology

The prototyping iterative in engineering disciplines, particularly in synthetic biology and metabolic engineering, is a technique used to effectively explore the solution space by creating new experiments based on insights gained from data from earlier prototypes. In this sense, the DBTL approach marries synthetic biology and automation engineering to create high-throughput biological solutions and support and speed up cell factory development.

The DBTL cycle seeks to satisfy specific design requirements for synthetic biology applications. It is centered on a biorefinery that will create a wide range of goods through more circular and sustainable economic methods. It also enhances research efficiency and accelerates the shift from problem-solving to innovation (Driessen, 2022; Gupta et al., 2024). Each stage was described by Hillson et al. (2019) and Driessen (2022) (see Figure 3.4), as summarized below:

Design: This phase includes the computational design of genetic components, circuits, regulatory and metabolic pathways, and entire genomes. It also states the intended target function and specifications.

Building: Entails connecting certain genetic components. In this stage, biological parts are combined in a designed chassis strain to create the desired cell factory. However,

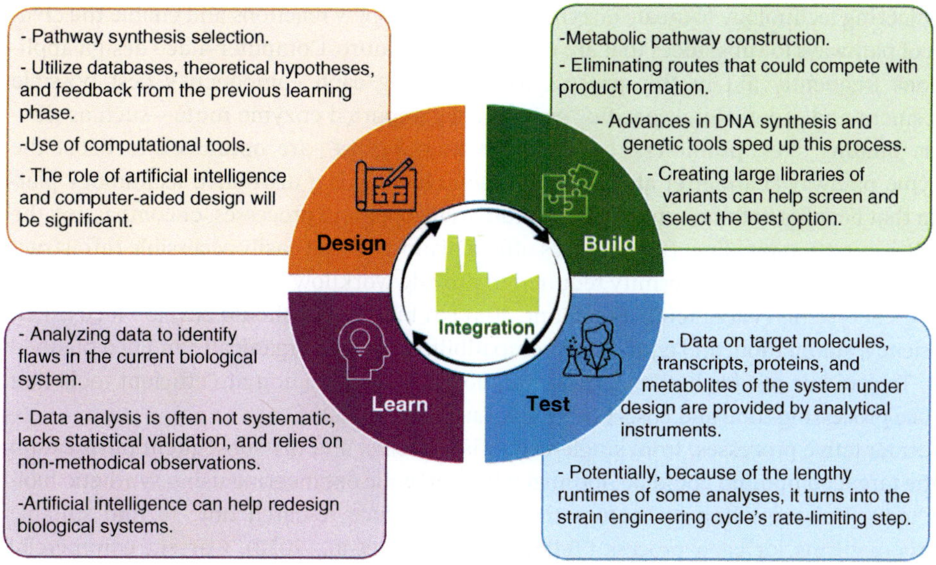

Figure 3.4 The biofoundry innovation system based on the Design-Build-Test-Learn cycle (DBTL-cycle) and its characteristics. Developing data analysis and machine learning methods is crucial in each stage.

accurately programming metabolic pathways is challenging due to the context-dependent nature of biological elements.

Test: Here are prototyped and tested in living cells (sometimes called "chasses") at various scales. Detailed analytical measurements (or "omics") of cellular components are also included. Different techniques such as next-generation sequencing, gas chromatography, liquid chromatography, mass spectrometry, Raman spectroscopy, and nuclear magnetic resonance are used for genomics, transcriptomics, proteomics, and metabolomics analysis. Potentially, because of the lengthy runtimes of some analyses, it turns into the strain engineering cycle's rate-limiting step.

Learn: Using the information gathered in T (Test) to guide the design process, L (Learn) applies modeling and computational learning technologies. The final phase involves analyzing data to identify current biological system flaws and improve metabolic models, the host organism, or the production pathway.

DBTL enables several rounds of design iterations for optimization. Furthermore, if inadequate performance is identified during later scale-up processes, feedback can be relayed to the biofoundry through scale-down simulators for its redesign (Philp, 2023).

3.3.3 Biofoundry and Process Integration in Biorefinery

Synthetic biology technology, which uses a fermentative recombinant microbial approach to fine chemical manufacturing, is gradually replacing traditional genetic engineering pathways of biocatalytic processes. This groundbreaking method can use enzyme engineering technology to create enzymes that catalyze new reactions and enable the creation of pathways to substances that are not present in nature. Computer-aided design applications frequently aid in this process by forecasting each component's best possible placement and order. Here, specific "parts" of the imported enzyme route—such as ribosome binding sites, promoters, and enzyme homologues—are optimized to boost the enzyme pathway (Banner et al., 2021). Synthetic biology is a disruptive technology platform that benefits both new and existing industrial goods and processes, encompassing the whole innovation pipeline. Biofoundries offer a distinctive and easily accessible infrastructure that promotes the uniformity required to provide workflow design and engineering of biological systems (Farzaneh & Freemont, 2021). A biofoundry should develop focused on efficiency, automation, and equipment compatibility, growing organically and in a planned way. To avoid or resolve bottlenecks, simulation and optimization are efficient tools that reduce processing time and costs (Tellechea-Luzardo et al., 2022).

Fermentative processes, from single-cell to large-scale, and the subsequent purification of the target compound could be optimized by metabolic engineering using synthetic biology tools. Before scaling up, automatization can reduce research time to find suitable microorganisms for each process (Tellechea-Luzardo et al., 2022). Current commercial microbial fermentations that use lignocellulose waste as a carbon source depend on costly pretreatment techniques to liberate plentiful sugars (like glucose) due to cellulose recalcitrancy. An alternate method that likely turns down feedstock pre-processing costs might be to use a consolidated bioprocessing (CBP) technique, in which fermentation, lignocellulose breakdown (saccharification), and the generation of enzymes are all carried out within a single microorganism. For this, secondary product biocatalytic pathways must either be integrated into naturally cellulolytic bacteria or current commercial strains should be engineered to include an extracellular localizing cellulolytic system. Advancements in CBP include engineered microbial chassis that utilize lignocellulose as their sole carbon source to produce valuable chemicals (Banner et al., 2021).

Process integration is visible in the bioethanol industry. A clear example is narrated by Chauvet and Lescieux-Katir (2024), who have a track record of the Bazancourt-Pomacle biorefinery near Reims, France. History began in 1953 when farmers converted a failing distillery into a sugar beet refinery. Now, it produces sucrose, bioethanol, starch, glucose, gluten, and various chemicals and cosmetics. This case in point highlights the importance of biorefineries in the bioeconomy and the integration of food and non-food production. Works like that must be replicated worldwide, given the urgent need to shift from a linear economy, which generates excessive single-use waste, to a circular economy that focuses on environmental, social, and economic well-being. Other bioethanol industries are focused on process integration, using emerging technologies for carbon capture and utilization (CCU). In addition to tackling significant greenhouse gas emissions, this is a step toward increasing the efficiency of carbon consumption and generating revenue. Nonetheless, because of the substantial carbon losses that occur during the production process, bioethanol output can yet be improved (Pavan, 2022). Metabolic engineering can be used to improve the co-utilization of multiple substrates to raise product yields and degrade inhibitory compounds in lignocellulose, thus biologically detoxifying biomass hydrolysate (Hillson et al., 2019).

To convert biomass efficiently, it is important to utilize its components to the maximum, minimize carbohydrate losses, and remove and convert lignin. Efforts are also being made to reduce the risk of oversaturating chemical markets and increase the value of traditionally underutilized biomass fractions to produce low-cost biofuels in the growing bioindustry. Additionally, research is focused on speeding up strain improvements and developing strong, scalable organisms for fuel and product manufacturing and diversification. The expanding capacity to manipulate biological processes necessitates a greater integration of computational techniques in advancing the DBTL cycle advocated by synthetic biology. Biofoundries and Cloud Laboratories will serve as pivotal components in facilitating this progression (Davis et al., 2018).

Biosynthesis of important chemical building blocks is achievable, replacing those traditionally sourced from petroleum refineries. Nevertheless, it usually demonstrates low yields. Natural systems can enhance productivity aided by metabolic engineering strategy. This involves gene identification and modification that participate in these compounds' synthesis. For example, Dai et al. (2023) achieved a higher 3-hydroxypropionic acid (3-HP) yield from derived sugars corn stover using *A. niger* transgenic strain after several DBTL cycles and optimizations of different cultures. This means a considerable rise of 200% beyond the maximum titers reached by the organism *A. pseudoterreus*. For their part, Adamczyk et al. (2023) analyzed a combination of genetic and metabolite data of *R. toruloides* that can be modified to control the profiles of four valuable chemicals—d-ribulose, d-arabitol, xylitol, and d-xylulose—produced from sustainable biomass feedstocks. Furthermore, the researchers consider that a comprehensive approach to understanding the metabolism of *R. toruloides* and related fungi will improve metabolic engineering for bioproducts by identifying rate-limiting steps and genes involved in enzymatic redundancy in key catabolic pathways.

For biorefinery development and process integration, simulation tools and life cycle analysis (LCA) are essential, as they allow for visualization of the best strategy to reduce production costs, increase profits, and reduce the negative environmental impact. For instance, Correia et al. (2024) evaluated bioethanol production from eucalyptus residues and corn stover through process simulation using Aspen Plus software, economic analysis, and LCA. To determine their economic impact on the minimal selling price, several scenarios were investigated, such as renewable energy source use, heat integration, molasses sales, and in-situ enzyme synthesis. The researchers concluded that combining *in situ* enzyme production with renewable energy is the most efficient way to lessen environmental consequences; this reduces CO_2 equivalent emissions for each biomass by 89%.

The ABF consortium, the USDOE-sponsored collaboration, combines sophisticated computational tools for biological engineering, process design and data analysis, and economic/sustainability modeling. They deliver integrated, bench-scale data for TEA/LCA and data for the Learn stage in the DBTL cycle and develop scaled fermentations to optimize titer, rate, and yield. Additionally, the operation creates, stores, and ships biomass hydrolysates while evaluating their performance against clean sugar methods (Benavides et al., 2021).

3.3.4 Global Expansion of Biofoundries: Innovation and Collaboration

Biofoundries are becoming increasingly popular globally, as academic and commercial facilities are set up in North America, Europe, and the Asia-Pacific regions. They can become key players in Industry 4.0, which is based on a circular economy. Chui et al. (2023) predicted that biology, computing, and automation intersection could impact the economy by up to $4 trillion annually in the next two decades. Expanding successful examples like the USDOE ABF and the Department of Defense BioMADE would boost the bioeconomy.

Creating public-private partnerships and research consortia for commodity fuel and chemical production will help address industry challenges and provide more equitable access to physical and virtual facilities. Additionally, there is a need for ongoing private sector input, interdisciplinary and international collaborations, and leveraging expertise to spur innovation globally. These consortiums can come from diverse fields, including information technology, mechanics, electronics, and biology. It is essential to take proactive steps to expedite the education of potential stakeholders who are not yet familiar with these technological advances. Universities, research institutions, and industry must be prepared to bring this new type of service, train specialists to develop and use automated laboratories, and expand biotechnological routes to boost the bioeconomy. Moreover, societal engagement, policy incentives, and economic benefits are needed to advocate this multidisciplinary collaboration.

3.4 Economic Competitiveness in the Production of Bioproducts of Commercial Interest

Coming from biomass, bioproducts have immense potential to transform fuel and chemical markets, but realizing this potential is contingent upon achieving economic viability. One of the advantages of biomass use as raw material is that compared with fossil feedstocks, biomass achieves a better balance between carbon dioxide emitted when burned and absorbed during feedstock growth (Guimarães et al., 2023). First-generation feedstocks (1G) are easy to process but compete with crops used for food products and come with higher environmental costs due to land and water use. Contrary to this, LCB (2G) comes from residues and does not compete with crops used for food products. However, it often demands intensive pretreatment to release sugars, which can offset some of its benefits. Despite advancements in LCB pretreatment like acid and enzymatic hydrolysis having risen to >80% glucose yield, it requires substantial capital investment, representing, for example, 45% of the total production cost in bioethanol manufacturing from sugarcane bagasse. Moreover, the need for robust microbial strains that can withstand industrial-scale pressures without compromising yield remains a technical bottleneck (Ajala et al., 2021). This is why LCA has become indispensable in optimizing bioproduct development, quantifying environmental impacts, and guiding design improvements. In the next section, we will analyze the techno-economic aspects, the LCA and the market value projections for four key bioproducts: bioethanol, lactic acid (LA), succinic acid, and xylitol.

3.4.1 Techno-Economic Analysis and Life Cycle Assessments for Sustainable Bioproducts

LCA provides a comprehensive platform for evaluating emissions, energy use, and resource efficiency across the life cycle of bioproducts. The integration of LCA with TEA allows for

Table 3.2 Comparative Analysis of Bioproducts Based on Their Environmental Performance and Economic Potential

Bioproduct	Main Applications	Environmental Performance	Economic Potential	References
Bioethanol	Fuel, Chemical Intermediate	Reduced GHG emissions, moderate energy use	Mature market, stable demand	Peng et al. (2023)
Lactic Acid	Bioplastics (Poli Lactic Acid), Food Additive	Low carbon footprint, high land use efficiency	High demand for bioplastics, supported by plastic bans	Munagala et al. (2021)
Succinic Acid	Bioplastics, Pharmaceuticals	Low GHG emissions, moderate water use	High market demand, scalable production	Gadkari et al. (2021); Klein et al. (2017)
Xylitol	Sweetener, Dental Care	Low environmental impact, high resource efficiency	Niche market, growing health awareness	Shaji et al. (2022); Vollmer et al. (2023)

the simultaneous evaluation and selection of more profitable and sustainable production routes. Feedstock selection and production technology heavily influence LCA outcomes. LCA studies emphasize the need for process optimization. Strategies such as energy recovery, process integration, and using renewable energy sources can significantly reduce the environmental footprint of biorefineries (Delgado-Arcaño et al., 2024; Ferraz & Pyka, 2023). Advances in computational modeling, including machine learning algorithms, are emerging as tools to optimize biorefinery design, making TEA and LCA more accurate and informative (Osman et al., 2024). Researchers have studied the environmental performance and economic potential of some bioproducts based on integrated TEA-LCA assessments. We will consider bioethanol, lactic acid, succinic acid, and xylitol. These products not only demonstrate strong market demand but also exhibit relatively low environmental impacts (Louw et al., 2023). The comparative analysis is shown in Table 3.2.

Bioethanol production demonstrates significant sustainability benefits and strong economic potential. Nevertheless, LCA is needed to show its environmental effects. Several researchers have analyzed the LCA of bioethanol production, but contrasting results between studies are not easy. Soleymani Angili et al. (2021) emphasized that variations in methodological assumptions, like system boundary, functional unit, and differing technical and agricultural factors among authors, complicate direct comparisons of LCA results. Most studies viewed pretreatment as a critical step in the bioconversion process. Besides, they highlighted that while bioethanol typically shows a reduction in global warming potential compared to fossil fuels, challenges persist in addressing impacts on human, water, air, marine, or terrestrial life, like ecotoxicity, acidification, and eutrophication, largely driven by feedstock production and pretreatment processes.

Peng et al. (2023) explored bioethanol production from rice straw using different pretreatment methods (sodium hydroxide (NaOH), choline chloride/lactic acid (DES-LA) and choline chloride/urea (DES-U)), demonstrating that NaOH pretreatment exhibits higher energy efficiency and lower greenhouse gas emissions than alternative methods evaluated. Besides, their study underscores the importance of further integrating waste valorization strategies, such as using lignin as an energy source, to enhance environmental and economic performance. Demichelis et al. (2020), however, take a comprehensive approach to assessing bioethanol biorefineries, concluding that second-generation bioethanol systems, although more sustainable, face technical and economic bottlenecks, particularly in feedstock logistics and pretreatment technologies. Their analysis shows that coproduct recovery is essential for improving economic feasibility, and they stress the need for tailored solutions to specific feedstocks.

The analysis of lactic acid production, in terms of TEA and LCA, presents a promising but complex picture. LA production in biorefineries, particularly those utilizing second-generation feedstocks like sugarcane bagasse (Munagala et al., 2021), macroalgae (Tong et al., 2024), or even glucose and xylose (Lee et al., 2024), demonstrates significant potential for improving economic viability and reducing environmental impacts. These feedstocks are abundant, renewable, and aligned with circular economic principles. Moreover, integrating LA production into existing sugar mills or biorefineries that co-produce bioethanol leverages synergistic resource utilization, reducing overall costs and waste (Munagala et al., 2021).

Biobased succinic acid (bio-SA) production integration into biorefineries regarding the economic and environmental approach reveals a promising yet challenging landscape. LCA indicates that the choice of LCB, such as apple pomace, positively impacts both the economy and the environment, especially due to higher solid loading, reduced alkali consumption, and enzyme loading. High fertilizer use, land impact, and energy-intensive separation increase environmental burdens. Improving yields and reducing separation energy requirements are crucial for sustainable biochemical valorization (González-García et al., 2018). Another common environmental inconvenience is the high energy consumption in the form of steam and heating required in feedstocks and product purification processes (Roja, 2024). Production technology also clearly interferes with economic and environmental aspects. Advanced process designs, such as membrane-based purification, have been explored to reduce environmental impacts and enhance cost-effectiveness in manufacturing bio-SA, presenting a green alternative that challenges traditional, energy-intensive methods. However, scaling up these technologies for commercial viability remains an obstacle (Kumar et al., 2020). Studies couple innovative methods that attenuate environmental impacts by utilizing carbon dioxide as a feedstock in fermentation processes (Nghiem & Senske, 2015). Overall, while bio-SA production has demonstrated clear environmental advantages, economic and process challenges still necessitate further technological advancements and strategic integration within biorefineries.

The LCA and TEA of xylitol production provide comprehensive insights into its environmental and economic feasibility. According to Vollmer et al. (2023), xylitol production through bioprocesses demonstrates significantly lower environmental impacts compared to traditional chemical routes. Utilizing wheat straw as feedstock, the biorefinery integrates value-added co-products like succinic acid, optimizing resource utilization and reducing waste. The study highlights moderate greenhouse gas emissions, with energy consumption in downstream processes identified as a critical factor. While the biorefinery process offers environmental advantages, its economic feasibility is heavily influenced by capital and operational expenses, highlighting the importance of technological optimization.

Similarly, Shaji et al. (2022) investigated a sugarcane bagasse-based xylitol biorefinery, revealing key economic and environmental trade-offs. Their analysis underscores the significance of fermentation efficiency and process integration in reducing costs. The LCA results indicate that electricity usage in the pretreatment and purification stages contributes significantly to the overall carbon footprint. Cost optimization through heat integration and reduction of fermentation times emerges as a pivotal improvement area. Santos et al. (2024) also emphasized xylitol's role as a high-value biochemical derived from waste biomass, offering dual benefits of waste valorization and circular bioeconomy enhancement. The studies highlight that while xylitol biorefineries are environmentally sustainable, overcoming technological and cost barriers remains critical for commercial scalability.

To fully realize the potential of these main components, broader global market insights are needed to encourage industrial-scale adoption and technological innovation, as discussed in the next section.

3.4.2 Market of Bioproducts: Insights and Challenges

The commercial viability of bioproducts is a critical factor that determines their success in the marketplace. The market value and the Compound Annual Growth Rate (CAGR) for the same products that were analyzed in the previous section: bioethanol, lactic acid, succinic acid, and xylitol—predicted by consultancy companies, are shown in Table 3.3. These bioproducts are positioned as key players in advancing the bioeconomy due to their wide-ranging applications and growing demand.

Bioethanol, with a market value of $80 billion in 2023 and a CAGR of 5.3% from 2024 to 2032, dominates the bioproduct landscape in terms of economic value (Fortune Business Insights, 2024). Its primary use as a biofuel and in the beverage industry drives this significant market size. Precedence Research (2024) notes that the ethanol market's growth is fueled by global efforts to reduce greenhouse gas emissions and the adoption of biofuel blending mandates. However, the ethanol market faces challenges from fluctuating fossil fuel prices, which affect its competitiveness, and land use concerns associated with first-generation feedstocks. Second-generation bioethanol derived from LCB presents a pathway to addressing these challenges while enhancing environmental sustainability (Market Research Future, 2024).

Table 3.3 Market Growth and Value Projections for Key Bioproducts

Bioproduct	CAGR	Market Value in 2023	References
Bioethanol	5.3% from 2024 to 2032	80 billion	Fortune Business Insights (2024)
Lactic acid	8.0% from 2024 to 2034	3.6 billion	Future Market Insights (2024)
Succinic acid	17.8% from 2024 to 2034	160 million	Transparency Market Research (2024)
Xylitol	3.7% from 2024 and 2032	957 million	Global Markets Insights (2024)

Lactic acid demonstrates significant growth potential with a CAGR of 8.0% from 2024 to 2034 and a market value of $3.6 billion by 2024. Its application spans the food and beverage, pharmaceutical, and biodegradable plastics sectors. The increasing demand for polylactic acid in sustainable packaging drives this growth, supported by regulatory bans on single-use plastics. However, the lactic acid market's expansion may face feedstock availability and cost challenges, as second-generation feedstocks and optimized fermentation processes are necessary to improve economic feasibility. According to Future Market Insights (2024), the lactic acid market is also witnessing advancements in low-pH fermentation technology to reduce costs and environmental impacts.

The succinic acid market, growing at an impressive CAGR of 17.8% from 2024 to 2034, reached $160 million in 2023. This rapid growth reflects its expanding application in producing biodegradable plastics, solvents, and pharmaceuticals. Transparency Market Research highlights that bio-based succinic acid is gaining momentum due to its sustainability and lower environmental footprint compared to its petroleum-based counterpart (Transparency Market Research, 2024). However, the market remains small compared to the other bioproducts analyzed in Table 3.3, partially due to its production costs and scalability challenges. Advances in microbial engineering and integration into biorefineries can further enhance their viability (Research and Markets, 2024).

Xylitol, with a market value of $957 million in 2023 and a CAGR of 3.7% from 2024 to 2032, exhibits slower growth than the other bioproducts shown in Table 3.3 but remains significant in the food and pharmaceutical industries. Its use as a natural sweetener and in dental care products drives steady demand. However, due to the feedstock variability, xylitol production's reliance on hemicellulose hydrolysates can pose challenges. Global Market Insights highlights advancements in ultrasound-assisted fermentation as emerging technology to improve efficiency and reduce production costs (Global Markets Insights, 2024; Verified Market Research, 2024).

In summary, the diverse growth trajectories of these bioproducts reflect their varying applications and production challenges. While lactic and succinic acids exhibit high growth potential driven by environmental policies and bioplastics demand, bioethanol dominates the market in sheer value due to its established role as a biofuel. Xylitol, despite slower growth, sustains relevance through niche applications in health and nutrition. However, all four bioproducts face shared challenges, including feedstock variability, technological scalability, and production costs. Overcoming these hurdles requires continued innovation in fermentation technologies, microbial engineering, and supply chain optimization. Together, these bioproducts illustrate the evolving landscape of the bioeconomy, underscoring the need for balanced approaches to economic design and environmental sustainability.

3.4.3 Market Growth, Cost Challenges, and Policy Drivers in Biorefineries

The biorefinery market was valued at US$143.2 billion in 2023 and is anticipated to grow at a CAGR of over 7.2% from 2024 to 2032. Governments support biofuels due to various factors, including economic growth, alternatives to fossil fuels, and the creation of employment opportunities (Global Markets Insights, 2024). By diversifying the product portfolio, biorefineries can improve profitability and reduce the risk associated with market fluctuations. However, achieving optimal integration requires sophisticated process design and supply chain management, which can be capital-intensive. In contrast to traditional fossil-based industries, biorefineries offer environmental benefits but have a major hurdle to achieving market competitiveness. Reducing the high production costs associated with bioprocess remains a huge challenge (Pérez-Almada et al., 2023). Several factors, including expensive feedstock supply chains, the complexity of biomass conversion processes, and the energy-intensive nature of some bio-based production methods, drive these costs. For instance, technologies such as enzymatic hydrolysis, used in the conversion of LCB, are effective but costly due to the need for specialized enzymes and extensive pretreatment stages (Basera et al., 2024).

Breakthroughs in microbial biotechnology and synthetic biology can potentially make bioproduct synthesis more cost-effective. Innovations in strain engineering and metabolic pathway optimization have shown promise in increasing yields and reducing energy demands, yet these technologies require significant investment in research and development (Abbate et al., 2023). This is how it happened with CBP has revolutionized the production of biofuels and chemicals by enabling the production of cellulolytic enzymes that hydrolyze biomass and ferment the resulting sugars in a single step. However, despite decades of research, CBP still suffers from low yields and is not yet ready for market launch (Philp, 2023).

Public policies and regulatory frameworks also play a crucial role in enhancing the competitiveness of bio-based alternatives. Government subsidies, tax incentives, and carbon pricing mechanisms can level the playing field, making bio-based chemicals more attractive to investors and consumers (Ali, 2023). For instance, the European Union's Renewable Energy Directive (RED) and Brazil's RenovaBio (ANP, 2024) programs have been instrumental in advancing biofuels and biochemicals production. The RED establishes a framework for promoting energy from renewable sources, setting binding targets for renewable energy consumption within the EU (Renewable Energy Directive, 2008). Similarly, RenovaBio aimed at reducing carbon intensity in transportation fuels by expanding biofuel use and creating a carbon credit market. Despite these initiatives, the lack of a unified global framework for bioeconomy regulations can create uncertainties for investors and impede innovation. Achieving a global consensus on bioeconomy regulations is essential to establishing a predictable environment that encourages investment and fosters innovation in sustainable bio-based industries. Engaging with the scientific community is crucial to address these challenges to expand awareness and client bases.

3.4.4 Biomanufacturing and Biofoundries: Addressing Technological and Operational Challenges

Bioproducts demand high levels of quality, purity, and effectiveness, but pursuing them necessitates removing challenges in downstream processing, which requires up to 80% of production costs. In this sense, it is important to consider how to improve the critical and expensive stages, such as purification. Many products are purified using chromatographic methods. Hence, controlling factors that could damage yields, such as cleaning and packing of columns, membrane materials, buffer composition, flow and elution rate, become crucial. Pandey et al. (2024) listed some procedures to improve the handling of chromatographic columns, enabling cost-effective purification. For instance, bioburden contamination can be reduced effectively by automated packing. Also, automatic pH and conductivity in-line adjustments minimize variability in common ion effects in buffer formulations.

Process simulations using mathematical modeling and statistical tools are crucial for accurate bioprocesses analysis, enabling real-time monitoring and control while ensuring high-quality bioproducts. Banner et al. (2021) mentioned automated tools for planning reaction series required to synthesize molecules (retrosynthetic pathways), such as Selenzyme, SelProm, RETROPATH 2.0, and PathPred. The latter can predict, for each catalytic step, the appropriate enzymes and design bespoke biocatalytic routes.

Biofoundries can reduce biorefinery production costs and then enhance the bioeconomy. Also, must prioritize developing a sustainable service model, building a core clientele, and diversifying funding sources to ensure long-term viability. Strengthening client engagement and addressing legal and technical obstacles will be essential for their growth and contribution to the bioeconomy (Holowko et al., 2020). Developing cell factories for large-scale bioprocessing remains a time-consuming and costly endeavor. According to Driessen (2022), it usually requires an investment exceeding US$50 million, spanning 6–8 years.

3.5 Conclusions

Modern biorefinery development capable of processing LCB is pivotal to fostering a sustainable bioeconomy. This chapter highlights innovative pretreatment and purification methods to obtain 2G sugars and other high-value compounds. While advancements in physical, chemical, physico-chemical and biological pretreatment offer new opportunities, high costs remain a challenge. It addressed biofoundries as a combination of technological advancements with a growing role in enabling automated, high-throughput processes that reduce costs, enhance efficiency, and expand biotechnological possibilities. The DBTL framework has proven essential in accelerating innovation and dealing with bottlenecks in biotechnological processes. Additionally, LCA and TEA have also been considered to evaluate the commercial viability of bioproducts. Notably, bioethanol, lactic acid, succinic acid, and xylitol were examined, with the first two showing the greatest promise based on their environmental performance, economic potential, and market value.

Lastly, biorefineries have the potential to be key drivers in the global shift toward a sustainable and circular economy, provided there is continued investment in innovation. Advancements in feedstock processing, synthetic biology, automation, and simulation technologies can enhance the economic competitiveness of bioproducts by lowering costs and reducing processing times. Moreover, fostering interdisciplinary collaboration, proactive stakeholder engagement, and a unified global regulatory framework is crucial for transforming the industrial landscape and contributing significantly to Industry 4.0.

References

Abbate, E., Andrion, J., Apel, A., Biggs, M., Chaves, J., Cheung, K., Ciesla, A., Clark-ElSayed, A., Clay, M., Contridas, R., Fox, R., Hein, G., Held, D., Horwitz, A., Jenkins, S., Kalbarczyk, K., Krishnamurthy, N., Mirsiaghi, M., Noon, K., ... Yerramsetty, K. (2023). Optimizing the strain engineering process for industrial-scale production of bio-based molecules. *Journal of Industrial Microbiology and Biotechnology*, *50*(1), 1–15. https://doi.org/10.1093/jimb/kuad025

Achinivu, E. C., Blankenship, B. W., Baral, N. R., Choudhary, H., Kakumanu, R., Mohan, M., Baidoo, E. E. K., Scown, C. D., George, A., Simmons, B. A., & Gladden, J. (2024). Biomass pretreatment with distillable ionic liquids for an effective recycling and recovery approach. *Chemical Engineering Journal*, *479*, 147824. https://doi.org/10.1016/J.CEJ.2023.147824

Adamczyk, P. A., Coradetti, S. T., & Gladden, J. M. (2023). Non-canonical d-xylose and l-arabinose metabolism via d-arabitol in the oleaginous yeast Rhodosporidium toruloides. *Microbial Cell Factories*, *22*(1), 145. https://doi.org/10.1186/s12934-023-02126-x

Ajala, E. O., Ighalo, J. O., Ajala, M. A., Adeniyi, A. G., & Ayanshola, A. M. (2021). Sugarcane bagasse: a biomass sufficiently applied for improving global energy, environment and economic sustainability. *Bioresources and Bioprocessing*. https://doi.org/10.1186/s40643-021-00440-z

Ali, J. (2023). Road map for sustainable and effective carbon pricing: bridging the gap of realities and ambitions. *Environmental Science and Pollution Research*, *30*(41), 94070–94080. https://doi.org/10.1007/s11356-023-29034-2

Alves De Oliveira, R., Alexandri, M., Komesu, A., Venus, J., Vaz Rossell, C. E., & Maciel Filho, R. (2020). Current advances in separation and purification of second-generation lactic acid. *Separation and Purification Reviews*, *49*(2), 159–175. https://doi.org/10.1080/15422119.2019.1590412

ANP. (2024). Sobre o RenovaBio. *Agência Nacional Do Petróleo, Gás Natural e Biocombustíveis*, 15–16. https://www.gov.br/anp/pt-br/assuntos/renovabio/sobre-o-renovabio

Antunes, F. A. F., Chandel, A. K., Terán-Hilares, R., Ingle, A. P., Rai, M., dos Santos Milessi, T. S., da Silva, S. S., & dos Santos, J. C. (2019). Overcoming challenges in lignocellulosic biomass pretreatment for second-generation (2G) sugar production: emerging role of nano, biotechnological and promising approaches. *3 Biotech*, *9*(6), 230. https://doi.org/10.1007/s13205-019-1761-1

Banner, A., Toogood, H. S., & Scrutton, N. S. (2021). Consolidated bioprocessing: Synthetic biology routes to fuels and fine chemicals. *Microorganisms*, *9*(5), 1079. https://doi.org/10.3390/microorganisms9051079

Baral, P., Kumar, V., & Agrawal, D. (2022). Emerging trends in high-solids enzymatic saccharification of lignocellulosic feedstocks for developing an efficient and industrially deployable sugar platform. *Critical Reviews in Biotechnology*, *42*(6), 873–891. https://doi.org/10.1080/07388551.2021.1973363

Basera, P., Chakraborty, S., & Sharma, N. (2024). Lignocellulosic biomass: Insights into enzymatic hydrolysis, influential factors, and economic viability. *Discover Sustainability*, *5*(1). https://doi.org/10.1007/s43621-024-00543-5

Benavides, P. T., Klein, B., & Davis, R. (2021, October). Economic and environmental assessment of biological conversions of agile BioFoundry (ABF) bio-derived chemicals.

Chandel, A. K., Antunes, Felipe, A. F., Terán-Hilares, R., Cota, J., Ellilä, S., Silveira, M. H. L., Santos, J. C., & Silva, S. S. (2018). Bioconversion of hemicellulose into ethanol and value-added products. In *Advances in sugarcane biorefinery* (1st ed., pp. 97–134). Elsevier.

Chaturvedi, T., Hulkko, L. S. S., Fredsgaard, M., & Thomsen, M. H. (2022). Extraction, isolation, and purification of value-added chemicals from lignocellulosic biomass. *Processes, 10*(9), 1752. https://doi.org/10.3390/pr10091752

Chauvet, J.-M., & Lescieux-Katir, H. (2024). From primary food processing industries to biorefineries: Example of bazancourt-pomacle. *Green Chemistry and Agro-Food Industry: Towards a Sustainable Bioeconomy, 313–338*. https://doi.org/10.1007/978-3-031-54188-9_13

Chui, M., Evers, M., Manyika, J., Zheng, A., & Nisbet, T. (2023). The bio revolution: Innovations transforming economies, societies, and our lives. In D. Araya & P. Marber (Eds.), *Augmented Education in the Global Age* (1st ed., pp. 48–74). Routledge.

del Contreras-Gámez, M., Galán-Martín, Á., Seixas, N., da Costa Lopes, A. M., Silvestre, A., & Castro, E. (2023). Deep eutectic solvents for improved biomass pretreatment: Current status and future prospective towards sustainable processes. *Bioresource Technology, 369*, 128396. https://doi.org/10.1016/j.biortech.2022.128396

Correia, B., Matos, H. A., Lopes, T. F., Marques, S., & Gírio, F. (2024). Sustainability assessment of 2G bioethanssol production from residual lignocellulosic biomass. *Processes, 12*(5), 987. https://doi.org/10.3390/pr12050987

Dai, Z., Pomraning, K. R., Deng, S., Kim, J., Campbell, K. B., Robles, A. L., Hofstad, B. A., Munoz, N., Gao, Y., Lemmon, T., Swita, M. S., Zucker, J. D., Kim, Y.-M., Burnum-Johnson, K. E., & Magnuson, J. K. (2023). Metabolic engineering to improve production of 3-hydroxypropionic acid from corn-stover hydrolysate in Aspergillus species. *Biotechnology for Biofuels and Bioproducts, 16*(1), 53. https://doi.org/10.1186/s13068-023-02288-1

Davis, R. E., Grundl, N. J., Tao, L., Biddy, M. J., Tan, E. C., Beckham, G. T., Humbird, D., Thompson, D. N., & Roni, M. S. (2018). Process design and economics for the conversion of lignocellulosic biomass to hydrocarbon fuels and coproducts: 2018 biochemical design case update. *Biochemical Deconstruction and Conversion of Biomass to Fuels and Products via Integrated Biorefinery Pathways*. https://doi.org/10.2172/1483234

Delgado-Arcaño, Y., Valmaña-García, O. D., da S., N., Neto, G. J. M., Guerra, G. N. F., & Pontes, L. A. M. (2024). Lignocellulosic biomass transformation processes into building blocks: Focus on simulation, optimization, and the life cycle assessment (LCA) techniques. *Green Energy and Technology, F2511*, 75–112. https://doi.org/10.1007/978-3-031-51601-6_4

Demichelis, F., Laghezza, M., Chiappero, M., & Fiore, S. (2020). Technical, economic and environmental assessement of bioethanol biorefinery from waste biomass. *Journal of Cleaner Production, 277*, 124111. https://doi.org/10.1016/j.jclepro.2020.124111

Densmore, D., Hillson, N. J., Klavins, E., Myers, C., Peccoud, J., & Stracquadanio, G. (2023). Introduction to the Special Issue on BioFoundries and Cloud Laboratories. *ACM Journal on Emerging Technologies in Computing Systems*, 19. https://doi.org/10.1145/3609485

Driessen, J. (2022). *Integration of biomass conversion, cell factory optimization and bioprocess design to innovate the valorization of lignocellulosic fibers*. Technical University of Denmark.

Farzaneh, T., & Freemont, P. S. (2021). Biofoundries are a nucleating hub for industrial translation. *Synthetic Biology, 6*(1). https://doi.org/10.1093/synbio/ysab013

Ferraz, D., & Pyka, A. (2023). Circular economy, bioeconomy, and sustainable development goals: A systematic literature review. *Environmental Science and Pollution Research, Lim, 2022*. https://doi.org/10.1007/s11356-023-29632-0

Ferreira, J. A., & Taherzadeh, M. J. (2019). Improving the economy of lignocellulose-based biorefineries with organosolv pretreatment. *Bioresource Technology, 299*. https://doi.org/10.1016/J.BIORTECH.2019.122695

Fortune Business Insights. (2024). *Bioethanol Market Size, Share & Industry Analysis, By Feedstock (Starch based, Sugar based, and Others), By Application (Transportation Fuel, Power Generation, Cosmetic, Pharmaceutical, and Others), and Regional Forecast, 2024-2032.*

Frisvold, G. B., Moss, S. M., Hodgson, A., & Maxon, M. E. (2021). Understanding the U.S. bioeconomy: A new definition and landscape. *Sustainability, 13*(4), 1627. https://doi.org/10.3390/su13041627

Future Market Insights. (2024). Lactic acid market outlook from 2024 to 2034. https://www.futuremarketinsights.com/reports/lactic-acid-market

Gadkari, S., Kumar, D., Qin, Z., Sze, C., Lin, K., & Kumar, V. (2021). Life cycle analysis of fermentative production of succinic acid from bread waste. *Waste Management, 126*, 861–871. https://doi.org/10.1016/j.wasman.2021.04.013

Global Markets Insights. (2024). Xylitol market size Xylitol. https://www.gminsights.com/industry-analysis/xylitol-market

González-García, S., Argiz, L., Míguez, P., & Gullón, B. (2018). Exploring the production of bio-succinic acid from apple pomace using an environmental approach. *Chemical Engineering Journal, 350*(June), 982–991. https://doi.org/10.1016/j.cej.2018.06.052

Guimarães, C. E. C., Neto, F. S., de Castro Bizerra, V., do Nascimento, J. G. A., Valério, R. B. R., de Sousa Junior, P. G., de Sousa Braz, A. K., Melo, R. L. F., de França Serpa, J., de Lima, R. K. C., Guimarães, A. P., de Souza, M. C. M., Lopes, A. A. S., de Sousa Rios, M. A., Desai, A. S., Bilal, M., Smułek, W., Jesionowski, T., & dos Santos, J. C. S. (2023). Sustainable bioethanol production from first- and second-generation sugar-based feedstocks: Advanced bibliometric analysis. *Bioresource Technology Reports, 23*, 101543. https://doi.org/10.1016/j.biteb.2023.101543

Gupta, A., Lee, S.-G., Sung, B.-H., Lee, D.-H., Cho, B.-K., Kim, D.-M., & Kim, H. (2024). Advancing biofoundry development: Strategies and challenges. *Biotechnology and Bioprocess Engineering, 29*(5), 771–778. https://doi.org/10.1007/s12257-024-00121-6

Hillson, N., Caddick, M., Cai, Y., Carrasco, J. A., Chang, M. W., Curach, N. C., Bell, D. J., Le Feuvre, R., Friedman, D. C., Fu, X., Gold, N. D., Herrgård, M. J., Holowko, M. B., Johnson, J. R., Johnson, R. A., Keasling, J. D., Kitney, R. I., Kondo, A., Liu, C., ... Freemont, P. S. (2019). Building a global alliance of biofoundries. *Nature Communications, 10*(1), 1–4. https://doi.org/10.1038/s41467-019-10079-2

Hoang, A. T., Nguyen, X. P., Duong, X. Q., Ağbulut, Ü., Len, C., Nguyen, P. Q. P., Kchaou, M., & Chen, W. H. (2023). Steam explosion as sustainable biomass pretreatment technique for biofuel production: Characteristics and challenges. *Bioresource Technology, 385*, 129398. https://doi.org/10.1016/J.BIORTECH.2023.129398

Holowko, M. B., Frow, E. K., Reid, J. C., Rourke, M., & Vickers, C. E. (2020). Building a biofoundry. *Synthetic Biology (Oxford, England), 6*(1). https://doi.org/10.1093/SYNBIO/YSAA026

Hu, N., Liu, X., Wei, S., Yao, J., Wang, W., Liu, B., Tang, T., Jiang, J., & Wang, L. (2024). Current status and future prospects of pretreatment for tobacco stalk lignocellulose. *Frontiers in Bioengineering and Biotechnology*, *12*, 1465419. https://doi.org/10.3389/FBIOE.2024.1465419/BIBTEX

Joshi, M., & Manjare, S. (2024). Chemical approaches for the biomass valorisation: A comprehensive review of pretreatment strategies. *Environmental Science and Pollution Research*, *31*(36), 48928–48954. https://doi.org/10.1007/s11356-024-34473-6

Klein, B. C., Silva, J. F. L., Junqueira, T. L., Rabelo, S. C., Arruda, P. V., Ienczak, J. L., Mantelatto, P. E., Pradella, J. G. C., Junior, S. V., & Bonomi, A. (2017). Process development and techno-economic analysis of bio-based succinic acid derived from pentoses integrated to a sugarcane biorefinery. *Biofuels, Bioproducts and Biorefining*, *11*(6), 1051–1064.

Kruschitz, A., & Nidetzky, B. (2020). Downstream processing technologies in the biocatalytic production of oligosaccharides. *Biotechnology Advances*, *43*. https://doi.org/10.1016/J.BIOTECHADV.2020.107568

Kumar, R., Basak, B., & Jeon, B. (2020). Sustainable production and purification of succinic acid: A review of membrane-integrated green approach. *Journal of Cleaner Production*, *277*, 123954. https://doi.org/10.1016/j.jclepro.2020.123954

Lecona-Vargas, C. S., Orsat, V., & Dumont, M. J. (2022). Carbohydrate-based biorefineries for the production of 5-hydroxymethylfurfural and 2,5-furandicarboxylic acid and their separation and purification methods. *Biomass Conversion and Biorefinery*, *14*(21), 26575–26595. https://doi.org/10.1007/S13399-022-03477-Y

Lee, Y., Kyu, N., Kim, C., Tran, V. G., Cao, M., Yoshikuni, Y., Zhao, H., & Jin, Y. (2024). Bioresource technology self-buffering system for cost-effective production of lactic acid from glucose and xylose using Acid-tolerant Issatchenkia orientalis. *Bioresource Technology*, *399*(March), 130641. https://doi.org/10.1016/j.biortech.2024.130641

Lobato-Rodríguez, Á., Gullón, B., Romaní, A., Ferreira-Santos, P., Garrote, G., & Del-Río, P. G. (2023). Recent advances in biorefineries based on lignin extraction using deep eutectic solvents: A review. *Bioresource Technology*, *388*, 129744. https://doi.org/10.1016/J.BIORTECH.2023.129744

Louw, J., Dogbe, E. S., Yang, B., & Görgens, J. F. (2023). Prioritisation of biomass-derived products for biorefineries based on economic feasibility: A review on the comparability of techno-economic assessment results. *Renewable and Sustainable Energy Reviews*, *188*(October). https://doi.org/10.1016/j.rser.2023.113840

Market Research Future. (2024). Global ethanol market overview. https://www.marketresearchfuture.com/reports/ethanol-market-7304

Munagala, M., Shastri, Y., Nalawade, K., Konde, K., & Patil, S. (2021). Life cycle and economic assessment of sugarcane bagasse valorization to lactic acid. *Waste Management*, *126*, 52–64. https://doi.org/10.1016/j.wasman.2021.02.052

Nghiem, N. P., & Senske, G. E. (2015). Capture of carbon dioxide from ethanol fermentation by liquid absorption for use in biological production of succinic acid. *Applied Biochemistry and Biotechnology*, *175*, 2104–2113. https://doi.org/10.1007/s12010-014-1369-1

NSF. (2024). *NSF invests in BioFoundries to drive advances across science and engineering*. U.S. National Science Foundation.

Osman, A. I., Fang, B., Zhang, Y., Liu, Y., Yu, J., & Farghali, M. (2024). Life cycle assessment and techno - economic analysis of sustainable bioenergy production: A review. *Environmental Chemistry Letters*, *22*(3). https://doi.org/10.1007/s10.11-023-01694-z

Pandey, K., Pandey, M., Kumar, V., Aggarwal, U., & Singhal, B. (2024). Bioprocessing 4.0 in biomanufacturing: Paving the way for sustainable bioeconomy. *Systems Microbiology and Biomanufacturing*, *4*(2), 407–424. https://doi.org/10.1007/S43393-023-00206-Y/FIGURES/6

Panić, M., Andlar, M., Tišma, M., Rezić, T., Šibalić, D., Cvjetko Bubalo, M., & Radojčić Redovniković, I. (2021). Natural deep eutectic solvent as a unique solvent for valorisation of orange peel waste by the integrated biorefinery approach. *Waste Management*, *120*, 340–350. https://doi.org/10.1016/J.WASMAN.2020.11.052

Patel, A., & Shah, A. R. (2021). Integrated lignocellulosic biorefinery: Gateway for production of second generation ethanol and value added products. *Journal of Bioresources and Bioproducts*, *6*(2), 108–128. https://doi.org/10.1016/J.JOBAB.2021.02.001

Pavan, M. (2022). The new biorefineries: Integration with new technologies for carbon capture and utilization to produce bioethanol. In C. R. Soccol, G. A. G. Pereira, C. G. Dussap, & L. P. de Souza Vandenberghe (Eds.), *Liquid Biofuels: Bioethanol. Biofuel and Biorefinery Technologies* (Vol. 12, pp. 457–487). https://doi.org/10.1007/978-3-031-01241-9_19

Peng, J., Xu, Y., Zhu, B., Yu, H., Li, B., & Xu, H. (2023). Life cycle assessment of bioethanol production by two methods of pretreatment of rice straw based on process simulation. *Industrial Crops and Products*, *191*(PB), 115810. https://doi.org/10.1016/j.indcrop.2022.115810

Pérez-Almada, D., Galán-Martín, Á., del M., C., & Castro, E. (2023). Integrated techno-economic and environmental assessment of biorefineries: Review and future research directions. *Sustainable Energy & Fuels*, *7*(17), 4031–4050. https://doi.org/10.1039/d3se00405h

Philp, J. (2023). Bioeconomy and net-zero carbon: Lessons from Trends in Biotechnology, volume 1, issue 1. *Trends in Biotechnology*, *41*(3), 307–322. https://doi.org/10.1016/j.tibtech.2022.09.016

Precedence Research. (2024). List of contents ethanol market size | Share and trends 2024 to 2034. https://www.precedenceresearch.com/ethanol-market

Renewable Energy Directive. (2008). Renewable energy directive. *European Wind Energy Conference and Exhibition*, *2008*(1), 32–38.

Research and Markets. (2024). Succinic acid - Global strategic business report. https://www.researchandmarkets.com/reports/1382364/succinic_acid_global_strategic_business_report?utm_source=GNE&utm_medium=PressRelease&utm_code=gvqmdw&utm_campaign=1980452+-+Global+Succinic+Acid+Strategic+Industry+Report+2023-2030%3A+Expansion+into+Emer

Roja, K. (2024). Sustainable production and scale-up of succinic acid from cyanobacteria. In *34th European Symposium on Computer Aided Process Engineering /15th International Symposium on Process Systems Engineering* (Vol. 53). Elsevier Masson SAS. 10.1016/B978-0-443-28824-1.50428-2

Santos, J. R., Rambo, M. K. D., & Scapin, E. (2024). Recent advances in xylitol production in biorefineries from lignocellulosic biomass: A review study. *Brazilian Journal of Environmental Sciences*, *56*(1), 1–12.

Sewsynker-Sukai, Y., Naomi David, A., & Gueguim Kana, E. B. (2020). Recent developments in the application of kraft pulping alkaline chemicals for lignocellulosic pretreatment: Potential beneficiation of green liquor dregs waste. *Bioresource Technology*, *306*. https://doi.org/10.1016/J.BIORTECH.2020.123225

Shaji, A., Shastri, Y., Kumar, V., Ranade, V. V., & Hindle, N. (2022). *Sugarcane bagasse valorization to xylitol: Techno-economic and life cycle assessment*. Society of Chemical Industry and John Wiley & Sons.

Shen, Y. (2024). Biomass pretreatment for steam gasification toward H2-rich syngas production – An overview. *International Journal of Hydrogen Energy, 66*, 90–102. https://doi.org/10.1016/J.IJHYDENE.2024.04.096

Silveira, M. H. L., Chandel, A. K., Vanelli, B. A., Sacilotto, K. S., & Cardoso, E. B. (2018). Production of hemicellulosic sugars from sugarcane bagasse via steam explosion employing industrially feasible conditions: Pilot scale study. *Bioresource Technology Reports, 3*, 138–146.

Singh, A., Tsai, M. L., Chen, C. W., Rani Singhania, R., Kumar Patel, A., Tambat, V., & Dong, C. D. (2023). Role of hydrothermal pretreatment towards sustainable biorefinery. *Bioresource Technology, 367*, 128271. https://doi.org/10.1016/J.BIORTECH.2022.128271

Soleymani Angili, T., Grzesik, K., & Kaltschmitt, M. (2021). Life cycle assessment of bioethanol production: A review of feedstock, technology and methodology.

Tellechea-Luzardo, J., Otero-Muras, I., Goñi-Moreno, A., & Carbonell, P. (2022). Fast biofoundries: Coping with the challenges of biomanufacturing. *Trends in Biotechnology, 40*(7), 831–842. https://doi.org/10.1016/J.TIBTECH.2021.12.006/ASSET/241FCC8E-6CB6-46C0-93F5-CB7ECE9B5172/MAIN.ASSETS/GR2.SML

Terrazas, M. (2024). UGA receives $18M NSF award to democratize glycoscience. @ UGAResearch Newsletter, University of Georgia.

Tong, K. T. X., Shi, I., Henry, T., Yew, C., Man, F., Lam, K., Lim, S., & Lee, K. T. (2024). *Advancement of biorefinery-derived platform chemicals from macroalgae: A perspective for bioethanol and lactic acid*. Biomass Conversion and Biorefinery. https://doi.org/10.1007/s13399-022-02561-7

Transparency Market Research. (2024). Global bio-based succinic acid market outlook 2034. https://www.transparencymarketresearch.com/bio-succinic-acid.html

Unyay, H., Perendeci, N. A., Piersa, P., Szufa, S., & Skwarczynska-Wojsa, A. (2024). Harnessing switchgrass for sustainable energy: Bioethanol production processes and pretreatment technologies. *Energies, 17*(19), 4812. https://doi.org/10.3390/EN17194812

Verified Market Research. (2024). Xylitol market size and forecast. https://www.verifiedmarketresearch.com/product/xylitol-market/

Verma, M. L., Kumar, P., & Chandel, H. (2023). Nanobiotechnological routes in lignocellulosic waste pre-treatment for bio-renewables production. *Recent Advances in Bio-Energy Research*, 23–34. https://doi.org/10.1007/978-981-99-5758-3_3

Vollmer, N. I., Gargalo, C. L., Gernaey, K. V., Olsen, S. I., & Sin, G. (2023). Life cycle assessment of an integrated xylitol biorefinery with value - added co – products. *The International Journal of Life Cycle Assessment, 28*, 1–14.

Woiciechowski, A. L., Dalmas Neto, C. J., de Souza, P., Vandenberghe, L., de Carvalho Neto, D. P., Novak Sydney, A. C., Letti, L. A. J., Karp, S. G., Zevallos Torres, L. A., & Soccol, C. R. (2020). Lignocellulosic biomass: Acid and alkaline pretreatments and their effects on biomass recalcitrance - Conventional processing and recent advances. *Bioresource Technology, 304*. https://doi.org/10.1016/J.BIORTECH.2020.122848

Yu, Y., Wu, J., Ren, X., Lau, A., Rezaei, H., Takada, M., Bi, X., & Sokhansanj, S. (2022). Steam explosion of lignocellulosic biomass for multiple advanced bioenergy processes: A review.

Renewable and Sustainable Energy Reviews, *154*, 111871. https://doi.org/10.1016/J.RSER.2021.111871

Zabochnicka, M., Krzywonos, M., Romanowska-Duda, Z., Szufa, S., Darkalt, A., & Mubashar, M. (2022). Algal biomass utilization toward circular economy. *Life*, *12*(10), 1480. https://doi.org/10.3390/LIFE12101480

Zhang, B., Liu, X., & Bao, J. (2023). High solids loading pretreatment: The core of lignocellulose biorefinery as an industrial technology – An overview. *Bioresource Technology*, *369*, 128334. https://doi.org/10.1016/J.BIORTECH.2022.128334

Ziaei-Rad, Z., & Pazouki, M. (2024). Cost-effective ionic liquid pretreatment of poplar wood for biofuel production: An optimization study using response surface methodology. *Industrial Crops and Products*, *219*, 119036. https://doi.org/10.1016/J.INDCROP.2024.119036

4

Lignin Valorization Is the Key for a Win–Win Situation in a Biomass Refinery

Lucas Ramos[1], Carina Prado[1], Maria Teresa Ferreira Ramos Raimundo[2], Uirajá C. M. Ruschoni[2], Vinícius Pereira Shibukawa[2], and Anuj Kumar Chandel[1]

[1] *Renewable Carbon and Biology Systems Laboratory-ReCABS, Department of Biotechnology, Engineering School of Lorena, University of São Paulo (EEL-USP), Lorena, São Paulo, Brazil*
[2] *Department of Biotechnology, Engineering School of Lorena, University of São Paulo (EEL-USP), Lorena, São Paulo, Brazil*

4.1 Introduction

The use of agro-industrial residues and byproducts has gained attention over the last decades due to their vast availability, low cost, carbon-rich content, and the increasingly evident effects of climate change (Mussatto et al., 2021). The world is pursuing ways to promote an energetic transition to a low-carbon economy, and the utilization of biomass to obtain drop-in and drop-out bioproducts is of utmost importance to achieve this goal and favor a circular bioeconomy (Jie et al., 2023; Shibukawa et al., 2023).

Lignocellulosic biomass is chemically well-suited for the production of carbon-based bioproducts, as its main composition is based on cellulose, hemicellulose, and lignin (Mussatto et al., 2021). Deconstructing the biomass fibers can yield a rich sugar stream that can be, for example, fermented to produce various products, for example, xylitol, ethanol, and pullulan (Chandel, et al., 2018a; Cruz-Santos et al., 2023; Prado et al., 2023).

Cellulose is the major carbohydrate fraction of lignocellulosic biomass, comprising 40–60% of its composition. It is a homopolymer made up of glucose units linked by β(1–4) bonds (Ashokkumar et al., 2022; Ilic et al., 2023). Hemicellulose makes up 20–40% of the biomass, and it is characterized as a heteropolymer. Its content varies according to the specific biomass but usually includes xylose, arabinose, mannose, glucose, and acetic and glucuronic acids (Ashokkumar et al., 2022; Chen et al., 2022).

The last major component of lignocellulosic biomass is lignin, comprising 10–25% of the biomass. It is characterized by being a highly complex, amorphous, branched, and polyphenolic macromolecule (Ashokkumar et al., 2022; Huang et al., 2022). The primary compounds within lignin are guaiacyl, syringyl and p-hydroxyphenyl units, with minor compounds including hydroxyphenylpropanoics, such as, tricin, hydroxycinnamaldehydes, and dihydroxycinnamic alcohols (Huang et al., 2022; Nelson & Cox, 2014). Lignin

Biofoundry Techniques for Biotechnology Applications, First Edition. Edited by Anuj Kumar Chandel.

forms alkyl/aryl-ester bonds with hemicellulose, creating lignin-carbohydrate complexes that inhibit enzyme access to cellulose (Huang et al., 2022). It provides cohesion between the fibers of lignocellulosic material, forming an insoluble three-dimensional network that ensures rigidity, mechanical support, and facilitates the transport of water and nutrients (Nelson & Cox, 2014).

Lignin ranks as the second most abundant macromolecule on Earth. Due to its characteristics, such as high calorific power, oxidation resistance, biodegradability, and high thermal stability, it is used in sectors such as pharmaceutical, energy production, chemistry, polymers, and resins (Liu et al., 2021; Nelson & Cox, 2014). However, its depolymerization to obtain small units of high added value still faces significant challenges, a challenge that has persisted since the 1930s. This difficulty is primarily due to the high heterogeneity of the macromolecule (da Cruz et al., 2022; Subbotina et al., 2021).

The facilities that process lignocellulosic biomasses are known as biorefineries. This concept is derived from petrochemical industries, where a diversity of products is obtained from a single raw material—oil for conventional refineries and biomasses for biorefineries (Siddiki et al., 2022). Unfortunately, second-generation biorefineries, i.e., facilities that process biomass, still face bottlenecks that hinder their economic viability, such as the high cost of enzyme cocktails, the pretreatment step, and the formation of inhibitory compounds during upstream stages (Mendes et al., 2018; Vasconcelos et al., 2020).

The carbohydrate fractions are already well utilized within a biorefinery context for producing bioproducts such as cello- and xylo-oligosaccharides, ethanol, biopolymers, xylitol, biosurfactants, and a range of other bioproducts (Chandel et al., 2018b; Cruz-Santos et al., 2023; Marcelino et al., 2020; Mendes et al., 2018; Prado et al., 2023; Shibukawa et al., 2023; Silveira et al., 2018; Vasconcelos et al., 2020). However, the lignin fraction remains underutilized and is more commonly sent to cogeneration system to be burned for steam and electricity production (Mendes et al., 2018; Shibukawa et al., 2023; Vasconcelos et al., 2020). Nevertheless, lignin shows high potential to produce high-value bioproducts, which could increase the biorefinery portfolio and thus enhance its economic sustainability (da Cruz et al., 2022; Shibukawa et al., 2023; Subbotina et al., 2021).

In this sense, research to develop and improve technologies to valorize the lignin fraction present in lignocellulosic biomasses is of utmost importance to achieve a circular bioeconomy within a biorefinery context. Lignin can replace fossil-based products, and its valorization contributes to valorize the whole biomass fractions. Additionally, the obtention of high-value compounds from lignin can improve economic aspects of second-generation biorefineries.

4.2 Lignin: Dispensable Source of Renewable Carbon

As is known, the second most abundant macromolecule on the planet is lignin, accounting for approximately 30% of the planet's non-fossil carbon (Luo & Abu-Omar, 2017). Until the middle of the last decade, literature indicated that lignin was an underutilized fraction, primarily used as fuel for burning and converting thermal energy into electrical energy for agro-industrial and pulp and paper facilities (Luo & Abu-Omar, 2017; Sun, 2020), rather than being used as a sustainable feedstock for aromatic chemical production (Hilares et al., 2020).

The estimated global industrial production of lignin is around 0.05 to 0.1 billion tons, derived from the pulping and biofuel industries, of which 95% is used to fuel boilers and generate electricity at \$0.08/kg of lignin (Luo & Abu-Omar, 2017; Sun, 2020; Windeisen & Wegener, 2016). The calorific value of different biomasses with varying lignin content ranges from 16 to 21 MJ/kg on a dry material basis (Mendu et al., 2011; Taylor et al., 2024), and the calorific value reported for lignin, by itself, was 25.5 MJ/kg (Luo & Abu-Omar, 2017; Yoo & Ragauskas, 2021). Thus, the lignin-burning process can save 3.8–7.6 billion dollars for industries with energy conversion generating 1.25–2.5 MJ.

In addition to the justification of the easy conversion of thermal energy into electrical energy generating monetary savings, the application of lignin as a raw material to yield fine chemicals and high-value-added products is hindered by several factors: (a) Variation in the composition of lignin's constituent monomers (Sun, 2020); (b) Amorphous organization of molecular chemistries (Sun, 2020); (c) Presence of modifications resulting from biomass fractionation processes (Sun, 2020); (d) Inefficient and economically unfeasible lignin conversion, separation, and purification technologies under development (Gaspar & Fardim, 2023; Luo & Abu-Omar, 2017); (e) Mechanical rigidity and chemical recalcitrance (Luo & Abu-Omar, 2017); (f) Molecular heterogeneity varies by biomass source, tissue, and aging (Gaspar & Fardim, 2023).

The method of obtaining and isolating lignin directly impacts the application of this macromolecule and was discussed in another section of this chapter (Brienza et al., 2024). Biorefineries and pulp and paper facilities overcome biomass recalcitrance by applying preprocessing such as pulping and pretreatments (Hilares et al., 2020; Windeisen & Wegener, 2016; Yoo & Ragauskas, 2021). However, these processes primarily focus on isolating and purifying carbohydrates, with lignin being a secondary residue in these current processes. The concept of "lignin-first biorefineries" arises from the development of processes that position lignin as a primary product in conventional industrial processes, reducing the effects of lignin condensation and modification, and demonstrating that it is possible to profit from lignin, contrary to paper and cellulose industry maxims that "one can make anything from lignin except money" (Brienza et al., 2024).

The market analysis developed by Dessbesell and collaborators (2020) categorizes potential lignin-based products as "high volume, low cost," exemplified by lignin's use in energy generation in biorefineries (previously covered in this topic); "low volume and high cost," such as the production of carbon fibers (Bajwa et al., 2019) and biomedical materials (Gaspar & Fardim, 2023); and "intermediate market volume and cost," where the market for bulk chemicals like phenols can be found (Tribot et al., 2019). This market analysis highlights the potential of applying lignin as a feedstock molecule to generate new products and meet existing and emerging market demands (Dessbesell et al., 2020).

In addition to the applications mentioned in the chart, lignin and its derivatives have been explored as food additives, dispersants, rheology modifiers, bioplastics, paper products, biocomposites, biofuel and bioenergy production and for biomedical applications (Bajwa et al., 2019; Brodin et al., 2017; Gaspar & Fardim, 2023). All of these applications leverage the natural chemical properties of lignin, applying this macromolecule in its native form or with specific chemical modifications resulting from the pretreatment and processing of the starting biomass (Brienza et al., 2024; Gaspar & Fardim, 2023). Unconventional monolignols, such as hydroxycinnamyl acetates, could be extracted from the phenolic fraction of biomass, potentially increasing the range of lignin-based by-products with higher added value (Sun, 2020).

Lignin also presents potential properties such as bactericidal effects, antioxidant activity, and ultraviolet radiation blocking (European Investment Bank, 2024 https://www.eib.org/en/stories/sustainable-solution-lignin; Gaspar & Fardim, 2023). For example, emerging industries demonstrate that it is possible to add value from this phenolic fraction of biomass, such as the Austrian company Lignovations, founded in 2021 (Lignovations, 2024 https://lignovations.com). The company's foundation is based on a patent for lignin fractionation that facilitates the subsequent manufacture of this material.

The trend in lignin use demonstrates the application of low molecular weight fractions or phenolic monomers of this macromolecule as building blocks for new materials or as components of existing polymeric materials (Tribot et al., 2019). An example of this application is the production of bioplastics and biopolymers from biomass, primarily lignin, such as polylactic acid (PLA), polyhydroxyalkanoates (PHA), and bio-based polyethylene (Brodin et al., 2017). The production routes for bioplastics from less costly biomass produce bio-based or bio-enriched polyphenol (PF) and polyurethane (PUR). On the other hand, the production of PHA, PLA, and PE involves fermentative and biotechnological routes which, when using lignin as a precursor, encounter challenges due to the antibiotic properties of this macromolecule and its precursor monomers (Brodin et al., 2017; Tribot et al., 2019).

Despite the advantages of using lignin's calorific power to generate energy through thermal processes, other properties of this macromolecule pave the way for developing new bioproducts. Using lignin in its native form requires less purity and has been explored mainly as an additive in established formulations to meet sustainability demands and reduce the carbon footprint of materials. However, employing lignin in fine chemistry and exploring its precursors as building blocks for new chemicals, polymers/resins, and carbon fibers requires the development of technologies from obtaining pure lignin fractions to processing and obtaining the final product, meeting the requirements of a non-polluting and green production chain. Although some of these materials do not meet the biodegradability requirement, the use of lignocellulosic biomass as a starting material reduces the carbon footprint and mitigates the environmental impacts of the high demand for plastic consumption, replacing fossil sources, i.e. BioPolyEthylene (BioPE) (Brodin et al., 2017; Tribot et al., 2019). PHA is the most prominent bioplastic material to replace fossil-based plastics; PHA is biodegradable and capable of forming molecules like carbon dioxide, methane, water, and biomass when left under mild environmental conditions (Sohn et al., 2022). Therefore, advancing the use of lignin as a raw material for new biobased products depends on the development of new isolation and manufacturing technologies for this potential material.

4.3 Lignin Chemistry

The composition of lignocellulosic materials consists of three primaries components: cellulose, hemicellulose and lignin. Different from the carbohydrate fractions, lignin has a complex, cross-linked polymeric structure that gives cell walls rigidity and recalcitrance. The chemical composition of lignin and the fraction of lignin present in plant biomass

Table 4.1 Main Sources of Lignin from Hardwood, Softwood and Grasses, and Their Composition in Terms of Lignin Percentage, S/G and H/G ratio

Source	Kind	% lignin	S/G	H/G	Reference
Bitternut hickory	Hardwood	23.71%	1.74	–	Obrzut et al. (2023)
White spruce	Softwood	28.12%	0.06	–	Obrzut et al. (2023)
Corn straw	Grass	15.60%	0.55	–	Obrzut et al. (2023)
Oat straw	Grass	12.90%	0.62	0.20	Rencoret et al. (2023)
Eucalyptus	Hardwood	–	2.00	0.00	Pereira et al. (2017)
Wheat straw	Grass	–	0.50	0.05	Pereira et al. (2017)
Spruce	Softwood	–	0.00	0.01	Pereira et al. (2017)
Birch	Hardwood	–	0.77	–	Faleva et al. (2021)
Eucalyptus	Hardwood	22.70%	3.63	0.09	Rencoret et al. (2011)

differ depending on the plant species and the specific morphological part of the plant. In general, lignin represents 10–30% by weight of plant biomass, with softwoods having the highest lignin content at around 30%, this fraction is lower in hardwoods, ranging from 20 to 25%, and even lower in grasses, from 10 to 15% (Li et al., 2015).

In terms of its chemical composition, lignin is an amorphous molecule, made up mainly of three primary units: *p*-coumaryl alcohols (4-hydroxycinnamyl), coniferyl alcohols (3-methoxy 4-hydroxycinnamyl) and sinapyl alcohols (3,5-dimethoxy 4-hydroxycinnamyl). These monolignols that make up the lignin molecule give rise to the p-hydroxyphenyl (H), guaiacyl (G) and syringyl (S) units, respectively, which are linked to each other by ether bonds and carbon–carbon bonds (Alam et al., 2024).

In addition to monolignols, other chemical species can also be present in the lignin molecule, and no plant contains lignin made up only of *p*-hydroxyphenyl (H), guaiacyl (G) and syringyl (S). Among the chemical compounds that constitute the structure of lignin are ferulic acid, ferulates, coniferaldehyde, sinapaldehyde, 5-hydroxyconiferyl alcohol and acylated monolignols containing acetate, *p*-hydroxybenzoate or *p*-coumarate fractions. In herbaceous raw materials, for example, lignin contains high levels of *p*-coumaric acid and ferulic acid units, which are linked to other units by ether and ester bonds (Li et al., 2015).

The proportion between the monolignols found in the lignin depends on the type of plant. The main sources of lignin and their chemical compositions are shown in Table 4.1. Gymnosperm lignin (softwood) has more guaiacyl units, while angiosperm lignin (hardwood) contains a mixture of guaiacyl and syringyl, and grass wood contains a mixture of all three monolignols. In this way, lignins can be classified according to their chemical composition into G-type lignin from softwoods, GS-type lignin from hardwoods and HGS-type lignin from grasses (Obrzut et al., 2023).

The types of chemical bonds present in the lignin molecule vary according to the type of plant and stage of development. In general, most of the bonds are of the carbon–carbon type. To determine the types of bonds, the carbon atoms in the aliphatic side chains of monolignols are designated as α, β, and γ, whereas the carbon atoms in the aromatic rings are numbered from 1 to 6. The predominant type of bond in lignin molecules is β-O-4, which consists of a bond formed between the β carbon of the aliphatic side chain and the oxygen atom attached to the C4 position of the aromatic fraction. This type of bond represents 43% of the bonds in softwoods and around 65% of the bonds in hardwoods. Other common bond types in lignin are β-β, β-5, α-O-4, 4-O-5, 5-5, α-O-γ and β-1 (Li et al., 2015).

The reactivity of lignin is directly related to its chemical composition and the types of bonds between the monomers. The β-O-4 bonds, the most abundant in lignin molecules, are among the easiest to cleave and are often the most studied targets in delignification methods. Another important characteristic of lignin molecules is the presence of carbon–carbon bonds, the quantity of which is directly related to the proportion of monolignols present. In softwoods containing G-type lignin, carbon–carbon bonds are more frequent than in hardwoods containing GS-type lignin. This difference in the number of carbon–carbon bonds is considered significant in the reactivity of lignin (Schutyser et al., 2018).

4.4 Lignin Extraction Methods

The valorization of the lignin fraction present in lignocellulosic materials is fundamental in the context of biorefineries, since it represents the most abundant non-carbohydrate component in the cell wall of plants and is still underutilized compared to the cellulose and hemicellulose fractions. In addition, lignin is the largest source of aromatic compounds available in nature, making it an interesting opportunity to obtain high value-added compounds. Therefore, an important stage in the process of valorizing lignin is the extraction of this component from lignocellulosic material (Ascencio et al., 2020; Brienza et al., 2024).

Various methods are available for extracting lignin from lignocellulosic materials, some of which are more commonly used industrially, such as the Kraft process, while others have been developed on a laboratory scale, such as organosolv extraction processes and using ionic liquids (Alam et al., 2024). Choosing the right extraction process is fundamental to ensuring its best utilization. It is important to consider characteristics such as extraction efficiency, sustainability of the process, and the chemical interactions of the reagents used with the lignin. Different extraction processes produce different types of lignin due to the various structural modifications that occur during the process, directly interfering with its properties and applications (Sang et al., 2023; Schutyser et al., 2018). Table 4.2 shows the main lignin extraction methods and their efficiency.

The Kraft process is the most widely used industrially to extract lignin from wood in pulp and paper production, which accounts for more than 80% of world production. This method uses sodium hydroxide and sulfide anions to delignify the wood at temperatures of 150–170 °C and removes over 90% of the lignin. Breaking the β-O-4 bonds in the lignin creates phenolic hydroxyl groups and reduces the molecular weight of the lignin. The recovered lignin (Kraft lignin) can be precipitated and extracted using mineral acids or carbon dioxide. The process can be carried out in batch or continuous systems, both of which are widely used. Despite the large amount of sulfur, only 1–2% of sulfide remains in the regenerated lignin. Although Kraft lignin is generally of poor quality and contains impurities such as sulfur and residual carbohydrates, it can be processed and modified for use in adhesives, binders, and biofuels (Schutyser et al., 2018).

Table 4.2 Principal Lignin Extraction Methods and Their Efficiency

Method	Lignocellulosic Material	Solvent	Reaction Conditions	Lignin Extraction (%)	Reference
Kraft	Pine	19% NaOH	170 °C, 150 min	89.59	Jardim et al. (2022)
Kraft	Acacia	16% NaOH	160 °C, 150 min	85.30	Jardim et al. (2022)
Kraft	Sweetgum	16% NaOH	160 °C, 150 min	95.37	Jardim et al. (2022)
Kraft	Eucalyptus	16% NaOH	160 °C, 150 min	95.56	Jardim et al. (2022)
Soda pulping	Hemp	16% NaOH	170 °C, 90 min	64.63	Gaynor et al. (2024)
Soda pulping	Miscanthus	80 g/kg NaOH	170 °C, 99 min	80.29	Kim and Um (2018)
Organosolv	Hazelnut skin	50% methanol	200 °C, 60 min	8.47	Oliva et al. (2021)
Organosolv	Sugar cane bagasse	75% ethanol	121 °C, 30 min	56.00	Schmatz and Brienzo (2022)
Ionic liquid	Eucalyptus	Ionic liquid	140 °C, 100 min, microwave-assisted	55.50	Sun et al. (2019)
Supercritical fluid	Eucalyptus	CO_2, ethanol and water (2:1)	180 °C, 15 MPa, 60 min	35.90	Jiang et al. (2020)

The extraction of lignin from non-wood lignocellulosic materials such as sugarcane bagasse, wheat straw, hemp, kenaf, and sisal is commonly carried out using soda pulping processes. This technique involves alkaline hydrolysis with sodium hydroxide at concentrations ranging from 13 to 16%, carried out in pressurized reactors that reach temperatures between 140 and 170 °C to isolate lignin from plant biomass. During the process, the α-aryl ether bonds within the phenolic structures of lignin are broken. Lignin is then precipitated from the solution by adding sulfuric acid until the black liquor reaches a pH of 2–3. The lignin can be recovered through centrifugation or filtration (Alam et al., 2024).

In addition to conventional pulping methods, more environmentally favorable lignin extraction methods have been developed, which require less energy and do not generate large amounts of waste, reducing the environmental impact of the process when compared to conventional methods. The supercritical fluid method, for example, makes it possible to extract lignin in a more sustainable way. This approach utilizes supercritical fluids, substances exposed to temperatures and pressures above their critical point. These fluids exhibit properties such as low viscosity, low dielectric constant, and high diffusivity, which enhance their capacity to permeate the lignin matrix effectively. Among them, carbon dioxide stands out as the most appropriate supercritical fluid for the pretreatment of lignocellulosic biomass, owing to its relatively low critical temperature (31.1 °C) and pressure (73.8 bar), which make the process more energy-efficient and environmentally benign. The lignin extraction process consists of adding ground vegetable biomass to an extraction tank along with carbon dioxide, under temperature and pressure conditions that exceed its critical point (Alam et al., 2024).

Another promising approach for lignin extraction is the Organosolv process, which employs a mixture of water and organic solvents, such as acetone, methanol, ethanol, or ethylene glycol, and can be carried out with a catalytic agent. This technique enables the removal of the lignin fraction by cleaving the α- and β-O-aryl ether linkages within its structure. As a result, the lignin obtained is typically of high purity, superior quality, and less condensed compared to the lignin produced through the Kraft process. (Alam et al., 2024).

4.5 Lignin Conversion Route

Lignin, the largest source of aromatic compounds available in nature, represents a promising raw material for the sustainable production of various chemicals, fuels, and materials. The structural heterogeneity and macromolecular characteristics of lignin are factors that hinder conversion processes. Despite this, strategies have been developed that make it possible to convert phenolic compounds derived from the lignin molecule into high value-added products such as plastics, adhesives, biofuels, polymer additives, and bioactive molecules (Liu et al., 2022).

Lignin can be converted by chemical or biological routes. Among the main chemical routes used are organic synthesis reactions, in which phenols, esters, and alcohols derived from lignin are converted into products of interest using chemical catalysts (Jadhav et al., 2024). The conversion of lignin via biological pathways, on the other hand, can be carried out by lignolytic microorganisms, which have developed metabolic pathways capable of converting lignin derivatives into chemical products. These microorganisms can metabolize different aromatic compounds derived from lignin through a set of pathways that generate various products, so the bioconversion of lignin represents a promising strategy to overcome limitations related to its heterogeneity (Liu et al., 2022).

4.6 Biological Routes

Lignin is a fundamental element of lignocellulosic biomass that serves as an underexploited resource in the expanding biofuels sector. Its conversion into useful products can be broadly categorized into biological and chemical routes, each with distinct mechanisms and end products. Biological conversion of lignin involves the use of microorganisms or enzymes to break down lignin into smaller, valuable compounds. These processes typically occur under milder conditions compared to chemical methods and can offer high specificity. Nevertheless, transforming lignin into value-added products is still difficult because of its complex and irregular composition. Numerous naturally occurring microbes have developed sophisticated metabolic and enzymatic pathways to break down and convert lignin. Significant efforts have been devoted to harnessing their potential for lignin valorization (Beckham et al., 2016; Chen & Wan, 2017).

4.6.1 Microbial Degradation

Fungi: White-rot fungi (such as *Phanerochaete chrysosporium*) are particularly effective at degrading lignin through the secretion of ligninolytic enzymes like lignin peroxidases, manganese peroxidases, and laccases. These fungi can convert lignin into aromatic monomers and other low-molecular-weight compounds. The fungal species *Calvatia fumosa* and *Flammulina velutipes* were able to remove 40% of the methoxy groups from Kraft lignin after three weeks. When two brown-rot fungi, *Postia placenta* (Mad-698-R) and *Gloeophyllum trabeum* (Mad-617-R), were used to degrade wood, the aromatic dihydroxy content in lignin increased due to demethylation (Chen & Wan, 2017; Zou et al., 2015).

Bacteria: Some bacterial strains, such as *Actinobacteria* and *Rhodococcus*, can also degrade lignin through their own set of enzymes. Bacteria can offer advantages in terms of faster growth rates and the ability to function under different environmental conditions. Recent research has shown that numerous bacterial species can degrade lignin. Their possible routes for lignin breakdown and valorization have been previously examined (Bugg et al., 2011; Tian et al., 2014). The *Rhodococcus genus* have been recognized as possessing lignin-degrading abilities. Among various *Rhodococcus species*, *R. jostii* RHA1 is the one that has been studied most. *R. jostii* RHA1 was first reported to be a polychlorinated biphenyl (PCB) degrader (Seto et al., 1995), and recently identified to be able to metabolize lignin (Ahmad et al., 2010, 2011).

4.6.2 Enzymatic Conversion

Lignin-degrading Enzymes: Enzymes such as laccases, peroxidases, and dioxygenases can break down lignin into smaller units. These enzymes can be produced through microbial fermentation or recombinant DNA technology. Lignin peroxidase (LiP) is a key oxidative biocatalyst responsible for lignin breakdown. This peroxide-activated enzyme catalyzes several oxidations through a mechanism based on the formation of aryl cation radicals (Chen & Wan, 2017; Kersten et al., 2012). Veratryl alcohol acts to stabilize the enzyme's catalytic conformation while operating as an electron-transfer agent that facilitates select LiP-driven reactions (Hammel et al., 2014). The catalytic cycle usually includes four steps: (a) the formation of compound I; (b) the reduction of compound I by non-phenolic aromatic substrate; (c) the formation of compound II; and (d) the completion of cycle by returning enzyme to the native ferric state. MnP, as a heme-containing peroxidase, uses Mn^{2+} as an electron donor, which is then oxidized to Mn^{3+} through peroxide-dependent oxidation (Dashtban et al., 2010; Hofrichter et al., 2002).

Metabolic Engineering: By genetically modifying microorganisms to overexpress specific enzymes, researchers can enhance lignin degradation and convert it into desired products like biofuels, bioplastics, and specialty chemicals. Such as the chelated Mn^{3+} functions as a low-molecular-weight, mobile redox agent that oxidizes phenolic compounds. The lignin breakdown process concurrently produces multiple lipid byproducts (Chen & Wan, 2017). The peroxidation of these lipids mediated by MnP is proposed to be a common mechanism for the degradation of non-phenolic structures (Chen & Wan, 2017).

4.6.3 Fermentation

One application of fermentation is in biofuel production. After the initial breakdown of lignin, the aromatic compounds derived from lignin can be further processed by engineered microbes into biofuels such as ethanol and butanol, as well as other valuable chemicals (Leonel et al., 2021). Another application of fermentation focuses on producing value-added chemicals. This process can convert lignin into a variety of high-value chemicals, including vanillin, adipic acid, and other aromatic compounds used in pharmaceuticals and fragrances.

4.7 Chemical Routes

Chemical conversion of lignin involves breaking down its complex structure through various chemical reactions, often requiring harsh conditions such as high temperatures and pressures, and the use of catalysts. Lignin depolymerization with selective bond cleavage is the major challenge for converting it into value-added chemicals (Pandey & Kim, 2011).

4.7.1 Thermal Decomposition

- **Pyrolysis**: Lignin is thermally decomposed in the absence of oxygen at high temperatures (300–700 °C), producing bio-oil, gases, and char. The bio-oil can be further refined into fuels and chemicals. Despite the fact that thermal degradation starts around 200 °C. The first linkages to be cleaved are the weakest C—O—C ether bonds (Agrizzi et al., 2024). The yields to pyrolytic liquids are generally around 40–60 wt% while gases represent 8–20 wt%. Several authors have investigated the experimental conditions, and to a lesser extent, their impact on the oil composition (Chen & Wan, 2017; Dias et al., 2023).
- **Gasification**: Lignin is converted into syngas (a mixture of hydrogen and carbon monoxide) at very high temperatures (>700 °C) in the presence of a controlled amount of oxygen or steam. Syngas can be used to produce methanol, hydrogen, and other chemicals through catalytic processes.

4.7.2 Catalytic Depolymerization

- **Hydrogenolysis**: Lignin is broken down in the presence of hydrogen and a catalyst (often a metal catalyst) to produce monomeric phenols and other aromatic compounds. This process can be conducted at moderate temperatures and pressures.
- **Oxidative Depolymerization**: Lignin is oxidized using oxidizing agents (e.g., oxygen, hydrogen peroxide) and catalysts to produce various aromatic aldehydes, acids, and ketones.

- **Solvolysis Organosolv Process**: Lignin is dissolved in organic solvents (e.g., ethanol, methanol) at elevated temperatures and pressures, often with the addition of acids or bases to enhance depolymerization. This method produces high-purity lignin and valuable chemicals. According to Sharma et al. (2004), when conversion of lignin was performed at 345 °C, the yields into liquids were still low and high quantities (20–30%) of condensation products, precursors of coke and chars, were formed after aliphatic cleavages.

Lignin is processed in water at high temperatures with catalysts to produce hydrogen, hydrocarbons, and other chemicals. In fact, based on conversion parameters from the literature, lignin can be depolymerized in aqueous solutions under subcritical or supercritical conditions at relatively low temperatures, typically ranging from 280 to 400 °C. This process usually takes from a few minutes to a few hours, under a total pressure of 20–25 MPa, with a water-to-lignin ratio generally varying from 2 to 50 (Pandey & Kim, 2011).

4.7.3 Electrochemical Conversion

Lignin is broken down through electrochemical reactions, often involving the use of electricity to drive the depolymerization process, producing various aromatic compounds. Both biological and chemical routes for lignin conversion offer unique advantages and challenges. Biological routes are generally more environmentally friendly and can offer high specificity but may suffer from slower reaction rates and lower yields. Chemical routes can achieve faster and more complete conversion of lignin but often require harsh conditions and can produce a wider range of byproducts. Advances in both fields are essential for the development of efficient and sustainable lignin valorization processes, contributing to the overall economic viability of biomass refineries and supporting the transition to a bio-based economy.

4.8 Lignin in the Pulp and Paper Industry

Pulp and paper industry (PPI) is one of the largest commodity industries in the world and holds significant influence over the global greenery share (Yadav et al., 2022). Different countries acknowledge PPI as a high-priority industrial sector, making it a major trade within a global economy (Haq et al., 2020; Mandal et al., 2022). It is also one of the most water-consuming industries globally, with an estimated consumption of about 905.8 million m^3/year, generating approximately 695.7 million m^3 of wastewater every year (Haq et al., 2020). However, modern facilities have the potential to decrease water consumption by a factor of 10 compared to older facilities. Despite this, more than 42% of total wastewater generated globally comes from PPI (Mandal et al., 2022).

The effluent and wastewater contain inorganic and organic contaminants, primarily generated from chlorine compounds, resins, lignins, and tannins (Yadav et al., 2022; Haq et al., 2020). The disposal of this effluent and wastewater poses a significant environmental issue due to their alkaline pH, high chemical oxygen demand, dark color, and suspended solids, all of which can lead to aquatic life damage (Mandal et al., 2022). The dark brownish appearance of the wastewater has an unpleasant aspect and negatively affects the photosynthesis process by hindering sunlight's penetration into the water (Haq et al., 2020; Mandal et al., 2022).

The dark brownish appearance is primarily due to lignin and lignin-derived compounds. PPI stands as one of the largest lignin, and lignin-derived compound producers globally (Yadav et al., 2022). These companies produce around 50–70 million tons of lignin each year, with a production projection of 225 million tons annually by 2030 (Haq et al., 2020; Yadav et al., 2022). This lignin was once considered a low-value byproduct; however, nowadays it has gained attention for its potential industrial applications (Haq et al., 2020; Liu et al., 2021; Subbotina et al., 2021; Yadav et al., 2022).

Different types of lignin are generated according to the pulping process employed. For example, the most common pulping processes are sulfate (kraft) and sulfite, originating kraft and sulfite lignin, respectively (Haq et al., 2020). Kraft process is the most common pulping process used worldwide, with kraft lignin accounting for 85% of the total production. Kraft lignin differs from raw lignin regarding phenolic hydroxyl groups (Yadav et al., 2022). Most kraft lignin is used for energy production; however, 2% of it is directed towards the generation of high-value products (Haq et al., 2020). The sulfite process is the second most common pulping method, producing sulfite lignin, also known as lignosulphonates. Annually, about 7 million tons of lignosulphonate are produced. Lignosulfonate is water-soluble, with elevated ash content and high molecular weight (Haq et al., 2020; Yadav et al., 2022). It is primarily used for the obtention of various high-value products, such as surfactants, cement additives, suspensions, and dispersant agents (Yadav et al., 2022).

Two other types of lignin worth mentioning are soda lignin and organosolv lignin. Soda lignin obtention derives from soda-anthraquinone pulping process and is normally produced from non-wood fibers, flax and straws. Due to the annual availability of these feedstocks, this type of lignin is produced at a low rate (Haq et al., 2020). The black liquor containing soda lignin is studied for energy recovery, and this lignin is closer to natural lignin compared to kraft lignin (Haq et al., 2020; Yadav et al., 2022). Organosolv lignin is obtained from a process that utilizes organic solvents for delignification, also known as organosolv pulping process. Organosolv lignin is relatively hydrophobic and characterized by low molecular weight and high purity, with minimal ash and carbohydrate content and no sulfur (Haq et al., 2020).

The color removal from wastewater is a good indicator of effective lignin removal. Current treatments in facilities primarily focus on eliminating lignin without converting it into high-value products (Haq et al., 2020; Yadav et al., 2022). Various treatments are used in PPI today, with a special emphasis on physicochemical processes. For instance, physical treatments are utilized for suspended solids removal. Chemical treatments include coagulation, precipitation, and adsorption. Recently, advanced oxidation processes and membrane techniques have gained attention (Mandal et al., 2022).

Although physicochemical treatments are the most employed in PPI, they have several drawbacks, such as the formation of by-products, high energy consumption, significant capital investment, and the generation of large quantities of sludge. This is one reason why biological treatments have gained attention, as they are eco-friendly and sustainable (Mandal et al., 2022).

The utilization of lignin and lignin derivatives to produce various bioproducts aligns with several modern concepts, including the United Nations' Sustainable Development Goals and the principles of a circular bioeconomy (Mandal et al., 2022). This approach generates value from by-products, reduces waste, and enhances the profitability of biorefineries. Furthermore, using lignin as a renewable source for different molecules can decrease reliance on fossil resources, as lignin-derived bioproducts can serve as drop-in or drop-out substitutes for fossil-based alternatives (Haq et al., 2020; Mandal et al., 2022; Shibukawa et al., 2023; Yadav et al., 2022). This can significantly impact industries such as plastics, fuels, fertilizers, chemical building blocks, lubricants, adhesives, and more (Haq et al. 2020).

4.9 Conclusion and Future Directions

Despite the technological challenges associated with applying lignin for the development of a chemical platform, advancements in lignin valorization are ongoing, resulting in lignin with lower impurities that is suitable for various applications. Although the number of companies operating in this sector is still small, there is a growing trend. In this context, lignin valorization represents a cleaner alternative that adds value to the production chain. Besides technological development and the standardization of production routes, political incentives are essential for the development of lignin-based chemicals and the circular economy in biorefineries. Tax and environmental incentives, increased investment in the sector, and the generation of carbon credits can encourage the development of the sector and the valorization of the production chain.

Acknowledgment

A.K. Chandel gratefully acknowledges the Brazilian National Council for Scientific and Technological Development (CNPq), Brazil for scientific productivity program (Process number: 304451/2025-8). A.K. Chandel gratefully acknowledges São Paulo State Funding Agency-FAPESP, Brazil (grant numbers: 2020/12559-6, 2022/13184-1, and 2023/08226-0) for the financial project.

References

Agrizzi, T., Oliveira, M. A., Faria, E. V., Santos, K. G., Xavier, T. P., & Lira, T. S. (2024). Assessing coconut shell pyrolysis: Biomass characterization, activation energy estimation, and statistical analysis of operating conditions. *Bioresource Technology Reports*, *26*, 101831.

Ahmad, M., Taylor, C. R., Pink, D., Burton, K., Eastwood, D., Bending, G. D., & Bugg, T. D. (2010). Development of novel assays for lignin degradation: Comparative analysis of bacterial and fungal lignin degraders. *Molecular Biosystems*, *6*(5), 815–821.

Ahmad, M., Roberts, J. N., Hardiman, E. M., Singh, R., Eltis, L. D., & Bugg, T. D. (2011). Identification of DypB from Rhodococcus jostii RHA1 as a lignin peroxidase. *Biochemistry*, *50*(23), 5096–5107.

Alam, M. M., Greco, A., Rajabimashhadi, Z., & Esposito Corcione, C. (2024). Efficient and environmentally friendly techniques for extracting lignin from lignocellulose biomass and subsequent uses: A review. *Clean. Mater.*, *13*(June), 100253.

Ascencio, J. J., Chandel, A. K., Philippini, R. R., & Silva, S. S. (2020). Comparative study of cellulosic sugars production from sugarcane bagasse after dilute nitric acid, dilute sodium hydroxide and sequential nitric acid-sodium hydroxide pretreatment. *Biomass Conversion and Biorefinery*, *10*, 813–822.

Ashokkumar, V., Venkatkarthick, R., Jayashree, S., Chuetor, S., Dharmaraj, S., Kumar, G., Chen, W., & Ngamcharussrivichai, C. (2022). Recent advances in lignocellulosic biomass for biofuels and value-added bioproducts—A critical review. *Bioresource Technology*, *344*. https://doi.org/10.1016/j.biortech.2021.126195

Bajwa, D. S., Pourhashem, G., Ullah, A. H., & Bajwa, S. G. (2019). A concise review of current lignin production, applications, products and their environment impact. *Industrial Crops and Products*, *139*. https://doi.org/10.1016/j.indcrop.2019.111526

Beckham, G. T., Johnson, C. W., Karp, E. M., Salvachúa, D., & Vardon, D. R. (2016). Opportunities and challenges in biological lignin valorization. *Current opinion in biotechnology*, *42*, 40–53.

Brienza, F., Cannella, D., Montesdeoca, D., Cybulska, I., & Debecker, D. P. (2024). A guide to lignin valorization in biorefineries: Traditional, recent, and forthcoming approaches to convert raw lignocellulose into valuable materials and chemicals. *RSC Sustainability*, *2*(1), 37–90. https://doi.org/10.1039/d3su00140g

Brodin, M., Vallejos, M., Opedal, M. T., Area, M. C., & Chinga-Carrasco, G. (2017). Lignocellulosics as sustainable resources for production of bioplastics—A review. *Journal of Cleaner Production*, *162*, 646–664. https://doi.org/10.1016/j.jclepro.2017.05.209

Bugg, T. D., Ahmad, M., Hardiman, E. M., & Singh, R. (2011). The emerging role for bacteria in lignin degradation and bio-product formation. *Current opinion in biotechnology*, *22*(3), 394–400.

Chandel, A. K., Garlapati, V. K., Singh, A. K., Antunes, F. A., & Silva, S. S. (2018a). The path forward for lignocellulosic biorefineries: Bottlenecks, solutions, and perspectives on commercialization. *Bioresource Technology*, *264*, 370–381. https://doi.org/10.1016/j.biortech.2018.06.004

Chandel, A. K., Antunes, F. A. F., Terán-Hilares, R., Cota, J., Ellilä, S., Silveira, M. H. L., Santos, J. C., & Silva, S. S. (2018b). Bioconversion of Hemicellulose into Ethanol and Value-Added Products. In *Advances in Sugarcane Biorefinery* (1st ed., pp. 97–134). Elsevier.

Chen, Z., & Wan, C. (2017). Biological valorization strategies for converting lignin into fuels and chemicals. *Renewable and Sustainable Energy Reviews*, *73*, 610–621.

Chen, Z., Wang, Y., Cheng, H., & Zhou, H. (2022). Hemicellulose degradation: An overlooked issue in acid deep eutectic solvents pretreatment of lignocellulosic biomass. *Industrial Crops and Products*, *187*. https://doi.org/10.1016/j.indcrop.2022.115335

da Cruz, M., Rodrigues, B., Ristic, A., Budnyk, S., Das, S., & Slabon, A. (2022). On the product selectivity in the electrochemical reductive cleavage of 2-phenoxyacetophenone, a lignin model compound. *Green Chemistry Letters and Reviews*, *15*(1), 153–161. https://doi.org/10.1080/17518253.2022.2025462

Cruz-santos, M. M., Antunes, F. A., Arruda, G. L., Shibukawa, V. P., Prado, C. A., Ortiz-silos, N., Castro-Alonso, M. J., Marcelino, P. R. F., & Santos, J. C. (2023). Production and applications of pullulans from lignocellulosic biomass: Challenges and perspectives. *Bioresource Technology*, *385*. https://doi.org/10.1016/j.biortech.2023.129460

Dashtban, M., Schraft, H., Syed, T. A., & Qin, W. (2010). Fungal biodegradation and enzymatic degradation genotypes and phenotypes among prokaryotes. *Appl Microbiol, 20*, 10–621.

Dessbesell, L., Paleologou, M., Leitch, M., Pulkki, R., & Xu, C. (C.) (2020). Global lignin supply overview and kraft lignin potential as an alternative for petroleum-based polymers. *Renewable and Sustainable Energy Reviews, 123*. https://doi.org/10.1016/j.rser.2020.109768

Dias, I. A., Aparício, R. R., Zakaluk, I. C. B., Naide, T. L., Cordeiro, L. A., Benedito, D. C. D., & de Cademartori, P. H. G. (2023). Wood-based materials for sustainable applications. In *Forest science: Sustainable processes and wood products* (pp. 25–50). Springer Nature Singapore: Singapore.

Faleva, A. V., Belesov, A. V., Kozhevnikov, A. Y., Falev, D. I., Chukhchin, D. G., & Novozhilov, E. V. (2021). Analysis of the functional group composition of the spruce and birch phloem lignin. *Int. J. Biol. Macromol., 166*, 913–922.

Gaspar, R., & Fardim, P. (2023). Lignin-based materials for emerging advanced applications. *Green and Sustainable Chemistry, 41*, 2452–2236. https://doi.org/10.1016/j.cogsc.2023.100834

Gaynor, J. G., Agwuncha, S. C., Smith, A., Gaynor, G., Harrington, M. J., & Lucia, L. (2024). Alkaline pretreatment and soda pulping of genetically improved hemp. *Ind. Crops Prod., 211*(February), 118181.

Hammel, Y. A., Dubois, M., Delatour, T., & Stadler, R. H. (2014). N, N-dimethylpiperidinium (mepiquat): Part 1. Formation in model systems and relevance to roasted food products. *Food Additives & Contaminants: Part A, 31*(2), 226–233.

Haq, I., Mazumder, P., & Kalamdhad, A. (2020). Recent advances in removal of lignin from paper industry wastewater and its industrial applications—A review. *Bioresource Technology, 312*. https://doi.org/10.1016/j.biortech.2020.123636

Hilares, R. T., Ramos, L., Ajaz Ahmed, M., Ingle, A. P., Kumar Chandel, A., da Silva, S. S., Choi, J. W., & dos Santos, J. C. (2020). Valorization of Lignin Into Value-Added Chemicals and Materials. In P. Avinash Ingle, A. Kumar Chandel, & S. Silvério da Silva (Eds.), *Lignocellulosic biorefining technologies* (1st ed.). Wiley.

Hofrichter, M. (2002). Review: Lignin conversion by manganese peroxidase (MnP). *Enzyme and Microbial Technology, 30*(4), 454–466.

Huang, C., Jiang, X., Shen, X., Hu, J., Tang, W., Wu, X., Ragauskas, A., Jameel, H., Meng, X., & Yong, Q. (2022). Lignin-enzyme interaction: A roadblock for efficient enzymatic hydrolysis of lignocellulosics. *Renewable an Sustainable Energy Reviews, 154*, 111822. https://doi.org/10.1016/j.rser.2021.111822

Ilic, N., Milic, M., Beluhan, S., & Brankovic, S. D. (2023). Cellulases: From lignocellulosic biomass to improved production. *Energies, 16*(8). https://doi.org/10.3390/en16083598

Jadhav, P., Bhuyar, P., Misnon, I. I., Rahim, M. H. A., & Roslan, R. (2024). Advancement of lignin into bioactive compounds through selective organic synthesis methods. *Int. J. Biol. Macromol., 276*(P2), 134061.

Jardim, J. M., Hart, P. W., Lucia, L. A., Jameel, H., & Chang, H. M. (2022). The effect of the kraft pulping process, wood species, and pH on lignin recovery from black liquor. *Fibers, 10*(2), 16.

Jiang, Y., Feng, Y., Lei, B., & Zhong, H. (2020). Impact mechanisms of supercritical CO2–ethanol–water on extraction behavior and chemical structure of eucalyptus lignin. *Int. J. Biol. Macromol., 161*, 1506–1515.

Jie, H., Khan, I., Alharthi, M., Zafar, M. W., & Saeed, A. (2023). Sustainable energy policy, socio-economic development, and ecological footprint: The economic significance of natural resources, population growth and industrial development. *Utilies Policy, 81*. https://doi.org/10.1016/j.jup.2023.101490

Kersten, P. J., Tien, M., Kalyanaraman, B., & Kirk, T. K. (2012). The ligninase of Phanerochaete chrysosporium generates cation radicals from methoxybenzenes. *Journal of Biological Chemistry, 260*, 2609–2612.

Kim, S. J., & Um, B. H. (2018). Delignification from Geodae-Uksae1 using soda-pulping followed by evaluation on recycling of liquid-liquid extraction solvent. *Biomass and Bioenergy, 109*, 23–30.

Leonel, L. V., Arruda, P. V., Chandel, A. K., Felipe, M. G. A., & Sene, L. (2021). *Kluyveromyces marxianus*: A potential biocatalyst of renewable chemicals and lignocellulosic ethanol production. *Critical Reviews in Biotechnology*. https://doi.org/10.1080/07388551.2021.1917505

Li, C., Zhao, X., Wang, A., Huber, G. W., & Zhang, T. (2015). Catalytic transformation of lignin for the production of chemicals and fuels. *Chemical reviews, 115*(21), 11559–11624.

Liu, H., Xu, T., Liu, K., Zhang, M., Liu, W., Li, W., Li, H., Du, H., & Si, C. (2021). Lignin-based electrodes for energy storage application. *Industrial Crops and Products, 165*, 113425.

Liu, Z. H., Li, B. Z., Yuan, J. S., & Yuan, Y. J. (2022). Creative biological lignin conversion routes toward lignin valorization. *Trends Biotechnol., 40*(12), 1550–1566.

Luo, H., & Abu-Omar, M. M. (2017). Chemicals from lignin. In M. A. Abraham (Ed.), *Reference module in earth systems and environmental sciences* (Vol. 3, pp. 573–585). Elsevier.

Mandal, D., Singh, G., Majumdar, S., & Chanda, P. (2022). Challenges in developing strategies for the valorization of lignin—A major pollutant of the paper mill industry. *Environmental Science and Pollution Research, 30*, 11119–11140. https://doi.org/10.1007/s11356-022-24022-4

Marcelino, P., Gonçalves, F., Jimenez, I., Carneiro, B., Santos, B., & da Silva, S. (2020). Sustainable production of biosurfactants and their applications. In A. Ingle, A. Chandel, & S. S. da Silva (Eds.), *Lignocellulosic biorefining technologies* (Vol. 1, pp. 159–183). Wiley. https://doi.org/10.1002/9781119568858.ch8

Mendes, F. M., Vasconcelos, M. H., Dias, M. O., Ferraz, A., Milagres, A. M., Santos, J. C., Jesus, C. D. F., Watanabe, M. D. B., Junqueira, T. L., & Bonomi, A. (2018). Alkaline sulfite pretreatment for integrated first and second generation ethanol production: A techno-economic assessment of sugarcane hybrids. *Biomass and Bioenergy, 119*, 314–321. https://doi.org/10.1016/j.biombioe.2018.10.005

Mendu, V., Harman-Ware, A. E., Crocker, M., Jae, J., Stork, J., Morton, S., Placido, A., Huber, G., & Debolt, S. (2011). Identification and thermochemical analysis of high-lignin feedstocks for biofuel and biochemical production. *Biotechnology for Biofuels, 4*. https://doi.org/10.1186/1754-6834-4-43

Mussatto, S. I., Yakamawa, C., van der Maas, L., & Dragone, G. (2021). New trends in bioprocesses for lignocellulosic biomass and CO2 utilization. *Renewable and Sustainable Energy Reviews, 152*. https://doi.org/10.1016/j.rser.2021.111620

Nelson, D., & Cox, M. (2014). *Princípios de bioquímica de Lehninger* (6th ed.). Porto Alegre: Artmed.

Obrzut, N., Hickmott, R., Shure, L., & Gray, K. A. (2023). The effects of lignin source and extraction on the composition and properties of biorefined depolymerization products. *RSC Sustain., 1*(9), 2328–2340.

Oliva, A., Tan, L. C., Papirio, S., Esposito, G., & Lens, P. N. L. (2021). Effect of methanol-organosolv pretreatment on anaerobic digestion of lignocellulosic materials. *Renew. Energy*, *169*, 1000–1012.

Pandey, M. P., & Kim, C. S. (2011). Lignin depolymerization and conversion: A review of thermochemical methods. *Chemical engineering & technology*, *34*(1), 29–41.

Pereira, A., Hoeger, I. C., Ferrer, A., Rencoret, J., Del Rio, J. C., Kruus, K., Rahikainen, J., Kellock, M., Gutiérrez, A., & Rojas, O. J. (2017). Lignin films from spruce, eucalyptus, and wheat straw studied with electroacoustic and optical sensors: Effect of composition and electrostatic screening on enzyme binding. *Biomacromolecules*, *18*(4), 1322–1332.

Prado, C. A., Cunha, M. L., Terán-hilares, R., Arruda, G. L., Antunes, F. A., Pereira, B., da Silva, S., & Santos, F. C. (2023). Hydrodynamic cavitation—Assisted oxidative pretreatment and sequential production of ethanol and xylitol as innovative approach for sugarcane bagasse biorefineries. *BioEnergy Research*. https://doi.org/10.1007/s12155-022-10555-6

Rencoret, J., Gutiérrez, A., Nieto, L., Jiménez-Barbero, J., Faulds, C. B., Kim, H., Ralph, J., Martínez, Á. T., & del Río, J. C. (2011). Lignin composition and structure in young versus adult eucalyptus globulus plants. *Plant Physiol.*, *155*(2), 667–682.

Rencoret, J., Marques, G., Rosado, M. J., Benito, J., Barro, F., Gutiérrez, A., & del Río, J. C. (2023). Variations in the composition and structure of the lignins of oat (Avena sativa L.) straws according to variety and planting season. *Int. J. Biol. Macromol.*, *242*(May), 124811.

Sang, Y., Chen, H., Khalifeh, M., & Li, Y. (2023). Catalysis and chemistry of lignin depolymerization in alcohol solvents–A review. *Catal. Today*, *408*, 168–181.

Schmatz, A. A., & Brienzo, M. (2022). Butylated hydroxytoluene improves lignin removal by organosolv pretreatment of sugarcane bagasse. *Bioenergy Res.*, *15*(1), 166–174.

Schutyser, W., Renders, T., Van Den Bosch, S., Koelewijn, S. F., Beckham, G. T., & Sels, B. F. (2018). Chemicals from lignin: An interplay of lignocellulose fractionation, depolymerisation, and upgrading. *Chem. Soc. Rev.*, *47*(3), 852–908.

Seto, M., Masai, E., Ida, M., Hatta, T., Kimbara, K., Fukuda, M., et al. (1995). Multiple polychlorinated biphenyl transformation systems in the Gram-Positive bacterium Rhodococcus sp. Strain RHA1. *Appl Environ Microbiol*, *61*, 4510–4513.

Sharma, R. K., Wooten, J. B., Baliga, V. L., Lin, X., Chan, W. G., & Hajaligol, M. R. (2004). Characterization of chars from pyrolysis of lignin. *Fuel*, *83*(11–12), 1469–1482.

Shibukawa, V. P., Ramos, L., Cruz-santos, M. M., Prado, C. A., Jofre, F. M., de Arruda, G. L., da Silva, S. S., Mussatto, S. I., & dos Santos, J. C. (2023). Impact of product diversification on the economic sustainability of second-generation ethanol biorefineries: A critical review. *Energies*, *16*(17). https://doi.org/10.3390/en16176384

Siddiki, S. Y. A., Mofijur, M., Kumar, P. S., Ahmed, S. F., Inayat, A., Kusumo, F., Badruddin, I. A., Khan, T. M. Y., Nghiem, L. D., Ong, H. C., & Mahlia, T. M. I. (2022). Microalgae biomass as a sustainable source for biofuel, biochemical and biobased value-added products: An integrated biorefinery concept. *Fuel*, *307*. https://doi.org/10.1016/j.fuel.2021.121782

Silveira, M. H. L., Chandel, A. K., Vanelli, B. A., Sacilotto, K. S., & Cardoso, E. B. (2018). Production of hemicellulosic sugars from sugarcane bagasse via steam explosion employing industrially feasible conditions: Pilot scale study. *Bioresour. Technol. Rep.*, *3*, 138–146.

Sohn, Y. J., Son, J., Lim, H. J., Lim, S. H., & Park, S. J. (2022). Valorization of lignocellulosic biomass for polyhydroxyalkanoate production: Status and perspectives. In *Bioresource technology* (Vol. 360). Elsevier Ltd. https://doi.org/10.1016/j.biortech.2022.127575

Subbotina, E., Rukkijakan, T., Marquez-Medina, M., Yu, X., Johnsson, M., & Samec, J. (2021). Oxidative cleavage of C–C bonds in lignin. *Nature Chemistry, 13*, 1118–1125.

Sun, R. C. (2020). Lignin source and structural characterization. *Chem Sus Chem, 13*(17), 4385–4393. https://doi.org/10.1002/cssc.202001324

Sun, Y. C., Liu, X. N., Wang, T. T., Xue, B. L., & Sun, R. C. (2019). Green process for extraction of lignin by the microwave-assisted ionic liquid approach: Toward biomass biorefinery and lignin characterization. *ACS Sustain. Chem. Eng.*, *7*(15), 13062–13072.

Taylor, M. J., Hornsby, K., Cheah, K. W., Hurst, P., Walker, S., & Skoulou, V. (2024). Repurposing lignin rich biorefinery waste streams into the next generation of sustainable solid fuels. *Sustainable Chemistry for the Environment, 7*. https://doi.org/10.1016/j.scenv.2024.100123

Tian, J. H., Pourcher, A. M., Bouchez, T., Gelhaye, E., & Peu, P. (2014). Occurrence of lignin degradation genotypes and phenotypes among prokaryotes. *Applied Microbiology and Biotechnology, 98*, 9527–9544.

Tribot, A., Amer, G., Abdou Alio, M., de Baynast, H., Delattre, C., Pons, A., Mathias, J. D., Callois, J. M., Vial, C., Michaud, P., & Dussap, C. G. (2019). Wood-lignin: Supply, extraction processes and use as bio-based material. *European Polymer Journal, 112*, 228–240. https://doi.org/10.1016/j.eurpolymj.2019.01.007

Vasconcelos, M. H., Mendes, F. M., Ramos, L., Dias, M. O. S., Bonomi, A., Jesus, C. D. F., Watanabe, M. D. B., Junqueira, T. L., Milagres, A. M. F., Ferraz, A., & Santos, J. C. (2020). Techno-economic assessment of bioenergy and biofuel in integrated sugarcane biorefinery: Identification of technological bottlenecks and economic feasibility of dilute acid pretreatment. *Energy, 199*, 117422.

Windeisen, E., & Wegener, G. (2016). Lignin as building unit for polymers. *Reference Module in Materials Science and Materials Engineering*. https://doi.org/10.1016/b978-0-12-803581-8.09240-7

Yadav, V., Kumar, A., Bilal, M., Nguyen, T., & Iqbal, H. (2022). Lignin removal from pulp and paper industry waste streams and its application. In R. Bhat, A. Kumar, T. Nguyen, & S. Sharma (Eds.), *Nanotechnology in paper and wood engineering: Fundamentals, challenges and applications* (Vol. 1, pp. 265–283). Elsevier. https://doi.org/10.1016/B978-0-323-85835-9.00019-2

Yoo, C. G., & Ragauskas, A. J. (2021). Opportunities and challenges of lignin utilization. In *ACS Symposium Series* (Vol. 1377, pp. 1–12). American Chemical Society. https://doi.org/10.1021/bk-2021-1377.ch001

Zou, L., Ross, B. M., Hutchison, L. J., Christopher, L. P., Dekker, R. F., & Malek, L. (2015). Fungal demethylation of Kraft lignin. *Enzyme and microbial technology, 73*, 44–50.

5

Sustainable Production of Advanced Alcohol-Based Biofuels in Biorefinery

From Alcohols to Sustainable Aviation Fuels

Danielle Matias Rodrigues[1,2,3], *Paula Zaghetto de Almeida*[2], *Allan H. Félix de Mélo*[2,4], *Juliana Velasco de Castro Oliveira*[2,5], *Ana Paula Jacobus*[2,6], *and Henrique Macedo Baudel*[2,3,7]

[1] *Postgraduate Program in Bioenergy, Faculty of Food Engineering, University of Campinas (UNICAMP), Campinas, São Paulo, Brazil*
[2] *SENAI Bom Retiro Biotecnologia, São Paulo, São Paulo, Brazil*
[3] *Bioenergy Department, America Biomass Technologies Division, Brix Brasil, São Paulo, Brazil*
[4] *Postgraduate Program in Food Engineering, Faculty of Food Engineering, University of Campinas (UNICAMP), Campinas, São Paulo, Brazil*
[5] *Postgraduate Program in Genetics and Molecular Biology, Institute of Biology, University of Campinas (UNICAMP), Campinas, São Paulo, Brazil*
[6] *São Paulo State University (Unesp), Institute for Research in Bioenergy, Rio Claro, São Paulo, Brazil*
[7] *Department of Chemical Engineering, Federal University of Pernambuco (UFPE), Recife, Pernambuco, Brazil*

5.1 Introduction

In the face of several challenges that the world economy faces today, such as climate change and greenhouse gas emissions, preservation of ecosystems and biodiversity, food security and availability of raw materials and energy; the bioeconomy represents an important advance towards a new economy and a new generation in industrial production. These challenges increase the pressure on the use of non-renewable resources, such as oil, while increasing the pressure for increasingly sustainable production and consumption in the context of consolidation of the local, regional and national-based bioeconomy. In principle, bioeconomy comprises all activities associated with the production of renewable biological resources, such as biomass from dedicated crops and agro-industrial residues, as well as the various ways in which such materials are transformed into different products of industrial interest. These resources are used directly in industrial processes or are transformed into high value-added products with biological bias (bio-based) and bioenergy. In this context, the production of advanced biofuels from biomass-based feedstocks, such as agro-industrial residues and dedicated energy crops, is relevant. This chapter presents relevant insights into the production of advanced alcohol-based biofuels.

Biofoundry Techniques for Biotechnology Applications, First Edition. Edited by Anuj Kumar Chandel.

5.2 Bioethanol

Energy sufficiency is critical to global socioeconomic development. Currently, approximately 84% of the global energy matrix is derived from fossil fuels such as coal, crude oil, and natural gas. (Institute Energy, 2024). About 45% of global CO_2 emissions are related to the transport sector (Ahn & Park, 2024). Environmental impacts, especially the increase in global temperature due to the greenhouse effect, have triggered the growing need to replace fossil-based projects with the production of bio-based analogs (Rautela et al., 2024). Research on biotechnological routes to produce renewable fuels, materials, and products has become crucial to foster techno-economic and sustainable alternatives, especially zero-carbon or even carbon-negative (Galván et al., 2024).

In the energy sector, especially related to liquid fuels, ethanol is the product with the most promising character, either as an additive, whole fuel, or building block (e.g., acetaldehyde, ethylene, 1,3-butadiene and ethyl acetate) (He et al., 2023). Globally, approximately 130 billion liters of ethanol are produced annually. Many factors have influenced the expansion of this alcohol as a biofuel: (a) ethanol can be produced by renewable feedstocks and; (b) its carbon neutral footprint. (c) lower toxicity compared to other alcohols (such as butanol); (d) it can be incorporated at varying levels into fuels; (e) it does not require modification of the current engines for its use. (f) it emits fewer pollutants than fossil fuels do (Samantaray et al., 2024).

Bioethanol production can be divided into four categories, or generations. The first generation (1G) is the most predominant, developed based on fermentable sugars from raw food materials, especially corn, sugarcane, and sugar beet (Samantaray et al., 2024). The crisis between food and fuel production triggered the second generation (2G) of bioethanol. The 2G process uses lignocellulosic materials, especially agro-industrial residues as substrates. The third generation (3G) uses photosynthetic organisms; such as microalgae; to obtain ethanol (Müller et al., 2023). Fourth-generation (4G) bioethanol is produced from renewable feedstocks that involve advanced metabolic engineering and carbon capture technologies (Shakelly et al., 2023).

Processes that have been gaining prominence are the so-called 1.5G processes, which are hybrid systems that combine first- and second-generation processes (Wu et al., 2020). The objective of these models is to unite the benefits of the consolidated industrial efficiency of 1G with the environmental sustainability and biomass availability of 2G. In addition, the tools of metabolic engineering, synthetic biology, and advanced enzymes are widely employed in these systems (Bermejo et al., 2020).

An example of a 1.5G system is an advanced process for producing bioethanol from enzyme-mediated sugarcane molasses, with or without the addition of lignocellulosic liquors (Saini et al., 2020). The application of enzymes, such as cellulases, dextranases, and amylases, in the context of ethanol production from sugarcane molasses has shown promising results. These enzymes play complementary roles in transforming complex sugars present in molasses into fermentable sugars. An estimated increase of between 12 and 15% in ethanol yield (Onishi et al., 2014). Considering the volume of production, the gain in yield represents a significant increase in the viability and profitability of this production system.

Two recent studies demonstrated advances in ethanol production from molasses using metabolically engineered *Saccharomyces cerevisiae*. First, a strain called NGTM-F1, developed by physical mutagenesis, was able to tolerate high concentrations of molasses (250 g/L total dissolved solids). The sugar conversion rate was 95.53%, with a yield of 1.86 g/L/h (Wang et al., 2024). In a second study, the replacement of the PHO4 gene in *S. cerevisiae* resulted in a genetically optimized strain, with a 5.30% increase in ethanol yield and a 12.5% reduction in fermentation time (Wu et al., 2020).

Bermejo et al. (2020) proposed a 1.5 process based on an alternative biorefinery, where the sucrose-rich stream from the 1G process is mixed with a cellobiose-rich stream sourced from partially hydrolyzed lignocellulosic liquor. The advantages described by the authors are reduced concentrations of inhibitors, metabolites that are typically produced during pretreatment and reduced availability of free glucose to contaminating microorganisms, and unwanted effects of glucose repression. To make the recommended process viable, an active transport system for sucrose and cellobiose, followed by hydrolysis in the cytosol via intracellular invertase and heterologously expressed beta-glucoside (BGL), was built using a cluster regularly interspaced short palindromic repeats (CRISPR) genome editing system.

Corn bioethanol production processes have also presented advanced 1.5G. This initiative aims to convert corn fiber and unconverted starch in vinasse into additional ethanol. Companies such as D3MAX and Edeniq have processes of this type, reporting a 15% increase in yield and a 20% reduction in energy consumption (Hoang & Nghiem, 2021).

One of the current challenges in ethanol 2G production is xylose fermentation. The most commonly used microorganism in ethanolic fermentation, *S. cerevisiae*, does not have the enzymes needed for xylose assimilation, whereas microorganisms that can use xylose for ethanol production are either not sufficiently robust to resist the harsh conditions of fermentation or do not have a good conversion rate of ethanol. As an alternative, the metabolic routes of the xylose isomerase pathway (bacterial) or xylose-oxidoreductase (fungal) strains (Figure 5.1) can be inserted into the *S. cerevisiae* industrial strains using molecular biology tools to allow xylose conversion to ethanol (Basso et al., 2023; Fernandes & Murray, 2010; Ruchala & Sibirny, 2021).

The theoretical maximum ethanol yield is greater in the isomerase pathway than in the xylose-oxidoreductase pathway, due to some carbon loss in the oxidative pentose phosphate pathway that is needed for NADPH supply. Another drawback of this pathway is the possible accumulation of xylitol when xylose reductase and xylose dehydrogenase are unbalanced. Some strategies of cautious aeration may help improve the production of ethanol over xylitol accumulation caused by redox balance and changes in yeast oxygen demands. The great advantage of the oxidoreductase pathway is higher xylose flux, which usually results in higher ethanol productivity than the isomerase pathway because of a thermodynamic limit with an unfavorable equilibrium between xylose and xylulose (Moysés et al., 2016; Zhu et al., 2021).

Transporters engineering are also important research targets that can benefit from a biofoundry approach. *S. cerevisiae* has 18 native glucose transporters that are classified based on their affinity for glucose. The transporters HXT7, HXT5, GAL2, HXT1, and HXT4 are xylose-permissive, and xylose uptake depends on xylose and glucose levels outside the cell. The *HXT7* and *HXT5* transporters are the main transporters responsible for xylose uptake when xylose is the only available carbon source or when co-fermentation with

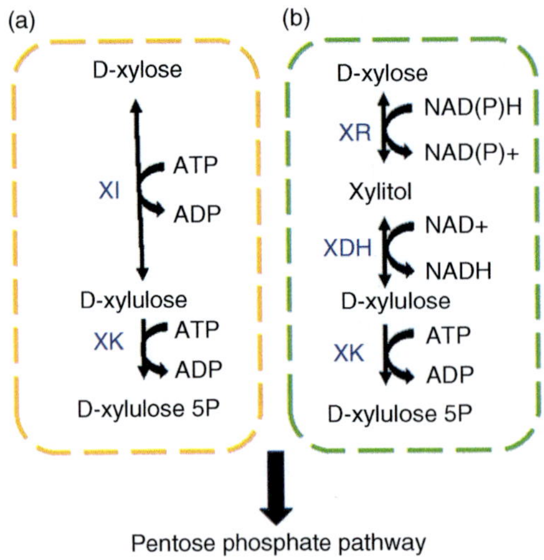

Figure 5.1 Native (a) Bacterial and (b) Fungal metabolism of D-xylose pathways to ethanol production. The enzyme abbreviations are: XI–xylose isomerase; XK–xylulose kinase; XR–xylose reductase; XDH–xylitol dehydrogenase. *Source:* Adapted from Baudel (2024).

low-concentration glucose occurs. On the other hand, when glucose is higher than 200 mM the transcription of *HXT7* and *HXT5* are repressed, and the transcription of *HXT1* is induced. When the glucose concentration drops below 2% in the co-fermentation media, *HXT4* is induced. Fermentation with low concentration of glucose ($\pm$0.5 g/L) can improve xylose uptake when compared to fermentations of pure xylose. Ideally, xylose and glucose should be consumed simultaneously to reduce fermentation time and improve productivity; however, the xylose uptake is significant only after glucose depletion (Gonçalves et al., 2014; Runquist et al., 2010).

5.3 Advanced Alcohol-Based Fuels

Despite the relevance and consolidation of bioethanol in the market, its application is limited to certain systems, requiring the development of new biofuels. Advanced biofuels are organic molecules suitable for use as whole or partial fuels and have superior properties to ethanol (low hygroscopicity, high energy density, and simple purification) and biodiesel (low cloud point) (Keasling et al., 2021). In addition, they must be used in engines already available. Advanced biofuels can be produced via microbial fermentation or the chemical conversion of biologically produced intermediates (Figure 5.2). Higher alcohols are suitable substitutes for gasoline and can be used to synthesize diesel and jet fuels (Rabinovitch-Deere et al., 2013). Metabolic engineering is a great challenge in creating or optimizing specific routes for each bioproduct to ensure the highest yields and the fewest number of by-products. Developing an optimized route of some bioprivileged molecules and chemically converting them to other products of interest is a strategy that overcomes some relevant limitations. One example is that hydrogenation, deoxygenation, and carbon coupling reactions in direct bonding with carbon are chemically well-elucidated reactions with excellent yields, while the use of biological systems in these Sustainable Aviation Fuels (SAFs) cases is practically unfeasible (Baudel, 2024; Huo & Shanks, 2020).

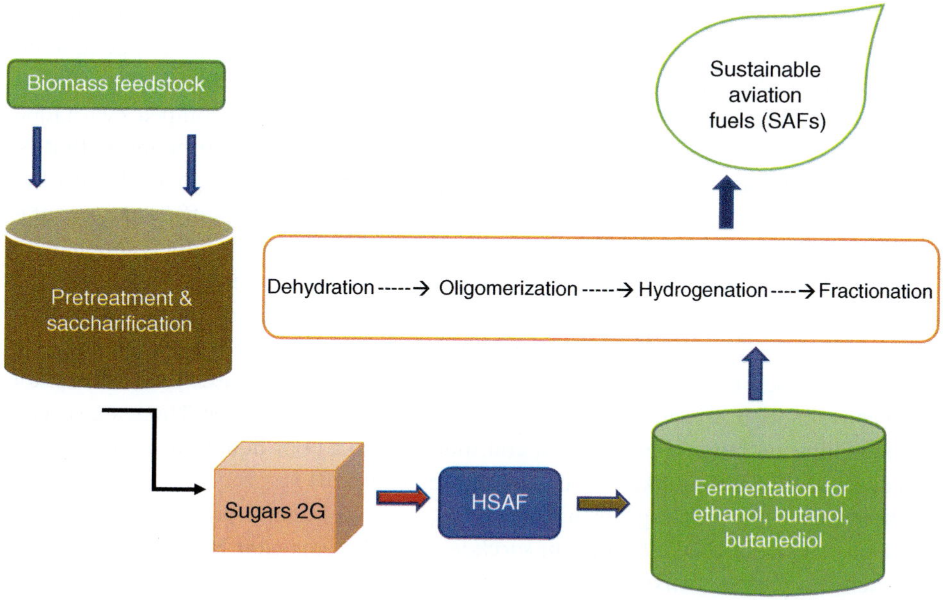

Figure 5.2 Sustainable production of advanced alcohol-based biofuels in biorefinery.

5.4 Biobutanol: The Biofoundry as a Tool to Optimize

Biofoundry can greatly accelerate biofuel yields and the development of economically viable bioprocesses. As the demand for decarbonization of the economy increases, several companies, such as LanzaJet and GEVO, have invested in butanol production. In this section, we discuss the scope of research on *n*-butanol (also known as 1-butanol, biobutanol, or butanol) production, particularly the biochemical routes and synthetic biology challenges, highlighting some points where the biofoundry approach is a key factor for better and faster results.

Butanol has interesting characteristics when compared to ethanol as a biofuel. It has 25% more energetic potential, is less corrosive, has lower water miscibility, can be blended in gasoline and diesel in higher proportions than ethanol and can even fully replace gasoline (Pugazhendhi et al., 2019). Another possibility is to convert from alcohol (ethanol or butanol) to aviation fuel, which is known as SAF. This process route has already been approved by ASTM (ASTM D7566), and LanzaJet opened this year (2024) the first SAF production facility in Georgia, USA (ATSM, 2019; LanzaJet, 2024).

Another example of industrial production of renewable isobutanol, used as a building block for many chemicals and biofuels, came from the company GEVO, and developed a process that starts with corn fermentation by a yeast developed by them. They claimed that their process is advantageous because of their low contamination rate and simple fermentation. As with other microorganisms, their yeast does not tolerate high isobutanol levels, so the company created a process to continuously remove alcohol during the fermentation, maintaining a low tank concentration (Ryan, 2023).

In the review of Guo et al. (2022), they summarized the main feedstocks used for butanol production via ABE fermentation with its benefits and challenges: The first-generation feedstock (starch or sucrose) results in a high butanol yield and supplies great amount of sugars through simple pretreatment processes, but on the other hand, competes with food sources and agricultural lands and has a high cost. Second-generation biomass, that is, lignocellulosic biomass, does not compete with food supply and has a high availability but requires complex pretreatment and detoxification processes to obtain sufficient fermentable sugars. The third-generation feedstock is algal biomass, which does not compete with land and food supply and has little to no lignin or fermentation inhibitors, but it is difficult to obtain sufficient fermentable sugars and results in costly downstream processes due to low butanol production. The fourth-generation feedstock is syngas; it has a greater CO_2 capture ability, does not require complex pretreatments and directly uses *Clostridia* sp. for butanol conversion; however, this technology still requires more studies. There are several parameters limiting butanol production, and there is also a poor liquid/gas mass transfer (Dutta et al., 2014; Gottumukkala et al., 2017; Wang et al., 2017).

In an example of modeling the biological response to butanol production, Assis de Oliveira et al. (2024), used *Clostridium saccharoperbutilacetonicum* N1–4 and the strain was fermented in various mixtures of substrates, focusing on second-generation feedstock. The model demonstrated a good conversion of pentoses up to 25% and tolerance to 0.33 mmol/L furfural and 0.03 mmol/L Hydroxy Methyl Furfural (HMF).

The native butanol synthesis pathways are found in bacteria from the genus *Clostridium*. Solvetogenic *Clostridium* species are characterized as anaerobic, non-pathogenic, rod-shaped, capable of forming spores, and performing biphasic fermentation. The first phase of the fermentation, known as acidogenesis, is characterized by the production of acetic acid, butyric acid, hydrogen, and carbon dioxide. The second phase, known as solventogenesis, is characterized by the production of butanol, acetone, isopropanol, and fermentation gases (Patakova et al., 2022; Xin et al., 2018).

Although there is a well-characterized native metabolism pathway for butanol production, using *Clostridium* species faces many challenges. Due to its high toxicity, the production of butanol is typically limited to concentrations below 13 g/L. The accumulation of butanol triggers cell sporulation and terminates fermentation. Additionally, the growth rate of *Clostridium* is very low, and the bioprocess is highly susceptible to phage contamination. Another drawback is that native *Clostridium* production requires more purification steps when compared to engineered strains because of the production of several by-products (Lin et al., 2023; Patakova et al., 2022). Significant efforts have been made on the genetic manipulation of microorganisms to enhance butanol production. Here we describe some strategies applied to *Clostridium sp., Saccharomyces cerevisiae,* and *Escherichia coli.*

5.4.1 *Clostridium* Pathway: Acetone-Butanol-Ethanol (ABE) Synthesis

This fermentation pathway is called ABE because the main products are acetone, butanol and ethanol, in a ratio of 3:6:1, and there is the generation of CO_2 and H_2 (Niglio et al., 2019). This fermentation is divided into two phases, the acidogenic and the solventogenic phase (Figure 5.3). The acidogenic phase is characterized by rapid cellular multiplication and the secretion of the carboxylic acids, acetate and butyrate, which cause the pH to drop

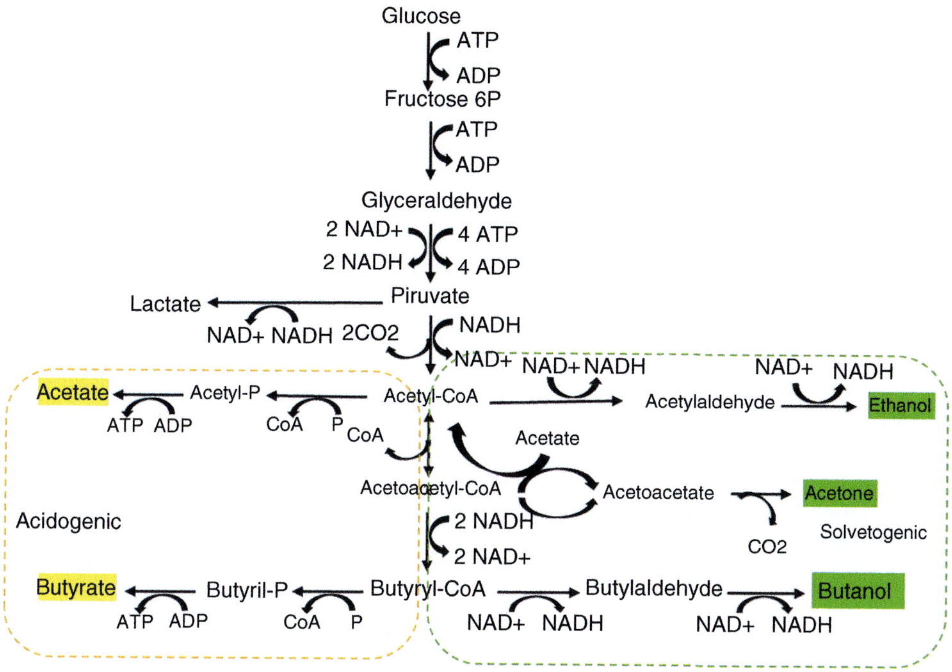

Figure 5.3 *Clostridium* sp. butanol pathway. The metabolic route is divided into the acidogenic phase (yellow dashed) and the solventogenic phase (green dashed) with the production of butanol.

approximately 4.5. The low pH acts as an inducer for the enzymes of solventogenic phase, triggering the onset of this second metabolic stage. The acids produced during the acidogenic phase can re-enter the cell and act as co-substrates for the solventogenic phase. As a result, the extracellular pH rises again. The main products of this phase are ethanol, acetone and butanol (Gheshlaghi et al., 2009).

The butanol production of *Clostridium* strains, unlike yeasts, can also be achieved from syngas via Wood–Ljungdahl pathway, which offers an advantage over expensive sugar—or starch-based substrates (Dürre, 2016).

The metabolism of pyruvate is considered a key factor for the conversion of the acids into the solvents and the subsequent redox reactions of NADH/NAD^+. In another example of a biofoundry approach, Quintero-Díaz et al. (2023) simulated ABE fermentation, showing that an increase in intracellular NADH concentration leads to a greater butanol production and selectivity over acetone. The study suggests that substrate consumption, and the production of butyryl-CoA from acetyl-CoA, should be the primary targets for metabolic engineering (Quintero-Díaz et al., 2023).

5.4.2 *S. cerevisiae*

S. cerevisiae is widely used for the production of bioproducts. It is a very versatile microorganism with well-developed genetic tools. *S. cerevisiae* naturally produces isobutanol and butanol in low titers; the butanol synthesis is still not well understood. Consequently, many

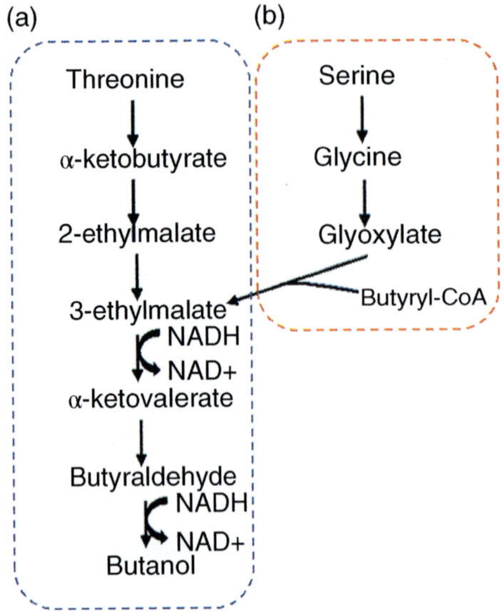

Figure 5.4 The figure shows simplified representations of (a) the endogenous threonine-based ketoacid pathway described by Si et al. (2014) and in (b) the glycine-butanol pathway as described by Branduardi et al. (2013). *Sources:* (a) Adapted from Si et al. (2014) and (b) Adapted from Branduardi et al. (2013).

strategies can be developed in a biofoundry context to improve this technology. In general, two strategies are commonly employed for butanol production: leveraging the amino acid uptake pathways shown (Figure 5.4) or expressing the *Clostridium* pathway (Figure 5.3). Another significant challenge is the low butanol tolerance of *S. cerevisiae* strains (Azambuja & Goldbeck, 2020; Branduardi et al., 2013; Si et al., 2014; Swidah et al., 2018).

The most promising strategy for butanol production appears to be the expression of exogenous (ABE) combined with the improvement of endogenous pathways in the same strain. Deleting any gene from the ABE pathway significantly impacts butanol production in engineered strains. Endogenous butanol production strongly correlates with glycine metabolism via an α-ketovalerate intermediate. *S. cerevisiae* can achieve titers greater than 2 g/L by supplementing with α-ketovalerate (Swidah et al., 2018). As titers increase, developing more tolerant strains becomes necessary. In the work of Azambuja et al. (2024) different strategies were tested to enhance the resistance of *S. cerevisiae* strains to butanol, including UV radiation and the adaptative laboratory evolution. Although some promising results were obtained, both strategies were labor-intensive and time-consuming. These processes could be significantly accelerated in a biofoundry approach through the automation of the workflow.

5.4.3 *E. coli*

E. coli is considered a cell factory due to its versatility and the extensive availability of molecular biology tools for editing and improving desired characteristics. Among the strategies described to improve the butanol production, those that combine multiple synthetic biology approaches are the most successful. To date, the highest butanol production across all microorganisms was achieved using an *E. coli* host, with a titer of 30 g/L. In this study, the Clostridial pathway was expressed with modifications, including the deletion of enzymes involved in acetate formation and the expression of a transenoyl-CoA reductase. These modifications aimed to direct the carbon flux exclusively to the butanol synthesis pathway. With this strategy, about 70–88% of the maximum theoretical production from glucose was achieved. Without these modifications on Clostridial pathway butanol production decreases by a factor of 10 (Shen et al., 2011).

According to Sheridan (2009), an economically viable bioprocess should achieve a titer of 100 g/L and a productivity greater than 2 g/L/h. Several key factors must be considered to ensure profitable production: the substrate should be inexpensive, sustainable, and easy to convert; competing pathways should be deleted or downregulated; *E. coli* strains need to be more tolerant to butanol; cofactor imbalance must be addressed; and the expression of the heterologous butanol pathway must be tightly regulated. (Ferreira et al., 2020).

5.5 Biofoundry Synthetic Biology Tools

Biofoundry tools can significantly accelerate the development of these technologies by enabling automation in screening, rapid testing, analytics, generation of new engineered strains, *in silico* modeling of new pathways and modification of routes for butanol production and carbon source utilization. Special attention is given to redox balance, butanol tolerance, and elimination of competing pathways (Banner et al., 2021).

The Design-Build-Test-Learn (DBTL) framework is a cyclical process (Figure 5.5), and a highly valuable approach in biofoundries. It integrates synthetic biology, machine learning, and automation, making a powerful tool for developing new strains for biofuel production. The first step is **Design**, where a prototype is created based on the initial ideas and project requirements. The next step is **Build,** where the strain is constructed using genetic engineering tools. This is followed by **Test,** where the strain's performance is evaluated using clear metrics, allowing for comparison across iterations. Finally, the **Learn** step involves analyzing the data and gaining insights from this design, enabling improvements in subsequent DBTL cycles (Opgenorth et al., 2019).

The integration of synthetic biology, machine learning and automation presents a multidisciplinary challenge but holds the potential to produce disruptive advancements both in computer science and biology. The predictive capabilities of machine learning models, combined with the synthetic biology tools, can probe and expand the mechanistic understanding, accelerate the DBTL cycles, and drive the development of more efficient strains (Carbonell et al., 2019).

As an example of testing biofoundry tools, Casini et al. (2018) completed a project in 90 days to produce 10 molecules previously unknown to the group. During the process, they constructed 1.2 Mb of DNA, tested 215 modified strains of *S. cerevisiae, E. coli,* and *Streptomyces* sp., developed two cell-free systems, and performed 690 assays designed specifically for these molecules.

The synthetic biology step of DBTL involves (a) the preparation and construction of the genetic parts, (b) multi-part assembly, and (c) construct verification. The construction of genetic parts can be achieved either by simply purchasing the desired sequences synthesized by a company or through a more labor-intensive process involving PCR reactions, digestions, and gel purifications. This process can be significantly accelerated by modern

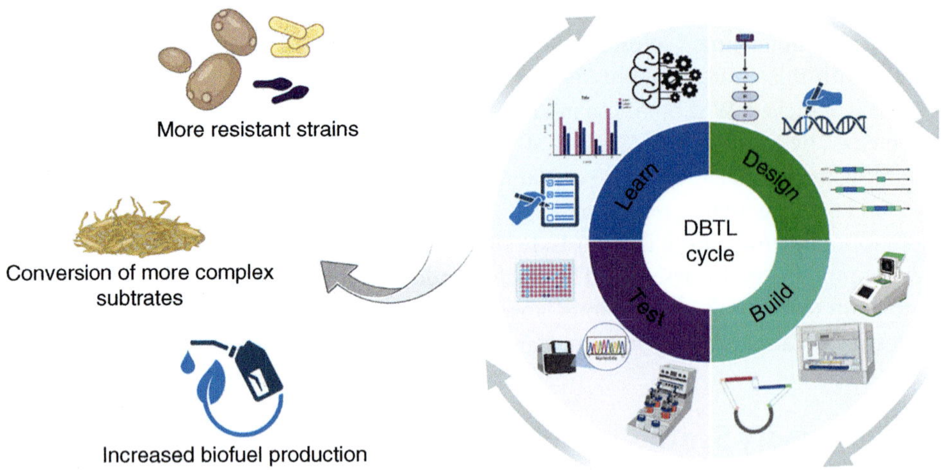

Figure 5.5 DBTL cyclic system to accelerate the research process on the development of biofuels. *Source:* Adapted from Baudel (2024).

equipment such as liquid handlers (Zhang et al., 2021a). The DNA assembly of the desired parts can be performed using various technologies, including Gibson, Golden Gate, and Ligase Cycling Reaction. This step can be integrated into automated pipelines; however, there is a pressing need for standardization of the process and the development of reliable protocols and metrics (Otero-Muras & Carbonell, 2021). The final process, construct verification, is expedited with the techniques such as qPCR (real-time PCR) and NGS (next-generation sequencing) techniques (Zhang et al., 2021a).

The effective testing and screening of the genetically modified strains can also be a time-consuming step, as it relies on cell culture, analysis of growth and product formation. Equipment such as multimode microplate readers can simplify this process, and new strategies and assays are continually being developed. For example, Luo et al. (2018) constructed a system called Synthetic Chromosome Rearrangement and Modification by LoxP-mediated Evolution (SCRaMbLE), which involved the introduction of thousands of symmetrical LoxP sites (loxPsym). This system enables extensive chromosome rearrangements and produces strains with high genetic diversity. The selection of mutants was achieved using a loxP-mediated switch of two markers named as ReSCuES. As a result, their study successfully identified several strains with high tolerance to ethanol, acetic acid, and heat.

To detect the butanol production on mutant strains more rapidly, Scheel and Lütke-Eversloh (2013) developed a high-throughput screening method based on a colorimetric reaction. This reaction involved an alcohol dehydrogenase-coupled conversion of nitroblue tetrazolium chloride (NBT) to insoluble formazan. Using this method, they screened 4,390 mutants with a false positive rate of only 2.2%. In another example of automation, a biosensor based on the transcription factor *akls* was developed for detection of C4 alcohols, such as butanol. The transcription factor recognizes the target product (in this case biofuel) and triggers the expression of a fluorescent protein, significantly simplifying the screening process (Bahls et al., 2022).

5.5.1 2,3-BDO

The biofoundry approach is a crucial strategy for producing non-conventional and/or promising biofuels, such as **2,3-Butanediol (2,3-BDO) or Butane-2,3-diol (its IUPAC name). In addition, 2,3 BDO** is a C4 chemical platform with significant potential across various other industries, including cosmetics, agriculture, pharmaceuticals, and chemicals. 2,3-DO is and colorless liquid that exists in three isomeric forms upon production: (2R,3R)-2,3-BDO (levo), (2R,3S)-2,3-BDO (meso), and (2S,3S)-2,3-BDO (dextro). These isomers exhibit different boiling points, ranging from 177 to 182 °C (Syu, 2001) as well as distinct physicochemical properties and unique efficacies (Song et al., 2019) as demonstrated in numerous studies. 2,3-BDO and its derivatives, such as methyl ethyl ketone and 1,3-butadiene (1,3-BD), can be directly used as fuel additives, or further converted into bio-jet fuels and alkanes through several steps (recently reviewed by Bai et al. (2023)).

Unfortunately, petroleum-based chemical catalysis for 2,3-BDO production necessitates costly separation and purification processes to achieve high-purity stereoisomers. As a result, 2,3-BDO is generally considered to have low industrial value and is produced in limited quantities. Consequently, the market price of 2,3-BDO remains high, averaging US$1,600/ton, which significantly limits its application potential (Shrivastav et al., 2013).

In recent years, the growing scarcity of fossil fuels and increasing environmental concerns have driven interest in biotechnologies that utilize renewable resources and microorganisms. Microbial biorefining of biomass to produce 2,3-BDO presents a sustainable and attractive solution, with the use of renewable biomass offering a promising pathway toward a carbon-neutral economy. Biofoundries could play a pivotal role in this transition, enabling high-throughput strain optimization and process development to enhance 2,3-BDO production. These facilities can accelerate genetic engineering, fermentation process improvements, and scale-up efforts, contributing to more sustainable and cost-effective processes (Amraoui et al. 2021).

Microbial production of 2,3-BDO is not a novel concept. The first documented bacterial production of 2,3-BDO dates back to 1906. During World War II, bio-based industrial production of 2,3-BDO was explored, particularly for its dehydrated derivative, 1,3-BD (Ji et al., 2011a). However, the development of more cost-effective petroleum-based methods for producing 1,3-BD reduced interest in these biological initiatives (Amraoui et al., 2021).

Fortunately, advancements in modern biotechnological tools, metabolic engineering, innovative technologies for utilizing carbon sources from non-cellulosic and lignocellulosic biomass, and improved fermentation methodologies have revitalized interest in 2,3-BDO (Ji et al., 2011b). Several microbial species, such as *Enterobacter cloacae*, *Klebsiella pneumoniae*, *Serratia marcescens*, *Paenibacillus polymyxa*, and *Bacillus licheniformis*, are known to naturally produce 2,3-BDO (Bai et al., 2023; Song et al., 2019). In a fed-batch culture, a genetically engineered strain of *E. cloacae* was able to synthesize up to 152 g/L of 2,3-BDO in a synthetic medium containing a mixture of glucose and xylose, the primary carbohydrate components of lignocellulosic hydrolysates (Li et al., 2015). The metabolic modifications introduced in this strain included enhanced expression of (2R,3R)-2,3-BDO dehydrogenase-encoding gene, disruption of meso-2,3-BDO dehydrogenase-encoding gene (to reduce the production of meso-BDH), and inactivation of the glucose transporter gene (*ptsG*) to overcome catabolite repression. Additionally, overexpression of the galactose permease gene (*galP*) was implemented, and genes associated with byproduct formation, such as *ldh* (leads to lactate production) and *frdA (succinid acid production)*, were knocked out. Using lignocellulosic hydrolysate derived from corn stover as the substrate, the strain produced 119.4 g/L of (2R,3R)-2,3-BDO with a purity exceeding 96.0% (Li et al., 2015).

In *S. marcescens*, a mutation induced by UV irradiation resulted in the deletion of the *swrW* gene, which encodes serrawettin W1, a surface-active exolipid responsible for excessive foam formation during fermentation (Zhang et al., 2010). This mutant strain, when cultivated in a fed-batch process with sucrose as the carbon source, produced 152 g/L of 2,3-BDO, although the isomer was not specified. Despite the promising results from species like *Serratia*, *Enterobacter*, and *Klebsiella (Chu et al.,* 2021*; Jo et al.,* 2024*; Jung et al.,* 2012*; Lu et al.,* 2024*; Ma et al.,* 2009*; Rathnasingh et al.,* 2016*; Zhang et al.,* 2010*).* which are considered efficient producers of 2,3-BDO for industrial scale-up, these genera are classified as biosafety level 2 due to their pathogenic characteristics. This classification significantly limits their industrial applications. While genetic engineering can remove virulence traits without compromising 2,3-BDO production (Huynh et al., 2015), ensuring the safety of these attenuated strains remains challenging, as pathogenicity often involves a wide range of genes.

An alternative approach involves using natural and non-pathogenic producers that are generally recognized as safe (GRAS). In this context, Song et al. (2019) isolated a strain of *B. licheniformis* with considerable potential for 2,3-BDO production. This strain underwent UV-induced random mutagenesis, resulting in a mucoid-free derivative strain suitable for genetic manipulation. In a two-stage fed-batch fermentation process, *B. licheniformis* produced a mixture of (2R,3S)- and (2R,3R)-2,3-BDO at concentrations of 63.4 g/L and 59.6 g/L, respectively, in an approximately 1:1 ratio. This *Bacillus* strain possesses two genes responsible for producing 2,3-BDO isomers: *levo-bdh*, which encodes the enzyme that synthesizes (2R,3R)-2,3-BDO, and *budC*, which produces (2R,3S)-2,3 BDO. To demonstrate the potential for controlling isomer production through genetic engineering, the *budC* gene was deleted. Although this led to a lower titer of 2,3-BDO (a total of 105 g/L), the stereoisomer ratio shifted significantly, achieving 91% selectivity for (2R,3R)-2,3-BDO, with concentrations of 96.3 g/L of (2R,3R)-2,3-BDO and 9.33 g/L of (2R,3S)-2,3-BDO.

Another biotechnological strategy is the use of non-natural producers of 2,3-BDO. As mentioned above, *S. cerevisiae* is classified as GRAS and represents an ideal cell factory for many industrial processes, including the production of food ingredients, pharmaceuticals, and fuels. To date, it is the best-studied eukaryote and is easily subjected to genetic manipulation. Additionally, it exhibits unique fermentation capabilities, resilience to various industrial processes, and resistance to harsh conditions, including osmotic stress and low pH (Parapouli et al., 2020).

Since it is a natural producer of ethanol, some genes related to ethanol production, such as *pdc* and *adh*, were disrupted. Additionally, its endogenous *bdh* gene was overexpressed, along with *alsS* and *alsD* from *Bacillus subtilis* (which promote the production of acetoin, a precursor of 2,3-BDO). As the disruption of ethanol pathways causes oxidative stress in *S. cerevisiae* cells, the *noxE* gene from *Lactococcus lactis* was also expressed to maintain redox balance under reduced ethanol production. The genetically engineered strain demonstrated an impressive titer of 178 g/L of 2,3-BDO when fermenting glucose. When cassava hydrolysate was used as carbon source, this strain produced 132 g/L of this biomolecule.

The use of GRAS microorganisms, such as *Bacillus* sp. and *S. cerevisiae*, combined with genetic engineering to precisely control the stereoisomer ratio of 2,3-BDO production, presents a significant advantage over traditional petroleum-based methods. This approach offers the flexibility to tailor processes to meet specific industrial needs. However, the recovery and separation of 2,3-BDO present several challenges due to its strong affinity for water, as well as complications arising from energy-intensive evaporation and multi-stage distillation processes (Gawal & Subudhi, 2023; Song et al., 2019).

The integration of biofoundries holds the key to overcoming these challenges by accelerating the development of cost-effective and scalable processes for 2,3-BDO production. By streamlining experimentation and optimization, biofoundries could significantly reduce development time, enhance process efficiency, and drive the industrial-scale production of 2,3-BDO as a sustainable, bio-based alternative.

5.6 Sustainable Aviation Fuel (SAF)

The aviation industry continues to experience significant global growth, including within the European Union (EU), but it faces increasing pressures to decarbonize (Chiaramonti, 2019). While other sectors, such as industry and electricity generation, have made progress in reducing their greenhouse gas (GHG) emissions, the transportation sector, particularly aviation, is moving in the opposite direction, with emissions on the rise (Raman et al., 2024). Aviation stands out as one of the most significant challenges to decarbonization due to the complexity of its operations and the limited availability of viable technological alternatives. Solutions like route optimization and improvements in the jet engines efficiency provide only incremental solutions. (Cavalett & Cherubini, 2018).

Additionally, the sustainable aviation fuel (SAF) value chain needs across several stages, from the cultivation and collection of raw materials to the final delivery of fuel to airlines (Martinez-Valencia et al., 2021). Raw materials, which can come from a variety of sources, such as vegetable oils, agricultural waste, biomass, and even CO_2, must be managed efficiently to ensure the economic viability of production (Emmanouilidou et al., 2023). The cost of cultivating, transporting, and processing raw materials remains significant component of the overall SAF cost, making it essential to develop technologies that can reduce expenses along the entire value chain (Becken & Mackey, 2017). Logistical efficiency, waste reduction, and the optimized use of biomass sources are critical factors for making SAF competitive with fossil fuels (Raman et al., 2024).

Within this context, one of promising routes for the production of SAF is the "Alcohol to Jet" (AtJ) pathway, which stands out as one of the most advanced and viable methods for the producing sustainable aviation fuels (Teixeira et al., 2024). This route involves converting alcohol, typically ethanol, into jet fuels, such as Jet-A, through catalytic dehydration and hydration processes (Kourkoumpas et al., 2024). The AtJ pathway is a flexible and effective solution, as ethanol is one of the most widely available feedstocks and can be produced from a variety of sources, including sugarcane, corn, and lignocellulosic biomass.

Additionally, ethanol is easier to transport and store compared to other SAF precursors, facilitating the integration of this technology into the global supply chain (Zahid et al., 2024).

The production of SAF via the AtJ route also benefits from the advancements in biocatalysis and microbial engineering technologies, which optimize the conversion of alcohol into high-performance compounds with chemical and physical properties comparable to conventional aviation kerosene (Geleynse et al., 2018). These technological developments are making the process increasingly efficient, with significant improvements in yield and reductions in production costs. However, for the ATJ route to become a large-scale solution, challenges must be addressed, including cost reduction and expansion of ethanol production infrastructure, and ensuring logistical efficiency throughout the value chain (Raman et al., 2024).

To achieve more efficient and sustainable production of SAF, it is essential to integrate innovative technologies, such as biofoundries, which offer promising solutions in biotechnology (Devi et al., 2024). Biofactories are advanced platforms that enable the engineering of microorganisms to produce renewable fuel precursors more efficiently using biological feedstocks (Zhang et al., 2021b). These technologies not only enhance large-scale production, but also help reduce the environmental impact of SAF production. Moreover, they enable adaptability to different biomass sources, increasing the flexibility of the process and making it more sustainable and adaptable to a variety of resources (Wang et al., 2023).

The transition to SAF, particularly through the alcohol-to-Jet route, requires a multipronged approach that includes substantial investments in infrastructure, value chain optimization, and continuous technological advancements, such as biofactories, to ensure that the final product meets the stringent safety and performance requirements required for use on aircraft (Köpke & Simpson, 2020). Overcoming these challenges will be crucial for aviation to align with global decarbonization goals, offering a sustainable and scalable alternative to fossil fuels and contributing to a greener and more efficient for air transport (Shahriar & Khanal, 2022).

5.7 Conclusion

Despite the relevance and consolidation of bioethanol in the market, its application is limited for some systems, requiring the development of new biofuels. Advanced biofuels are organic molecules suitable for use as whole or partial fuels that have complementary properties to ethanol (low hygroscopicity, high energy density, and simple purification) and biodiesel (low cloud point) as transportation fuels. Alcohol-based advanced biofuels have emerged over the last years as candidates for use as lower-carbon sustainable aviation and maritime fuels (SAF and SMF, respectively). Such molecules can be produced from microbial fermentation or chemical conversion of biologically produced intermediates. Higher alcohols are suitable substitutes for gasoline and can be used to synthesize diesel and jet fuel via ATJ (Alcohol-to-Jet) routes. Metabolic engineering has a great challenge to create or optimize specific routes for each bioproduct in order to ensure the highest yields and the fewest number of by-products. Developing optimized routes for the production of biomolecules and selectively converting them via hydrogenation, deoxygenation, and carbon coupling reactions to products of interest such as advanced biofuels is a strategy that overcomes some relevant limitations, while reaching decarbonization targets for the transport sector and industries.

References

Ahn, H., & Park, E. (2024). The impact of consumers' sustainable electronic-word-of-mouth in purchasing sustainable mobility: An analysis from online review comments of e-commerce. *Research in Transportation Business and Management*, *52*(December 2023), 101086. https://doi.org/10.1016/j.rtbm.2023.101086

Amraoui, Y., Narisetty, V., Coulon, F., Agrawal, D., Chandel, A. K., Maina, S., Koutinas, A., & Kumar, V. (2021). Integrated fermentative production and downstream processing of 2,3-butanediol from sugarcane bagasse-derived xylose by mutant strain of *Enterobacter ludwigii*. *ACS Sustainable Chemistry and Engineering*. https://doi.org/10.1021/acssuschemeng.1c03951

Assis de Oliveira, R., Zetty Arenas, A. M., Plácido, J., & Tovar, L. P. (2024). Integrated kinetic modeling of acetone-butanol-ethanol fermentation by Clostridium saccharoperbutilacetonicum N1-4: Co-utilization strategy with cane molasses, biomass-derived sugars, and furaldehydes. *Biomass and Bioenergy*, *191*, 107435. https://doi.org/10.1016/j.biombioe.2024.107435

ATSM. (2019). Specification for aviation turbine fuel containing synthesized hydrocarbons. In ATSM international (Ed.), *2019* (Vol. D7566). ATSM International.

Azambuja, S. P. H., de Mélo, A. H. F., Bertozzi, B. G., Inoue, H. P., Egawa, V. Y., Rosa, C. A., Rocha, L. O., Teixeira, G. S., & Goldbeck, R. (2024). Performance of Saccharomyces cerevisiae strains against the application of adaptive laboratory evolution strategies for butanol tolerance. *Food Research International*, *190*, 114637. https://doi.org/10.1016/j.foodres.2024.114637

Azambuja, S. P. H., & Goldbeck, R. (2020). Butanol production by Saccharomyces cerevisiae: Perspectives, strategies and challenges. *World Journal of Microbiology and Biotechnology*, *36*(3), 48. https://doi.org/10.1007/s11274-020-02828-z

Bahls, M. O., Platz, L., Morgado, G., Schmidt, G. W., & Panke, S. (2022). Directed evolution of biofuel-responsive biosensors for automated optimization of branched-chain alcohol biosynthesis. *Metabolic Engineering*, *69*, 98–111. https://doi.org/10.1016/j.ymben.2021.10.014

Bai, Y., Feng, H., Liu, N., & Zhao, X. (2023). Biomass-derived 2,3-butanediol and its application in biofuels production. *Energies*, *16*(15). https://doi.org/10.3390/en16155802

Banner, A., Toogood, H. S., & Scrutton, N. S. (2021). Consolidated bioprocessing: Synthetic biology routes to fuels and fine chemicals. *Microorganisms*, *9*(5), 1079. https://doi.org/10.3390/microorganisms9051079

Basso, T. P., Procópio, D. P., Petrin, T. H. C., Giacon, T. G., Jin, Y.-S., Basso, T. O., & Basso, L. C. (2023). Engineering xylose fermentation in an industrial yeast: continuous cultivation as a tool for selecting improved strains. *Letters in Applied Microbiology*, *76*(7). https://doi.org/10.1093/lambio/ovad077

Baudel, H. M. (2024). *Biobased SAF (sustainable aviation biofuels): Opportunities and challenges*. Sertãozinho, Brazil: Fenasucro & Agrocana.

Becken, S., & Mackey, B. (2017). What role for offsetting aviation greenhouse gas emissions in a deep-cut carbon world? *Journal of Air Transport Management*, *63*, 71–83. https://doi.org/10.1016/J.JAIRTRAMAN.2017.05.009

Bermejo, P. M., Raghavendran, V., & Gombert, A. K. (2020). Neither 1G nor 2G fuel ethanol: Setting the ground for a sugarcane-based biorefinery using an iSUCCELL yeast platform. *FEMS Yeast Research*, *20*(4), 1–15. https://doi.org/10.1093/FEMSYR/FOAA027

Branduardi, P., Longo, V., Berterame, N. M., Rossi, G., & Porro, D. (2013). A novel pathway to produce butanol and isobutanol in Saccharomyces cerevisiae. *Biotechnology for Biofuels*, *6*(1), 68. https://doi.org/10.1186/1754-6834-6-68

Carbonell, P., Radivojevic, T., & García Martín, H. (2019). Opportunities at the intersection of synthetic biology, machine learning, and automation. *ACS Synthetic Biology*, *8*(7), 1474–-1477. https://doi.org/10.1021/acssynbio.8b00540

Casini, A., Chang, F.-Y., Eluere, R., King, A. M., Young, E. M., Dudley, Q. M., Karim, A., Pratt, K., Bristol, C., Forget, A., Ghodasara, A., Warden-Rothman, R., Gan, R., Cristofaro, A., Borujeni, A. E., Ryu, M.-H., Li, J., Kwon, Y.-C., Wang, H., ... Gordon, D. B. (2018). A pressure test to make 10 molecules in 90 days: External evaluation of methods to engineer biology. *Journal of the American Chemical Society*, *140*(12), 4302–4316. https://doi.org/10.1021/jacs.7b13292

Cavalett, O., & Cherubini, F. (2018). Contribution of jet fuel from forest residues to multiple Sustainable Development Goals. *Nature Sustainability*, *1*(12), 799–807. https://doi.org/10.1038/s41893-018-0181-2

Chiaramonti, D. (2019). Sustainable aviation fuels: The challenge of decarbonization. *Energy Procedia*, *158*, 1202–1207. https://doi.org/10.1016/J.EGYPRO.2019.01.308

Chu, W., Jiang, K., Lu, P., Xu, Y., Yang, J., Wei, X., Li, L., Liu, S., Wu, Y., Wang, S., & Zhao, H. (2021). Metabolic regulation and optimization of oxygen supply enhance the 2,3-butanediol yield of the novel Klebsiella sp. isolate FSoil 024. *Biotechnology Journal*, *16*(11). https://doi.org/10.1002/biot.202100279

Devi, N. B., Pugazhenthi, G., & Pakshirajan, K. (2024). Synthetic biology approaches and bioseparations in syngas fermentation. *Trends in Biotechnology*. https://doi.org/10.1016/J.TIBTECH.2024.07.008

Dürre, P. (2016). Butanol formation from gaseous substrates. *FEMS Microbiology Letters*, *363*(6), fnw040. https://doi.org/10.1093/femsle/fnw040

Dutta, K., Daverey, A., & Lin, J.-G. (2014). Evolution retrospective for alternative fuels: First to fourth generation. *Renewable Energy*, *69*, 114–122. https://doi.org/10.1016/j.renene.2014.02.044

Emmanouilidou, E., Mitkidou, S., Agapiou, A., & Kokkinos, N. C. (2023). Solid waste biomass as a potential feedstock for producing sustainable aviation fuel: A systematic review. *Renewable Energy*, *206*, 897–907. https://doi.org/10.1016/j.renene.2023.02.113

Fernandes, S., & Murray, P. G. (2010). Metabolic engineering for improved microbial pentose fermentation. *Bioengineered Bugs*, *1*(6), 424–428. https://doi.org/10.4161/bbug.1.6.12724

Ferreira, S., Pereira, R., Wahl, S. A., & Rocha, I. (2020). Metabolic engineering strategies for butanol production in *Escherichia coli*. *Biotechnology and Bioengineering*, *117*(8), 2571–2587. https://doi.org/10.1002/bit.27377

Galván, M. J., Badin, F., Cabrera, M., Martinez, D., & Dantur, A. (2024). GHG emissions intensity analysis. Case study: Bioethanol plant with cogeneration and partial CO_2 recovery. *Energy for Sustainable Development*, *83*(August). https://doi.org/10.1016/j.esd.2024.101598

Gawal, P. M., & Subudhi, S. (2023). Advances and challenges in bio-based 2,3-BD downstream purification: A comprehensive review. *Bioresource Technology Reports*, *24*, 101638. https://doi.org/10.1016/J.BITEB.2023.101638

Geleynse, S., Brandt, K., Garcia-Perez, M., Wolcott, M., & Zhang, X. (2018). The alcohol-to-jet conversion pathway for drop-in biofuels: Techno-economic evaluation. *ChemSusChem*, *11*(21), 3728–3741. https://doi.org/10.1002/cssc.201801690

Gheshlaghi, R., Scharer, J. M., Moo-Young, M., & Chou, C. P. (2009). Metabolic pathways of clostridia for producing butanol. *Biotechnology Advances*, *27*(6), 764–781. https://doi.org/10.1016/j.biotechadv.2009.06.002

Gonçalves, D. L., Matsushika, A., de Sales, B. B., Goshima, T., Bon, E. P. S., & Stambuk, B. U. (2014). Xylose and xylose/glucose co-fermentation by recombinant Saccharomyces cerevisiae strains expressing individual hexose transporters. *Enzyme and Microbial Technology*, *63*, 13–20. https://doi.org/10.1016/j.enzmictec.2014.05.003

Gottumukkala, L. D., Haigh, K., & Görgens, J. (2017). Trends and advances in conversion of lignocellulosic biomass to biobutanol: Microbes, bioprocesses and industrial viability. *Renewable and Sustainable Energy Reviews*, *76*, 963–973. https://doi.org/10.1016/j.rser.2017.03.030

Guo, Y., Liu, Y., Guan, M., Tang, H., Wang, Z., Lin, L., & Pang, H. (2022). Production of butanol from lignocellulosic biomass: Recent advances, challenges, and prospects. *RSC Advances*, *12*(29), 18848–18863. 10.1039/D1RA09396G

He, L., Zhou, B. C., Sun, D. H., Li, W. C., Lv, W. L., Wang, J., Liang, Y. Q., & Lu, A. H. (2023). Catalytic conversion of ethanol to oxygen-containing value-added chemicals. *ACS Catalysis*, *13*(17), 11291–11304. https://doi.org/10.1021/acscatal.3c01481

Hoang, T. D., & Nghiem, N. (2021). Recent developments and current status of commercial production of fuel ethanol. *Fermentation*, *7*(4). https://doi.org/10.3390/fermentation7040314

Huo, J., & Shanks, B. H. (2020). Bioprivileged molecules: Integrating biological and chemical catalysis for biomass conversion. *Annual Review of Chemical and Biomolecular Engineering*, *11*, 63–85. https://doi.org/10.1146/annurev-chembioeng-101519-121127

Huynh, D. T. N., Kim, A. Y., Seol, I. H., Jung, S., Lim, M. C., Lee, J. A., Jo, M. R., Choi, S. J., Kim, B., Lee, J., Kim, W., & Kim, Y. R. (2015). Inactivation of the virulence factors from 2,3-butanediol-producing Klebsiella pneumoniae. *Applied Microbiology and Biotechnology*, *99*(22), 9427–9438. https://doi.org/10.1007/s00253-015-6861-1

Ji, X. J., Huang, H., & Ouyang, P. K. (2011a). Microbial 2,3-butanediol production: A state-of-the-art review. *Biotechnology Advances*, *29*(3), 351–364. https://doi.org/10.1016/J.BIOTECHADV.2011.01.007

Ji, X. J., Huang, H., & Ouyang, P. K. (2011b). Microbial 2,3-butanediol production: A state-of-the-art review. *Biotechnology Advances*, *29*(3), 351–364. https://doi.org/10.1016/J.BIOTECHADV.2011.01.007

Jo, M. H., Ju, J. H., Heo, S. Y., Son, C. B., Jeong, K. J., & Oh, B. R. (2024). High production of enantiopure (R,R)-2,3-butanediol from crude glycerol by Klebsiella pneumoniae with an engineered oxidative pathway and a two-stage agitation strategy. *Microbial Cell Factories*, *23*(1). https://doi.org/10.1186/s12934-024-02480-4

Jung, M. Y., Ng, C. Y., Song, H., Lee, J., & Oh, M. K. (2012). Deletion of lactate dehydrogenase in Enterobacter aerogenes to enhance 2,3-butanediol production. *Applied Microbiology and Biotechnology*, *95*(2), 461–469. https://doi.org/10.1007/s00253-012-3883-9

Keasling, J., Garcia Martin, H., Lee, T. S., Mukhopadhyay, A., Singer, S. W., & Sundstrom, E. (2021). Microbial production of advanced biofuels. *Nature Reviews Microbiology*, *19*(11), 701–715. https://doi.org/10.1038/s41579-021-00577-w

Köpke, M., & Simpson, S. D. (2020). Pollution to products: recycling of 'above ground' carbon by gas fermentation. *Current Opinion in Biotechnology*, *65*, 180–189. https://doi.org/10.1016/J.COPBIO.2020.02.017

Kourkoumpas, D. S., Sagani, A., Hull, A., Hull, A., Karellas, S., & Grammelis, P. (2024). Life cycle assessment of an innovative alcohol-to-jet process: The case for retrofitting a bioethanol plant for sustainable aviation fuel production. *Renewable Energy*, *228*, 120512. https://doi.org/10.1016/J.RENENE.2024.120512

LanzaJet. (2024, January 24). LanzaJet celebrates grand opening of freedom pines fuels plant, the world's first ethanol to sustainable aviation fuel production facility. https://www.lanzajet.com/news-insights/lanzajet.celebrates-grand-opening-of-freedom-pines-fuels-plant-the-worlds-first-ethanol-to-sustainable-aviation-fuel-production-facility

Li, L., Li, K., Wang, Y., Chen, C., Xu, Y., Zhang, L., Han, B., Gao, C., Tao, F., Ma, C., & Xu, P. (2015). Metabolic engineering of Enterobacter cloacae for high-yield production of enantiopure (2R,3R)-2,3-butanediol from lignocellulose-derived sugars. *Metabolic Engineering*, *28*, 19–27. https://doi.org/10.1016/J.YMBEN.2014.11.010

Lin, Z., Cong, W., & Zhang, J. (2023). Biobutanol production from acetone–butanol–ethanol fermentation: Developments and prospects. *Fermentation*, *9*(9), 847. https://doi.org/10.3390/fermentation9090847

Lu, P., Bai, R., Gao, T., Chen, J., Jiang, K., Zhu, Y., Lu, Y., Zhang, S., Xu, F., & Zhao, H. (2024). Systemic metabolic engineering of Enterobacter aerogenes for efficient 2,3-butanediol production. *Applied Microbiology and Biotechnology*, *108*(1). https://doi.org/10.1007/s00253-023-12911-8

Luo, Z., Wang, L., Wang, Y., Zhang, W., Guo, Y., Shen, Y., Jiang, L., Wu, Q., Zhang, C., Cai, Y., & Dai, J. (2018). Identifying and characterizing SCRaMbLEd synthetic yeast using ReSCuES. *Nature Communications*, *9*(1), 1930. https://doi.org/10.1038/s41467-017-00806-y

Ma, C., Wang, A., Qin, J., Li, L., Ai, X., Jiang, T., Tang, H., & Xu, P. (2009). Enhanced 2,3-butanediol production by Klebsiella pneumoniae SDM. *Applied Microbiology and Biotechnology*, *82*(1), 49–57. https://doi.org/10.1007/s00253-008-1732-7

Martinez-Valencia, L., Garcia-Perez, M., & Wolcott, M. P. (2021). Supply chain configuration of sustainable aviation fuel: Review, challenges, and pathways for including environmental and social benefits. *Renewable and Sustainable Energy Reviews*, *152*, 111680. https://doi.org/10.1016/j.rser.2021.111680

Moysés, D., Reis, V., Almeida, J., Moraes, L., & Torres, F. (2016). Xylose fermentation by Saccharomyces cerevisiae: Challenges and prospects. *International Journal of Molecular Sciences*, *17*(3), 207. https://doi.org/10.3390/ijms17030207

Müller, C., Scapini, T., Rempel, A., Abaide, E. R., Camargo, A. F., Nazari, M. T., Tadioto, V., Bonatto, C., Tres, M. V., Zabot, G. L., Colla, L. M., Treichel, H., & Alves, S. L. (2023). Challenges and opportunities for third-generation ethanol production: A critical review. *Engineering Microbiology*, *3*(1). https://doi.org/10.1016/j.engmic.2022.100056

Niglio, S., Marzocchella, A., & Rehmann, L. (2019). Clostridial conversion of corn syrup to Acetone-Butanol-Ethanol (ABE) via batch and fed-batch fermentation. *Heliyon*, *5*(3), e01401. https://doi.org/10.1016/j.heliyon.2019.e01401

Onishi, V., Moraes, F., Olivo, J., Zanin, G. M., Onishi, V. C., Olivo, J. E., & Moraes, F. F. (2014). Enzymatic hydrolysis of sugarcane molasses as pre-treatment for bioethanol production. *Article in International Sugar Journal.* https://www.researchgate.net/publication/253643971

Opgenorth, P., Costello, Z., Okada, T., Goyal, G., Chen, Y., Gin, J., Benites, V., de Raad, M., Northen, T. R., Deng, K., Deutsch, S., Baidoo, E. E. K., Petzold, C. J., Hillson, N. J., Garcia Martin, H., & Beller, H. R. (2019). Lessons from two design–build–test–learn cycles of dodecanol production in *Escherichia coli* aided by machine learning. *ACS Synthetic Biology, 8*(6), 1337–1351. https://doi.org/10.1021/acssynbio.9b00020

Otero-Muras, I., & Carbonell, P. (2021). Automated engineering of synthetic metabolic pathways for efficient biomanufacturing. *Metabolic Engineering, 63*, 61–80. https://doi.org/10.1016/j.ymben.2020.11.012

Parapouli, M., Vasileiadis, A., Afendra, A. S., & Hatziloukas, E. (2020). Saccharomyces cerevisiae and its industrial applications. *AIMS Microbiology, 6*(1), 1–31). AIMS Press. https://doi.org/10.3934/microbiol.2020001

Patakova, P., Branska, B., Vasylkivska, M., Jureckova, K., Musilova, J., Provaznik, I., & Sedlar, K. (2022). Transcriptomic studies of solventogenic clostridia, Clostridium acetobutylicum and Clostridium beijerinckii. *Biotechnology Advances, 58*, 107889. https://doi.org/10.1016/j.biotechadv.2021.107889

Pugazhendhi, A., Mathimani, T., Varjani, S., Rene, E. R., Kumar, G., Kim, S.-H., Ponnusamy, V. K., & Yoon, J.-J. (2019). Biobutanol as a promising liquid fuel for the future—Recent updates and perspectives. *Fuel, 253*, 637–646. https://doi.org/10.1016/j.fuel.2019.04.139

Quintero-Díaz, J. C., Mendoza, D. F., & Avignone-Rossa, C. (2023). NADH-based kinetic model for acetone-butanol-ethanol production by Clostridium. *Frontiers in Bioengineering and Biotechnology, 11.* https://doi.org/10.3389/fbioe.2023.1294355

Rabinovitch-Deere, C. A., Oliver, J. W. K., Rodriguez, G. M., & Atsumi, S. (2013). Synthetic biology and metabolic engineering approaches to produce biofuels. *Chemical Reviews, 113*(7), 4611–4632. https://doi.org/10.1021/cr300361t

Raman, R., Gunasekar, S., Dávid, L. D., Rahmat, A. F., & Nedungadi, P. (2024). Aligning sustainable aviation fuel research with sustainable development goals: Trends and thematic analysis. *Energy Reports, 12*, 2642–2652. https://doi.org/10.1016/J.EGYR.2024.08.076

Rathnasingh, C., Park, J. M., Kim, D., Song, H., & Chang, Y. K. (2016). Metabolic engineering of Klebsiella pneumoniae and in silico investigation for enhanced 2,3-butanediol production. *Biotechnology Letters, 38*(6), 975–982. https://doi.org/10.1007/s10529-016-2062-y

Rautela, A., Chatterjee, R., Yadav, I., & Kumar, S. (2024). A comprehensive review on engineered microbial production of farnesene for versatile applications. *Journal of Environmental Chemical Engineering, 12*(2), 112398. https://doi.org/10.1016/j.jece.2024.112398

Ruchala, J., & Sibirny, A. A. (2021). Pentose metabolism and conversion to biofuels and high-value chemicals in yeasts. *FEMS Microbiology Reviews, 45*(4). https://doi.org/10.1093/femsre/fuaa069

Runquist, D., Hahn-Hägerdal, B., & Rådström, P. (2010). Comparison of heterologous xylose transporters in recombinant Saccharomyces cerevisiae. *Biotechnology for Biofuels, 3*(1), 5. https://doi.org/10.1186/1754-6834-3-5

Ryan, C. (2023). An overview of Gevo's biobased isobutanol production process isobutanol production process 1. https://gevo.com/wp-content/uploads/2023/03/Gevo-Whitepaper-Overview-of-Gevos-Biobased-Isobutanol-Production-Process.pdf

Saini, S., Chandel, A. K., & Sharma, K. K. (2020). Past practices and current trends in recovery and purification of first-generation ethanol: A learning curve for lignocellulosic ethanol. *Journal of Cleaner Production.*, *268*, 122357.

Samantaray, B., Mohapatra, S., Mishra, R. R., Behera, B. C., & Thatoi, H. (2024). Bioethanol production from agro-wastes: A comprehensive review with a focus on pretreatment, enzymatic hydrolysis, and fermentation. *International Journal of Green Energy*, *21*(6), 1398–1424. https://doi.org/10.1080/15435075.2023.2253871

Scheel, M., & Lütke-Eversloh, T. (2013). New options to engineer biofuel microbes: Development and application of a high-throughput screening system. *Metabolic Engineering*, *17*, 51–58. https://doi.org/10.1016/j.ymben.2013.03.002

Shahriar, M. F., & Khanal, A. (2022). The current techno-economic, environmental, policy status and perspectives of sustainable aviation fuel (SAF). *Fuel*, *325*, 124905. https://doi.org/10.1016/J.FUEL.2022.124905

Shakelly, N., Pérez-Cardona, J. R., Deng, S., Maani, T., Li, Z., & Sutherland, J. W. (2023). Comparative life cycle assessment of bioethanol production from different generations of biomass and waste feedstocks. *Procedia CIRP*, *116*, 630–635. https://doi.org/10.1016/j.procir.2023.02.106

Shen, C. R., Lan, E. I., Dekishima, Y., Baez, A., Cho, K. M., & Liao, J. C. (2011). Driving forces enable high-titer anaerobic 1-butanol synthesis in Escherichia coli. *Applied and Environmental Microbiology*, *77*(9), 2905–2915. https://doi.org/10.1128/AEM.03034-10

Sheridan, C. (2009). Making green. *Nature Biotechnology*, *27*(12), 1074–1076. https://doi.org/10.1038/nbt1209-1074

Shrivastav, A., Lee, J., Kim, H. Y., & Kim, Y. R. (2013). Recent insights in the removal of klebseilla pathogenicity factors for the industrial production of 2,3-butanediol. *Journal of Microbiology and Biotechnology*, *23*(7), 885–896. https://doi.org/10.4014/jmb.1302.02066

Si, T., Luo, Y., Xiao, H., & Zhao, H. (2014). Utilizing an endogenous pathway for 1-butanol production in Saccharomyces cerevisiae. *Metabolic Engineering*, *22*, 60–68. https://doi.org/10.1016/j.ymben.2014.01.002

Song, C. W., Park, J. M., Chung, S. C., Lee, S. Y., & Song, H. (2019). Microbial production of 2,3-butanediol for industrial applications. *Journal of Industrial Microbiology and Biotechnology*, *46*(11), 1583–1601. https://doi.org/10.1007/s10295-019-02231-0

Swidah, R., Ogunlabi, O., Grant, C. M., & Ashe, M. P. (2018). n-Butanol production in S. cerevisiae: Co-ordinate use of endogenous and exogenous pathways. *Applied Microbiology and Biotechnology*, *102*(22), 9857–9866. https://doi.org/10.1007/s00253-018-9305-x

Syu, M. J. (2001). Biological production of 2,3-butanediol. *Applied Microbiology and Biotechnology*, *55*(1), 10–18. https://doi.org/10.1007/s002530000486

Teixeira, A. T., da Silva, A. C. M., Cavalcante, R. M., & Young, A. F. (2024). Process simulation and economic evaluation of the Alcohol-to-Jet production of sustainable aviation fuel in the Brazilian context. *Energy Conversion and Management*, *319*, 118947. https://doi.org/10.1016/J.ENCONMAN.2024.118947

Wang, W. Y., Wang, B. P., Su, H. S., Wei, M. M., Wei, Y. T., & Niu, F. X. (2024). Key role of K+ and Ca2+ in high-yield ethanol production by S. Cerevisiae from concentrated sugarcane molasses. *Microbial Cell Factories*, *23*(1), 1–13. https://doi.org/10.1186/s12934-024-02401-5

Wang, Y., Ho, S.-H., Yen, H.-W., Nagarajan, D., Ren, N.-Q., Li, S., Hu, Z., Lee, D.-J., Kondo, A., & Chang, J.-S. (2017). Current advances on fermentative biobutanol production using third generation feedstock. *Biotechnology Advances*, *35*(8), 1049–1059. https://doi.org/10.1016/j.biotechadv.2017.06.001

Wang, Y., Wang, Q., Sha, A., Ren, K., Hu, M., Xu, M., Zhang, X., & Rao, Z. (2023). Advances in synthetic biology techniques and industrial applications of Corynebacterium glutamicum. *Fermentation*, *9*(8). https://doi.org/10.3390/fermentation9080729

Wu, R., Chen, D., Cao, S., Lu, Z., Huang, J., Lu, Q., Chen, Y., Chen, X., Guan, N., Wei, Y., & Huang, R. (2020). Enhanced ethanol production from sugarcane molasses by industrially engineered: Saccharomyces cerevisiae via replacement of the PHO4 gene. *RSC Advances*, *10*(4), 2267–2276. https://doi.org/10.1039/c9ra08673k

Xin, F., Yan, W., Zhou, J., Wu, H., Dong, W., Ma, J., Zhang, W., & Jiang, M. (2018). Exploitation of novel wild type solventogenic strains for butanol production. *Biotechnology for Biofuels*, *11*(1), 252. https://doi.org/10.1186/s13068-018-1252-3

Zahid, I., Nazir, M. H., Chiang, K., Christo, F., & Ameen, M. (2024). Current outlook on sustainable feedstocks and processes for sustainable aviation fuel production. *Current Opinion in Green and Sustainable Chemistry*, *49*, 100959. https://doi.org/10.1016/J.COGSC.2024.100959

Zhang, J., Chen, Y., Fu, L., Guo, E., Wang, B., Dai, L., & Si, T. (2021a). Accelerating strain engineering in biofuel research via build and test automation of synthetic biology. *Current Opinion in Biotechnology*, *67*, 88–98. https://doi.org/10.1016/j.copbio.2021.01.010

Zhang, J., Chen, Y., Fu, L., Guo, E., Wang, B., Dai, L., & Si, T. (2021b). Accelerating strain engineering in biofuel research via build and test automation of synthetic biology. *Current Opinion in Biotechnology*, *67*, 88–98. https://doi.org/10.1016/J.COPBIO.2021.01.010

Zhang, L., Sun, J., Hao, Y., Zhu, J., Chu, J., Wei, D., & Shen, Y. (2010). Microbial production of 2,3-butanediol by a surfactant (serrawettin)- deficient mutant of Serratia marcescens H30. *Journal of Industrial Microbiology and Biotechnology*, *37*(8), 857–862. https://doi.org/10.1007/s10295-010-0733-6

Zhu, Y., Zhang, J., Zhu, L., Jia, Z., Li, Q., Xiao, W., & Cao, L. (2021). Minimize the xylitol production in saccharomyces cerevisiae by balancing the xylose redox metabolic pathway. *Frontiers in Bioengineering and Biotechnology*, *9*. https://doi.org/10.3389/fbioe.2021.639595

6

Biomanufacturing of Smart Packaging Materials, Cosmetics, Therapeutics, and Nanomaterials Through Lignocellulosic Biorefinery Framework

Sounak Maitra, Muskaan Sethi, Prisha Inani, Palak Shrivastava, C. Shriya, and Samuel Jacob

Department of Biotechnology, College of Engineering and Technology, Faculty of Engineering and Technology, School of Bioengineering, SRM Institute of Science and Technology, Kattankulathur, Chengalpattu Dist., Tamil Nadu, India

6.1 Introduction

Economic growth of a country relies on many factors; among these, energy plays a pivotal role. A report by the International Energy Agency (IEA, 2011) indicates that global primary energy. Courtney and Dorman (2003) stated that current crude oil reserves are expected to be depleted within 40–70 years, while natural gas reserves are anticipated to be exhausted within the next 50 years. In addition, CO_2 emissions due to energy generation from fossil fuels will rise by 20% and the global average temperature is expected to rise from 1.4 to 5.8 °C by the year 2100 (Dow & Downing, 2006). Fossil fuels are finite in nature and are considered as depleting resources. Thus, strong efforts are oriented to search for new energy sources like renewable fuels, where the resources from they are derived are not significantly depleted by their utilization.

Recently, renewable energy is progressively substituting conventional fuels in four different areas, such as electricity generation, transportation, heating, and rural energy services (Appels et al., 2011). Biorefinery products are derived from biomass conversion by microbial intervention, physical, chemical, and physiochemical processing. The sources of biomass include the biodegradable portion of municipal wastes, industrial solid/liquid effluents, forestry and agricultural energy crops, i.e., lignocellulosics (Angenent et al., 2004; Silveira et al., 2018).

The biorefinery products should have the following criteria to be justified as sustainable and alternative:

- Provide net energy gain
- Environmentally friendly
- Cost effective and economically feasible
- Larger manufacturing capacities should not disrupt the food chain/web

Biofoundry Techniques for Biotechnology Applications, First Edition. Edited by Anuj Kumar Chandel.

CO_2 released during the manufacturing of these biorefinery products is the same CO_2 that the plants have assimilated during photosynthesis to produce organic biomass; therefore, the use of bioproducts is regarded as CO_2 neutral (Jorgensen, 2009).

The biomanufacturing of smart packaging materials, cosmetics, medicines, and nanomaterials via a lignocellulosic biorefinery framework signifies a revolutionary method for sustainable material production (Silva et al., 2012). This chapter examines the feasibility of employing agricultural leftovers, including maize stover, paddy straw, and potato trash, as feedstocks for the extraction and conversion of biopolymers. Corn stover, abundant in cellulose and hemicellulose, is a significant resource for generating biodegradable products that help alleviate plastic pollution (Velásquez, 2023). Paddy straw has been effectively utilized in multiple applications, including activated carbon synthesis, demonstrating its adaptability and promise for waste valorization (Jagaba et al., 2022). Potato waste, rich in starch, can be converted into polylactic acid (PLA), a commonly utilized biopolymer in packaging owing to its advantageous mechanical qualities and biodegradability (Bharathi et al., 2025). The incorporation of agricultural leftovers into biomanufacturing processes improves resource efficiency and adheres to circular economy principles, facilitating material reuse and waste reduction (Asim et al., 2019). This chapter seeks to illustrate the paths for generating sustainable materials from lignocellulosic biomass, addressing environmental concerns while fostering economic growth and innovation in the packaging, cosmetics, and therapeutic sectors.

6.2 Lignocellulosic Raw Materials and Their Potential as Industrial Raw Materials

6.2.1 Corn Wastes

Corn waste, consisting of byproducts from the cultivation and processing of corn (*Zea Mays* L.), encompasses several elements, including cobs, husks, stalks, and leaves. Each of these byproducts possesses unique chemical compositions and possible applications, especially in biofuel generation and animal feed. Corn cobs consist mainly of cellulose (40–60%), hemicellulose (20–30%), and lignin (15–20%) (Norvell & Nghiem, 2018). The substantial cellulose and hemicellulose content in corn cobs makes them appropriate for biofuel production, particularly for cellulosic ethanol, which is obtained through the fermentation of these polysaccharides (Sun & Cheng, 2002; Illukpitiya et al., 2012) (Figure 6.1).

Moreover, crushed corn cobs can provide a significant amount of protein and energy in animal feed, improving the nutritional composition of cattle diets (Rosentrater et al., 2003). The economic feasibility of employing maize cobs for biofuel production is considerable, as it can fulfill cellulose demands at a reduced cost while mitigating environmental repercussions linked to waste disposal (Illukpitiya et al., 2012). Corn husks have a chemical composition comprising cellulose (40–45%), hemicellulose (30–35%), and lignin (5–10%) (Wang & Padua, 2010). The absorbent characteristics of maize husks render them very beneficial for animal bedding, enhancing hygiene and comfort in cow management (Pleasants et al., 2001). The use of maize husks in this manner offers a sustainable bedding solution while successfully managing waste, therefore diminishing the environmental impact of corn production (Rosentrater et al., 2003). The treatment and processing of corn waste can result in improved

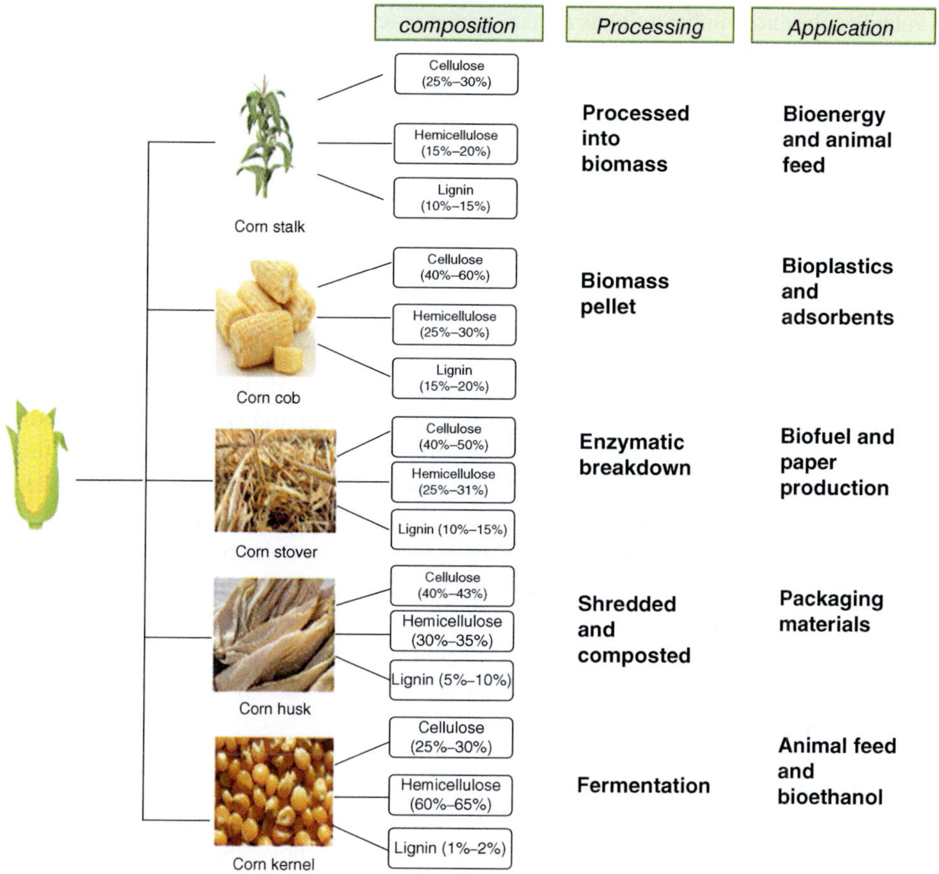

Figure 6.1 Different types of corn waste and its applications. *Source:* sudowoodo / Adobe Stock Photos.

economic and environmental benefits. Optimal pretreatment procedures can enhance cellulose accessibility in maize stalks, hence increasing yields of fermentable sugars essential for biofuel generation (Jeoh et al., 2007). Furthermore, incorporating maize wastes into animal feed systems might diminish dependence on traditional feed sources, thus fostering a circular economy in agricultural operations (Reis & Hu, 2022).

6.2.2 Sugarcane and Sugar Beet Residues

Every year, sugarcane and sugar beet cultivation produce large amounts of garbage worldwide. The most prominent residues generated are discussed in the following subsections.

6.2.2.1 Bagasse

With an annual production of almost 700 million tons, Brazil, India, and Thailand are the main producers of sugarcane bagasse, a waste product of the sugar-alcohol industry. The physicochemical properties of sugarcane bagasse are fascinating. First, its high levels of cellulose and volatile substances indicate it has a tremendous potential for producing liquid fuels (De Miranda & Fonseca, 2020).

Second, it has low sulfur and ash concentrations, which distinguishes its characteristics from many other fuels and is something that the world has been working to lower, especially in the instance of fossil fuels (Wani et al., 2023; Miranda et al., 2021). Sugar beet bagasse (pulp) is made of 20–30% cellulose and 18–36% hemicellulose, 20–32% pectin, 7–15% protein, and 1–4% lignin (Kyzy & Mazhitova, 2023; Mustafa et al., 2023). Sugarcane bagasse's high cellulose content allows for the extraction of cellulose, nanocellulose, and fibers for use in textile and food packaging applications (Mahmuda & Anannyab, 2021).

6.2.2.2 Molasses

Sugarcane and sugar beet molasses, which are byproducts of processing sugar crops, have a high sugar content. Molasses is created when sugar is prepared and crystallized repeatedly (Jamir et al., 2021). Molasses contains around 50% sucrose and 80% dry substance, consisting of a significant amount of antioxidants, vitamins, and minerals (Kyzy & Mazhitova, 2023). Although lignocellulosic bagasse and sugar beet pulp are the main non-food byproducts of producing sucrose from sugarcane and sugar beet, respectively, these products don't have a large amount of sugar (Kumar et al., 2020; Vladu et al., 2021; Tayyab et al., 2023). The importance of employing by-products for their antioxidant, antibacterial, and other nutritional qualities is emphasized by the possible health advantages of molasses (Jamir et al., 2021). Because of this, using molasses directly as a fermentation substrate holds great promise for future biotechnological advancement.

6.2.2.3 Vinasse

Vinasse, or spent wash, is a liquid residue that occurs during bioethanol production from sugarcane and sugar beet. The fermentation of sugarcane juice, molasses, and syrup results in the production of ethanol and vinasse. These are separated by distillation (Hoarau et al. (2018); Nakashima et al., 2020). Vinasse production was estimated to reach 1742 billion liters by 2024. There are high concentrations of nutrients, along with potassium, phosphorus, and nitrogen, and heavy metals like zinc, barium, and copper, present in vinasse (Meghana & Shastri, 2020).

6.2.2.4 Wastewater from the Sugar Industry

Large amounts of water are used to wash away chemicals used in the processing of sugar. Millhouses, sulfur furnaces, boiling houses, and lime hydrator units primarily dispose of water at the end of the process. This wastewater has heavy concentrations of biological oxygen demand (BOD) and various biomolecules like cellulose and sugars (Razia et al., 2020) (Table 6.1; Figure 6.2).

6.2.3 Paddy Processing Wastes

Paddy processing waste includes several byproducts produced during rice milling and processing, such as rice straw, husk, bran, and stubble. Rice straw is distinguished by its substantial cellulose concentration, which varies from 30% to 40%, and hemicellulose content of 20–25% (Feng et al., 2021). This composition indicates substantial potential for biofuel production; nevertheless, the high silica (10%–15%) and ash content (15–20%) hinder direct combustion

Table 6.1 Valorization of Sugarcane and Sugar Beet Waste and Their Yield

Waste Residue	Microorganism	Product	References
Bagasse	*Saccharomyces cerevisiae, Candida tropicalis, Actinobacillus succinogenes*	Paper, cardboard, textile fiber, bioethanol, biochar, xylitol, succinic acid	Sayed et al. (2022); Razia et al. (2020); Ungureanu et al. (2022); Alokika et al. (2021); Iwuozor et al. (2022)
Molasses	*S. cerevisiae, Scheffersomyces stipites, Zymomonas mobilis*	Fertilizers, animal feed	Razia et al. (2020); Jamir et al. (2021); Zhang et al. (2021)
Vinasse	Acetogenic bacteria, Methanogenic archaea	Fertilizers, biogas	Marafon et al. (2020); Miranda et al. (2021); Montiel-Rosales et al. (2022)
Waste water	Enterobacteriaceae, *Staphylococcus, Syntrophomonas wolinii, Methanosarcina*	Biogas	Tshemese et al. (2023); Ungureanu et al. (2022)

(Hossain et al., 2018). The elevated silica content may result in operational difficulties in combustion systems, as it might generate clinkers and diminish efficiency (Ullal et al., 2022). Rice husk, a byproduct, consists of roughly 25–35% cellulose, 18–22% hemicellulose, and 20–25% lignin (Aliah, 2023; Samosir et al., 2022). This composition renders rice husk especially appropriate for the fabrication of bio-based composites and particle boards, as lignin functions as a natural binder (Aliah, 2023). Moreover, the elevated silica content in rice husk, which may attain 20% (Muzni et al., 2020), augments its use in sectors like construction, where silica is an essential element in the production of glass, ceramics, and silicon chips (Bach et al., 2019; Borah et al., 2023). The extraction of silica from rice husk has been investigated using several methods, such as combustion and chemical treatments, which can produce high-purity silica appropriate for industrial uses (Phoohinkong & Kitthawee, 2014; Wahyuningsih et al., 2022). Rice bran contains lower levels of cellulose (10–15%) and hemicellulose (5–10%) but is abundant in oil (15–20%) and protein (10–15%), according to Feng et al. (2021). The nutritional composition of rice bran renders it an ideal choice for edible oil extraction and as an additive in animal feed (Hossain et al., 2018). The use of rice bran enhances economic value and mitigates waste in rice processing activities. The components of paddy processing waste, specifically rice straw, husk, and bran, provide prospects for sustainable application. The elevated cellulose and hemicellulose levels in rice straw and husk facilitate biofuel production and composite materials, while the nutritional properties of rice bran augment its utility in the food and feed sectors (Singh et al., 2016). Efficient management and processing of these wastes can yield substantial environmental and economic advantages.

6.2.4 Potato Processing Wastes

Potatoes serve multiple functions beyond being a vegetable solely for domestic culinary use. Additionally, they are processed into starch for other industrial purposes. The

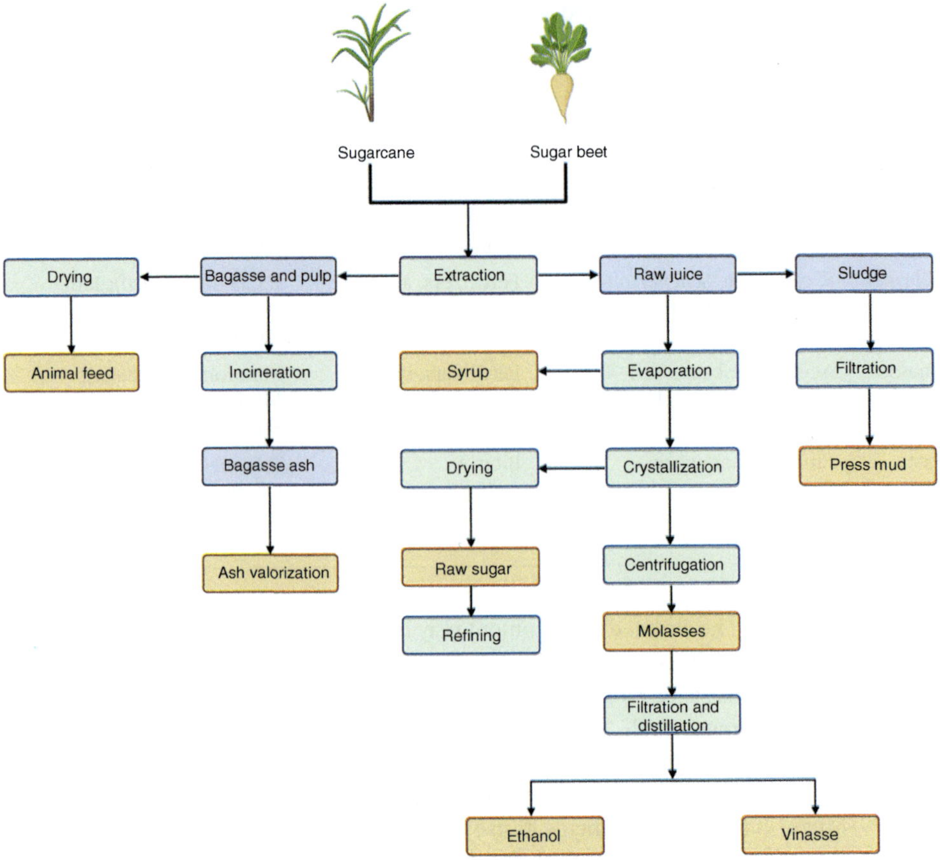

Figure 6.2 Processing of sugarcane and sugar beet and applications of byproducts.

worldwide consumption of potatoes is transitioning from fresh potatoes to value-added, processed food products (Mironescu, 2011; Brar et al., 2017; Statista, 2024). The food processing and manufacturing sector generates food losses and waste throughout the production phase due to factors such as damage during transportation, inadequate transport systems, storage issues, processing losses or contamination, and unsuitable packaging. Efficient management and disposal of these byproducts are essential to prevent environmental pollution. Moreover, the effective utilization of these byproducts can boost industry revenue (Girotto et al., 2015).

Due to the extensive processing of potatoes, the wastes obtained from industry are diverse. These wastes are divided into four categories:

- Potato peels
- Potato starch
- Potato protein
- Potato wastewater

6.2.4.1 Potato Peels

Potato peels are a byproduct of the food processing industry that can be used as a cost-effective and valuable raw material to enhance value, extract products, and produce economically significant components like dietary fiber, biopolymers, organic antioxidants, and nutritious food additives (Javed et al., 2019). Potato peel processing provides value-added products that have diverse applications. These include biosorbents, used for treating effluents and various sources of contamination, like metals, pigments, colors, biocomposites, and packaging materials. In addition, processed potato peels can be used for dietary fiber and medications. These materials can potentially produce energy in the form of biogas, biohydrogen, and biodiesel. They create biopolymer films and cellulose nanocrystals, which are applied in industrial processes such as paper production, as food ingredients, and in pet and animal food (Piyaporn, 2015; Elgarahy et al., 2023). Industrial processing generates an annual output of 70 to 140 thousand metric tons of peels globally (Abebaw, 2020).

6.2.4.2 Potato Starch from Processing Wastes

Starch is quantitatively the most dominant storage carbohydrate on Earth and is synthesized mostly in plants and some cyanobacteria (Apriyanto et al., 2022). The primary crops cultivated for industrial starch production are cereals, particularly maize, followed by wheat and potatoes (De Baere, 1999). Starch is easily obtained from potatoes because no steeping or germ removal is necessary; after the disintegration of potatoes in wet mills, pulp and proteins are separated in the same way as in the obtaining of cornstarch. Potato starch is a finely ground powder that lacks flavor but has a pleasant texture when consumed. It has a higher thickness compared to wheat and maize starches and enhances the taste of the final product. It serves as a viscosity enhancer in sauces and stews and as a cohesive agent in cake mixes, doughs, biscuits, and ice cream. Crushed potatoes are heated in Eastern Europe and Scandinavia to convert the starch in them into fermentable sugars. These sugars are then used in the production of alcoholic beverages, such as vodka and akvavit (Mironescu, 2011).

6.2.4.3 Potato Protein

In the industrial manufacturing of potato starch, a by-product known as potato fruit juice (PFJ) is produced (Schoenbeck et al., 2013). Potato proteins are categorized into three groups: patatin, protease inhibitors, and high molecular weight (MW) proteins, comprising roughly 40%, 50%, and 10% of the total protein content, respectively (Pots et al., 1999). Patatin, a tuber storage protein, possesses a unique structure that imparts remarkable functional and bioactive characteristics, rendering it a valuable component for various uses in the food sector. This category of glycoproteins consists of isoforms with molecular weights between 40 and 45 kDa and an isoelectric point ranging from 4.5 to 5.2 (Alting et al., 2011; Fu et al., 2020; Pots et al., 1999).

6.2.4.4 Potato Wastewater

Potato processing requires large amounts of water. Raw potatoes need cleaning before processing. In addition, operations such as potato peeling, dicing, or cutting cause the cells to release starch, which is often rinsed off using fresh water, resulting in large quantities of starch-loaded wastewater (Waterleau, 2024). The wastewater produced by the potato starch industry contains high concentrations of potassium salt, potato proteins, soluble starch, and other organic components (Bharathi et al., 2025). One can compare the amount of waste produced by a city with 200,000 residents to the amount produced during potato processing, such as washing, cutting, or hydro-transporting potatoes (Hongbin Li et al., 2020). Because of the high chemical and biological demand for oxygen and suspended solids, it is also very difficult to repurpose potato processing water (PPW) in manufacturing. To obtain high yields of highly concentrated protein, the protein recovery from PPW was examined using seven combinations of sedimentation, centrifugation, paper filtration, microfiltration (MF), and ultrafiltration (UF) processes (Dabestani et al., 2017) (Table 6.2; Figure 6.3).

6.2.5 Oil Processing Industry Residues

Residues from oil processing, especially those derived from palm oil and olive oil production, are abundant in lignocellulosic constituents, rendering them significant for bioenergy and biomaterials. Palm kernel shells (PKS), a byproduct of palm oil extraction, comprise roughly 30–50% cellulose and 25–30% lignin (Turan, 2023). The elevated cellulose content of PKS renders it an exceptional option for bioethanol conversion, a sustainable and renewable energy source. Moreover, the lignin in PKS can be employed to generate charcoal or activated carbon, hence augmenting its economic worth (Medouni-Haroune et al., 2017). Olive pomace, the solid byproduct of olive oil extraction, contains approximately 30–50% cellulose and 25–30% lignin (Difonzo et al., 2020). The lignocellulosic composition of olive pomace facilitates its utilization in diverse applications, such as bioenergy production via combustion or gasification (Valenti et al., 2017). Additionally, olive pomace can be

Table 6.2 Valorization of Potato Waste and Their Yield

Waste Residue	Product	Microorganism	References
Potato peel	Bioethanol; lactic acid (LA); biogas	*Saccharomyces cerevisiae* (yeast); *Lactobacillus delbrueckii; Methanobacterium, Methanosarcina*	Claassen et al. (1999); John et al. (2006); Bouallagui et al. (2004); Wu (2016)
Potato wastewater	Biofertilizer; single-cell protein; probiotics; citric acid (CA)	*Azotobacter, Bacillus subtilis; Candida utilis, S. cerevisiae; Lactobacillus plantarum; Aspergillus niger (fungus)*	Nasseri et al. (2011); Ranadheera et al. (2014); Soccol et al. (1994)
Potato starch	Lactic acid production, bioethanol, single-cell protein (SCP), biohydrogen.	*Lactococcus lactis, Zymomonas mobilis.*	Bhanwar et al. (2014); Sampaio et al., (2020)
Potato protein	Protein fortifier, bioplastics, adhesives.	*Lactobacillus* spp.; *Bacillus* spp.; *Aspergillus niger; Trichoderma reesei.*	Bhutto et al. (2024)

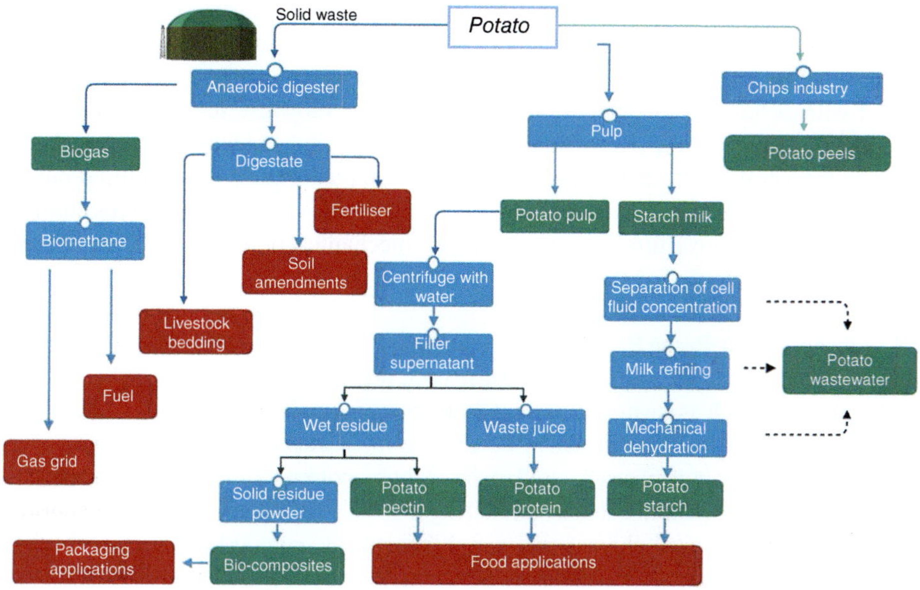

Figure 6.3 Processing of potatoes and industrial applications of their byproducts. *Source:* w-ings/ Getty Images.

subjected to anaerobic digestion to generate biogas, which acts as an alternative renewable energy source (Valero-Muñoz et al., 2013). The extraction of cellulose and lignin from olive pomace facilitates the creation of bioplastics and biocomposites, consequently advancing the development of sustainable materials (Touil, 2023). The use of these residues not only resolves waste management challenges but also conforms to the ideals of a circular economy by transforming trash into valuable resources. The valorization of olive pomace via gasification technology has been suggested as a means to optimize energy recovery from this byproduct (Lacolla et al., 2019). Recent research has emphasized the potential of deriving bioactive chemicals from depleted olive pomace, hence increasing its value (Barreca & Praticò, 2018; Difonzo et al., 2020). In a nutshell, both PKSs and olive pomace have significant opportunities for sustainable energy production and material development. Their high lignocellulosic content facilitates various conversion processes, making them crucial for the development of bioenergy and biomaterials in sustainable agricultural practices.

6.3 Smart Packaging Materials

6.3.1 Starch and Lignocellulose-Based Biopolymers

6.3.1.1 Starch-Based Biopolymer

Starch-based biopolymers occupy a significant place in the polymer industry for replacing plastics derived from petrochemicals. Their non-toxicity, biocompatible nature, and comparable mechanical and degradation properties make them suitable candidates for various applications, including the packaging sector. Starch derived from a variety of agricultural waste is converted to biocomposites by chemical modification in the presence of plasticizers and other chemical moieties. These biopolymers are biodegradable; however, they are associated with certain disadvantages, such as brittleness, moisture sensitivity, and poor thermal and mechanical properties (Agarwal et al., 2023). Biodegradable plastic has been synthesized from chitosan and starch waste derived from potato peels. The acquired membranes demonstrated notable biodegradability characteristics, validated by various characterization techniques. These films exhibited superior mechanical properties and accelerated decomposition (Taha et al., 2024).

6.3.1.2 Lignocellulosic-Based Biopolymer

Once sugar is extracted from sugarcane or sugar beet, organic waste streams, including fermentable substrates such as sugars, acids, and aromatics, are produced. To create biodegradable food packaging materials with active or sensing molecules embedded in biopolymer matrices, microbial metabolism can be used to direct those molecules toward platform chemicals or biopolymers (Jayasekara et al., 2022; Bahcegul et al., 2020). Lignocellulosic waste can enhance the physicochemical characteristics and increase the compostability of packaging materials. Hence, it can be used as an additive and a raw material (Ballesteros et al., 2018; Jahangiri, 2024). Bagasse and other lignocellulosic materials are abundant in sugars, acids, and aromatics that can be fermented by microorganisms. These precursors can then be used to create a variety of value-added products. The development of metabolic engineering methods makes it possible to turn microbes into intelligent cell factories that can produce a wide range of desired substances in an environmentally friendly manner. The approach primarily entails creating artificial metabolic pathways and genetically modifying capable microbial hosts under those pathways to maximize the production of targeted products. Food packaging composed of biodegradable plastic has been produced to lessen the negative effects of synthetic plastics generated from petroleum (Jayasekara et al., 2022).

6.3.2 PLA, PHA, and PHB

Growing environmental apprehensions about conventional plastics have encouraged the development of bioplastics as viable and eco-friendly substitutes. PLA, polyhydroxyalkanoates (PHA), and polyhydroxybutyrate (PHB) are notable bioplastics that are manufactured by microbial fermentation. These materials possess both biodegradability and distinctive characteristics that render them appropriate for smart packaging applications (Auras et al., 2004).

6.3.2.1 Polylactic Acid (PLA)

PLA is a thermoplastic polymer that may be broken down by natural processes. It is commonly derived by fermenting sugars obtained from renewable sources, including maize starch, sugarcane, and other plant materials. The manufacture of PLA by microorganisms involves the transformation of lactic acid monomers, which are generated through the process of fermentation, into PLA by a polymerization reaction (Madhavan Nampoothiri et al., 2010). Lignocellulosic biomass, such as corn stover, wheat straw, and sugarcane bagasse, is essential for the sustainable manufacture of PLA.

The agricultural residues mostly consist of cellulose, hemicellulose, and lignin, which can be transformed into fermentable sugars using several methods. Corn stover, comprising the residual stalks, leaves, and cobs remaining after corn harvest, is one of the most abundant agricultural residues in the world. It is rich in cellulose and hemicellulose, which makes it highly suitable for fermentation. Utilizing corn stover for PLA production reduces agricultural waste and adds value to a by-product that would otherwise be left to decompose or be burned, thus contributing to a circular economy. Sugarcane bagasse refers to the fibrous remains that are left behind once the juice has been extracted from sugarcane. It is a significant type of plant material made up of lignocellulose that can be found abundantly in tropical and subtropical areas. Sugarcane bagasse, commonly used as a fuel source in sugar mills, can also be utilized in the manufacturing of PLA as a valuable alternative application, contributing to sustainability (Pandey et al., 2000).

6.3.2.2 Polyhydroxyalkanoates (PHA)

PHA are a type of biodegradable polyester that is synthesized by microorganisms as intracellular carbon and energy storage compounds. PHAs are produced by bacteria when they are exposed to situations where nutrients are scarce and carbon sources are abundant (Anderson & Dawes, 1990).

In the production of PHAs, lignocellulosic biomass such as rice husk and sugarcane bagasse plays a crucial role as a source of carbon. The cellulose- and hemicellulose-rich agricultural leftovers undergo pretreatment to liberate fermentable sugars via hydrolysis. Bacteria consume these sugars to create PHAs. For example, rice husk, which is a byproduct of rice milling, and sugarcane bagasse, which is the fibrous residue remaining after extracting juice from sugarcane, offer a plentiful and renewable carbon source for microbial fermentation. Furthermore, maize stover, which refers to the residual stalks, leaves, and cobs remaining after corn harvest, is also a substantial source of lignocellulosic feedstock employed in the manufacturing of PHA. Utilizing these leftovers not only improves the sustainability of PHA production but also aids in the valorization of agricultural waste, hence promoting a circular economy. PHAs display a wide range of mechanical properties, varying from stiff plastics to flexible elastomers, based on the makeup of their monomers. These materials are both biocompatible and biodegradable, which makes them well-suited for use in packaging applications (Raza et al., 2018). They are used in the production of packaging films, containers, and coatings. Their ability to be engineered for specific properties makes them suitable for smart packaging (Kourmentza et al., 2017) (Figure 6.4).

6.3.2.3 Polyhydroxybutyrate (PHB)

Bacterial fermentation of carbohydrates or lipids synthesizes PHB, the most common form of PHA. PHB is renowned for its exceptional crystallinity and biodegradability, rendering it a compelling substitute for traditional plastics (Anderson & Dawes, 1990). The production process of PHB utilizes bacteria such as *Ralstonia eutropha* or *Cupriavidus necator*. These microorganisms amass PHB as granules inside their cells when they are subjected to situations where nutrients are limited and there is an excess of carbon sources (McAdam et al., 2020). The qualities of PHB are comparable to those of polypropylene, exhibiting excellent tensile strength and barrier properties. Furthermore, it can naturally decompose and originate from sustainable sources (Raza et al., 2018; Dariš & Knez 2020). The

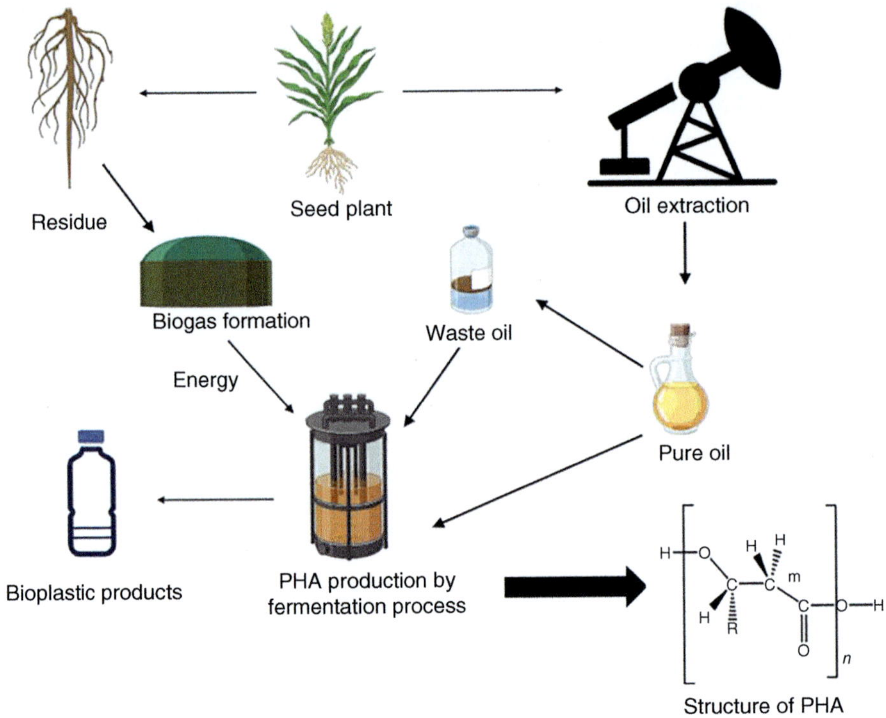

Figure 6.4 Production of polyhydroxyalkanoates from renewable sources.

application of lignocellulosic biomass—such as corn stover, rice husk, and sugarcane bagasse—significantly impacts PHB's mechanical and chemical properties. Rice husk contains a significant amount of silica and cellulose. The presence of silica can influence the efficiency of hydrolysis and the production of fermentable sugars, which in turn can affect the purity and crystallinity of the resulting PHB. This can impact the brittleness and tensile strength of PHB films and packaging materials (Kourmentza et al., 2017).

6.4 Cosmetics and Therapeutics

6.4.1 Active Pharmaceutical Components from Bioresources

Active pharmaceutical ingredients (APIs) are the essential constituents in pharmaceutical formulations that elicit the intended therapeutic effects. Historically, APIs have been generated via chemical synthesis, extraction from natural sources, or biotechnological techniques. There is a growing interest in employing lignocellulosic biomass, including maize stalks, corn cobs, and sugarcane bagasse, as renewable resources for the biotechnological synthesis of APIs due to concerns over environmental sustainability (Rahman et al., 2020; Wang, 2024). Agricultural residues contain high levels of cellulose, hemicellulose, and lignin, which can be transformed into important precursors for APIs via microbial fermentation and enzymatic methods (Rahman et al., 2020). The transformation of lignocellulosic biomass into API precursors entails multiple stages. The biomass must first undergo pretreatment to extract fermentable sugars, including glucose and xylose, from the cellulose and hemicellulose constituents (Rahman et al., 2020). This step is essential since it improves the accessibility of these sugars for later fermentation. Subsequent to sugar release, microbial fermentation is utilized, wherein particular microbes, such as *Escherichia coli* and *Saccharomyces cerevisiae*, metabolize these sugars to generate valuable compounds, including ethanol and lactic acid (Wang, 2024). The production of lactic acid is significant due to its extensive use in numerous industries, including medicines, owing to its market value and versatility (Wang, 2024). Furthermore, the employment of lignocellulosic biomass for the manufacturing of APIs not only mitigates sustainability concerns but also offers a cost-efficient substitute to conventional sources of APIs. The plentiful availability of lignocellulosic materials, along with progress in biotechnological methods, establishes this strategy as a viable prospect for the future of pharmaceutical production (Rahman et al., 2020; Wang, 2024). The incorporation of biorefineries capable of transforming agricultural waste into useful chemicals and energy carriers highlights the potential of lignocellulosic biomass in the pharmaceutical industry (Gómez-Cruz et al., 2022). Lactic acid, an organic acid, plays a crucial role as an intermediate in the production of APIs. It acts as a precursor for several pharmaceutical products, including biodegradable polymers like PLA that are utilized in drug administration systems (Nampoothiri et al., 2010).

6.4.2 Bio-Oil as a Resource for the Cosmetics Industry

Bio-oil, derived from the thermal conversion of biomass, has attracted considerable interest as a flexible and environmentally reliable chemical source. Bio-oil is a complex mixture of many organic chemicals, such as phenols, aldehydes, ketones, furans, acids, and alcohols (Kumar et al., 2022). It is mostly made from lignocellulosic materials through processes like pyrolysis. It is commonly a dark-brown, thick liquid with a strong and unpleasant pungent smell. The chemical makeup of the substance is very diverse and is influenced by the type of substrate employed and the specific circumstances in which pyrolysis takes place (Machado et al., 2022).

As a multifaceted liquid product obtained from the rapid pyrolysis of biomass, it comprises numerous key components that enhance its applicability across diverse uses. Phenolic chemicals, furans, organic acids, sugars, and alcohols are pivotal components.

Phenolic compounds are essential components of bio-oil, augmenting its antioxidant characteristics, which are advantageous in numerous applications, including the development of resins and adhesives (Zhang et al., 2021). Catechol and guaiacol can react with other bio-oil constituents to synthesize polymers, rendering them significant for the fabrication of resilient materials (Banerjee & Mushrif, 2017; Zhang et al., 2021). Moreover, phenolic compounds are progressively employed in cosmetic formulations for their capacity to mitigate the detrimental impacts of free radicals, thereby safeguarding skin cells and enhancing skin health. They are crucial components in anti-aging creams, serums, and skincare routines aimed at protecting the skin from environmental harm (Zhang et al., 2021).

Table 6.3 Types of APIs and Their Therapeutic Use

Type of Active Pharmaceutical Ingredients	Description	Therapeutic Use	References
Corn bract polysaccharides (CBPs)	Extracted using ultrasound-assisted enzymatic extraction; composed of various sugars	Antioxidant, Anti-Diabetic	Liu et al. (2022)
Corn straw fiber	Combined with waste FRP in wood plastic composites	Mechanical enhancement, biodegradable packaging	Chen et al. (2017)
Carboxymethyl cellulose (CMC) from corn peel	Derived from cellulose conversion; biocompatible	Drug formulations	Rojsitthisak et al. (2017)
Compost from paddy husk	Co-composted with chicken manure and clinoptilolite zeolite	Soil quality enhancement and agricultural applications	Latifah et al. (2015)
Bio-char from paddy waste	High surface area and activation potential	Medicinal adsorbents; bioremediation	Jatav et al. (2021)
Bio-adsorbents from paddy waste	Removal of heavy metals from contaminated environments	Environmental clean-up, functional food ingredients	Ziarati et al. (2020)

Furans, a significant class of chemicals in bio-oil, are noted for their exceptional stability and reactivity. They can undergo chemical transformation into extremely valuable molecules, rendering them significant for the manufacture of biofuels and other chemicals (Banerjee & Mushrif, 2017). 5-Hydroxymethylfurfural (HMF), a furan derivative, possesses potential as a sustainable alternative to traditional fuels (Rivilli et al., 2012). The adaptability of furans in chemical reactions highlights their significance in the composition of bio-oil (Table 6.3).

Organic acids, including acetic acid and lactic acid, are significant components of bio-oil and have potential applications in the manufacture of biodegradable polymers and other synthetic goods (Meng et al., 2015). These acids promote the exfoliation of dead skin cells, hence increasing skin renewal, texture, and appearance. Lactic acid is extensively utilized in chemical peels and exfoliating scrubs for its effectiveness in refining the skin's texture, diminishing fine lines, and evening skin tone (Meng et al., 2015). Organic acids in bio-oil boost its chemical reactivity and utility in cosmetic compositions. Furthermore, the sugars and alcohols contained in bio-oil can be subjected to fermentation methods to produce bioethanol and several biochemical organic compounds (Hita et al., 2020). This capability underscores the promise of bio-oil as a renewable resource for biofuel production, hence enhancing sustainable energy solutions. The fermentation of sugars obtained from bio-oil can yield valuable compounds applicable in diverse industrial sectors (Sun et al., 2022).

In summary, the essential constituents of bio-oil—comprising phenolic compounds, furans, organic acids, sugars, and alcohols—are crucial to its efficacy and adaptability.

Their varied applications in cosmetics, materials science, and biofuel production demonstrate the potential of bio-oil as a sustainable resource within the chemical industry.

6.4.3 Application of Bio-Oils in the Cosmetics Industry

Bio-oils are mostly included in lotions and creams to augment moisturization and enhance the texture of the skin. Their emollient characteristics contribute to the softening and refined texture of the skin. Concentrated bio-oil formulations can target specific skin concerns, such as aging, hyperpigmentation, and uneven skin tone, providing potent active ingredients in a lightweight format. Bio-oils provide significant conditioning benefits by improving hair luster and manageability while also reducing frizz and split ends. Furthermore, they can protect hair from harmful environmental effects.

6.5 Bio-Nanotechnology Through Biomass

Biogenic techniques for nanoparticle production provide a more convenient and eco-friendly approach for using non-toxic and green raw materials for specific tailoring of nanomaterials according to the requirements. The synthesis can be carried out at ambient pressures and temperatures, minimizing the need for energy-intensive equipment and the risk of accidents. Also, green synthesis of nanoparticles is carried out in a single step without the need of multiple procedures, thereby generating low toxic wastes at every step. Water is preferably used since it qualifies as a universal solvent, negating the noxious effects of other reagents and providing a greater yield of nanomaterials (Álvarez-Chimal & Ángel Arenas-Alatorre, 2023).

A study investigated the synthesis of CuO nanoparticles (NPs) from the leaf extract of *T. chebula* using water as a solvent, which was further used for both esterification and transesterification of *T. chebula* seed oil to produce 96.6% biodiesel due to the presence of balanced active sites on the surface of the NPs to enable methanol to carry out the reaction. The improved catalytic activity is attributed to the formation of aggregated spherical morphologies and greater surface areas for CuO NPs, which provide a greater degree of interaction between the triglycerides of the feedstock and the catalytic active sites. With lower HC and CO emissions, the diesel engine performed satisfactorily when biodiesel test blends were used in place of diesel fuel (Yatish et al., 2021).

A comparative study was conducted with green synthesized iron oxide nanoparticles, NaOH, and HCl for the transesterification of pre-treated wet algal biomass, which depicted 81%, 48%, and 64% of biodiesel yields respectively. The green synthesis of iron oxide NPs was carried out using leaf extract of *Hibiscus rosa-sinensis* for which water was used as the solvent, which was non-toxic. The total Global Warming Potential (GWP) contribution for the generation of biodiesel in the transesterification process utilizing HCl as the catalyst was 1.22 kg CO_2 equivalent, whereas the total GWP contribution utilizing Fe_2O_3 nanocatalyst was 1.14 kg CO_2 equivalent, resulting in a 6.55% drop in comparison to the HCl process. It is predicted that the total eutrophication potential contribution to the synthesis of biodiesel using HCl and Fe_2O_3 catalyst is

6.349 kg and 5.947 kg phosphate equivalent. In addition to indicating sustainability, the 16.81% drop in phosphate equivalent with Fe_2O_3 also suggests its superiority for biodiesel production as compared to conventional mechanisms (Banerjee et al., 2019).

6.6 Conclusion

There is a pressing need to shift to alternate and renewable bioresources for fuels, chemicals and materials to curb the shortage of oil supply, balance the economic inflation due to fuel crisis and combat climate change. Biorefinery is gaining global prominence due to its wider application and availability of know-how process technologies for industrial production. In addition, these technologies serve to be environment friendly, non-hazardous, biodegradable, and they do not produce toxic intermediates that pose a threat to public health. The demand for these bio-based product markets surged higher due to various policies and laws enforced by different nations to combat climate change and to achieve the target of net zero emissions. This chapter provided an overview of some of the salient products of wide applications using waste resources through a circular bioeconomy approach.

References

Abebaw, G. (2020). Review on: Its potentials and application of potato peel (waste). *Journal of Aquaculture and Livestock Production, 1*, 1–4. https://doi.org/10.47363/JALP/2020(1)104

Agarwal, S., Singhal, S., Godiya, C. B., & Kumar, S. (2023). Prospects and applications of starch based biopolymers. *International Journal of Environmental Analytical Chemistry, 103*(18), 6907–6926. https://doi.org/10.1080/03067319.2021.1963717

Aliah, H. (2023). Influence of sieve size on calorific value and proximate properties of bio-briquette composites. *Journal of Ecological Engineering, 24*(7), 25–34. https://doi.org/10.12911/22998993/163309

Alokika, A., Kumar, A., Kumar, V., & Singh, B. (2021). Cellulosic and hemicellulosic fractions of sugarcane bagasse: Potential, challenges and future perspective. *International Journal of Biological Macromolecules, 169*, 564–582. https://doi.org/10.1016/j.ij.iomac.2020.12.175

Alting, A. C., Pouvreau, L., Giuseppin, M. L. F., & Van Nieuwenhuijzen, N. H. (2011). Potato proteins. In *Handbook of food proteins* (pp. 316–334). Woodhead Publishing. https://doi.org/10.1533/9780857093639.316

Álvarez-Chimal, R., & Ángel Arenas-Alatorre, J. (2023). Green synthesis of nanoparticles: A biological approach. In K. J. Shah (Ed.), *Green Chemistry for Environmental Sustainability—Prevention-Assurance-Sustainability (P-A-S) Approach*. IntechOpen. https://doi.org/10.5772/intechopen.1002203

Anderson, A. J., & Dawes, E. A. (1990). Occurrence, metabolism, metabolic role, and industrial uses of bacterial polyhydroxyalkanoates. *Microbiological Reviews, 54*(4), 450–472. https://doi.org/10.1128/mr.54.4.450-472.1990

Angenent, L. T., Karim, K., Al-Dahhan, M. H., Wrenn, B. A., & Domíguez-Espinosa, R. (2004). Production of bioenergy and biochemicals from industrial and agricultural wastewater. *Trends in Biotechnology*, *22*(9), 477–485. https://doi.org/10.1016/j.tibtech.2004.07.001

Appels, L., Lauwers, J., Degrève, J., Helsen, L., Lievens, B., Willems, K., Van Impe, J., & Dewil, R. (2011). Anaerobic digestion in global bio-energy production: Potential and research challenges. *Renewable and Sustainable Energy Reviews*, *15*(9), 4295–4301. https://doi.org/10.1016/j.rser.2011.07.121

Apriyanto, A., Compart, J., & Fettke, J. (2022). A review of starch, a unique biopolymer–Structure, metabolism and in planta modifications. *Plant Science*, *318*, 111223. https://doi.org/10.1016/j.plantsci.2022.111223

Asim, N., Badiei, M., Alghoul, M., Mohammad, M., Fudholi, A., Akhtaruzzaman, M., ... Sopian, K. (2019). Biomass and industrial wastes as resource materials for aerogel preparation: opportunities, challenges, and research directions. *Industrial & Engineering Chemistry Research*, *58*(38), 17621–17645. https://doi.org/10.1021/acs.iecr.9b02661

Auras, R., Harte, B., & Selke, S. (2004). An overview of polylactides as packaging materials. *Macromolecular Bioscience*, *4*(9), 835–864. https://doi.org/10.1002/mabi.200400043

Bach, Q., Vu, C., Vu, H., & Nguyen, D. (2019). Enhancing mode i and ii interlaminar fracture toughness of carbon fiber-filled epoxy-based composites using both rice husk silica and silk fibroin electrospun nanofibers. *High Performance Polymers*, *31*(9–10), 1195–1203. https://doi.org/10.1177/0954008319840404

Bahcegul, E., Bahçegül, E., & Özkan, N. (2020). 3D printing of hemicellulosic biopolymers extracted from lignocellulosic agricultural wastes. *ACS Applied Polymer Materials*, *2*(7), 2622–2632. https://doi.org/10.1021/acsapm.0c00256

Ballesteros, L. F., Michelin, M., Vicente, A. A., Teixeira, J. A., & Cerqueira, M. Â. (2018). Use of lignocellulosic materials in bio-based packaging. In *Lignocellulosic Materials and Their Use in Bio-based Packaging*, Briefs in Molecular Science. Springer. https://doi.org/10.1007/978-3-319-92940-8_5

Banerjee, A., & Mushrif, S. (2017). Reaction pathways for the deoxygenation of biomass-pyrolysis-derived bio-oil on ru: A dft study using furfural as a model compound. *ChemCatChem*, *9*(14), 2828–2838. https://doi.org/10.1002/cctc.201700036

Banerjee, S., Rout, S., Banerjee, S., Atta, A., & Das, D. (2019). Fe2O3 nanocatalyst aided transesterification for biodiesel production from lipid-intact wet microalgal biomass: A biorefinery approach. *Energy Conversion and Management*, *195*, 844–853. https://doi.org/10.1016/j.enconman.2019.05.060

Barreca, F., & Praticò, P. (2018). Post-occupancy evaluation of buildings for sustainable agri-food production—a method applied to an olive oil mill. *Buildings*, *8*(7), 83. https://doi.org/10.3390/buildings8070083

Bhanwar, S., Singh, A., & Ganguli, A. (2014). Effective conversion of industrial starch waste to L-Lactic acid by *Lactococcus lactis* in a dialysis sac bioreactor. *Annals of Microbiology*, *64*, 1447–1452. https://doi.org/10.1007/s13213-013-0754-2

Bharathi, S. D., Baldia, A., Aktas, E., Roy, D. D., Dubey, K. K., Garlapati, V. K., Kumar, V., Kundu, D., & Jacob, S. (2025). Systematic valorisation and circular bioeconomy prospects from potato wastes: A review. *Bioresource Technology Reports*, *32*, 102327. https://doi.org/10.1016/j.biteb.2025.102327

Bhutto, R. A., Hira Bhutto, N., Khanal, S., Wang, M., Iqbal, S., Fan, Y., & Yi, J. (2024). Potato protein as an emerging high-quality: Source, extraction, purification, properties (functional, nutritional, physicochemical, and processing), applications, and challenges using potato protein. *Food Hydrocolloids*, *157*, 110415. https://doi.org/10.1016/j.foodhyd.2024.110.15

Borah, M., Kalita, B., Jose, S., & Baruah, S. (2023). Fabrication of hydrophobic surface on eri silk/wool fabric using nano silica extracted from rice husk. *Silicon*. https://doi.org/10.1007/s12633-023-02568-3

Bouallagui, H., Torrijos, M., Godon, J. J., Moletta, R., Ben Cheikh, R., Touhami, Y., Delgenes, J. P., & Hamdi, M. (2004). Two-phases anaerobic digestion of fruit and vegetable wastes: bioreactors performance. *Biochemical Engineering Journal*, *21*(2), 193–197. https://doi.org/10.1016/j.bej.2004.05.001

Brar, A., Bhatia, A. K., Pandey, V., & Kumari, P. (2017). Biochemical and phytochemical properties of potato: A review. *Chemical Science Review and Letters*, *6*, 117–129.

Chen, J., Teng, Z., & Wu, J. (2017). Recycling of waste FRP and corn straw in wood plastic composite. *Polymer Composites*, *38*, 2140–2145. https://doi.org/10.1002/pc.23789

Claassen, P., van Lier, J., Lopez Contreras, A., van Niel, E. W. J., Sijtsma, L., Stams, A. J. M., de Vries, S. S., & Weusthuis, R. A. (1999). Utilisation of biomass for the supply of energy carriers. *Applied Microbiology and Biotechnology*, *52*, 741–755. https://doi.org/10.1007/s002530051586

Courtney, B., & Dorman, D. (2003). *Worldwide fossil fuels*. Louisiana: Department of Chemistry, Louisiana State University.

Dabestani, S., Arcot, J., & Chen, V. (2017). Protein recovery from potato processing water: Pre-treatment and membrane fouling minimization. *Journal of Food Engineering*, *195*, 85–96. https://doi.org/10.1016/j.jfoodeng.2016.09.013

Dariš, B., & Knez, Ž. (2020). Poly(3-hydroxybutyrate): Promising biomaterial for bone tissue engineering. *Acta Pharmaceutica*, *70*(1), 1–15. https://doi.org/10.2478/acph-2020-0007

De Baere, H. (1999). Starch policy in the European community. *Starch-Stärke*, *51*(6), 189–193. https://doi.org/10.1002/(SICI)1521-379X(199906)51:6<189::AID-STAR189>3.0.CO;2-G

De Miranda, E. E., & Fonseca, M. F. (2020). Sugarcane: Food production, energy, and environment. In *Sugarcane Biorefinery, Technology and Perspectives* (pp. 67–88). https://doi.org/10.1016/B978-0-12-814236-3.00004-4

Difonzo, G., Troilo, M., Squeo, G., Pasqualone, A., & Caponio, F. (2020). Functional compounds from olive pomace to obtain high-added value foods—A review. *Journal of the Science of Food and Agriculture*, *101*(1), 15–26. https://doi.org/10.1002/jsfa.10478

Dow, K., & Downing, T. (2006). *The atlas of climate change: Mapping the world's greatest challenge*. Los Angeles: University of California Press.

El Sayed, A., Hamad, K., Eid, Y., Alaa, A., Hassan, A., Mohsen, A., Safwat, A., Atta, K., Benbella, M., & Abdel Tawab, S. (2022). Production of paper using chemical pulping process of sugarcane bagasse. *International Journal of Industry and Sustainable Development*. https://doi.org/10.21608/ijisd.2022.145854.1013

Elgarahy, A. M., Eloffy, M. G., Alengebawy, A., El-Sherif, D. M., Gaballah, M. S., Elwakeel, K. Z., & El-Qelish, M. (2023). Sustainable management of food waste; Pre-treatment strategies, techno-economic assessment, bibliometric analysis, and potential utilizations: A systematic review. *Environmental Research*, *225*, 155558. https://doi.org/10.1016/j.envres.2023.115558

Feng, Y., Tian, S., Du, K., & Wang, X. (2021). Effects of steam explosion pretreatment on the extraction of xylooligosaccharide from rice husk. *BioResources*, *16*(4), 6910–6920. 10.15376/biores.16.4.6910-6920

Fu, Y., Liu, W. N., & Soladoye, O. P. (2020). Towards potato protein utilisation: Insights into separation, functionality and bioactivity of patatin. *International Journal of Food Science & Technology*, *55*(6), 2314–2322. https://doi.org/10.1111/ijfs.1434

Girotto, F., Alibardi, L., & Cossu, R. (2015). Food waste generation and industrial uses: A review. *Waste Management*, *45*, 32–41. https://doi.org/10.1016/j.wasman.2015.06.00

Gómez-Cruz, I., Romero, I., Contreras, M., Labidi, J., Hernández-Ramos, F., Roseiro, L., ... Carvalheiro, F. (2022). Combined extraction and ethanol organosolv fractionation of exhausted olive pomace for bioactive compounds. *Advanced Sustainable Systems*, *6*(4). https://doi.org/10.1002/adsu.202100361

Hita, I., Cordero-Lanzac, T., Kekäläinen, T., Okafor, O., Rodríguez-Mirasol, J., Cordero, T., ... Castaño, P. (2020). In-depth analysis of raw bio-oil and its hydrodeoxygenated products for a comprehensive catalyst performance evaluation. *ACS Sustainable Chemistry & Engineering*, *8*(50), 18433–18445. https://doi.org/10.1021/acssuschemeng.0c05533

Hoarau, J., Caro, Y., Grondin, I., & Petit, T. (2018). Sugarcane vinasse processing: Toward a status shift from waste to valuable resource. A review. *Journal of Water Process Engineering*, *24*, 11–25. https://doi.org/10.1016/j.jwpe.2018.05.003

Hossain, S., Mathur, L., & Roy, P. (2018). Rice husk/rice husk ash as an alternative source of silica in ceramics: A review. *Journal of Asian Ceramic Societies*, *6*(4), 299–313. https://doi.org/10.1080/21870764.2018.1539210

Illukpitiya, P., Gopalakrishnan, C., & Ratnayake, R. (2012). Economic competitiveness of bio-fuel production from underutilized ethanol feedstock in Sri Lanka. *Journal of Natural Resources Policy Research*, *4*(2), 91–102. https://doi.org/10.1080/193909IEA459.2012.673853

International Energy Agency (IEA). (2011). World Energy Outlook 2011 Report. https://www.iea.org/reports/world-energy-outlook-2011.

Iwuozor, K. O., Emenike, E. C., Ighalo, J. O., Omoarukhe, F. O., Omuku, P. E., & Adeniyi, A. G. (2022). A review on the thermochemical conversion of sugarcane bagasse into biochar. *Cleaner Materials*, *6*, 100162. https://doi.org/10.1016/j.clema.2022.10.162

Jagaba, A., Kutty, S., Noor, A., Affam, A., Ghfar, A., Usman, A., ... Soja, U. (2022). Parametric optimization and kinetic modelling for organic matter removal from agro-waste derived paper packaging biorefinery wastewater. *Biomass Conversion and Biorefinery*, *14*(2), 2235–2252. https://doi.org/10.1007/s13399-022-02431-2

Jahangiri, F. (2024). Sustainable biodegradable coatings for food packaging: challenges and opportunities. *Green Chemistry*, *26*(9), 4934–4974. https://doi.org/10.1039/d3gc02647g

Jamir, L., Kumar, V., Kaur, J., Kumar, S., & Singh, H. (2021). Composition, valorization and therapeutical potential of molasses: A critical review. *Environmental Technology Reviews*, *10*(1), 131–142. https://doi.org/10.1080/21622515.2021.1892203

Jatav, H. S., Rajput, V. D., Minkina, T., Singh, S. K., Chejara, S., Gorovtsov, A., Barakhov, A., Bauer, T., Sushkova, S., Mandzhieva, S., Burachevskaya, M., & Kalinitchenko, V. P. (2021). Sustainable approach and safe use of biochar and its possible consequences. *Sustainability*, *13*(18), 10362. https://doi.org/10.3390/su131810362

Javed, A., Ahmad, A., Tahir, A., Shabbir, U., Nouman, M., & Hameed, A. (2019). *Potato peel waste—Its nutraceutical, industrial and biotechnological applications*. https://doi.org/10.3934/agrfood.2019.3.807

Jayasekara, S., Dissanayake, L., & Jayakody, L. N. (2022). Opportunities in the microbial valorization of sugar industrial organic waste to biodegradable smart food packaging materials. *International Journal of Food Microbiology, 377*, 109785. https://doi.org/10.1016/j.ijfoodmicro.2022.109785

Jeoh, T., Ishizawa, C., Davis, M., Himmel, M., Adney, W., & Johnson, D. (2007). Cellulase digestibility of pretreated biomass is limited by cellulose accessibility. *Biotechnology and Bioengineering, 98*(1), 112–122. https://doi.org/10.1002/bit.21408

John, R. P., Nampoothiri, K. M., & Pandey, A. (2006). Solid-state fermentation for l-lactic acid production from agro wastes using *Lactobacillus delbrueckii. Process Biochemistry, 41*, 759–763. https://doi.org/10.1016/j.procbio.2005.09.013

Jorgensen, P. J. (2009). *Biogas: Green Energy* (2nd ed.). Plan Energy and Researcher for a Day - Faculty of Agricultural Sciences, Aarhus University.

Kourmentza, C., Plácido, J., Venetsaneas, N., Burniol-Figols, A., Varrone, C., Gavala, H. N., & Reis, M. A. M. (2017). Recent advances and challenges towards sustainable polyhydroxyalkanoate (PHA) production. *Bioengineering, 4*(2), 55. https://doi.org/10.3390/bioengineering4020055

Kumar, R., Strezov, V., Weldekidan, H., He, J., Singh, S., Kan, T., & Dastjerdi, B. (2020). Lignocellulose biomass pyrolysis for bio-oil production: A review of biomass pre-treatment methods for production of drop-in fuels. *Renewable and Sustainable Energy Reviews, 123*, 109763. https://doi.org/10.1016/j.rser.2020.109763

Kumar, V., Bansal, V., Madhavan, A., Kumar, M., Sindhu, R., Awasthi, M. K., Binod, P., & Saran, S. (2022). Active pharmaceutical ingredient (API) chemicals: A critical review of current biotechnological approaches. *Bioengineered, 13*(2), 4309–4327. https://doi.org/10.1080/21655979.2022.2031412

Kyzy, A. M., & Mazhitova, A. (2023). Biotechnological valorization of sugar beet wastes into value-added products. *MANAS Journal of Engineering, 11*(1), 136–144. 10.51354/mjen.1280304

Lacolla, G., Fortunato, S., Nigro, D., Pinto, M., Mastro, M., Caranfa, D., ... Cucci, G. (2019). Effects of mineral and organic fertilization with the use of wet olive pomace on durum wheat performance. *International Journal of Recycling of Organic Waste in Agriculture, 8*(S1), 245–254. https://doi.org/10.1007/s40093-019-00295-7

Latifah, O., Ahmed, O. H., Susilawati, K., & Majid, N. M. (2015). Compost maturity and nitrogen availability by co-composting of paddy husk and chicken manure amended with clinoptilolite zeolite. *Waste Management & Research: The Journal for a Sustainable Circular Economy*. 2015 *33*(4), 322–331. https://doi.org/10.1177/0734242X15576771

Li, H., Zeng, X., Shi, W., Zhang, H., Huang, S., Zhou, R., & Qin, X. (2020). Recovery and purification of potato proteins from potato starch wastewater by hollow fiber separation membrane integrated process. *Innovative Food Science & Emerging Technologies, 63*, 102380.

Liu, Y., Li, Y., Niu, N., Xie, Y., Zhang, W., Dong, S., Pu, G., Liu, C., Jiang, C., Cai, M., & Zhang, Y. (2022). Ultrasound-assisted enzymatic extraction of polysaccharides from waste corn bract: process optimization, characterization, antioxidant and anti-diabetic potentials. *American Journal of Biochemistry and Biotechnology, 18*(2), 238–251. https://doi.org/10.3844/ajbbsp.2022.238.251

Machado, H., Cristino, A. F., Orišková, S., & Galhano Dos Santos, R. (2022). Bio-oil: The next-generation source of chemicals. *Reactions, 3*(1), 118–137. https://doi.org/10.3390/reactions3010009

Madhavan Nampoothiri, K., Nair, N. R., & John, R. P. (2010). An overview of the recent developments in polylactide (PLA) research. *Bioresource Technology, 101*(22), 8493–8501. https://doi.org/10.1016/j.biortech.2010.05.092

Mahmuda, M. A., & Anannyab, F. R. (2021). Sugarcane bagasse—A source of cellulosic fiber for diverse applications. *Heliyon, 7*(8), e07771. https://doi.org/10.1016/j.heliyon.2021.e07771

Marafon, A. C., Salomon, K. R., Amorim, E. L. C., & Peiter, F. C. (2020). Use of sugarcane vinasse to biogas, bioenergy, and biofertilizer production. *Sugarcane Biorefinery, Technology and Perspectives*, 179–194. https://doi.org/10.1016/B978-0-12-814236-3.00010-X

McAdam, B., Brennan Fournet, M., McDonald, P., & Mojicevic, M. (2020). Production of polyhydroxybutyrate (PHB) and factors impacting its chemical and mechanical characteristics. *Polymers, 12*(12), 2908. https://doi.org/10.3390/polym12122908

Medouni-Haroune, L., Zaïdi, F., Medouni-Adrar, S., Roussos, S., Azzouz, S., Desseaux, V., ... Kecha, M. (2017). Selective isolation and screening of actinobacteria strains producing lignocellulolytic enzymes using olive pomace as substrate. *Iranian Journal of Biotechnology, 15*(1), 74–77. 10.15171/ijb.1278

Meghana, M., & Shastri, Y. (2020). Sustainable valorization of sugar industry waste: Status, opportunities, and challenges. *Bioresource Technology, 303*, 122929. https://doi.org/10.1016/j.biortech.2020.122929

Meng, J., Moore, A., Tilotta, D., Kelley, S., Adhikari, S., & Park, S. (2015). Thermal and storage stability of bio-oil from pyrolysis of torrefied wood. *Energy & Fuels, 29*(8), 5117–5126. https://doi.org/10.1021/acs.energyfuels.5b00929

Miranda, N. T., Motta, I. L., Maciel Filho, R., & Maciel, M. R. W. (2021). Sugarcane bagasse pyrolysis: A review of operating conditions and products properties. *Renewable and Sustainable Energy Reviews, 149*, 111394. https://doi.org/10.1016/j.rser.2021.111394

Mironescu, M. (2011). Investigations on wastewaters at potato processing and starch recovery and characterisation. *Journal of Agroalimentary Processes and Technologies, 17*(2), 134–138.

Montiel-Rosales, A., Montalvo-Romero, N., García-Santamaría, L. E., Sandoval-Herazo, L. C., Bautista-Santos, H., & Fernández-Lambert, G. (2022). Post-industrial use of sugarcane ethanol vinasse: A systematic review. *Sustainability, 14*(18), 11635. https://doi.org/10.3390/su141811635

Mustafa, G., Arshad, M., Bano, I., & Abbas, M. (2023). Biotechnological applications of sugarcane bagasse and sugar beet molasses. *Biomass Conversion and Biorefinery, 13*, 1489–1501. https://doi.org/10.1007/s13399-020-01141-x

Muzni, N., Jamil, N., Faizul, C., & Arif, W. (2020). Effect of acid leaching on different state of rice husk. *Materials Science Forum, 1010*, 532–537. https://doi.org/10.4028/http://www.scientific.net/msf.1010.532

Nakashima, R. N., Orrego, D. F., Velásquez, H. I., & Junior, S. D. O. (2020). Sugarcane bagasse and vinasse conversion to electricity and biofuels: An exergoeconomic and environmental assessment. *International Journal of Exergy, 33*(1), 44. https://doi.org/10.1504/IJEX.2020.109623

Nampoothiri, K. M., Nair, N. R., & John, R. P. (2010). An overview of the recent developments in polylactide (PLA) research. *Bioresource Technology, 101*, 8493–8501. https://doi.org/10.1016/j.biortech.2010.05.092

Nasseri, A. T., Rasoul-Amini, S., Morowvat, M. H., & Ghasemi, Y. (2011). Single Cell Protein: Production and Process. *American Journal of Food Technology, 6*, 103–116.

Norvell, K., & Nghiem, N. (2018). Soaking in aqueous ammonia (saa) pretreatment of whole corn kernels for cellulosic ethanol production from the fiber fractions. *Fermentation, 4*(4), 87. https://doi.org/10.3390/fermentation4040087

Pandey, A., Soccol, C. R., Nigam, P., Soccol, V. T., Vandenberghe, L. P. S., & Mohan, R. (2000). Biotechnological potential of agro-industrial residues. II: Cassava bagasse. *Bioresource Technology*, *74*(1), 81–87. https://doi.org/10.1016/S0960-8524(99)00143-1

Phoohinkong, W., & Kitthawee, U. (2014). Low-cost and fast production of nano-silica from rice husk ash. *Advanced Materials Research*, *979*, 216–219. https://doi.org/10.4028/http://www.scientific.net/amr.979.216

Piyaporn, K. (2015). Extraction and characterization of cellulose nanocrystals produced by acid hydrolysis from corn husk. *Materials and Minerals*. https://doi.org/10.14456/JMMM.2015.3

Pleasants, J., Hellmich, R., Dively, G., Sears, M., Stanley-Horn, D., Mattila, H., ... Jones, G. (2001). Corn pollen deposition on milkweeds in and near cornfields. *Proceedings of the National Academy of Sciences*, *98*(21), 11919–11924. https://doi.org/10.1073/pnas.211287498

Pots, A. M., Gruppen, H., de Jongh, H. H., van Boekel, M. A., Walstra, P., & Voragen, A. G. (1999). Kinetic modeling of the thermal aggregation of patatin. *Journal of Agricultural and Food Chemistry*, *47*(11), 4593–4599. https://doi.org/10.1021/jf990191t

Rahman, M., Mondal, M., Yeasmin, M., Sayeed, M., & Hossain, A. (2020). Conversion of lignocellulosic corn agro-waste into cellulose derivative and its potential application as pharmaceutical excipient. *Processes*, *8*(6), 711. https://doi.org/10.3390/pr8060711

Ranadheera, R. D. S., Vidanarachchi, J. K., Naumovski, N., & Ajlouni, S. (2014). Effect of dairy probiotic combinations on in vitro gastrointestinal tolerance, intestinal epithelial cell adhesion and cytokine secretion. *Journal of Functional Foods*, *10*, 369–379.

Raza, Z. A., Abid, S., & Banat, I. M. (2018). Polyhydroxyalkanoates: Characteristics, production, recent developments, and applications. *International Biodeterioration & Biodegradation*, *126*, 45–56. https://doi.org/10.1016/j.ibiod.2017.10.001

Razia, M., V., U. M. N., & Sivaramakrishnan, S. (2020). Agro-based sugarcane industry wastes for production of high-value bioproducts. *Biovalorisation of Wastes to Renewable Chemicals and Biofuels*, 303–316. https://doi.org/10.1016/B978-0-12-817951-2.00016-X

Reis, C., & Hu, B. (2022). Volumetric scale-up of a packed-bed ion-exchange system to extract phytate from thin stillage. *Membranes*, *12*(2), 230. https://doi.org/10.3390/membranes12020230

Rivilli, P., Alarcón, R., Isasmendi, G., & Perez, J. (2012). Stepwise isothermal fast pyrolysis (sifp). part ii. sifp of peanut shells—Antifungal properties of phenolic fractions. *BioResources*, *7*(1), 112–117. https://doi.org/10.15376/biores.7.1.112-117

Rojsitthisak, P., Khunthon, S., Noomun, K., & Limpanart, S. (2017). Response surface method to optimize the preparation of carboxymethyl cellulose from corn peel agricultural waste. *ScienceAsia*, *43*(1), 8–14. https://doi.org/10.2306/scienceasia1513-1874.2017.43.008

Rosentrater, K., Richard, T., Bern, C., & Flores, R. (2003). Economic simulation modeling of reprocessing alternatives for corn masa byproducts. *Resources, Conservation and Recycling*, *39*(4), 341–367. https://doi.org/10.1016/s0921-3449(02)00175-1

Samosir, G., Sariati, M., Sitepu, I., Ritonga, E., Samosir, A., & Martgrita, M. (2022). Optimisation of rice husks' lignocellulose pre-treatment methods and use of central composite design in citric acid fermentation: A review. *IOP Conference Series: Earth and Environmental Science*, *1097*(1), 012015. https://doi.org/10.1088/1755-1315/1097/1/012015

Sampaio, S. L., Petropoulos, S. A., Alexopoulos, A., Heleno, S. A., Santos-Buelga, C., Barros, L., & Ferreira, I. C. F. R. (2020). Potato peels as sources of functional compounds for the food industry: a review. *Trends in Food Science & Technology*. https://doi.org/10.1016/j.tifs.2020.07.015

Schoenbeck, I., Graf, A. M., Leuthold, M., Pastor, A., Beutel, S., & Scheper, T. (2013). Purification of high value proteins from particle containing potato fruit juice via direct capture membrane adsorption chromatography. *Journal of Biotechnology, 168*(4), 693–700. https://doi.org/10.1016/j.jbiotec.2013.09.018

Silva, S. S., Chandel, A. K., Wickramasinghe, S. R., & González, J. M. D. (2012). Fermentative production of value-added products from lignocellulosic biomass. *BioMed Research International, 2012*, 826162.

Silveira, M. H. L., Vanelli, B. A., & Chandel, A. K. (2018). Second generation ethanol production: Potential biomass feedstock, engineering of feedstock, engineering of microorganisms for improved cellulose and ethanol production, Biomass deconstruction, process configurations and new process schemes, Techno-economic analysis. In *Advances in Sugarcane Biorefinery* (1st ed., pp. 135–152). Elsevier.

Singh, R., Srivastava, M., & Shukla, A. (2016). Environmental sustainability of bioethanol production from rice straw in India: A review. *Renewable and Sustainable Energy Reviews, 54*, 202–216. https://doi.org/10.1016/j.rser.2015.10.005

Soccol, C. R., Marin, B., Raimbault, M., & Lebeaultm, M. (1994). Potential of solid state fermentation for production of L(+)-lactic acid by *Rhizopus oryzae*. *Applied Microbiology and Biotechnology, 41*, 286–290. https://doi.org/10.1007/BF00221220

Srivastava, A., Srivasatva, A., Singh, A., Singh, P., Verma, S., Vats, M., ... Sagadevan, S. (2022). Biopolymers as renewable polymeric materials for sustainable development—An overview. *Polimery, 67*(5), 185–196. 10.14314/polimery.2022.5.1

Statista. (2024). Production volume of potato across India from financial year 2015 to 2022, with an estimate for 2023. https://www.statista.com/statistics/1038959/india-production-of-potato

Sun, Y., & Cheng, J. (2002). Hydrolysis of lignocellulosic materials for ethanol production: A review. *Bioresource Technology, 83*(1), 1–11. https://doi.org/10.1016/S0960-8524(01)00212-7

Sun, J., Shao, S., Hu, X., Li, X., & Zhang, H. (2022). Synthesis of oxygen-containing precursors of aviation fuel via carbonylation of the aqueous bio-oil fraction followed by c–c coupling. *ACS Sustainable Chemistry & Engineering, 10*(33), 11030–11040. https://doi.org/10.1021/acssuschemeng.2c03379

Taha, T. H., Abu-Saied, M. A., Elnouby, M., Hashem, M., Alamri, S., Desouky, E. E., & Morsy, K. (2024). Profitable exploitation of biodegradable polymer including chitosan blended potato peels' starch waste as an alternative source of petroleum plastics. *Biomass Conversion and Biorefinery, 14*(1), 207–215. https://link.springer.com/article/10.1007%2Fs13399-021-02244-9

Tayyab, M., Wakeel, A., Mubarak, M. U., Artyszak, A., Ali, S., Hakki, E. E., Mahmood, K., Song, B., & Ishfaq, M. (2023). Sugar beet cultivation in the tropics and subtropics: Challenges and opportunities. *Agronomy, 13*, 1213. https://doi.org/10.3390/agronomy13051213

Touil, I. (2023). Development and mechanical assessment of corn flour and olive pomace reinforced bioplastics. *Annales De Chimie Science Des Matériaux, 47*(4), 193.

Tshemese, Z., Deenadayalu, N., Linganiso, L. C., & Chetty, M. (2023). An overview of biogas production from anaerobic digestion and the possibility of using sugarcane wastewater and municipal solid waste in a South African context. *Applied System Innovation, 6*(1), 13. https://doi.org/10.3390/asi6010013

Turan, S. (2023). Hot-air-assisted radiofrequency drying of olive pomace and its effect on the quality of olive pomace oil. *Food, 12*(18), 3515. https://doi.org/10.3390/foods12183515

Ullal, V., Shivaramu, H., & Aveen, K. (2022). Rice husk ash utilization, composition and properties: A brief review. *Journal of Metals Materials and Minerals, 32*(4), 37–46. 10.55713/jmmm.v32i4.1544

Ungureanu, N., Vlăduț, V., & Biriș, S.-Ștefan. (2022). Sustainable valorization of waste and by-products from sugarcane processing. *Sustainability, 14*(17), 11089. https://doi.org/10.3390/su141711089

Valenti, F., Arcidiacono, C., Chinnici, G., & Cascone, G. (2017). Quantification of olive pomace availability for biogas production by using a gis-based model. *Biofuels, Bioproducts and Biorefining, 11*(5), 784–797. https://doi.org/10.1002/bbb.1784

Valero-Muñoz, M., Martín-Fernández, B., Ballesteros, S., Fuente, E., Quintela, J., Lahera, V., ... Heras, N. (2013). Protective effect of a pomace olive oil concentrated in triterpenic acids in alterations related to hypertension in rats: Mechanisms involved. *Molecular Nutrition & Food Research, 58*(2), 376–383. https://doi.org/10.1002/mnfr.201300256

Velásquez, M. (2023). Aprovechamiento de residuos del maíz (Zea mays) en la industria: revisión bibliográfica. *Revista Rivar, 11*(31), 212–229. 10.35588/rivar.v10i31.5980

Vladu, M., Tudor, V. C., Mărcuță, L., Mihai, D., & Tudor, A. D. (2021). Study on the production and valorization of sugar beet in the European Union. *Romanian Agricultural Research, 38*.

Wahyuningsih, K., Yuliani, S., & Diono, W. (2022). Improvement of the characteristics of silica particles extracted from rice husks using chemical–physical method. *Journal of Physics Conference Series, 2190*(1), 012011. https://doi.org/10.1088/1742-6596/2190/1/012011

Wang, C. (2024). Detoxification of corn stover prehydrolysate by different biochars and its effect on lactic acid fermentation. *RSC Advances, 14*(7), 4315–4323. https://doi.org/10.1039/d3ra08055b

Wang, Y., & Padua, G. (2010). Formation of zein microphases in ethanol–water. *Langmuir, 26*(15), 12897–12901. https://doi.org/10.1021/la101688v

Wani, A. K., Rahayu, F., Fauziah, L., & Suhara, C. (2023). Advances in safe processing of sugarcane and bagasse for the generation of biofuels and bioactive compounds. *Journal of Agriculture and Food Research, 12*, 100549. https://doi.org/10.1016/j.jafr.2023.100549

Waterleau. (2024) Energy from wastewater in the potato and starch industry. https://www.waterleau.com/en/projects/potato-starch-industry#:~:text=Potato%20processing%20requires%20large%20amounts,quantities%20of%20starch%2Dloaded%20wastewater.

Wu, D. (2016). Recycle technology potato peel waste processing: a review. *Procedia Environmental Science., 31*, 103–107.

Yatish, K. V., Prakash, R. M., Ningaraju, C., Sakar, M., GeethaBalakrishna, R., & Lalithamba, H. S. (2021). Terminalia chebula as a novel green source for the synthesis of copper oxide nanoparticles and as feedstock for biodiesel production and its application on diesel engine. *Energy, 215*, 119165. https://doi.org/10.1016/j.energy.2020.119165

Zhang, S., Wang, J., & Jiang, H. (2021). Microbial production of value-added bioproducts and enzymes from molasses, a by-product of sugar industry. *Food Chemistry, 346*, 128860. https://doi.org/10.1016/j.foodchem.2020.128860

Ziarati, P., Moradi, D., & Vambol, V. (2020). Bioadsorption of heavy metals from the pharmaceutical effluents, contaminated soils and water by food and agricultural waste: a short review. *Labour Protection Problems in Ukraine, 36*(2), 3–7. https://doi.org/10.36804/nndipbop.36-2.2020.3-7

7

White Biotechnology for Skincare

Unveiling the Power of Bioactives for the Cosmetic Industry

Samatha Paladugu[1], *Sarepalli Sai Sathwik*[1], *and Mamatha Potu*[2]

[1] *Prameela Life Sciences Private Limited, Hyderabad, Telangana, India*
[2] *9MSS UK Ltd., Aberystwyth, Ceredigion, UK*

7.1 Introduction

The modern cosmetic industry is being transformed by the increasing demand for sustainable, effective, and science-backed solutions. Products designed with cutting-edge biotechnology-based ingredients to harness the power of nature are becoming popular among consumers. The shift toward advanced biotechnology in skincare, haircare, and personal care goes beyond a trend; it offers a solution to the limitations of traditional cosmetics, which often rely on harsh chemicals. Currently, consumers are very concerned about the ingredients in their skincare, haircare, beauty, and personal care products. They now seek products that are both effective and gentle on the skin and the environment. This demand is driving research and development, leading to the discovery of novel formulations and ingredients that are both effective and sustainable (Zappelli et al., 2016). White biotechnology, emphasizing sustainability and creativity, is at the forefront of this change, promising revolutionary bioactive components with bio-ferments transforming the cosmetic industry (Pérez-Rivero & López-Gómez, 2023). The cosmetic sector is investing in innovative fermented bioactives that are effective due to their biological properties, without harming nature. White biotechnology aligns with the "clean beauty" trend by ensuring cosmetics have minimal impact on the environment and human health (Mellou et al., 2019). Probiotics and fermentation-derived substances, such as bioferments and biosurfactants, are leading the way in replacing chemical-based active substances. These ingredients are highly demanded in the rapidly expanding organic cosmetics industry (Klimek-Szczykutowicz et al., 2024). Probiotic cosmetic formulations or fermented raw materials certified by recognized certification bodies like the United States Food and Drug Administration (US FDA), ECOCERT, and COSMetic Organic and Natural Standard (COSMOS) are considered to be organic and natural (Fonseca-Santos et al., 2015).

The white biotechnology and bioeconomy industries are now focused on improving the production of bioactive compounds essential to the cosmetics industry (Pérez-Rivero &

Biofoundry Techniques for Biotechnology Applications, First Edition. Edited by Anuj Kumar Chandel.

López-Gómez, 2023). Advances in biotechnology have led to the creation of next-generation biological ingredients and microbial pigments from microbial ferments and plant extracts (Sanchez-Munoz et al., 2019). These ingredients are known for their wide range of beneficial properties, such as antioxidant, anti-irritant, anti-aging, skin-firming, hydrating, probiotic, topical microbiome-enhancing, soothing, whitening, and conditioning effects (Herman & Herman, 2023). Eco-friendly ingredients like plant extracts, fermented ingredients, and microbial-based bioactives are gaining popularity due to their distinct benefits (Majchrzak et al., 2022; Martínez-Ruiz et al., 2022). Bioferment ingredients are rich in active compounds like amino acids, peptides, organic acids, short-chain fatty acids (SCFAs), and antioxidants, offering targeted skincare, haircare, and beauty solutions for diverse consumer needs (Gupta et al., 2019). Recent innovations in biotechnology have enabled advanced extraction techniques and downstream processing methods. These methods enhance the activity of bioactive compounds, increasing their concentrations and effectiveness in final formulations. Innovative formulations like emulsions, serums, essences, gels, lotions, mist sprays, masks, scrubs, sunscreens, shampoos, conditioners, and many more can be developed using these natural bio-ferment ingredients combined with natural base formulations (Majchrzak et al., 2022). The natural base improves the absorption of bioactive compounds, providing skincare, haircare, and personal care benefits. Another strategy involves incorporating biomolecules into nanostructures, lipid dispersions, or gels to improve absorption, efficacy, and stability. Expanding projects on plant-originated compounds, particularly from forests also holds promise with white biotechnology, utilizing innovative botanical formulation applications. In summary, integrating white biotechnology into skincare, and haircare is a move toward more sustainable and efficient cosmetic solutions. Throughout this chapter, we will examine biotechnology-derived elements like fermented bioactives, and microbial derivatives, highlighting their potential to revolutionize cosmetics and meet consumers' evolving needs.

The global cosmetics market size is expected to reach $760 billion by 2034, up from $429 billion in 2022, holding 4.10% of the probiotic cosmetics market share (Cosmetics Market Size, Share, and Trends 2024 to 2034, n.d.). The probiotic skincare market is projected to reach $22.91 billion in 2030 with a 5.71% compound annual growth rate (CAGR) from $17.36 billion in 2025 (Probiotic Cosmetic Products Market Size & Share Analysis—Growth Trends & Forecasts [2025–2030], n.d.). The market size of cosmetic raw materials, including both synthetic and natural materials (active ingredients, aesthetics, formulation ingredients, etc.) was $27.9 billion in 2023 and is expected to reach $37.97 billion by 2032, holding a 2% share of the sustainable natural ingredient market share (https://www.businessresearchinsights.com/market-reports/cosmetic-raw-materials-market-100628, n.d.). The increasing demand for sustainable cosmetics has driven the sustainable raw material market to grow at a CAGR of 6.9%. Based on this growth rate, the sustainable natural ingredient market is projected to reach $1 billion by 2030, up from $642 million in 2022, highlighting rising demand and market opportunities for both probiotic-based cosmetics and natural raw materials in the green cosmetic sector. (Natural Cosmetics Ingredients Market by Ingredient Type [Emollients, Bio Surfactants, Sugar Polymers, Natural Preservatives, Others], Application [Skin Care, Hair Care, Oral Care, Make-up] and Region—Global Forecast to 2030). This market trend, particularly in the cosmetics segment, is due to consumer awareness of sustainability and the environmental and societal impacts of products, resulting in a surge in demand for sustainable and natural ingredient products. Recent surveys indicate that more than 60% of cosmetic consumers want to shift towards sustainable products to mitigate environmental impacts. Nearly 75–80% of respondents are highly focused on choosing sustainable products, with the majority expressing interest in purchasing cosmetics from eco-friendly brands.

This increasing demand for sustainable products has prompted many cosmetic companies to adopt green practices in their choice of ingredients, formulation design, and packaging. Most cosmetic companies now prioritize reducing their carbon footprint per product, minimizing environmental impact, using biodegradable resources, mitigating pollution, and contributing to the circular economy in their operations (Martins & Marto, 2023).

From a consumer perspective, there is a definitive shift in trends regarding product expectations. Consumers increasingly seek products promoting healthier skin while avoiding toxic chemicals that may lead to health problems. Transparency in ingredient disclosure is essential, allowing consumers to be aware of what they are applying to their skin and enabling them to make informed choices. Key ingredients of interest include sustainable bio-fermented hyaluronic acid, vitamin C, peptides, retinol, niacinamide, UV filters, and ceramides.

Furthermore, many consumers are environmentally conscious and prefer sustainable packaging made from recycled or biodegradable materials, aiming to minimize the use of plastics. The assurance of cruelty-free practices is also a significant consideration. Affordability remains a crucial factor, as consumers balance quality and cost when making their purchase decisions.

7.2 Fermented Bioactives

Fermentation is a natural metabolic process that breaks down organic molecules, such as carbohydrates and proteins, through microorganisms like bacteria, yeast, and fungi into simpler compounds. The procedure results in biochemical transformation, typically producing valuable end products (Majchrzak et al., 2022). Fermentation enhances the bioavailability and potency of nutrients, making them more effective in absorption (Sivamaruthi et al., 2018). When microbes are offered a mixture of substrates for fermentation, they can produce a wide range of end products. They are known for their rapid growth, large-scale cultivation, and the ability to use sustainable and cost-effective substrates. Their versatility enables them to successfully convert a wide range of raw materials into usable metabolites, making them indispensable contributors to the synthesis of bioactive substances used in skincare, haircare, beauty, and other industries. In white biotechnological skincare, natural extracts and components from plants are fermented to produce bioactive molecules that improve the health of the scalp and skin (Pérez-Rivero & López-Gómez, 2023).

Fermentation is an innovative procedure in green cosmetic research that yields high-efficacy ingredients. It involves developing prebiotics from natural substrates, probiotics from microorganisms, and postbiotics from fermented metabolites. The procedure begins with the upstream process, which involves selecting unique sustainable substrates and suitable microorganisms. The fermentation process development includes choosing the appropriate microorganisms, optimizing fermentation parameters, and scaling up the technology. Downstream processing involves purification, extraction, and quantification of the end product. The final stage focuses on developing unique formulations. The overall process involves the following stages:

i) ***Selection of Substrates and Microorganisms***
The first stage involves selecting natural substrates rich in bioactive compounds such as minerals, vitamins, amino acids, and peptides. These substrates support a nutrient-rich habitat for the microorganisms involved, providing a foundation for fermentation. As shown in Figure 7.1, the next crucial step is the selection of suitable microorganisms, including bacteria, yeast, and fungi. Specific factors to consider include genus and species, strain safety and stability, fermentation requirements, substrate compatibility, and the desired end products.

ii) ***Optimization of Fermentation Parameters and Scaleup***
After the selection of substrates and microorganisms, the fermentation process requires optimizing several key parameters to ensure effective growth and metabolism. As illustrated in Figure 7.1 segment (ii), essential parameters include temperature control, aeration, agitation, pH control, seed preparation, and seed percentage. Initially, these parameters must be optimized using robust design of experiments (DOE). Each parameter should be controlled and optimized according to the robustness of the process to achieve concentrated bioactive compounds. Advanced statistical optimization methodologies, such as the Taguchi method, can be employed to design these experiments (Nagarjun et al., 2005; Rao et al., 2008). After optimization, the process should be scaled up from lab-scale fermentations to pilot scale and finally to demonstration scale.

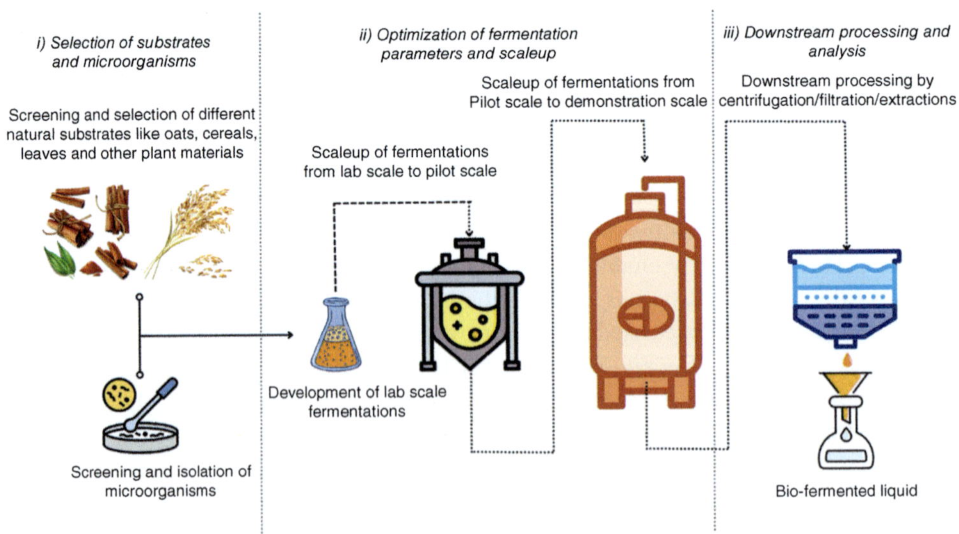

Figure 7.1 Process development of bio-fermentation.

iii) ***Downstream Processing and Analysis***
Post-fermentation, the process advances to downstream processing, where the byproducts are extracted and purified as shown in Figure 7.1 segment (iii). During this stage, bioactive compounds are extracted in their most effective form to preserve their activity and efficacy. The final stage involves analytical confirmation and end-product stability testing, which are crucial to ensuring the product's stability, efficacy, and safety in cosmetic applications.

iv) ***Bio-Ferment-Based Probiotic Skincare Formulations***
The fermented product is incorporated into various natural skincare and haircare formulations. In skin care, the end bioferment can be added to serums, lotions, sunscreens, and creams. For haircare, it can be integrated into shampoos, conditioners, serums, and similar products, as depicted in Figure 7.2. During the formulation stage, interactive studies should be considered between the ferment, bioactive compounds, and the other base ingredients. Final formulations undergo rigorous stability, safety, and efficacy analyses. The stability will be analyzed and studied based on temperature stability, pH, viscosity, heavy metal concentration, microbial stability, and pathogenic contamination. Safety assessments are conducted using non-cruelty irritation tests. Efficacy is assessed through non-cruelty methods such as graft studies, tests on human volunteers, and 3D-skin models made of fibroblast cell lines. This comprehensive approach ensures that products developed through bio-fermentation are both effective and safe for application.

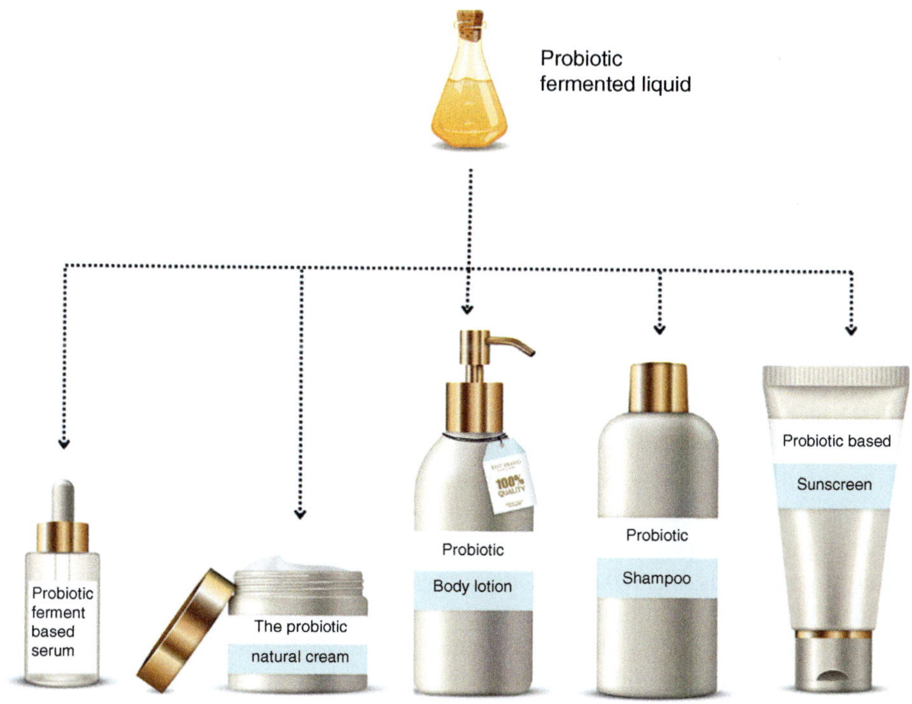

Figure 7.2 Bio-ferment-based probiotic skincare formulations.

7.3 Innovative Approaches in Green Bio-ferment Cosmetic Formulations

In recent years, probiotics and bio-ferments in skincare, haircare, and cosmetic products have gained much demand. This trend is driven by increasing studies demonstrating the beneficial effects of these substances on human health. Probiotics, sometimes known as "good bacteria," and bio ferments—nutrient-rich fermented substances—are essential for preserving a healthy skin microbiome and enhancing the overall vitality of the skin (Dou et al., 2023). In the rapidly evolving cosmetics, personal care, and beauty industry, the definition of probiotics has expanded beyond traditional ingredients to include microbial lysates, filtrates, ferments, lyophilized cells, tyndallized ferments, and cell wall components (Puebla-Barragan & Reid, 2021). New fermentation science-based technologies can improve the utilization potential of probiotics by incorporating innovative bio-ferments into skincare, haircare, and personal care formulations. This shift emphasizes the importance of selecting beneficial microorganisms, such as bacteria, yeast, fungi, and algae; optimizing fermentation conditions; and utilizing a variety of sustainable substrates to produce unique metabolite profiles.

The incorporation of new bio-ferments is revolutionizing the future of cosmetics. These ferments are made by fermenting natural substrates of plants with useful microorganisms. This gives rise to unique bioactive compounds like amino acids, peptides, vitamins, minerals, organic acids, alkaloids, flavonoids, and other secondary metabolites with promising health benefits. Lab-scale research on different bioferments shows the potential skincare and haircare benefits. For example, cereals like oats, barley, rice, and quinoa, when fermented with different strains like *Lacticaseibacillus casei* and *Saccharomyces cerevisiae*, are found to decrease oxidative stress, hyperpigmentation, and impart both anti-wrinkle and antioxidant benefits, along with imparting skin barrier protection (Anton Pallin, 2015; Carciochi et al., 2016; Chen et al., 2024; Cho et al., 2024; Kim Su Yeong, 2021; Lee et al., 2022; Miyoshi et al., 2023; Mo et al., 2022). Fruits like *Cornus mas* and passion fruit, fermented with symbiotic bacteria and *S. cerevisiae*, are found to inhibit collagenase and elastase, as well as protect the skin from UV radiation (Fang et al., 2023; Zagórska-Dziok et al., 2023). Flowers like jasmine, when treated with *Lactobacillus rhamnosus*, impart UV protection (Ho et al., 2021). Leaves and leaf extracts, like yerba mate tea and *aloe vera* gel, fermented with cultures like *Acetobacter* and *Lactobacillus plantarum*, show anti-wrinkle, antioxidant, anti-inflammatory, and wound-healing activities (Shi et al., 2020; Ziemlewska et al., 2021). Stems of plants like dendrobium orchids, when fermented with *Lactobacillus paracasei* and *Lactobacillus acidophilus*, also impart wound healing and anti-inflammatory benefits (Lin et al., 2024). Roots, especially black and red ginseng, ferment with *S. cerevisiae* and *Lactobacillus brevis* to impart anti-wrinkle and antioxidant properties (Ghosal et al., 2021; Nam et al., 2019; Puebla-Barragan & Reid, 2021) (Table 7.1).

For hair care, oil seeds, like flax seeds, are fermented with *Lactobacillus* or *S. cerevisiae*, stimulating hair growth and inhibiting hair fall, whereas fruit extracts, like Trapa japonica, fermented with *Bacillus subtilis,* also stimulate hair growth. Together, these ferments, from a range of substrates and tapped through white biotechnology are a sustainable and very effective means of developing personal care products (Ghosal et al., 2021; Nam et al., 2019; Puebla-Barragan & Reid, 2021) (Table 7.1).

Table 7.1 Skincare and Haircare Activities of Bio-Ferments

Substrate	Plant Variety	Microorganisms	Activities	References
Skincare				
Cereals	Oats	*L. casei*	Reduces oxidative stress and hyperpigmentation	Cho et al. (2024); Lee et al. (2022)
	Barley	*L. plantarum, L. paraplantarum* and *L. brevis*	Anti-wrinkle and antioxidant activities	Kim Su Yeong (2021)
	Rice	*Lactobacillus* sp, *S. cerevisiae, A. oryzae*	Protection of skin barrier	Chen et al. (2024); Miyoshi et al. (2023); Mo et al. (2022)
	Quinoa	*S. cerevisiae*	Antioxidant	Carciochi et al. (2016)
Fruits	*Cornus mas* fruit	*S. cerevisiae* and *Symbiotic bacteria*	Inhibits collagenase and elastase	Zagórska-Dziok et al. (2023)
	Passion fruit peel (*Passiflora edulis*)	*S. cerevisiae*	Protects from UV radiation	Fang et al. (2023)
Flowers	Jasmine	*L. rhamnosus*	Protects from UV radiation	Ziemlewska et al. (2021)
Leaves and leaf extracts	Yebra mate	*Acetobacter, Gluconobacter, Saccharomyces*	Anti-wrinkle, antioxidant, and has anti-inflammatory activities	Ziemlewska et al. (2021); Shi et al. (2020)
	Aloe vera gel	*L. plantarum*	Contains wound healing activity	
Stem	Dendrobium orchids (*Dendrobiumofficinale*)	*L. paracasei* and *L. acidophilus*	Contains wound healing activity and Anti-inflammatory activities	Lin et al. (2024)
Roots	Black Ginseng	*S. cerevisiae*	Contains antiaging and antioxidant properties	Pham et al. (2017); Lee et al. (2012)
	Red Ginseng	*L. brevis*	Contains antiaging and antioxidant properties	
Haircare				
Oil seeds	Flax seeds	*Lactobacillus or Saccharomyces cerevisiae*	Hair growth activity and reduces hair fall	Ghosal et al. (2021)
Fruit extract	Trapa japonica	*B. subtilis*	Hair growth activity	Ghosal et al. (2021); Nam et al. (2019)

Substrates made of oats, barley, rice, wheat, corn, millet, sorghum, rye, and buckwheat are rich sources of bioactive compounds like phenolic acids, flavonoids, dietary fiber, carotenoids, tocols, phytosterols, gamma-oryzanol, phytic acid, and many more. These compounds are responsible for many benefits, like antioxidant properties, radical scavenging, antioxidant properties, and inhibition of pathogenic organisms. Rice extracts, rice bran, and fermented rice extracts can control the growth of pathogens, even those responsible for topical infections (Farahanah et al., 2025).

Advanced biorefinery processes use new pretreatment methods to break down tough plant materials. These materials are then fermented with suitable organisms to produce valuable molecules (Pérez-Rivero & López-Gómez, 2023). Cutting-edge extraction techniques like ultrafiltration, purification, and natural deep eutectic solvents (NaDES) extraction help in maintaining the efficacy and stability of sensitive bioactive molecules (Freitas et al., 2024; Villa et al., 2024). Despite these breakthroughs, there are still gaps in research and innovation that present opportunities for customized skincare and haircare solutions. Additionally, the cosmetic regulations of the European Union, the United States, and other countries have been changing their guidelines by regulating the usage of harsh chemicals and banning severe chemicals (Ferreira et al., 2022). As a result, cosmetic companies are shifting towards more sustainable formulations. This has increased the demand for eco-friendly and natural formulations. Regulatory actions will likely accelerate the integration of these bio-ferment-based products, which are ideally suited for eco-friendly applications and are safe for the skin. This industry shift not only aligns with environmental goals but also enhances product safety and efficacy.

7.4 Green Bio Ferments

Bio ferments are highly concentrated end products obtained from the fermentation of various natural and plant-based raw materials using microbial species such as *Lactobacillus, Bifidobacterium, Saccharomyces, Bacillus, Galactomyces*, lactic acid-producing bacteria, and other probiotic strains (Martins et al., 2023). The fermentation process produces bioactive compounds with significant benefits for skincare and haircare formulations. The bioprocess begins with pretreatment procedures, followed by extracting powders and extracts from plant materials with water. This extract is then fermented by distinct microbial strains with specific metabolic profiles. Advanced white biotechnology procedures guide both upstream and downstream processing phases to induce optimal production profiles of the bioactive compounds. These bioactive components significantly enhance cosmetic product efficacy, providing better skincare and haircare benefits. Developing and implementing probiotics and ferments in skincare has gained considerable momentum compared to previous traditional methods. Such advancements in biotechnology offer substantial benefits for skincare, haircare, and personal care (Bhatia et al., 2024; Majchrzak et al., 2022). Although the research is still in its early stages, the general theory behind the use of fermented substances in skincare is that they will improve the overall balance of skin microbiota, boost absorption, and increase effectiveness, all of which will result in better skin (Klimek-Szczykutowicz et al., 2024; Lyu, 2024).

Type of bio-ferments based on innovative approachs and popular ferments.

i) ***Lactobacillus Ferment***
Lactobacillus is a GRAS (Generally Recognized as Safe) probiotic species capable of fermenting versatile sustainable substrates, making it highly suitable for cosmetic formulations (Jung et al., 2019). Lactobacillus ferment can be developed in the form of lysates, filtrates, active, and inactive forms making them comprehensive bio ferments packed with essential nutrients (Jung et al., 2019). Various Lactobacillus strains, including *L. acidophilus, L. rhamnosus, L. casie, L. paracasei, L. plantarum, L. bulgaricus, L. reuteri, L. delbrueckii, L. fermentum, L. helviticus, L. salivarius,* and others, can be employed to formulate raw materials for skincare final products (Tóth & Németh, 2022). These bacteria can ferment sustainable raw materials, from simple ones like glucose to complex carbohydrates and proteins. Lactobacillus ferments can improve skin moisture levels, exfoliate dead skin, maintain optimal skin pH, strengthen the skin barrier, provide significant anti-aging effects, stimulate cell proliferation, aid in wound healing, and enhance cell turnover in the skin (Lim et al., 2022).

ii) ***Saccharomyces Ferment***
Saccharomyces ferment is a bio-ferment component derived from the yeast *S. cerevisiae*, which is well-known for its ability to boost skin brightness (Gaspar et al., 2008). These organisms can ferment and metabolize various substrates, including simple to complex carbohydrates, proteins, salts, and cellulosic materials. Post-fermentation, these ferments generate distinct metabolites with various skin benefits (Lyu, 2024). Saccharomyces ferment contains enzymes that improve skin tone and brightness, enhance mineral delivery to the skin, promote moisture retention, and aid in the exfoliation of dead skin and skin rejuvenation. This demonstrates the multifunctional advantages of *S. cerevisiae* in skin care applications (Hassan et al., 2020; Jo et al., 2022).

iii) ***Bifida Ferment***
Bifidobacterium ferments, including species such as *Bifidobacterium longum, Bifidobacterium breve, Bifidobacterium adolescentis, Bifidobacterium bifidum*, and *Bifidobacterium lactis*, etc., are widely used in cosmetics due to their skincare benefits. These ferments are typically produced industrially using inorganic techniques and genetically engineered strains. Bifidobacterium ferments provide several benefits, such as strengthening the skin barrier, increasing moisture retention, providing anti-aging effects by reducing wrinkles and fine lines, promoting cell renewal, lowering skin sensitivity and irritation, and preserving a healthy skin microbiome (Otsuka et al., 2022; Wang et al., 2023). Ferments from *Bifidobacterium* are generally added to serums, creams, and lotions to improve skin health, although eco-friendly products are not currently available (Pérez-Rivero & López-Gómez, 2023).

iv) ***Lactococcus Ferments***
Lactococcus ferments such as *Lactococcus lactis, Lactococcus cremoris, Lactococcus garvieae*, and *Lactococcus piscium* are increasingly used in skincare products due to their postbiotic effects (Hausmann et al., 2019). Currently, no technology exists on a commercial scale to provide sustainable fermentation for these ferments. The Lactococcus ferment-based formulations that are available in the market are typically produced using inorganic fermentation techniques or probiotic powders and lysates. Despite the

lack of sustainable production methods, Lactococcus ferments are used in serums, creams, and other formulations to improve skin barrier function and reduce skin sensitivity. Their ability to promote skin regeneration and provide anti-aging properties makes these ferments valuable ingredients in modern skincare products (Otsuka et al., 2022).

Some other organisms of *Bacillus species, Galactomyces, Altromonas, Pseudoaltromonas, Acetobacter, Chlorella vulgaris, Nitrosomonas eutrophia, Enterococcus faecalis, Vitreoscilla filiformis, Streptococcus thermophilus, Thermus thermophillus,* and *Vibrio alginolyticus* can be used in the development of innovative bio-ferments (Klimek-Szczykutowicz et al., 2024; Otsuka et al., 2022). While these organisms offer numerous skincare benefits, they still lack sustainable production methods (Majchrzak et al., 2022). These ferments cannot be considered natural bio-ferments until they follow sustainable procedures and use natural substrates. However, these organisms have the potential to ferment natural substrates and unlock innovative bioactive compounds for the cosmetic market in the future.

In the evolving landscape of skincare, sustainable fermented raw materials are gaining prominence for their natural efficacy and environmental benefits. Leading the charge, **Prameela Life Sciences Private Limited, India**, developed **KINMATI**®, a premium Himalayan cereal-based dual-fermented raw material with Lactobacillus. This cutting-edge method ensures high nutrient content, delivering various skin benefits without involving any chemical processing. **KINMATI** contains effective bioactive compounds such as organic acids, amino acids, gamma-aminobutyric acid (GABA), peptides, antioxidants, vitamins, and minerals.

When fermented with various substrates and microbial species, they deliver many novel molecules that significantly impact skincare, haircare, and beauty products. Further exploration of distinctive substrates, especially those from forest plants with high lignocellulosic content, may create new opportunities for innovative and sustainable skincare ingredients. Lignocellulosic materials, abundant in forest biomass, can be converted into highly potent bioactive compounds through advanced fermentation processes. These ingredients demonstrate a commitment to environmentally friendly, chemical-free bioprocessing. They leverage the inherent fermentation ability of several microbial species to produce effective skincare raw materials. This approach not only promotes skin health but also aligns with the growing trend toward sustainable and ethically sourced cosmetic ingredients.

7.5 Active Compounds from Bioferments

7.5.1 Organic Acids

Lactic acid is the major organic acid found in probiotic bacterial bio-ferment that is formed during natural substrate-based fermentation processes (Abedi & Hashemi, 2020). These organisms can ferment carbohydrates into different organic compounds. Besides lactic acid, citric acid, acetic acid, and succinic acid are other organic acids that are formed during bio-fermentation and can be utilized in cosmetic formulas (Punia Bangar et al., 2022;

Yaqi Wang et al., 2021). While these acids other than lactic acid are formed in lesser amounts, they contribute to better overall efficacy by delivering complementary benefits. Nonetheless, lactic acid is still the most vital ingredient because it is produced abundantly during carbohydrate fermentation. Lactic acid is highly useful in skin care. It is a good gentle exfoliant that removes dead skin cells and brings out brighter skin. It also aids in regulating skin pH, stimulates collagen production, keeps the skin hydrated, inhibits melanogenesis, and minimizes the visibility of fine lines and wrinkles (Feng et al., 2024; Huang et al., 2020; Thueson et al., 1998). Citric acid enhances skin texture and brightness, unclogs pores, assists in increased cell turnover, lightens dark spots, and balances pigmentation. Due to its astringent action, it assists in balancing oil and sebum production (Ditre et al., 1996; Yamamoto et al., 2006). Acetic acid is known for its capacity to balance pH and deliver mild exfoliation. It also acts as an astringent, tightening and firming the skin, inhibiting oil production, and reducing irritation, making it great for acne-prone skin (Elhage et al., 2022; Kapuścińska & Nowak, 2015; Măgeruşan et al., 2023). Succinic acid provides antimicrobial, anti-inflammatory, and moisturizing advantages. It is specifically beneficial for acne, soothing redness on the skin, and providing antioxidant advantages. It also enhances skin elasticity and firmness (Bonnet et al., 2022; Green et al., 2009; Yanhan Wang et al., 2014; Saxena & Yadav, 2020).

The utilization of organic acids from bio-fermentation in cosmetics is in line with the clean beauty trend that focuses on products that have biologically active ingredients that are not only effective but also safe for the environment. Through white biotechnology the cosmetic industry can generate innovative products that respond to customers' growing demands for eco-friendly and high-performance beauty products.

7.5.2 Amino Acids

Microorganisms, including *Lactobacillus, Bacillus,* and *Saccharomyces* species, can ferment natural substrates like rice, soybeans, and oats, causing substantial metabolic changes without the need for chemical extraction procedures or genetic modifications. These microbes metabolize the substrates, resulting in the creation and synthesis of organic forms of amino acids. The synthesis of amino acids involves several biochemical reactions. These biochemical transformations are significantly dependent on major metabolic pathways such as arginine deiminase, glutamate decarboxylase, branched chain amino acid pathway, aromatic amino acid metabolism, and the D-amino acid synthesis process (Sanchez et al., 2018). Natural fermentation processes create essential and non-essential amino acids such as alanine, arginine, citrulline, GABA, tryptophan, valine, leucine, isoleucine, proline, glycine, serine, threonine, methionine, aspartic acid, glutamic acid, phenylalanine, and ornithine, which are produced as post-biotics. D-amino acids like D-alanine, D-glutamate, and D-aspartate are also generated, of which some exist in the components of bacterial cell walls (Kobayashi, 2019). Free amino acids enhance the efficiency and efficacy of fermented haircare and skincare products and hence are a sustainable solution to conditioning and renewing skin and hair. Such an approach also conforms to green cosmetics principles since it employs the inherent metabolic potential of non-GMO microbes, thereby stimulating the generation of novel and potent formulations.

7.5.2.1 The Function of Amino Acids in Skin and Hair Care

Amino acids are important in ensuring the efficacy of fermented skincare and haircare formulations Table 7.2.

Skincare Benefits

- ***Collagen Synthesis***: Amino acids such as glutamine, hydroxyproline, lysine, glycine, proline, and branched-chain amino acids (isoleucine, leucine, and valine) combined with arginine and glutamine promote collagen formation, enhancing skin firmness and elasticity (Albaugh et al., 2017; Karna et al., 2001; Murakami et al., 2012; Noguchi & Djerassi, 2009).
- ***Expression of Filaggrin***: Histidine, which is transformed into urocanic acid, is essential for the expression of filaggrin, a protein crucial for skin barrier function (Hsu et al., 2011; Junichi Koyama et al., 1983).
- ***Glycosaminoglycan Synthesis***: Glutamine boosts the synthesis of glycosaminoglycans, which are vital for maintaining skin hydration and elasticity (Noguchi & Djerassi, 2009).
- ***Hydration and Moisturizing***: Amino acids like glycine, alanine, arginine, glutamic acid, aspartic acid, isoleucine, leucine, valine, threonine, and serine provide excellent hydration and moisturizing properties (Ahsan, 2019; Miyamoto et al., 2024; Wu, 2020).
- ***Skin Protectors***: Aspartic acid and glutamic acid, along with their derivatives, protect the skin from environmental stressors (Morichi et al., 1963; Wu, 2020).
- ***Antiperspirants***: Glycine is effective in reducing perspiration (Henkel AG and Co KGaA, 2017).
- ***Wound Healing and Skin Ulcer Treatment***: Glutamine, arginine, glycine, lysine, proline, leucine, and the combination of lysine and arginine aid in wound healing and skin ulcer treatment (Arribas-López et al., 2021; Wu, 2020).
- ***Anti-inflammatory***: Glutamine exhibits anti-inflammatory properties, soothing irritated skin (Jin et al., 2012).
- ***Antioxidant***: Methionine, cysteine, histidine, arginine, citrulline, and glutathione act as powerful antioxidants, protecting the skin from oxidative damage (Bin et al., 2017; Noguchi & Djerassi, 2009; Wu, 2020).
- ***Tyrosinase Inhibition***: Cysteine and glutathione suppress melanogenesis, while phenylalanine and tyrosine contribute to both eumelanin and pheomelanin production, other factors influencing skin pigmentation (Jara et al., 1988).
- ***Reduction of Fine Lines and Wrinkles***: Hydroxyproline, hyaluronic acid, proline, leucine, and branched-chain amino acids (leucine, glycine, and proline) help reduce the appearance of fine lines and wrinkles (Ohara et al., 2010).
- ***Prevention of Acne***: Lysine demonstrates effectiveness in preventing acne (Leignadier et al., 2023).
- ***Precursor for Melatonin***: Tryptophan is a precursor for melatonin, a hormone that protects the skin against oxidative stress (Slominski et al., 2018).

Haircare Benefits

- ***Hair Repairing and Anti-Hair Fall***: Lysine and isoleucine amino acids help in hair growth and strengthening by being building blocks of protein. Arginine is vital for healthy hair growth. It helps in improving blood circulation to the scalp, thereby promoting hair growth and preventing hair fall (Oshimura et al., 2007).

Table 7.2 Properties and Functions of Amino Acids in Skincare and Haircare Formulations

Properties in Skincare	**Amino Acid**	**References**
Collagen synthesis	Glutamine + Hydroxyproline	Karna et al. (2001); Noguchi and Djerassi (2009)
	Lysine Glycine Proline	Albaugh et al. (2017)
	Branched-chain amino acids: (Isoleucine-Leucine-Valine) + Arg, Gln, or Pro	Murakami et al. (2012)
Expression of filaggrin	Histidine to Urocanoic acid	Hsu et al. (2011); Junichi Koyama et al. (1983)
Glycosaminoglycan synthesis	Glutamine	Noguchi and Djerassi (2009)
Hydration and moisturizing	Glycine, alanine, arginine, glutamic acid, aspartic acid, isoleucine, leucine, valine, threonine, and serine.	Ahsan (2019); Miyamoto et al. (2024); Wu (2020)
Skin protectors	Aspartic acid, Glutamic acid, and their derivatives.	Morichi et al. (1963); Noguchi and Djerassi (2009); Wu (2020)
Antiperspirants	Glycine	Henkel AG and Co KGaA (2017)
Wound healing and skin ulcer treatment.	Glutamine Arginine Glycine Lysine Proline Leucine (Lysine+Arginine)	Arribas-López et al. (2021); Wu (2020)
Anti-inflammatory	Glutamine	Jin et al. (2012)
Antioxidant	Methionine Cysteine Histidine Arginine Citrulline Glutathione	Bin et al. (2017); Noguchi and Djerassi (2009); Wu (2020)
Tyrosinase inhibition (suppression of melanogenesis)	Cysteine + Glutathione	Jara et al. (1988)
Precursor for melanin	*Eumelanin:* (Phenylalanine or Tyrosine) *Pheomelanin:* (Phenylalanine or Tyrosine + Cysteine)	Wu (2020)

(*Continued*)

Table 7.2 (Continued)

Properties in Skincare	Amino Acid	References
Reduction of fine lines and wrinkles	Hydroxyproline and Hyaluronic acid (Proline + Leucine) *Branched-chain amino acids:* (Leu-Gly-Pro)	Wu (2020); Ohara et al. (2010)
Prevention of acne	Lysine	Leignadier et al. (2023)
Precursor for melatonin (hormone involved in skin protection against oxidative stress)	Tryptophan	Slominski et al. (2018)
Amino acids in hair care		
Hair repairing shampoo and conditioner	Lysine and Isoleucine	Kathed (2018); Oshimura et al. (2007); Wu et al. (2011)
Anti-hair fall compositions	Arginine	
Hair spray with restoration properties	Glutamic acid, arginine, serine, cystine, hydroxyproline	
Hair relaxing components	Cysteine and its derivatives	
Hair coloring creams	Arginine	

- ***Hair Restoration Properties***: Glutamic acid, arginine, serine, cystine, and hydroxyproline are mainly responsible for maintaining hair texture, enhancing the blood flow to the scalp, and providing natural moisturization to the hair, mainly hydroxyproline strengthens hair by improving collagen production (Kathed, 2018; Wu et al., 2011).
- ***Hair Coloring***: Arginine will minimize hair damage and improve hair elasticity and shine (Oshimura et al., 2007).

By utilizing the natural metabolic capabilities of non-GMO microorganisms, these fermented amino acids align with the principles of green cosmetics, promoting the development of innovative, effective, and environmentally friendly skincare and haircare products.

7.5.3 GABA

Gamma-aminobutyric acid is a non-proteinaceous amino acid. Lactic acid bacteria (LAB), *Escherichia coli,* and *Corynebacterium glutamicum* are natural microorganisms capable of producing GABA through the action of their specific glutamic acid decarboxylase (GAD) enzyme. These microorganisms are employed in industrial processes to produce GABA on a commercial scale (Kook & Cho, 2013). It can also be found in a variety of fruits and vegetables, including soybean, but its natural concentration is quite low, typically ranging from 0.03 to 2.00 μmol/g-wet weight, which is known to be difficult to have biological activity (Wei et al., 2022).

In plant sources, it is developed by the decarboxylase reaction of glutamic acid. Therefore, the amount of GABA is proportional to the amount of glutamic acid present, and adding glutamic acid is necessary to achieve higher concentrations of GABA. Researchers are focusing on producing higher concentrations of GABA using microorganisms by fermenting sustainable substrates containing L-glutamic acid, monosodium glutamate, glucose, starch, glycerol, xylose, and lignocellulose materials.

Lactic acid-producing organisms include *Lactobacillus brevis, L. buchneri, L. delbrueckii subs. Bulgaricus, L. plantarum, L. helveticus, Streptococcus thermophilus, Lactococcus lactics, Bifidobacterium, Enterococcus, Leuconostoc, Pediococcus, Propionibacterium,* and *Weisella*, which can produce GABA (Kook & Cho, 2013; Milon et al., 2024). In efforts to maximize the production of GABA, some LAB and other microorganisms have been considered. *L. brevis* yields the highest productivity in dual-stage fermentation, 1.3 g/L/h, which is far beyond other approaches like co-culture fermentation (0.018 g/L/h) and substrate enhancement (0.38 g/L/h). Similarly, *L. futsaii* also provides satisfactory results using fed-batch fermentation under 0.31 g/L/h. On the other hand, other LAB strains such as *L. plantarum, L. rhamnosus,* and *L. paracasei* provide relatively lower GABA productivity with varied fermentation processes between 0.003 and 0.033 g/L/h.

Among the non-LAB microorganisms, *C. glutamicum* is noteworthy with a productivity of 1.2 g/L/h employing fed-batch fermentation and 0.42 g/L/h in lignocellulose fermentation. *B. subtilis* and *Kluyveromyces marxianus* have moderate productivity with co-culture fermentation (0.16 g/L/h) and substrate augmentation (0.07 g/L/h), respectively. The above observations bring into perspective the enhanced efficiency of dual-stage and fed-batch fermentations to produce GABA with greater efficiency, of which *L. brevis* and *C. glutamicum* were found to be the most effective under these conditions. The thorough study presented herein helps select proper microbial strains and appropriate methods of fermentation for obtaining maximal yields of GABA in biotechnological applications (Milon et al., 2024).

GABA is gaining popularity in skincare for its ability to relax facial muscles, reduce fine lines, and improve skin hydration. The topical application of GABA is known for its calming properties, helping soothe irritation and enhancing the skin's overall appearance. As people age, their levels of GABA decline, which results in wrinkles, a graying of the skin, and a loss of elasticity. An external application of GABA can relax tense muscles, significantly reducing expression lines and crow's feet for a youthful and refreshed appearance. GABA also enhances skin elasticity by increasing the expression of Type-1 collagen and decreasing matrix metalloproteinase (MMP-1) levels, thereby promoting the production of tropoelastin, fibrillin-1, fibrillin-2, and fibulin-5 (Ferrillo et al., 2019).

Prameela Life Sciences developed Kinmati® by following starch-based dual fermentation. The compact bio-fermented liquid is rich in gamma-aminobutyric acid. Kinmati is a natural probiotic cosmetic raw material that can be easily infused into various skincare formulations. The final formulation will deliver the effective form of GABA as a potential.

7.5.3.1 Efficacy of Lactobacillus-Fermented GABA on Dermal Fibroblasts

Everspring Co., Ltd. and Kyung Hee University of Youngin have discovered that commercially available GABA and *Lactobacillus sakei* ferment extract have shown potential benefits for human dermal fibroblasts. In a recent efficacy study, researchers discovered that *Lactobacillus*-fermented GABA exhibited significant benefits on dermal fibroblasts. When treated with a concentration of 150 μg/mL for 72 hours, there was an increase in skin cell proliferation by 35%. The expression of MMP-1 was also decreased by nearly 75% under the

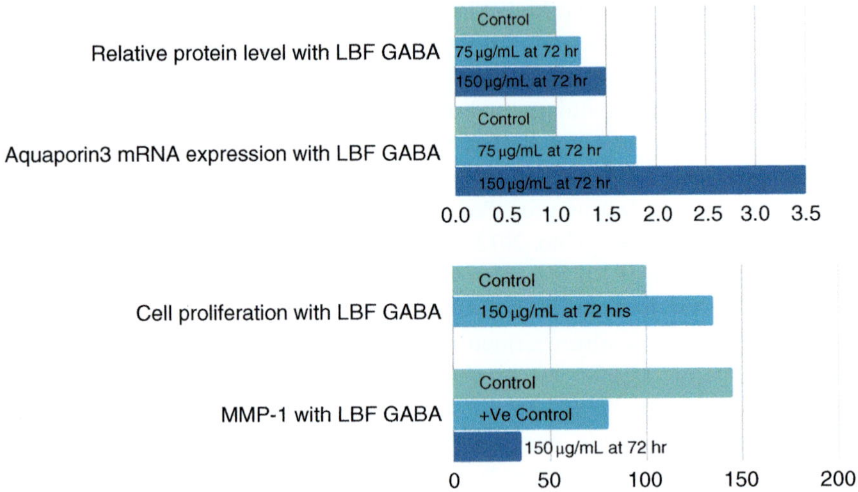

Figure 7.3 Relative protein level, aquaporin3 mRNA expression, cell proliferation, and (metalloproteinases-1) MMP-1 expression with lactobacillus-fermented GABA on human fibroblast cell lines.

same conditions (Figure 7.3). In addition, relative protein content in human fibroblasts increased by 25% when administered 75 μg/mL and by 50% with 150 μg/mL, both for 72 hours. Also, aquaporin3 mRNA expression elevated almost twofold at 75 μg/mL and threefold at 150 μg/mL following 72 hours of treatment. These findings highlight the potential of Lactobacillus-fermented GABA in enhancing skin health and promoting cellular activities (Table 7.3) (Zhao et al., 2023).

7.5.4 Peptides

Peptides are short chains comprising 2–50 amino acids bonded together by peptide bonds. The presence of peptides is found in a wide spectrum of natural sources: plants, animals, and microorganisms. Biofermented peptides are gaining more attention in the skincare, cosmetics, and beauty industries because of their natural origin and multifunctional benefits. Some of the most notable examples include soy-fermented oligo peptides such as (Val-Ala-His-Ile-Asn-Val-Gly-Lys), (Tyr-Val-Trp-Lys), (Ser-Tyr), and (Gly-Tyr) that are obtained by fermenting soy with Bacillus and Lactobacillus. These peptides have strong antioxidant and anti-inflammatory properties, making them useful for protecting the skin from oxidative stress and inflammation (Chatterjee et al., 2018). Cereal fermentation-derived peptides consist of proteins including glutelin, glutelin precursors, globulins, prolamin, albumin, glycine-histidine-lysine, and glutathione (Glu-Cys-Gly). They are generated from the fermentation of cereals like rice and oats with Bacillus and Lactobacillus (Ferri et al., 2017; Park et al., 2020). These peptides stimulate the proliferation of keratinocytes, hence improving the integrity of the skin barrier through an increase in the expression of integrin. It also inhibits melanogenesis; it provides antioxidant effects, contributing to an even skin tone as well as protection against oxidative damage. Jiupei (fermented grains)

Table 7.3 Probiotic Fermented Peptides, Their Sequences, Preparation Methods, and Skincare Activities

Name of the Peptide	Peptide Sequence	Preparation	Skincare Activity
Green fermented peptides			
Soy-fermented oligo peptides	(Val-Ala-His-Ile-Asn-Val-Gly-Lys) (Tyr-Val-Trp-Lys) (Ser-Tyr) (Gly-Tyr)	Bacillus/ Lactobacillus fermentations	Antioxidant and anti-inflammatory
Cereals fermentation-derived peptides	Glutelin Glutelin precursors Globulins Prolamin Albumin Glycine-Histidine-Lysine Glutathione (Glu-Cys-Gly)	Bacillus/ Lactobacillus fermentations	The proliferation of keratinocytes with increased expression of integrin. Suppression of melanogenesis and Antioxidant
Jiupei (fermented grains)	Val-Asn-Pro Tyr-Gly-Asp	Fermentation	Antioxidant activity
Goat milker	Ser-Ala-Glu-Glu-Gln-Leu-His-Ser-Met Lys Ile-Ala-Lys-Tyr-Ile-Pro-Ile-Gln-Tyr-Val Leu-Ser-Arg Glu-Ala-Leu-Glu-Lys-Phe-Asp-Lys	Fermentation	Antioxidant activity

Val-Asn-Pro and Tyr-Gly-Asp, which are produced during grain fermentation, are some of the peptides produced with such high antioxidant activity. These assist in neutralizing free radicals as well as protecting the skin from environmental damage (Jiang et al., 2019). Goat milk fermented peptides produced through goat milk fermentation include Ser-Ala-Glu-Glu-Gln-Leu-His-Ser-Met-Lys and Ile-Ala-Lys-Tyr-Ile-Pro-Ile-Gln-Tyr-Val-Leu-Ser-Arg and Glu-Ala-Leu-Glu-Lys-Phe-Asp-Lys. These have robust antioxidant activity for rejuvenation and protection of the skin (Ngoc et al., 2023; Panchal et al., 2020) (Table 7.3).

7.5.4.1 Types of Peptides

There are four different types of peptides

1) ***Signal Peptides***: Stimulate collagen and elastin production, enhancing skin firmness and elasticity.

2) ***Carrier Peptides***: Deliver trace elements like copper and magnesium essential for wound healing and enzymatic processes.
3) ***Neurotransmitter Inhibitory Peptides***: Relax facial muscles, reducing the appearance of fine lines and wrinkles.
4) ***Enzyme Inhibitory Peptides (Bio-fermented Peptides)***: Prevent collagen breakdown by inhibiting enzymes, maintaining skin integrity and elasticity (Husein el Hadmed & Castillo, 2016).

Signal peptides, carrier peptides, and neurotransmitter inhibitory peptides are synthesized by solid-phase peptide synthesis. This involves chemical-based synthesis and inorganic extractions. Enzyme inhibitory peptides are the biofermented peptides that are synthesized by a probiotic-based natural fermentation process. However, the most environment-friendly method for synthesizing peptides is through fermentation (Gomes et al., 2020).

In skincare, there are so many benefits when using peptides:

i) ***Promoting Collagen Formation***: Bio-fermented peptides from Lactobacillus enhance collagen production, improving skin firmness and elasticity; hence, the skin becomes firmer (An et al., 2023).
ii) ***Wound Healing:*** Peptides from Bacillus promote cell migration and proliferation, expediting wound healing (Lima & Pedriali Moraes, 2018).
iii) ***Anti-aging***: Increase collagen and elastin production, reducing fine lines and wrinkles.
iv) ***Moisturizing:*** The peptides enhance skin hydration and barrier function.
v) ***Antioxidant Effects***: Protect skin from oxidative stress, reducing damage caused by free radicals (Aguilar-Toalá et al., 2019; An et al., 2023; Apone et al., 2019; Lima & Pedriali Moraes, 2018; Ngoc et al., 2023).

Bio-fermented peptides can be used in a variety of topical skincare formulas, such as serums, creams, and masks. The addition of fermented peptides in skincare products makes the skin softer, more hydrated, and anti-aging while giving protection against environmental factors. This technology is in complete harmony with green cosmetics, which fosters the design of effective skincare products that cause no harm to the environment. To leverage the natural metabolic capacity of non-GMO microorganisms, fermented peptides hold onto their sources and also retain potent skincare benefits. Such a method, therefore, fulfills the demands of consumers to increasingly produce sustainable and environmentally friendly cosmetic products that have both high safety and efficacy (Gomes et al., 2020).

In summary, peptides from various natural sources undergoing bio-fermentation carry multiple benefits ranging from antioxidation, anti-inflammatory effects, enhanced skin barrier function and even skin tone, such that they can be considered precious added substances in the formulation of skincare and cosmetic products (Ngoc et al., 2023).

7.5.5 Antioxidant Substances

Glutathione is one of the key antioxidants found in lactic acid fermentation as well as bio-fermentation done in plants. It is an antioxidant thiol and a low-molecular-weight tripeptide molecule. Glutathione is prepared from a mixture of glutamate, cysteine, and glycine. When bacteria are carrying out fermentation to create lactic acid, they use a complex substrate to generate glutathione. It is an antioxidant that may reduce inflammation caused by the oxidizing effect of reactive oxygen species (ROS) and decrease oxidative stress. It helps in various skin benefits, including antiaging by reducing wrinkles and fine lines, works as an antioxidant, and contains skin-whitening properties (Cui et al., 2024; Etnawati et al., 2019; Watanabe et al., 2014; Weschawalit et al., 2017).

7.5.6 Short-Chain Fatty Acids

Lactic acid-producing microorganisms, such as *Lactobacilli* and *Bifidobacteria*, produce SCFAs like acetate, butyrate, and propionate when fermenting indigestible forms of carbohydrates. Probiotic organisms ferment dietary fibers, indigestible starch materials, lignocellulose materials, etc. to create SCFAs. SCFAs are subsequently used in several medicinal applications. These SCFAs are mainly essential to maintain cutaneous homeostasis. Introducing SCFAs in cosmetics is a novel and cutting-edge strategy with encouraging potential. Prameela Life Sciences has advanced this concept with its specialized ingredient, Kinmati.

Kinmati, developed by Prameela Life Sciences, is made using starch-based fermentation. This process results in a compact bioactive solution containing SCFAs that can be easily infused into various skincare formulations.

Effectiveness of SCFAs in Skincare

- ***Hydration and Skin Health***: Bio-fermented SCFAs provide better hydration for the skin than standard fatty acids. They increase skin moisture levels, which benefits general skin health (H. Park et al., 2021).
- ***Inflammation Management***: SCFAs can successfully regulate skin inflammation by inducing T-cell regulators. This results in a balanced inflammatory response in the skin (Schwarz et al., 2017).
- ***Microbiome Balance***: SCFAs promote the growth of beneficial microbes on the skin by acting as prebiotics for the skin's beneficial microbes, resulting in a balanced and healthy skin microbiome. This balance is important for preserving skin health and minimizing inflammatory reactions (Park et al., 2021).
- ***Regulation of Inflammatory Markers***: SCFAs regulate inflammatory indicators to minimize skin inflammation. This regulation is especially advantageous for diseases like psoriasis and allergic contact dermatitis (Egawa et al., 2017).
- ***Context-Dependent Effects***: The effects of SCFAs are dependent on various factors, including their type, concentration, and delivery method. Different SCFAs have unique qualities and advantage.

Specific Benefits

- Acetate increases the diversity of skin microbes and general skin health.
- Butyrate has anti-inflammatory actions and may reduce inflammation linked to psoriasis.
- Propionate will enhance the hydration and barrier function of the skin.

The use of SCFAs in skincare formulations is a promising development in dermatology, and Prameela Life Sciences' Kinmati is at the forefront of this innovation. Kinmati combines the benefits of SCFAs with an easy-to-formulate bioactive solution, and it shows a direct impact on skin health when used in skincare applications. SCFAs promote moisture, manage inflammation, and balance the skin microbiome, making them an effective remedy for a variety of skin issues. More research and development into the best delivery techniques and formulations of SCFAs will unlock their full potential in skincare applications.

7.6 Application of Bioferments in Skincare

All the bioactive compounds produced during the development of natural ferments are very effective based on the recent *in vitro* and *in vivo* efficacy studies.

7.6.1 Reducing Wrinkles and Signs of Aging

The natural aging process leads to visible signs like wrinkles and fine lines eventually, leading to signs of early aging. Probiotic bio-ferments can be applied to the skin by using them as cosmetic formulation ingredients to reduce the signs of aging and wrinkles. It can be blended with any organic topical base formulation. The bioactive compounds boost the production of procollagen type-1 and reduce enzymes called matrix metalloproteinases, mainly MMP-1, MMP-2, and MMP-9. These enzymes are mainly involved in the degradation of extracellular matrix proteins collagen, fibronectin, and elastin, which will eventually lead to the development of wrinkles, fine lines, and crow's feet lines (Majchrzak et al., 2022).

The topical application of fermented bioactive compounds like amino acids and the non-proteinaceous amino acid called GABA with a natural base formulation will improve the structural integrity and elasticity of the skin through increased type-1 collagen synthesis. GABA also stimulates the assembly of structural proteins like tropoelastin, fibrillin-1, fibrillin-2, and fibulin-5, which provide structural integrity to the skin and elasticity and inhibit the activity of collagenase and elastase. The fermented GABA also relaxes tensed muscles, eliminating deep lines of expression, crow's feet, and wrinkles to make the skin look younger (Zhao et al., 2023).

Bioferments also stimulate the production of hyaluronic acid for the hydration of the skin and enhance proteins like aquaporin 3 and filaggrin, which are important for maintaining the hydration and structural integrity of the skin (Klimek-Szczykutowicz et al., 2024). Applying all these fermented bioactive compounds, in turn, will improve human skin health by reducing the visible effects of aging signs with improved elasticity. This all-natural fermentation can be one solution for sustainable products in skincare against aging for future applications (Dou et al., 2023; Klimek-Szczykutowicz et al., 2024).

7.6.2 Strengthening Skin Barrier

Natural probiotic bio-ferments developed especially using cereals like barley, oats, and rice can significantly enhance the skin barrier (Miyoshi et al., 2023). This is achieved by topical application of probiotic ferments, which increases the levels of tight junction proteins such as claudin-1, claudin-3, claudin-4, zonula occludens-1, and occludin in the sub-dermal layers (Yang et al., 2022). These are some of the major proteins that prevent the breakdown of the skin. Bioactive compounds of *Lactobacillus, Acetobacter, Leuconostoc, Gluconobacter, Saccharomyces,* and *Bifidobacterium* have the potential to increase skin moisture, cell growth, and junctions between the different layers of the skin (Jung et al., 2019; Sultana et al., 2013).

Studies indicate that bioactive compounds of probiotic ferments enhance collagen production and minimize sagging of the skin. Alpha hydroxy acids help maintain the skin's optimal pH level and hydration levels, which are vital for a healthy microbiome (Salgaonkar et al., 2022). Lactic acid will exfoliate the dead skin cells at regular periods, leaving healthy epidermal layers. SCFAs and amino acids from the fermented bioactives nourish the skin microbiome. The healthy microbiome protects the skin from pollution and pathogens by working as a protective shield on the skin (Klimek-Szczykutowicz et al., 2024; Otsuka et al., 2022). Amino acids and peptides provide nourishment for both outer and inner skin layers of the skin by thickening the layers of the dermis and epidermis, making the epidermal barrier resistant to aggressors. The healthy skin barrier will minimize transdermal water loss, which will help in maintaining optimal skin hydration levels. Consequently, the cereal ferments of various probiotic microorganisms improve hydration levels of the skin, collagen concentrations, and structural strength, thus providing a natural and healthy skin barrier (da Silva Vale et al., 2023; Klimek-Szczykutowicz et al., 2024).

7.6.3 Reducing Inflammation

LAB such as *Lactiplantibacillus plantarum, Limosilactobacillus reuteri, Lacticaseibacillus rhamnosus, Lactobacillus helveticus,* and *L. sakei,* alongside other beneficial strains like *B. subtilis, Saccharomyces*, and *Bifidobacterium* produce various anti-inflammatory compounds (LeBlanc et al., 2008). Lactic acid, SCFAs, and exopolysaccharides are anti-inflammatory compounds derived from probiotic fermentation processes. These compounds effectively block several pro-inflammatory compounds like cytokines, chemokines, and oxidative stress markers (Brandi et al., 2020). An increase in these pro-inflammatory compounds shows typical impacts on skin health that lead to the cause of skin inflammatory conditions like contact dermatitis, eczema (atopic dermatitis), psoriasis, acne, rosacea, and some infections (Narla & Silverberg, 2020).

Probiotic ferments of *L. plantarum, L. reuteri, L. casei,* and many other strains have effectively reduced the levels of nitric oxide (NO), thereby reducing inflammatory responses (Choi et al., 2017). Bio-ferments contain several molecules that contribute to their anti-inflammatory effects. These include SCFAs that elicit anti-inflammatory activity by suppressing pro-inflammatory responses, consequently reducing the secretion of pro-inflammatory cytokines (Xiao et al., 2023). Lactic acid reduces inflammation. Anti-inflammatory bioactive peptides resulting from fermentation may result in the inhibition of various enzymes implicated in the inflammatory cascade. Glutathione is produced in the fermentation process, which acts as an antioxidant that scavenges ROS and decreases oxidative stress, thereby lowering inflammation. Amino acids like tryptophan and arginine contain anti-inflammatory properties. This research further supports the possibility of harnessing fermentable substrates and probiotic ferments to develop advanced skincare treatments that can diminish inflammation and promote healthy skin (Cui et al., 2024; Klimek-Szczykutowicz et al., 2024).

7.6.4 Helping Wound Healing

Recent advances in skincare have highlighted the potential of probiotic bio-ferments on wound healing and skin protection. According to the research, topical application of GABA synthesized during *lactobacillus* fermentation helps speed up wound healing. Along with that, it reduces the overgrowth of the epidermis by activating GABA(A) receptors in the skin. This activation promotes fibroblast migration, cell proliferation, inhibition of matrix metalloproteinase-1 (MMP-1), and increases skin moisture by improving filaggrin and aquaporin 3 levels in skin cells. Animal studies also indicated that GABA therapies speed up wound healing as they lower inflammation. The stimulation of epidermal growth, thereby making it a viable component for improving skin health and hydration (Zhao et al., 2023).

Amino acids also play an important role in wound healing, especially in collagen synthesis and skin repair. Key amino acids such as arginine, lysine, and leucine are vital for cell signaling and cell proliferation during healing. Studies have been able to prove that dressings infused with the mentioned amino acids can accelerate the closure of a wound in aged rats. These amino acids have modulated the inflammatory responses, increased the distribution of fibroblasts, and enhanced collagen production (Arribas-López et al., 2021; Noguchi & Djerassi, 2009; Wu, 2020).

Ongoing research on wound healing has demonstrated the benefits of topical application of probiotics at various stages of healing. Topical application of probiotic bacteria such as *L. plantarum* and *L. salivarius* has demonstrated effective responses to speed up wound closure in skin models (Brandi et al., 2020). Based on the recent reports published on In vitro studies on wound healing by topical application of probiotics species like *Lactobacillus* VITSAMJ1 and *L. plantarum* etc. and ferments of probiotic species and subspecies including *L. acidophilus, L. casei, L. plantarum, L. salivarius, L. reuteri, Bacillus coagulans, B. bifidum*, etc. have been applied on skin models. To treat wound infections and protect from pathogenic adherence during the wound healing process, probiotic species of *L. rhamnosus, L. acidophilus, L. casei, L. reuteri, L. plantarum, B. bifidum,* etc. and probiotic ferments of *L. rhamnosus, L. reuteri, L. salivarius, L. delbrueckii, L. brevis, L. acidophilus, L. plantarum, L. casei, B. animalis, B. lactis, B. longum,* etc. have been applied on skin models, tested on skin pathogens, etc. (Bădăluță et al., 2024).

LAB have shown considerable wound healing, anti-inflammatory, and antibacterial effects on damaged skin cells. These probiotics and their bioactive compounds influence the wound environment through immunological regulation, antimicrobial activity, and the release of growth factors (Brandi et al., 2020; Klimek-Szczykutowicz et al., 2024; Panagiotou et al., 2023).

7.6.5 Fighting Acne

Recent studies on natural ferments in fighting acne have indicated that probiotics and their bioactive compounds, like organic acids and peptides, have great potential to treat skin conditions like acne and pathogenic infections of the skin. Different types of probiotic bacteria and yeast, such as *Lactobacillus, Bifidobacterium, Staphylococcus, Lactococcus, and Saccharomyces,* can significantly reduce inflammation in the skin and eliminate pathogenic microorganisms.

Some of the lactic acid-producing bacteria possess antimicrobial activities against common cutaneous pathogens *Staphylococcus aureus, Pseudomonas aeruginosa, Micrococcus luteus, Cutibacterium acnes,* and *Propionibacterium acnes* (Mottin & Suyenaga, 2018).

The bioactives of probiotics, including lactic acid, bacteriocins, and other peptides, play the largest role in the inhibition of skin pathogens. They also help with the modulation of immune responses in keratinocytes and epithelial cells. For example, *Staphylococcus epidermidis* extracts from glycerol fermentation act as a natural defense against acne. Meanwhile, *Lactococcus* and *Enterococcus* also help in reducing acne. Bio-ferments derived from species of *Lactobacillus* and *Bifidobacterium* are effective in treating acne and other skin infections (Salgaonkar et al., 2022; Goodarzi et al., 2020; Klimek-Szczykutowicz et al., 2024).

7.7 KINMATI: The Advanced Probiotic Biofermented Raw Material for Skincare

KINMATI is a probiotic-based bio-ferment developed by Prameela Life Sciences Private Limited. It is obtained through the fermentation of a premium Himalayan cereal variety by *Lactobacillus*. This innovative bio-ferment is created using dual fermentation technology. The post-fermented liquid contains bioactive substances such as organic acids, antioxidants, amino acids, GABA, peptides, SCFAs, and fermented *Lactobacillus*. These components collectively contribute to KINMATI's potential in skincare.

Organic acids and antioxidants in KINMATI moisturize the skin and protect against oxidative stress. The available amino acids and peptides are essential for collagen synthesis and skin repair, increasing skin elasticity and firmness. GABA in KINMATI helps support the synthesis of tight junction proteins, supports the skin's microbiota, and reduces wrinkles and other signs of aging. SCFAs and Lactobacillus fermented extract will nourish and improve the skin microbiota and strengthen the skin barrier.

Prameela Life Sciences®, a private limited company in India, has developed an advanced probiotic-based natural skincare formulation called AKSHATA using KINMATI as the primary raw material. KINMATI was featured in the 2024 edition of "The Product Innovation Guide," published by "Cosmetics Business" Magazine, UK ("Product Innovation 2024," 2024). KINMATI as a raw material and AKSHATA as the final skincare formulation were launched into the market at the prestigious event, Global Bio India 2024 in New Delhi, India, organized by the Department of Biotechnology (DBT) and the Biotechnology Industry Research Assistance Council (BIRAC), Government of India (*KINMATI and AKSHATA Product launch at Global Bio India* 2024, *Delhi*, 2024).

7.8 Future of Bio-ferments, Active Ingredients, and Green Formulations

The future of sustainable bio-ferments in skincare will mainly depend on advanced white biotechnology, specifically the fermentation of natural plant-based substrates with wild, non-GMO, and safe microorganisms. Microbes produce a variety of useful active compounds/metabolites (primary and secondary metabolites) by fermenting diverse materials such as flower extracts, barks, leaves, lignocellulosic substances, and various plant extracts. The post-fermented compounds include unique bioactive molecules. Future innovations are projected to focus on the use of novel plant substrates sourced from rare and premium sources, expanding the range of effective, environmentally friendly, and green cosmetic formulations. Optimizations in process development and cost management can be achieved by implementing advanced statistical optimization techniques during the streamlining of the technology. Using cost-effective substrates, and natural fermentations following optimal parameters designed by statistical approach will help in the commercialization of the end product in high quality at low cost. Statistical optimization techniques like Taguchi methodology can be applied at various stages of the process development (Rao et al., 2008).

Compared to typical chemical-based cosmetics, plant-based extracts offer great skincare benefits without the use of synthetic chemicals. Natural ingredients, including *aloe vera*, jojoba oil, single-pressed carrier oils, floral hydrosols, plant-based vitamins, natural sugar polyols, polyol substitutes, natural emulsifiers, and butter, are very beneficial to the skin. Advancements in biotechnology and green chemistry will explore the discovery of innovative compounds, effective ingredients, and sustainable raw materials for the cosmetic industry. As consumers become increasingly aware of their choice of ingredients in their product formulations and focus on environmental and ethical impacts, the demand for sustainable skincare products is already rising. Innovations in packaging, such as biodegradable or refillable options, will further minimize environmental plastic and carbon footprint.

7.8.1 Increasing Demand for Eco-Friendly Ingredients

With growing awareness of environmental and social impacts, the cosmetics industry is experiencing a high demand for natural and sustainable ingredients. To satisfy customer demands and legal requirements, businesses are using eco-friendly products and cutting-edge preservation methods.

This demand is expected to propel the natural products market from $642 million to $1 billion by 2030, at an annual rate of 6.9% (*Natural Cosmetics Ingredients Market by Ingredient Type (Emollients, Bio Surfactants, Sugar Polymers, Natural Preservatives, Others), Application (Skin Care, Hair Care, Oral Care, Make-up) and Region—Global Forecast to 2030*, n.d.). Over 40% of consumers now prefer natural ingredients in their cosmetic products, and patents and publications on natural cosmetics have increased from 10 to 40 over the past 10 years.

7.8.2 Shift to Natural Emollients, Solvents, Surfactants, Thickeners, Exfoliators, Fragrances, Colourants, and Antioxidants

Innovations in biotechnology and green chemistry have made it easier to substitute natural oils for petrochemicals. Natural emollients currently dominate the market, accounting for more than 35% of cosmetic formulations. Traditional emollients like petrolatum, mineral oil, squalene, cetyl alcohol, and octyldodecanol can be replaced with vegetable oils, microbial ferments, and sustainably sourced plant materials. Conventional harsh/inorganic and petroleum-derived solvents can be replaced with green solvents like ionic liquids, NaDES, deep eutectic solvents (DES), and solvents derived from microbial fermentation. Anionic &

cationic surfactants can be replaced with non-ionic compounds like alkyl polyglycosides (APGs), sucrose esters, polyglycerol-based glycosides, olive oil esters, coconut oil esters, and some phospholipids.

Synthetic thickeners can be replaced with naturally derived thickeners and bio-ferment-based thickeners like xanthan gum, gelatin, chitosan, carrageenan, cellulose derivatives, pullulan, etc. Microplastic-based physical exfoliators can be replaced with granules obtained by milling plant materials like rice, nut shells, husks, etc. Inorganic pigments or colors can be replaced with microbial fermentation-derived pigments using safe organisms. Synthetic fragrances can be replaced with natural essential oils, plant extracts, and microbial extracts. Synthetic antioxidants like butylated hydroxytoluene and butylated hydroxyanisole can be replaced with vitamins, fermentation, and plant-based phenolic compounds.

7.8.3 Safer Preservation Methods

Health concerns over traditional preservatives have led to the adoption of green technologies. These include natural identical preservatives such as sodium benzoate, potassium sorbate, benzyl alcohol, and salicylic acid, as well as natural preservatives extracted from plant materials, such as tea, essential oils, cinnamon, and turmeric, to provide antimicrobial functionality. Harsh preservatives like phenoxyethanol (a phenol derivative), methylparaben (a paraben ester), imidazolidinyl urea, and diazolidinyl urea (formaldehyde donors) can be replaced with organic certified options such as Geogard® ECT (benzyl alcohol, salicylic acid, glycerin, and sorbic acid composition) and Sensicare® C 2010 (sodium benzoate and potassium sorbate composition). These preservatives are suitable for oil-in-water O/W and water-in-oil W/O emulsions and other organic formulations. These are compatible with formulations at a wide range of pH from 4 to 10. Multifunctional organic-sourced ingredients are incorporated into the formulations to prevent microbial contamination and spoilage. All of these alternatives address safety-related issues and conform to existing regulations and consumer demands.

7.8.4 Balancing Efficacy and Stability with NaDES

To assure the efficacy and stability of natural ingredients and sensitive active compounds, harsh solvents like propylene glycol, butylene glycol, ethanol, glycerin, and other petroleum-derived solvents used during extractions are gradually replaced with NaDES. The use of NaDES solves solubility and stability issues in the development of natural formulations (Freitas et al., 2024; Villa et al., 2024). By using these extractions, we can achieve low toxicity in raw materials and final formulations. Developing natural formulations with green chemistry is possible with NaDES extractions.

7.8.5 Sustainability Commitments of Industry Leaders

Leading cosmetic companies are setting some ambitious sustainability goals. For example, they aim to reduce their greenhouse gas emissions by 50% per product by 2030. They are also targeting the sustainable sourcing of raw materials, minimizing water consumption and waste production, increasing reliance on renewable energy, and achieving zero plastic pollution. These goals show that the industry is committed to sustainability, with market leaders pushing eco-friendly solutions (Martins & Marto, 2023).

7.9 Conclusion

White biotechnology is changing the cosmetic industry by introducing bioactive compounds and ferments that significantly benefit skin and scalp health. Conventional probiotic-based cosmetic formulations available in the market by cosmetic industries are prepared by infusing ingredients like microbial lysates, extracts, and filtrates. Raw material ingredients for probiotic cosmetics are available in the form of powders, lysates, and extracts. These probiotic ingredients are developed by the inorganic fermentation processes, due to inadequate process development, they do not contain natural fermented bioactive compounds. Even the lab-scale research on different natural bio-ferments still lacks scalable technology, commercial viability, efficacy in their activity, stability, and suitability with different cosmetic formulations. However, the development of these natural bio-ferments as commercially viable liquid cosmetic raw materials is a complex process, requiring expertise in microbiology, fermentation science, biotechnology techniques, expertise in scaleup knowledge, downstream processing, formulation science, safety and efficacy studies, knowledge in regulatory testing, documentation and licensing, etc.

The global cosmetic market is anticipated to grow from $432 billion in 2022 to $760 billion by 2034, with the probiotic cosmetics market holding a 4.10% share. The probiotic market, valued at $17 billion in 2024, is projected to reach $23 billion by 2034, growing at a CAGR of 5.7%, highlighting a significant market gap. Simultaneously, the cosmetic raw material market is projected to grow from $30 billion in 2024 to $40 billion by 2032, holding a 2% share of the natural cosmetic raw material market. The natural raw material market, valued at $600 million in 2024, is projected to grow to $1 billion by 2032 with a 7% CAGR, indicating rising demand for both probiotic-based cosmetics and natural raw materials. There is a significant market gap for probiotic cosmetics and raw materials, presenting substantial opportunities for the green cosmetic sector.

Developing bio-ferment cosmetic raw materials and final natural cosmetic formulations requires significant investments and collaborative efforts among microbiologists, biotechnologists, and organic chemists. These collaborative efforts are essential to progress from research to commercialization. Post-commercialization, there will be many challenges in regulations, packaging, large-scale manufacturing, and more. The main challenge includes quality control: evaluating the availability, efficacy, and activity of these molecules in the final product is challenging due to the lack of advanced testing and analytical facilities.

7.9.1 Regulatory Challenges

There are numerous challenges associated with cosmetic regulations. Each country has its own set of rules that include formulation standards, ingredient usage, packaging, labeling, product claims, testing, and toxicological analysis (Chauhan et al., 2022). This poses a significant challenge for small skincare brands aiming for a global market reach. Additionally, in developing probiotic-bioferment-based skincare solutions, the lack of specific guidelines in some regions complicates the process, despite growing market demand. Sourcing alternatives free from harsh chemicals and organic ingredients also presents a challenge in streamlining the supply chain.

7.9.2 Challenges in Sustainable Packaging

The European Council of the EU has signed new regulations on the usage of single-use plastic packaging. The new EU Packaging and Packaging Waste Regulation (PPWR) aims to achieve 15% waste prevention by 2040. This waste prevention focuses on shifting to plastic alternatives like paper and glass packaging. Innovations in biotechnology can pave the way for developing plastic alternatives with bio-fermented end products like polylactic acid, pullulan, seaweed extracts, and more. As we move towards sustainable packaging, focusing on net-zero, plastic-free, and carbon-free solutions, innovations in formulation and alternative methods of packaging should be explored. These innovations will underscore the principles behind sustainable and eco-friendly skincare products, alleviating the challenges involved.

7.9.3 Manufacturing Challenges

Developing formulations based on probiotic skincare presents significant manufacturing challenges. Proper microbial handling requires facilities that comply with Good Manufacturing Practices (GMP) and established Standard Operating Procedures (SOPs) and guidelines in process development. Additionally, proper regulatory oversight from authorities is necessary in this aspect.

7.9.4 Challenges for Biotech Skincare Startups

New startups face difficulties in developing a skincare brand due to challenges in obtaining government support, including licensing, investment, and comprehensive mentoring. This mentoring encompasses scientific guidance, business strategies, financial planning, and administrative tasks such as corporate documentation and company maintenance. A comprehensive understanding of these challenges is required to launch an innovative biotechnology-based formulation, but new startups often lack knowledge and support in these areas.

Despite these limitations, the future of sustainable skincare looks bright. New white biotechnology and sustainable practices will lead to innovative skincare solutions that improve skin health and promote a more responsible approach to natural beauty. Using natural ingredients from organic sources and sustainable methods can create effective skincare routines for a wide variety of consumer needs. Continuous research and development in white biotechnology hold immense promise for the future of skincare, offering effective sustainable alternatives to traditional chemical-based products. Ongoing innovation in biotechnology will ultimately lead to transformative changes for both the skincare industry and consumers.

7.9.5 From a Consumer Perspective

Consumers have become very keen on selecting natural products to avoid toxic chemicals that raise health concerns. They expect transparency in ingredients, cruelty-free practices in product testing, and sustainable packaging made from recycled or biodegradable materials. Due to the increasing market demand for active ingredients, consumers are keen on innovative and sustainable bio-fermented compounds such as vitamin C, peptides, retinol, hyaluronic acid, niacinamide, UV filters, ceramides, amino acids, and many more.

Acknowledgments

This research was supported and funded by the Department of Biotechnology (DBT), the Biotechnology Industry Research Assistance Council (BIRAC), the Government of India, and the Society for Innovation and Entrepreneurship (SINE IIT Bombay). Our products, KINMATI and AKSHATA™, were launched by DBT/BIRAC at Global Bio India 2024, New Delhi. We acknowledge our attribution to "Freepik" for vectors and icons.

References

Abedi, E., & Hashemi, S. M. B. (2020). Lactic acid production – Producing microorganisms and substrates sources-state of art. *Heliyon*, *6*(10), e04974. https://doi.org/10.1016/j.heliyon.2020.e04974

Aguilar-Toalá, J. E., Hernández-Mendoza, A., González-Córdova, A. F., Vallejo-Cordoba, B., & Liceaga, A. M. (2019). Potential role of natural bioactive peptides for development of cosmeceutical skin products. *Peptides*, *122*, 170170. https://doi.org/10.1016/j.peptides.2019.170170

Ahsan, H. (2019). Immunopharmacology and immunopathology of peptides and proteins in personal products. *Journal of Immunoassay and Immunochemistry*, *40*(4), 439–447. https://doi.org/10.1080/15321819.2019.1629590

Albaugh, V. L., Mukherjee, K., & Barbul, A. (2017). Proline precursors and collagen synthesis: biochemical challenges of nutrient supplementation and wound healing. *The Journal of Nutrition*, *147*(11), 2011–2017. https://doi.org/10.3945/jn.117.256404

An, J., Tsopmejio, I. S. N., Wang, Z., & Li, W. (2023). Review on extraction, modification, and synthesis of natural peptides and their beneficial effects on skin. *Molecules*, *28*(2), 908. https://doi.org/10.3390/molecules28020908

Apone, F., Barbulova, A., & Colucci, M. G. (2019). Plant and microalgae derived peptides are advantageously employed as bioactive compounds in cosmetics. *Frontiers in Plant Science*, *10*. https://doi.org/10.3389/fpls.2019.00756

Arribas-López, E., Zand, N., Ojo, O., Snowden, M. J., & Kochhar, T. (2021). The effect of amino acids on wound healing: A systematic review and meta-analysis on arginine and glutamine. *Nutrients*, *13*(8), 2498. https://doi.org/10.3390/nu13082498

Bădăluță, V. A., Curuțiu, C., Dițu, L. M., Holban, A. M., & Lazăr, V. (2024). Probiotics in wound healing. *International Journal of Molecular Sciences*, *25*(11), 5723. https://doi.org/10.3390/ijms25115723

Bhatia, L., Sarangi, P. K., Singh, A. K., Srivastava, R. K., & Chandel, A. K. (2024). Pre-, pro-, and postbiotics development from vegetable, fruit, and lignocellulosic biomass: A perspective. *Food Bioscience*, *61*(2024), 1–16.

Bin, P., Huang, R., & Zhou, X. (2017). Oxidation resistance of the sulfur amino acids: Methionine and cysteine. *BioMed Research International*, *2017*, 1–6. https://doi.org/10.1155/2017/9584932

Bonnet, I., Gault, M., & Andre-Frei, V. (2022). 226 Biotic complex to improve acne-prone skin. *Journal of Investigative Dermatology*, *142*(12), S219. https://doi.org/10.1016/j.jid.2022.09.237

Brandi, J., Cheri, S., Manfredi, M., Di Carlo, C., Vita Vanella, V., Federici, F., Bombiero, E., Bazaj, A., Rizzi, E., Manna, L., Cornaglia, G., Marini, U., Valenti, M. T., Marengo, E., & Cecconi, D. (2020). Exploring the wound healing, anti-inflammatory, anti-pathogenic and proteomic effects of lactic acid bacteria on keratinocytes. *Scientific Reports*, *10*(1), 11572. https://doi.org/10.1038/s41598-020-68483-4

Carciochi, R. A., Galván-D'Alessandro, L., Vandendriessche, P., & Chollet, S. (2016). Effect of germination and fermentation process on the antioxidant compounds of quinoa seeds. *Plant Foods for Human Nutrition*, *71*(4), 361–367. https://doi.org/10.1007/s11130-016-0567-0

Chatterjee, C., Gleddie, S., & Xiao, C.-W. (2018). Soybean bioactive peptides and their functional properties. *Nutrients*, *10*(9), 1211. https://doi.org/10.3390/nu10091211

Chauhan, A., Chauhan, S. B., & Nainwani, S. (2022). The review on cosmetic regulation: Quality education. *ECS Transactions*, *107*(1), 241–248. https://doi.org/10.1149/10701.0241ecst

Chen, M., Sun, Y., Zhu, L., Li, L., & Zhao, Y. (2024). Study on the skincare effects of red rice fermented by Aspergillus oryzae *in vitro*. *Molecules*, *29*(9), 2066. https://doi.org/10.3390/molecules29092066

Cho, H., Yang, J., Kang, J. Y., & Kim, K. E. (2024). Inhibitory effects of fermented sprouted oat extracts on oxidative stress and melanin overproduction. *Antioxidants*, *13*(5), 544. https://doi.org/10.3390/antiox13050544

Choi, J.-S., Jang, D. B., Cho, H. K., Lee, B.-B., Ha, Y.-M., & Choi, I. S. (2017). Anti-erythema and anti-edema activities of lactic acid bacteria extracts in mice. *Toxicology and Environmental Health Sciences*, *9*(1), 82–87. https://doi.org/10.1007/s13530-017-0307-3

Cosmetics Market Size, Share, and Trends 2024 to 2034. (n.d.). Retrieved from https://www.precedenceresearch.com/cosmetics-market

Cui, X., Mi, T., Xiao, X., Zhang, H., Dong, Y., Huang, N., Gao, P., Lee, J., Guelakis, M., & Gu, X. (2024). Topical glutathione amino acid precursors protect skin against environmental and oxidative stress. *Journal of the European Academy of Dermatology and Venereology*, *38*(S3), 3–11. https://doi.org/10.1111/jdv.19717

da Silva Vale, A., de Melo Pereira, G. V., de Oliveira, A. C., de Carvalho Neto, D. P., Herrmann, L. W., Karp, S. G., Soccol, V. T., & Soccol, C. R. (2023). Production, formulation, and application of postbiotics in the treatment of skin conditions. *Fermentation*, *9*(3), 264. https://doi.org/10.3390/fermentation9030264

Ditre, C., Griffin, T., Murphy, G., Sueki, H., Telegan, B., Johnson, W., Yu, R., & Van Scott, E. V. (1996). *Effects of alpha-hydroxy acids on photoaged skin: a pilot clinical, histologic, and ultrastructural study*. Journal of American Academy of Dermatology (Vol. 34, pp. 187–195).

Dou, J., Feng, N., Guo, F., Chen, Z., Liang, J., Wang, T., Guo, X., & Xu, Z. (2023). Applications of probiotic constituents in cosmetics. *Molecules*, *28*(19), 6765. https://doi.org/10.3390/molecules28196765

Egawa, G., Honda, T., & Kabashima, K. (2017). SCFAs control skin immune responses via increasing Tregs. *Journal of Investigative Dermatology*, *137*(4), 800–801. https://doi.org/10.1016/j.jid.2016.12.022

Elhage, K. G., St. Claire, K., & Daveluy, S. (2022). Acetic acid and the skin: A review of vinegar in dermatology. *International Journal of Dermatology*, *61*(7), 804–811. https://doi.org/10.1111/ijd.15804

Etnawati, K., Adiwinarni, D. R., Susetiati, D. A., Sauchi, Y., & Ito, H. (2019). The efficacy of skin care products containing glutathione in delivering skin lightening in Indonesian women. *Dermatology Reports*. https://doi.org/10.4081/dr.2019.8013

Fang, J., Sun, Q., Wang, Z., Song, Z., Geng, J., Wang, C., Li, M., & Wang, D. (2023). Enhancement of human epidermal cell defense against UVB damage by fermentation of Passiflora edulis sims peel with Saccharomyces cerevisiae. *Nutrients*, *15*(3), 501. https://doi.org/10.3390/nu15030501

Farahanah, I., Sahudin, S., Mohsin, H. F., Ariffin, S. A., & Aminuddin, L. D. (2025). Understanding investigational perspective of antioxidant and antibacterial properties of rice. *Rice Science*, *32*(1), 15–31. https://doi.org/10.1016/j.rsci.2024.10.004

Feng, X., Shang, J., Gu, Z., Luo, X., Chen, Y., & Liu, Y. (2024). Lactic acid chemical peeling in skin disorders. *Clinical, Cosmetic and Investigational Dermatology*, *17*, 901–909. https://doi.org/10.2147/CCID.S455700

Ferreira, M., Matos, A., Couras, A., Marto, J., & Ribeiro, H. (2022). Overview of cosmetic regulatory frameworks around the world. *Cosmetics*, *9*(4), 72. https://doi.org/10.3390/cosmetics9040072

Ferri, M., Graen-Heedfeld, J., Bretz, K., Guillon, F., Michelini, E., Calabretta, M. M., Lamborghini, M., Gruarin, N., Roda, A., Kraft, A., & Tassoni, A. (2017). Peptide fractions obtained from rice by-products by means of an environment-friendly process show *in vitro* health-related bioactivities. *PLoS One*, *12*(1), e0170954. https://doi.org/10.1371/journal.pone.0170954

Ferrillo, M., Vastarella, M., Cantelli, M., Mazzella, C., & Fabbrocini, G. (2019). Instrumental, clinical and subjective evaluation of the efficacy of a cosmetic treatment for home use. *Journal of Cosmetic and Laser Therapy*, *21*(4), 190–195. https://doi.org/10.1080/14764172.2018.1502454

Fonseca-Santos, B., Corrêa, M. A., & Chorilli, M. (2015). Sustainability, natural and organic cosmetics: Consumer, products, efficacy, toxicological and regulatory considerations. *Brazilian Journal of Pharmaceutical Sciences*, *51*(1), 17–26. https://doi.org/10.1590/S1984-82502015000100002

Freitas, D. S., Rocha, D., Santos, J., Noro, J., Tavares, T. D., Teixeira, M. O., Araújo, D., Castro, J., Silva, S., Felgueiras, H. P., Ribeiro, A., Venâncio, A., Cavaco-Paulo, A., & Silva, C. (2024). Developing translucent emulsions using sustainable NADES-based extracts. *Journal of Molecular Liquids*, *415*, 126413. https://doi.org/10.1016/j.molliq.2024.126413

Gaspar, L. R., Camargo, F. B., Gianeti, M. D., & Maia Campos, P. M. B. G. (2008). Evaluation of dermatological effects of cosmetic formulations containing Saccharomyces cerevisiae extract and vitamins. *Food and Chemical Toxicology*, *46*(11), 3493–3500. https://doi.org/10.1016/j.fct.2008.08.028

Ghosal, S., Bhattacharyya, D. K., & Bhowal, J. (2021). Effect of *Saccharomyces cerevisiae* fermentation process on the phenolic content, flavonoid content and antioxidant properties of flaxseeds. In *Advances in bioprocess engineering and technology* (pp. 119–131). https://doi.org/10.1007/978-981-15-7409-2_12

Gomes, C., Silva, A. C., Marques, A. C., Sousa Lobo, J., & Amaral, M. H. (2020). Biotechnology applied to cosmetics and aesthetic medicines. *Cosmetics*, *7*(2), 33. https://doi.org/10.3390/cosmetics7020033

Goodarzi, A., Mozafarpoor, S., Bodaghabadi, M., & Mohamadi, M. (2020). The potential of probiotics for treating acne vulgaris: A review of literature on acne and microbiota. *Dermatologic Therapy*, *33*(3). https://doi.org/10.1111/dth.13279

Green, B. A., Yu, R. J., & Van Scott, E. J. (2009). Clinical and cosmeceutical uses of hydroxyacids. *Clinics in Dermatology*, *27*(5), 495–501. https://doi.org/10.1016/j.clindermatol.2009.06.023

Gupta, P. L., Rajput, M., Oza, T., Trivedi, U., & Sanghvi, G. (2019). Eminence of microbial products in cosmetic industry. *Natural Products and Bioprospecting*, *9*(4), 267–278. https://doi.org/10.1007/s13659-019-0215-0

Hausmann, C., Hertz-Kleptow, D., Zoschke, C., Wanjiku, B., Wentzien-Odenthal, A., Kerscher, M., & Schäfer-Korting, M. (2019). Reconstructed human epidermis predicts barrier-improving effects of Lactococcus lactis emulsion in humans. *Skin Pharmacology and Physiology*, *32*(2), 72–80. https://doi.org/10.1159/000495255

Henkel AG and Co KGaA. (2017). Antiperspirant cosmetics comprising specific proteins from legumes of the species soybean and including no aluminum and/or zirconium halides and/or hydroxy halides.

Herman, A., & Herman, A. P. (2023). Biological activity of fermented plant extracts for potential dermal applications. *Pharmaceutics*, *15*(12), 2775. https://doi.org/10.3390/pharmaceutics15122775

Ho, C., Ng, S., Chuang, H., Wen, S., Kuo, C., Mahalakshmi, B., Huang, C., & Kuo, W. (2021). Extracts of Jasminum sambac flowers fermented by Lactobacillus rhamnosus inhibit H_2 O_2 - and UVB-induced aging in human dermal fibroblasts. *Environmental Toxicology*, *36*(4), 607–619. https://doi.org/10.1002/tox.23065

Hsu, C.-Y., Henry, J., Raymond, A.-A., Méchin, M.-C., Pendaries, V., Nassar, D., Hansmann, B., Balica, S., Burlet-Schiltz, O., Schmitt, A.-M., Takahara, H., Paul, C., Serre, G., & Simon, M. (2011). Deimination of human filaggrin-2 promotes its proteolysis by calpain 1. *Journal of Biological Chemistry*, *286*(26), 23222–23233. https://doi.org/10.1074/jbc.M110.197400

https://www.businessresearchinsights.com/market-reports/cosmetic-raw-materials-market-100628. (n.d.).

Huang, H.-C., Lee, I. J., Huang, C., & Chang, T.-M. (2020). Lactic acid bacteria and lactic acid for skin health and melanogenesis inhibition. *Current Pharmaceutical Biotechnology*, *21*(7), 566–577. https://doi.org/10.2174/1389201021666200109104701

Husein el Hadmed, H., & Castillo, R. F. (2016). Cosmeceuticals: Peptides, proteins, and growth factors. *Journal of Cosmetic Dermatology*, *15*(4), 514–519. https://doi.org/10.1111/jocd.12229

Jara, J. R., Aroca, P., Solano, F., Martinez, J. H., & Lozano, J. A. (1988). The role of sulfhydryl compounds in mammalian melanogenesis: The effect of cysteine and glutathione upon tyrosinase and the intermediates of the pathway. *Biochimica et Biophysica Acta (BBA) - General Subjects*, *967*(2), 296–303. https://doi.org/10.1016/0304-4165(88)90023-2

Jiang, Y., Zhao, D., Sun, J., Luo, X., Li, H., Sun, X., & Zheng, F. (2019). Analysis of antioxidant effect of two tripeptides isolated from fermented grains (Jiupei) and the antioxidative interaction with 4-methylguaiacol, 4-ethylguaiacol, and vanillin. *Food Science & Nutrition*, *7*(7), 2391–2403. https://doi.org/10.1002/fsn3.1100

Jin, Z. W., Kim, H.-K., Lee, C.-H., Jung, S.-W., Shin, S.-J., Im, S.-Y., Cho, B.-H., & Lee, H.-K. (2012). Glutamine suppresses dinitrophenol fluorobenzene-induced allergic contact dermatitis and itching: Inhibition of contact dermatitis by glutamine. *Journal of Dermatological Science*, *67*(2), 88–94. https://doi.org/10.1016/j.jdermsci.2012.05.012

Jo, D.-U., Chin, Y.-W., Kim, Y., Kim, K.-T., Kim, T.-W., & Lim, T.-G. (2022). By-product of Korean liquor fermented by Saccharomyces cerevisiae exhibits skin whitening activity. *Food Science and Biotechnology*, *31*(5), 587–596. https://doi.org/10.1007/s10068-022-01062-7

Jung, Y.-O., Jeong, H., Cho, Y., Lee, E.-O., Jang, H.-W., Kim, J., Nam, K. T., & Lim, K.-M. (2019). Lysates of a probiotic, Lactobacillus rhamnosus, can improve skin barrier function in a reconstructed human epidermis model. *International Journal of Molecular Sciences*, *20*(17), 4289. https://doi.org/10.3390/ijms20174289

Kapuścińska, A., & Nowak, I. (2015). Use of organic acids in acne and skin discolorations therapy. *Postępy Higieny i Medycyny Doświadczalnej*, *69*, 374–383. https://doi.org/10.5604/17322693.1145825

Karna, E., Miltyk, W., Wołczyński, S., & Pałka, J. A. (2001). The potential mechanism for glutamine-induced collagen biosynthesis in cultured human skin fibroblasts. *Comparative Biochemistry and Physiology Part B: Biochemistry and Molecular Biology*, *130*(1), 23–32. https://doi.org/10.1016/S1096-4959(01)00400-6

Kathed, A. (2018). Multisource radiofrequency for fractional skin resurfacing: Significant improvement treating acne scars, skin rejuvenation and stretch marks. *Journal of Cosmetology & Trichology*, *4*. https://doi.org/10.4172/2471-9323-C1-008

Kim, S. Y., Shin, S. Y., Kim, J. H., Lee, H. J., & Kim, K. M. (2021). Antioxidant and moisturizing effects of fermented Hordeum vulgare L. extract and heat-killed probiotics by Lactobacillus lactic acid bacteria isolated from kimchi. *Korean Journal of Microbiology*, *57*, 30–38.

KINMATI and AKSHATA Product launch at Global Bio India 2024, Delhi. (2024). Retrieved from https://www.prameelalifesciences.com/blogs/news

Klimek-Szczykutowicz, M., Błońska-Sikora, E. M., Kulik-Siarek, K., Zhussupova, A., & Wrzosek, M. (2024). Bioferments and biosurfactants as new products with potential use in the cosmetic industry. *Applied Sciences*, *14*(9), 3902. https://doi.org/10.3390/app14093902

Kobayashi, J. (2019). d-Amino acids and lactic acid bacteria. *Microorganisms*, *7*(12), 690. https://doi.org/10.3390/microorganisms7120690

Kook, M.-C., & Cho, S.-C. (2013). Production of GABA (gamma amino butyric acid) by lactic acid bacteria. *Korean Journal for Food Science of Animal Resources*, *33*(3), 377–389. https://doi.org/10.5851/kosfa.2013.33.3.377

Koyama, J., Horii, I., & Kawasaki, K. (1983). *Free amino acids of stratum corneum as a biochemical marker to evaluate dry skin* (Vol. 35, pp. 183–195). Journal of the Society of Cosmetic Chemists.

LeBlanc, J., de LeBlanc, A., Perdigon, G., Miyoshi, A., Rochat, T., Bermudez-Humaran, L., Langella, P., Sesma, F., & Azevedo, V. (2008). Anti-inflammatory properties of lactic acid bacteria: Current knowledge, applications and prospects. *Anti-Infective Agents in Medicinal Chemistry*, *7*(3), 148–154. https://doi.org/10.2174/187152108784911287

Lee, H.-S., Kim, M.-R., Park, Y., Park, H. J., Chang, U. J., Kim, S. Y., & Suh, H. J. (2012). Fermenting red ginseng enhances its safety and efficacy as a novel skin care anti-aging ingredient: *In vitro* and animal study. *Journal of Medicinal Food*, *15*(11), 1015–1023. https://doi.org/10.1089/jmf.2012.2187

Lee, J. K., Kim, I., Jeon, E.-K., Ha, J.-H., Hwang, C.-W., Kim, J.-C., Yang, W.-S., Choi, H., Kim, H.-D., & Kim, C.-H. (2022). Bacterially converted oat active ingredients enhances antioxidative and anti-UVB photoaging activities. *Evidence-based Complementary and Alternative Medicine*, *2022*, 1–12. https://doi.org/10.1155/2022/1901564

Leignadier, J., Drago, M., Lesouhaitier, O., Barreau, M., Dashi, A., Worsley, O., & Attia-Vigneau, J. (2023). Lysine-dendrimer, a new non-aggressive solution to rebalance the microbiota of Acne-Prone skin. *Pharmaceutics*, *15*(8), 2083. https://doi.org/10.3390/pharmaceutics15082083

Lim, H.-W., Huang, Y.-H., Kyeong, G., Park, M., & Lim, C.-J. (2022). Comparative insights into the skin beneficial properties of probiotic *Lactobacillus* isolates of skin origin. *BioMed Research International*, *2022*(1). https://doi.org/10.1155/2022/7728789

Lima, T., & Pedriali Moraes, C. (2018). Bioactive peptides: Applications and relevance for cosmeceuticals. *Cosmetics*, *5*(1), 21. https://doi.org/10.3390/cosmetics5010021

Lin, H., Cheng, Y., Liang, L., Qin, X., Dong, X., Guo, Y., Yu, Q., Zhang, G., & Hu, X. (2024). Effects of the lactobacillus fermentation on dendrobium officinale polysaccharide: Structural, physicochemical properties and *in vitro* bioactivities. *Food Bioscience*, *61*, 105003. https://doi.org/10.1016/j.fbio.2024.105003

Lyu, Y. (2024). Microorganisms and microbial metabolites: The next frontier in skincare products. *Theoretical and Natural Science*, *59*(1), 222–227. https://doi.org/10.54254/2753-8818/59/20241412

Măgeruşan, Ş. E., Hancu, G., & Rusu, A. (2023). A comprehensive bibliographic review concerning the efficacy of organic acids for chemical peels treating acne vulgaris. *Molecules*, *28*(20), 7219. https://doi.org/10.3390/molecules28207219

Majchrzak, W., Motyl, I., & Śmigielski, K. (2022). Biological and cosmetical importance of fermented raw materials: An overview. *Molecules*, *27*(15), 4845. https://doi.org/10.3390/molecules27154845

Martínez-Ruiz, M., Martínez-González, C. A., Kim, D.-H., Santiesteban-Romero, B., Reyes-Pardo, H., Villaseñor-Zepeda, K. R., Meléndez-Sánchez, E. R., Ramírez-Gamboa, D., Díaz-Zamorano, A. L., Sosa-Hernández, J. E., Coronado-Apodaca, K. G., Gámez-Méndez, A. M., Iqbal, H. M. N., & Parra-Saldivar, R. (2022). Microalgae bioactive compounds to topical applications products—A review. *Molecules*, *27*(11), 3512. https://doi.org/10.3390/molecules27113512

Martins, A. M., & Marto, J. M. (2023). A sustainable life cycle for cosmetics: From design and development to post-use phase. *Sustainable Chemistry and Pharmacy*, *35*, 101178. https://doi.org/10.1016/j.scp.2023.101178

Mellou, F., Varvaresou, A., & Papageorgiou, S. (2019). Renewable sources: Applications in personal care formulations. *International Journal of Cosmetic Science*, *41*(6), 517–525. https://doi.org/10.1111/ics.12564

Milon, R. B., Hu, P., Zhang, X., Hu, X., & Ren, L. (2024). Recent advances in the biosynthesis and industrial biotechnology of Gamma-amino butyric acid. *Bioresources and Bioprocessing*, *11*(1), 32. https://doi.org/10.1186/s40643-024-00747-7

Miyamoto, K., Munakata, Y., Fujii, K., Summer, W., Yang, L., Sudarsana, S., & Furue, M. (2024). Six amino acids among natural moisturizing factors responsible for skin hydration: Improvement and anti-aging of skin by Galactomyces ferment filtrate-PiteraTM containing skin moisturizer. *Journal of Cosmetics, Dermatological Sciences and Applications*, *14*(2), 113–127. https://doi.org/10.4236/jcdsa.2024.142008

Miyoshi, Y., Itoh, M., Ayaki, S., Mii, T., Matsuno, K., Tokuyama, T., Tokuyama, T., Tokuyama, T., & Dainichi, T. (2023). 702 Rice ferment filtrate increases the production of hyaluronic acid, natural moisturizing factors, and ceramides by keratinocytes and improves the barrier function of the skin. *Journal of Investigative Dermatology*, *143*(5), S121. https://doi.org/10.1016/j.jid.2023.03.711

Mo, Q., You, S., Fu, H., Wang, D., Zhang, J., Wang, C., & Li, M. (2022). Purification and identification of antioxidant peptides from rice fermentation of Lactobacillus plantarum and their protective effects on UVA – Induced oxidative stress in skin. *Antioxidants*, *11*(12), 2333. https://doi.org/10.3390/antiox11122333

Morichi, T., Irie, R., Yano, N., & Kembo, H. (1963). Protective effect of glutamic acid and related compounds on bacterial cells subjected to freeze-drying. *The Journal of General and Applied Microbiology*, *9*(2), 149–161. https://doi.org/10.2323/jgam.9.149

Mottin, V. H. M., & Suyenaga, E. S. (2018). An approach on the potential use of probiotics in the treatment of skin conditions: Acne and atopic dermatitis. *International Journal of Dermatology*, *57*(12), 1425–1432. https://doi.org/10.1111/ijd.13972

Murakami, H., Shimbo, K., Inoue, Y., Takino, Y., & Kobayashi, H. (2012). Importance of amino acid composition to improve skin collagen protein synthesis rates in UV-irradiated mice. *Amino Acids*, *42*(6), 2481–2489. https://doi.org/10.1007/s00726-011-1059-z

Nagarjun, P. A., Rao, R. S., Rajesham, S., & Rao, L. V. (2005). Optimization of lactic acid production in SSF by Lactobacillus amylovorus NRRL B-4542 using Taguchi methodology. *Journal of Microbiology (Seoul, Korea)*, *43*(1), 38–43.

Nam, G.-H., Jo, K.-J., Park, Y.-S., Kawk, H. W., Yoo, J.-G., Jang, J. D., Kang, S. M., Kim, S.-Y., & Kim, Y.-M. (2019). Bacillus/Trapa japonica fruit extract ferment filtrate enhances human hair follicle dermal papilla cell proliferation via the Akt/ERK/GSK-3β signaling pathway. *BMC Complementary and Alternative Medicine*, *19*(1), 104. https://doi.org/10.1186/s12906-019-2514-8

Narla, S., & Silverberg, J. I. (2020). Multimorbidity and mortality risk in hospitalized adults with chronic inflammatory skin disease in the United States. *Archives of Dermatological Research*, *312*(7), 507–512. https://doi.org/10.1007/s00403-020-02043-8

Natural Cosmetics Ingredients Market by Ingredient Type (Emollients, Bio Surfactants, Sugar Polymers, Natural Preservatives, Others), Application (Skin Care, Hair Care, Oral Care, Make-up) and Region - Global Forecast to 2030. (n.d.). Retrieved from https://www.marketsandmarkets.com/Market-Reports/natural-cosmetics-ingredients-market-117267500.html

Ngoc, L. T. N., Moon, J.-Y., & Lee, Y.-C. (2023). Insights into bioactive peptides in cosmetics. *Cosmetics*, *10*(4), 111. https://doi.org/10.3390/cosmetics10040111

Noguchi, A., & Djerassi, D. (2009). Amino acids and peptides: Building blocks for skin proteins. In *Nutritional cosmetics* (pp. 287–317). Elsevier. https://doi.org/10.1016/B978-0-8155-2029-0.50022-3

Ohara, H., Ichikawa, S., Matsumoto, H., Akiyama, M., Fujimoto, N., Kobayashi, T., & Tajima, S. (2010). Collagen-derived dipeptide, proline-hydroxyproline, stimulates cell proliferation and hyaluronic acid synthesis in cultured human dermal fibroblasts. *The Journal of Dermatology*, *37*(4), 330–338. https://doi.org/10.1111/j.1346-8138.2010.00827.x

Oshimura, E., Abe, H., & Oota, R. (2007). Hair and amino acids: The interactions and the effects. *Journal of Cosmetic Science*, *58*(4), 347–357.

Otsuka, A., Moriguchi, C., Shigematsu, Y., Tanabe, K., Haraguchi, N., Iwashita, S., Tokudome, Y., & Kitagaki, H. (2022). Fermented cosmetics and metabolites of skin microbiota—A new approach to skin health. *Fermentation*, *8*(12), 703. https://doi.org/10.3390/fermentation8120703

Pallin, A. (2015). *Fermentation of barley flour with Lactobacillus reuteri*. Swedish University of Agricultural Sciences.

Panagiotou, D., Filidou, E., Gaitanidou, M., Tarapatzi, G., Spathakis, M., Kandilogiannakis, L., Stavrou, G., Arvanitidis, K., Tsetis, J., Gionga, P., Shrewsbury, A., Manolopoulos, V., Kapoukranidou, D., Lasithiotakis, K., Kolios, G., & Kotzampassi, K. (2023). Role of Lactiplantibacillus plantarum UBLP-40, Lactobacillus rhamnosus UBLR-58 and Bifidobacterium longum UBBL-64 in the wound healing process of the excisional skin. *Nutrients*, *15*(8), 1822. https://doi.org/10.3390/nu15081822

Panchal, G., Hati, S., & Sakure, A. (2020). Characterization and production of novel antioxidative peptides derived from fermented goat milk by L. fermentum. *WT*, *119*, 108887. https://doi.org/10.1016/j.lwt.2019.108887

Park, H., Arellano, K., Lee, Y., Yeo, S., Ji, Y., Ko, J., & Holzapfel, W. (2021). Pilot study on the forehead skin microbiome and short chain fatty acids depending on the SC functional index in Korean cohorts. *Microorganisms*, *9*(11), 2216. https://doi.org/10.3390/microorganisms9112216

Park, J., Sung, J. M., Choi, Y.-S., & Park, J.-D. (2020). Effect of natural fermentation on milled rice grains: Physicochemical and functional properties of rice flour. *Food Hydrocolloids*, *108*, 106005. https://doi.org/10.1016/j.foodhyd.2020.106005

Pérez-Rivero, C., & López-Gómez, J. P. (2023). Unlocking the potential of fermentation in cosmetics: A review. *Fermentation*, *9*(5), 463. https://doi.org/10.3390/fermentation9050463

Probiotic Cosmetic Products Market Size & Share Analysis - Growth Trends & Forecasts (2025 - 2030). (n.d.). Retrieved from https://www.mordorintelligence.com/industry-reports/probiotic-cosmetic-products-market

Product Innovation 2024. (2024). Cosmetics business. Retrieved from https://flickread.com/edition/html/index.php?pdf=661e839461fb6&dm_i=8EU,8LJK4,550Y76,ZMQ2O,1#20

Puebla-Barragan, S., & Reid, G. (2021). Probiotics in cosmetic and personal care products: Trends and challenges. *Molecules*, *26*(5), 1249. https://doi.org/10.3390/molecules26051249

Punia Bangar, S., Suri, S., Trif, M., & Ozogul, F. (2022). Organic acids production from lactic acid bacteria: A preservation approach. *Food Bioscience*, *46*, 101615. https://doi.org/10.1016/j.fbio.2022.101615

Rao, R. S., Kumar, C. G., Prakasham, R. S., & Hobbs, P. J. (2008). The Taguchi methodology as a statistical tool for biotechnological applications: A critical appraisal. *Biotechnology Journal*, *3*(4), 510–523. https://doi.org/10.1002/biot.200700201

Salgaonkar, N., Kadamkode, V., Kumaran, S., Mallemala, P., Christy, E., Appavoo, S., Majumdar, A., Mitra, R., & Dasgupta, A. (2022). Glycerol fermentation by skin bacteria generates lactic acid and upregulates the expression levels of genes associated with the skin barrier function. *Experimental Dermatology*, *31*(9), 1364–1372. https://doi.org/10.1111/exd.14604

Sanchez, S., Rodríguez-Sanoja, R., Ramos, A., & Demain, A. L. (2018). Our microbes not only produce antibiotics, they also overproduce amino acids. *The Journal of Antibiotics*, *71*(1), 26–36. https://doi.org/10.1038/ja.2017.142

Sánchez-Muñoz, S., Mariano, S. G., de Oliveira, L. M., Mura, B. F., Verma, M. L., da Silva, S. S., & Chandel, A. K. (2019). Production of fungal and bacterial pigments and their applications. In M. L. Verma & A. K. Chandel (Eds.), *Biotechnological production of bioactive compounds*. Amsterdam, The Netherlands: Elsevier Press.

Saxena, V., & Yadav, K. (2020). Glycolic acid, lactic acid, mandelic acid, salicylic acid, citric acid, gluconolactone: Skin exfoliators in combination therapy of acne vulgaris. *International Journal of Research in Engineering, Science and Management*, *3*(10), 54–55. https://doi.org/10.47607/ijresm.2020.334

Schwarz, A., Bruhs, A., & Schwarz, T. (2017). The short-chain fatty acid sodium butyrate functions as a regulator of the skin immune system. *Journal of Investigative Dermatology*, *137*(4), 855–864. https://doi.org/10.1016/j.jid.2016.11.014

Shi, Y., Wei, K., Lu, J., Wei, J., Hu, X., & Chen, T. (2020). A clinic trial evaluating the effects of Aloe Vera fermentation gel on recurrent Aphthous stomatitis. *Canadian Journal of Infectious Diseases and Medical Microbiology*, *2020*, 1–9. https://doi.org/10.1155/2020/8867548

Sivamaruthi, B. S., Chaiyasut, C., & Kesika, P. (2018). Cosmeceutical importance of fermented plant extracts: A short review. *International Journal of Applied Pharmaceutics*, *10*(4), 31. https://doi.org/10.22159/ijap.2018v10i4.26355

Slominski, A. T., Hardeland, R., Zmijewski, M. A., Slominski, R. M., Reiter, R. J., & Paus, R. (2018). Melatonin: A cutaneous perspective on its production, metabolism, and functions. *Journal of Investigative Dermatology*, *138*(3), 490–499. https://doi.org/10.1016/j.jid.2017.10.025

Sultana, R., McBain, A. J., & O'Neill, C. A. (2013). Strain-dependent augmentation of tight-junction barrier function in human primary epidermal keratinocytes by lactobacillus and bifidobacterium lysates. *Applied and Environmental Microbiology*, *79*(16), 4887–4894. https://doi.org/10.1128/AEM.00982-13

Thueson, D. O., Chan, E. K., Oechsli, L. M., & Hahn, G. S. (1998). The roles of pH and concentration in lactic acid-induced stimulation of epidermal turnover. *Dermatologic Surgery*, *24*(6), 641–645. https://doi.org/10.1111/j.1524-4725.1998.tb04221.x

Tóth, P., & Németh, Á. (2022). Investigation and characterisation of new eco-friendly cosmetic ingredients based on probiotic bacteria ferment filtrates in combination with alginite mineral. *Processes*, *10*(12), 2672. https://doi.org/10.3390/pr10122672

Villa, C., Caviglia, D., Robustelli della Cuna, F. S., Zuccari, G., & Russo, E. (2024). NaDES application in cosmetic and pharmaceutical fields: An overview. *Gels*, *10*(2), 107. https://doi.org/10.3390/gels10020107

Wang, R., Yan, S., Ma, X., Zhao, J., Han, Y., Pan, Y., & Zhao, H. (2023). The pivotal role of *Bifida Ferment Lysate* on reinforcing the skin barrier function and maintaining homeostasis of skin defenses *in vitro*. *Journal of Cosmetic Dermatology*, *22*(12), 3427–3435. https://doi.org/10.1111/jocd.15831

Wang, Y., Kuo, S., Shu, M., Yu, J., Huang, S., Dai, A., Two, A., Gallo, R. L., & Huang, C.-M. (2014). Staphylococcus epidermidis in the human skin microbiome mediates fermentation to inhibit the growth of Propionibacterium acnes: Implications of probiotics in acne vulgaris. *Applied Microbiology and Biotechnology*, *98*(1), 411–424. https://doi.org/10.1007/s00253-013-5394-8

Wang, Y., Wu, J., Lv, M., Shao, Z., Hungwe, M., Wang, J., Bai, X., Xie, J., Wang, Y., & Geng, W. (2021). Metabolism characteristics of lactic acid bacteria and the expanding applications in food industry. *Frontiers in Bioengineering and Biotechnology*, *9*. https://doi.org/10.3389/fbioe.2021.612285

Watanabe, F., Hashizume, E., Chan, G. P., & Kamimura, A. (2014). Skin-whitening and skin-condition-improving effects of topical oxidized glutathione: A double-blind and placebo-controlled clinical trial in healthy women. *Clinical, Cosmetic and Investigational Dermatology*, *267*. https://doi.org/10.2147/CCID.S68424

Wei, K., Guo, C., Zhu, J., Wei, Y., Wu, M., Huang, X., Zhang, M., Li, J., Wang, X., Wang, Y., & Wei, X. (2022). The whitening, moisturizing, anti-aging activities, and skincare evaluation of selenium-enriched mung bean fermentation broth. *Frontiers in Nutrition*, *9*. https://doi.org/10.3389/fnut.2022.837168

Weschawalit, S., Thongthip, S., Phutrakool, P., & Asawanonda, P. (2017). Glutathione and its antiaging and antimelanogenic effects. *Clinical, Cosmetic and Investigational Dermatology*, *10*, 147–153. https://doi.org/10.2147/CCID.S128339

Wu, G. (2020). *Amino acids in nutrition and health* (Vol. 1265). Cham: Springer International Publishing. https://doi.org/10.1007/978-3-030-45328-2

Wu, G., Bazer, F. W., Burghardt, R. C., Johnson, G. A., Kim, S. W., Knabe, D. A., Li, P., Li, X., McKnight, J. R., Satterfield, M. C., & Spencer, T. E. (2011). Proline and hydroxyproline metabolism: Implications for animal and human nutrition. *Amino Acids*, *40*(4), 1053–1063. https://doi.org/10.1007/s00726-010-0715-z

Xiao, X., Hu, X., Yao, J., Cao, W., Zou, Z., Wang, L., Qin, H., Zhong, D., Li, Y., Xue, P., Jin, R., Li, Y., Shi, Y., & Li, J. (2023). The role of short-chain fatty acids in inflammatory skin diseases. *Frontiers in Microbiology*, *13*. https://doi.org/10.3389/fmicb.2022.1083432

Yamamoto, Y., Uede, K., Yonei, N., Kishioka, A., Ohtani, T., & Furukawa, F. (2006). Effects of alpha-hydroxy acids on the human skin of Japanese subjects: The rationale for chemical peeling. *The Journal of Dermatology*, *33*(1), 16–22. https://doi.org/10.1111/j.1346-8138.2006.00003.x

Yang, F., Zhou, Z., & Guo, M. (2022). Validation of the tight junction promotion and skin barrier enhancement by Saccharomyces rice ferment filtrate. *Cosmetic Science*, *73*, 201–212.

Zagórska-Dziok, M., Ziemlewska, A., Mokrzyńska, A., Nizioł-Łukaszewska, Z., Sowa, I., Szczepanek, D., & Wójciak, M. (2023). Comparative study of cytotoxicity and antioxidant, anti-aging and antibacterial properties of unfermented and fermented extract of Cornus mas L. *International Journal of Molecular Sciences*, *24*(17), 13232. https://doi.org/10.3390/ijms241713232

Zappelli, C., Barbulova, A., Apone, F., & Colucci, G. (2016). Effective active ingredients obtained through biotechnology. *Cosmetics*, *3*(4), 39. https://doi.org/10.3390/cosmetics3040039

Zhao, H., Park, B., Kim, M.-J., Hwang, S.-H., Kim, T.-J., Kim, S.-U., Kwon, I., & Hwang, J. S. (2023). The effect of γ-aminobutyric acid intake on UVB- induced skin damage in hairless mice. *Biomolecules & Therapeutics*, *31*(6), 640–647. https://doi.org/10.4062/biomolther.2023.085

Ziemlewska, A., Nizioł-Łukaszewska, Z., Bujak, T., Zagórska-Dziok, M., Wójciak, M., & Sowa, I. (2021). Effect of fermentation time on the content of bioactive compounds with cosmetic and dermatological properties in Kombucha Yerba Mate extracts. *Scientific Reports*, *11*(1), 18792. https://doi.org/10.1038/s41598-021-98191-6

8

Biotechnological Advancements in Lactic Acid Bacteria Fermentation

Metabolic Pathways and Metabolite Profiles

Samatha Paladugu[1,2], *Sarepalli Sai Sathwik*[1,2], *and Sreelatha Beemagani*[2]

[1]*Prameela Life Sciences Private Limited, Hyderabad, Telangana, India*
[2]*Department of Microbiology, Chaitanya (Deemed to be University), Hyderabad, Telangana, India*

8.1 Introduction

The most common applications of Lactobacillus fermentation are the production of lactic acid and biomass for use in food, medicine, and agriculture (Huang et al., 2023a, b; de Oliveira et al., 2021). In recent years, the metabolic profiles of Lactobacillus fermentation have received garnered significant interest (Gänzle, 2015; Kieliszek et al., 2021; Lee et al., 2019; Tang et al., 2023; Wang et al., 2021). Recent advances in the taxonomic nomenclature of the Lactobacillus family have made it possible to identify metabolic characteristics across three levels of evolution. First, the genus-level feature for hexoses is determined by the homofermentative or heterofermentative fermentation pathways. Second, the current classification facilitates the identification of genus-specific traits. Third, it identifies metabolic properties or coevolution with other bacteria. *Lactobacillaceae* and *Leuconostacaceae* have been merged into a single family, *Lactobacillaceae*. This reclassification resulted in approximately 300 species from seven genera and two families being reclassified into the *Lactobacillaceae* family, which now consists of 31 genera and 23 new genera comprising organisms previously considered to be Lactobacillus species (Qiao et al., 2022).

Lactobacillaceae is a bacterium capable of fermenting carbohydrates to produce organic acids, primarily lactic acid, which helps to maintain acidic pH and prevent the growth of pathogens (Gänzle, 2015; Wang et al., 2021). It can also break down proteins into simpler peptide structures, creating free amino acids. Alongside organic acids, EPS, vitamins, antioxidants, neurotransmitters, bacteriocins, short-chain fatty acids (SCFAs), and polyols, it generates a wide spectrum of volatile and nonvolatile chemicals. This review focuses on the metabolism of Lactobacillus bacteria (Kieliszek et al., 2021; Wang et al., 2021).

Lactobacillus, a multipurpose probiotic, is known for its dairy fermentations, which can ferment various substrates including polysaccharides, oligosaccharides, disaccharides, and monosaccharides. These lactic acid bacteria (LABs) can be classified into two groups: those that produce lactic acid using starch and those that do not. Starch-using LABs produce

Biofoundry Techniques for Biotechnology Applications, First Edition. Edited by Anuj Kumar Chandel.
 Published 2026 by John Wiley & Sons Ltd.

lactic acid through the enzyme amylase, breaking down amylose and amylopectin groups, resulting in starch degradation and fermentation (Petrova et al., 2013). Nonstarch-utilizing organisms use oligosaccharides and monosaccharides but cannot completely break down starch. Instead, they convert starch into glucose, fructose, hydrolysates of starch, cyclodextrins, derivatives of maltodextrins, and some other derivatives of starch through enzymatic or chemical conversions (Petrova & Petrov, 2012; Petrova et al., 2013).

Lactobacillus amylovorous, Amylolactobacillus amylophilus, Lactobacillus acidophilus, L. amylolyticus, L. amylotrophicus, L. fermentum, L. plantarum, L. paraplantarum, L. pentosus, L. manihotivorans, and *L. paracasei* are robust starch-utilizing and lactic acid-producing organisms. They ferment complex carbohydrates like wheat flour, corn flour, and rice, which are nature's energy reserves (Abedi & Hashemi, 2020). These substrates undergo transformative metabolic processes, producing nonvolatile compounds like free amino acids, simpler carbohydrates, peptides, organic acids, vitamins, and fatty acids, and volatile compounds like alcohols, benzene derivatives, furan derivatives, acids, aldehydes, esters, ketones, hydrocarbons, and terpenes (Kieliszek et al., 2021). Volatile compounds contribute to the aroma, while nonvolatile compounds influence the taste of fermented byproducts (Wang et al., 2021). Lactobacillus-based starch fermentation is a traditional practice in various foods and beverages. Fermentation of natural substrates with LAB enhances nutritional value and bioavailability of metabolites and bioactives, contributing to healthcare. These bioactive compounds contain antifungal compounds and regulate mycotoxin detoxification. LAB generates organic acids like citric acid, which are useful in degrading aflatoxins, thereby improving the food, skincare, and healthcare industries (Sionek et al., 2023).

8.2 Metabolism of Carbohydrates (Mono, Di, Oligo, and Polysaccharides)

Lactobacillus ferments carbohydrates in neutral to acidic environments. During the fermentation process, lactic acid is the primary product, formed through two distinct pathways. According to Liu et al. (Liu, 2003), the first mechanism is homofermentation, in which lactobacillus primarily produces lactic acid from carbohydrates. Heterofermentation is the second process, in which lactobacillus breaks down carbohydrates into lactic acid, acetic acid, ethanol, and carbon dioxide.

8.2.1 Homofermentation

For the fermentation of food and feed, LAB, such as Lactobacillus strains, have been employed extensively (Giraffa et al., 2010). They can perform homofermentation, which converts carbohydrates to lactic acid as the main product, making them ideal for creating a range of fermented foods and drinks. During homofermentation, the glycolysis pathway converts pentoses like arabinose and xylose and hexoses like glucose, fructose, lactose, and galactose into lactic acid (Behera et al., 2018). The volatile compounds that give fermented foods their sour, acidic, and unique scent are produced during this process. Additionally, it aids in extending the shelf life and safety of fermented foods (Wang et al., 2021). Multiple selection criteria will be used in a screening method to better identify Lactobacillus strains with optimal homofermentation ability. Homolactic fermentations are carried out by species of Lactobacillus and Lactococcus.

8.2.2 Heterofermentation

Heterolactic fermentation is another important metabolic activity in lactobacillus, where carbohydrates from lignocellulose materials are converted into the compounds lactic acid, acetic acid, ethanol, diacetyl acetoin, and CO_2 by the phosphogluconate and phosphoketolase pathways (Abdel-Rahman et al., 2011; Gänzle, 2015). Lactobacillus can employ a variety of carbon sources and produce a greater range of end products by using the heterolactic fermentation pathway. To determine if fermentation is homo or hetero, a CO_2 generation test can be utilized. Heterofermentative lactobacillus include *Lactobacillus brevis*, *Lactobacillus fermentum*, and *Lactobacillus reuteri*. Leuconostoc, Weissella, and Oenococcus are the three main species that do heterolactic fermentation (Abedi & Hashemi, 2020).

8.3 Monosaccharides

Glucose is a vital substrate and a primary source of energy in bacterial fermentation processes (Behera et al., 2018). As part of its glucose metabolism, LAB uses the Embden–Meyerhof path to turn glucose into pyruvate. Lactic acid is a byproduct of the pyruvate after lactate dehydrogenase processes (Gänzle, 2015). Since LAB can primarily convert glucose into lactic acid, acetic acid, ethanol, and carbon dioxide, heterofermentation yields a greater range of end products than homofermentation (Gänzle, 2015). This metabolic pathway allows lactobacilli to efficiently consume glucose and generates a range of metabolites that improve the flavor and aroma profiles of fermented foods (Bintsis, 2018; Wang et al., 2021). Galactose will be transformed by the lactobacillus into lactic acid in lactic acid fermentations via the tagatose or Leloir pathway, and it will be turned into lactic acid, acetic acid, and ethanol in heterolactic fermentations via the Leloir pathway (Iskandar et al., 2019; Turner & Martley, 1983). During the process of homolactic fermentation, bacteria metabolize monosaccharides like fructose, it first phosphorylates the fructose through fructose kinase to create fructose 1-phosphate. Pyruvate is then transformed to pyruvate via the Embden–Meyerhof–Parnas (EMP) route, and ultimately, pyruvate is reduced to lactic acid. Through the hexose monophosphate (HMP) pathway, bacteria produce lactic acid, acetic acid, ethanol, and carbon dioxide during heterolactic fermentation (Kang et al., 2013; Wu & Zhao, 2019). In homolactic fermentations, where one galactose molecule is directly metabolized to two molecules of lactic acid. Galactose molecules are metabolized into glucose 6-phosphate during heterolactic fermentations, whereupon they enter the glycolytic pathway and yield pyruvate. Following that, pyruvate is transformed into acetic acid, lactic acid, and ethanol (Grossiord et al., 1998; White et al., 1955).

In homolactic fermentation, LAB will ferment xylose to lactic acid. Xylose isomerase isomerizes xylose to xylulose, which is then phosphorylated to xylulose 5 phosphate by the xylulose kinase enzyme. The final products of xylulose 5 phosphate conversion are acetic and lactic acids. Lactobacillus species can also ferment arabinose, a pentose sugar commonly found in plant cells. During fermentation, lactobacillus converts arabinose into lactic acid through the pentose phosphate pathway. While in heterofermentation, it gets converted into lactic acid, acetic acid, ethanol, and CO_2 (Abedi & Hashemi, 2020; Gobbetti et al., 2000).

Table 8.1 Metabolism of Monosaccharides by Lactic Acid Bacteria

Disaccharides	Organism	Enzyme	Homolactic Fermentation	Heterolactic Fermentation
Sucrose	*Lb. delbrueckii sp. delbrueckii, Lb. rhamnosus, Halolactibacillus halophilus, Lb. reuteri, Lactococcus lactis, Lb. gasseri, Lb. plantarum and Lb. paracasei.*	Invertase	Lactate	Lactate/ethanol and CO_2
Lactose	*Lb. delbrueckii sp. bulgaricus, Lb. helveticus sp., Lb. rhamnosus, Lb. bulgaricus, Str. Thermophilus, Lb. plantarum and L. Lactis.*	Lactase	Lactate	Lactate, ethanol, and CO_2
Maltose	*Lb. helveticus sp. and L. lactis.*	Maltase	Lactate	Lactate/ethanol and CO_2
Cellobiose	*Lb. delbrueckii sp. bulgaricus*	Cellobiase	Lactate	–

Overall, the homofermentative and heterofermentative properties of Lactobacillus species allow them to efficiently utilize a wide spectrum of hexoses and pentoses including glucose, fructose, galactose, xylose, and arabinose (Table 8.1).

8.4 Disaccharides

Sugar moieties including sucrose, lactose, maltose, and cellobiose can all be fermented by LAB (Abdel-Rahman et al., 2011; Wu & Zhao, 2019). Two simple sugars are combined to create these molecules. Enzymes such as lactase, cellobiase, maltase, and invertase convert these disaccharides. During homofermentation, lactic acid is produced from each of these substrates. Heterolactic fermentation produces lactic acid, ethanol, and carbon dioxide as mentioned in Table 8.2 (Abedi & Hashemi, 2020; Calabia et al., 2011; Kawai et al., 2020).

Table 8.2 Metabolism of Disaccharides by Lactic Acid-Producing Bacteria

Oligosaccharides	Organisms	Short-Chain Fatty Acids
Fructooligosaccharides and inulin	*Lb. casei, Lb. helveticus, L. lactis,*	Lactic acid Propionic acid Acetic acid
	Lb. mali, Lb. paraplantarum, Lb. paracasei, Lb. acidophilus, Lb. reuteri, B. brevis.	Lactic acid Butyric acid Propionic acid Acetic acid

8.5 Oligosaccharides

Oligosaccharides like fructooligosaccharides and inulin will be converted to SCFAs like lactate, acetate, butyrate, and propionate molecules (Renye et al., 2021). Species including lactobacillus, enterococcus, and bifidobacterium are capable of fermenting oligosaccharides as mentioned in Table 8.3.

8.6 Polysaccharides and Indigestible Carbohydrates

Polysaccharides are called long-chain carbohydrate moieties that are bonded to one another via glycosidic linkages inside the glycan chain. Similar to α- or β-glycosidic bonds, these linkages usually occur between identical or nonidentical monosaccharides such as fructan, inulin, cellulose, hemicellulose, and starch. These polysaccharides can be created by combining homopolysaccharides (HoPS), which are combinations of identical monosaccharides, or heteropolysaccharides (HePS), which are combinations of HoPS that are not identical. LAB use a variety of hydrolases, including pullulanases, neopullulanases, amylopullulanases, maltogenic amylases, glycosidase, etc., to break down these complex compounds (Petrova et al., 2013). The metabolism of starch substrates by lactic acid-producing strains requires robust design of experiments, statistical optimizations, and validation in their process development. This will improve robustness in the technology development in a low cost (Rao et al., 2008).

The clade of polysaccharide fermentation includes a variety of lactic acid-producing species. According to the Kyoto Encyclopedia of Genes and Genomes (KEGG) pathway, (Gänzle, 2015) these organisms will metabolize the polysaccharides into monosaccharides or simple sugars using two catabolic processes for starch fermentation:

a) Hydrolysis of starch yields dextrin and subsequently glucose.
b) The breakdown and transformation of terminal glucose residues into α-D-glucose 1-phosphate, a phosphorylated form of glucose.

These are fermented into a variety of organic acids after being broken down into glucose and simple sugars. LAB species such as *Lb. amylophilus, Lb. amylovorous, Lb. acidophilus, Lb. amylolyticus, Lb. amylotrophicus, Lb. fermentum, Lb. plantarum, Lb. paraplantarum, Lb. pentosus, Lb. manihotivorans, Lb. paracasei,* etc., can ferment polysaccharides by utilizing polysaccharide-degrading enzymes (Nagarjun et al., 2005; Petrova & Petrov, 2012; Petrova et al., 2013; Wang et al., 2021).

8.7 Indigestible Starch/Resistant Starch

Indigestible starch also called resistant starch (RS) is a type of starch found in potatoes, rice, lotus seeds, unripened bananas, and many more. In the gut, it can be fermented by gut microbes like *Bifidobacterium adolescentis* and *Ruminococcus spp.*, producing three

Table 8.3 Metabolism of Fructooligosaccharides and Inulin by Lactic Acid-Producing Bacteria

	Hexose Sugars			Pentose Sugars	
Monosaccharide	**Glucose[a]**	**Fructose[b]**	**Galactose[c]**	**Arabinose[d]**	**Xylose[e]**
Homolactic fermentation	Lactic acid	Lactic acid	Lactic acid	Lactic acid	Lactic acid
Pathway	EMP	EMP	Tagatose or Leloir	Pentose phosphate	Phosphoketolase or pentose phosphate
Heterolactic fermentation	Lactic acid, acetic acid, ethanol and CO_2	Lactic acid, acetic acid, ethanol and CO_2	Lactic acid, acetic acid and ethanol	Lactic acid, acetic acid, ethanol and CO_2	Lactic acid and acetic acid
Pathway	Phosphoketolase	Mannitol dehydrogenase, phosphoketolase	Leloir	Phosphoketolase	Phosphoketolase

[a] *Lb. delbrueckii, Lb. helveticus sp. milano, Lb. rhamnosus, Lb. casei, Lb. bulgaricus, Lb. lactis sp., Lb. plantarum, Lb. amylophilus, Lb. zeae, Lb. salivarius, Lb. coryniformis, E. faecalis, Lb. pentosus, Lb. paracasei. Brevis, Sporolactobacillus sp. strain.*

[b] *Lb. rhamnosus and Lb. delbrueckii, Lb. panis, Leuconostoc citreum, Levilactobacillus brevis.*

[c] *Lb. bulgaricus, Lb. pentosus, Lb. plantarum, L. lactis.*

[d] *Lb. plantarum, E. faecium, P. acidilactici.*

[e] *Lb. pentosus, Lb. brevis, Leuconostoc lactis, Lb. delbrueckii sp. bulgaricus, Lb. lactis, Lb. plantarum, E. mundrii, Lb. rhamnosus, Weissella sp.*

Table 8.4 Different Varieties of Resistant Starch

Resistant Starch Variety	Availability
Type 1	Type 1 RS is a naturally occurring kind of resistant starch found in grain, seed, and legume cell walls
Type 2	Type 2 RS is a natural kind of resistant starch that can be found in starchy foods such as unripe bananas and uncooked potatoes
Type 3	Type 3 RS is a naturally occurring kind of resistant starch found in rice and potatoes, but it is created through heating/cooking and immediate cooling
Type 4	Type 4 RS is an artificial form of resistant starch where this RS is made by chemical processes
Type 5	Type 5 RS is an artificial form of resistant starch where the amylose content of the RS bounds with fat, protein, and other carbohydrates and making it more nondigestible

SCFAs: acetate, butyrate, and propionate (Crost et al., 2018; Jung et al., 2019; Ze et al., 2012). There are five different types of RS. The RS variety includes Type 1 of RS, found in grains, seeds, and legumes; Type 2 of RS is found in starchy foods like raw potatoes and unripened bananas; Type 3 is found in rice and potatoes formed by heating/cooking and cooling; Type 4 in the artificial form of RS made through chemical processes; and Type 5 is a synthetic form where amylose content binds with fat, protein, and carbohydrates, making it nondigestible, as shown in Table 8.4.

Recent advancements show that RS Types 2, 3, and 4 can be fermented by various lactic acid-producing organisms like *Lb. casei, Lb. plantarum, Lb. paracasei, Lb. acidophilus, Lb. delbrueckii, Lb. bulgaricus, B. longum*, etc. resulting in the production of SCFAs like acetate, butyrate, and propionate as shown in Figure 8.1. This RS content has prebiotic effects by improving the shelf life and viability of the probiotic strains for commercial use (Tekin & Dincer, 2023; Wang et al., 2022).

8.8 Metabolism of Nitrogen Source (Proteins)

Lactic acid bacteria undergo rapid intracellular metabolism during fermentation, requiring numerous free amino acids for growth and development. If availability is insufficient, LAB initiates proteolytic metabolism, varying based on species and subspecies (Chopin, 1993; Sanchez et al., 2018).

a) *Lacticaseibacillus casei* requires around 10 free amino acids: including Leucine, Valine, Arginine, Isoleucine, Phenylalanine, Glutamate, Aspartate, Tyrosine, Tryptophan, and Cysteine.
b) *Lactobacillus plantarum* requires around seven free amino acids: including Leucine, Valine, Methionine, Isoleucine, Phenylalanine, Tryptophan, and Glutamate.

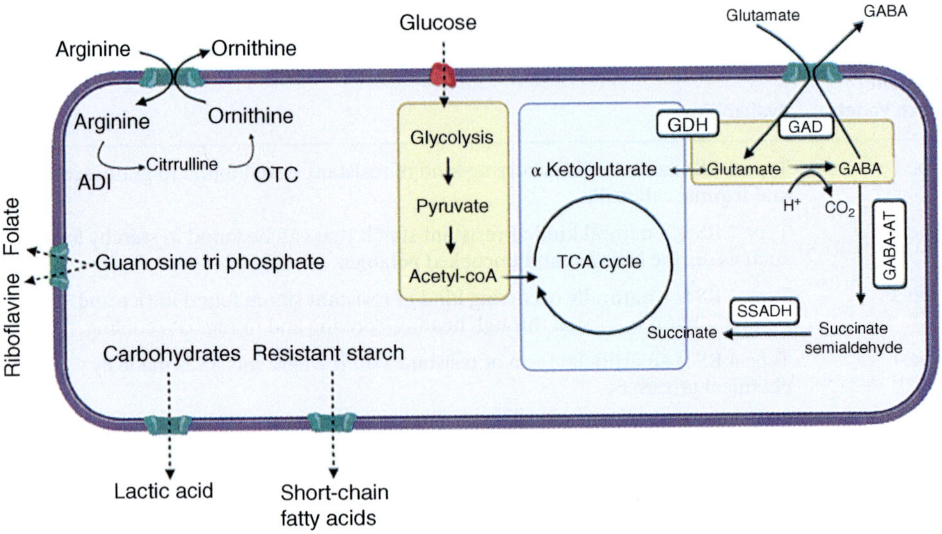

Figure 8.1 Mechanism of digestion of carbon and nitrogen sources and production of organic acid, GABA, vitamins, short-chain fatty acids, and ornithine. ADI: arginine deiminase; GABA: gamma amino butyric acid; GAD: glutamate decarboxylase; GDH: glutamate dehydrogenase; OTC: ornithine transcarbamylase; SSADH: succinic semialdehyde dehydrogenase; TCA: tricarboxylic acid cycle.

c) *Lactobacillus helveticus* requires around 15 free amino acids: including Leucine, Valine, Histidine, Isoleucine, Arginine, Methionine, Phenylalanine, Proline, Threonine, Lysine, Aspartate, Tyrosine, Cysteine, Tryptophan, and Glutamate.
d) *Lactococcus species* requires around seven free amino acids: including Valine, Leucine, Isoleucine, Histidine, Methionine, Glycine, and Glutamic acid.

A series of steps is involved in the protein metabolism to meet the amino acid requirement during the fermentation process. Step 1: Degradation of large protein molecules and formation of oligopeptides with proteinases. Step 2: Peptides are transported via three separate transport systems: oligopeptide, dipeptide, and tripeptide (Opp, DtpP, and DtpT). Step 3: The transported peptides will be digested by different peptidases such as aminopeptidases, dipeptidyl peptidases, tripeptidyl peptidases, carboxypeptidases, peptidyl dipeptidases, dipeptidases, omega peptidases, endopeptidases, and proline-specific peptidases (Kieliszek et al., 2021; Wang et al., 2021).

In the dairy industry, LABs will digest casein protein to match its requirements of nitrogen and amino acids. In the food industry, LAB digests different protein sources to meet the requirements of free amino acids. Lactobacillus can ferment other food materials like fruits, vegetables, flour, etc., (Sionek et al., 2023) and digest protein molecules like albumins, globulins, gliadins, glutelins, and prolamins (Park et al., 2020).

LAB strains, specifically *Lacticaseibacillus casei*, can hydrolyze protein allergens in wheat sourdough during fermentation. This helps reduce allergenicity caused by proteinaceous foods.

In the skincare industry, LAB ferments from natural substrates like flour, fruits, vegetables, millet, and rice can produce beneficial substances like peptides, metabolites, and free amino acids that are easily absorbed into the skin, providing beneficial effects (Noguchi & Djerassi, 2009). This fermentation process contributes to the development of effective skincare products (Burke et al., 1966; Fernandes et al., 2023).

8.8.1 Metabolism of Amino Acids

During acid stress, LAB shifts its metabolism from carbohydrate fermentation to amino acid utilization, producing various compounds like flavored, aromatic, nonproteinaceous, biogenic amines, and dextrorotatory amino acids (Gänzle, 2015; Sionek et al., 2023; Wang et al., 2021). The synthesis of these compounds depends on strain specificity and substrate specificity. The branched-chain amino acid metabolism, aromatic amino acid metabolism, glutamate decarboxylase (GAD) pathway, and arginine deiminase (ADI) pathway are the four primary amino acid metabolism processes in which LABs are implicated. These pathways produce ammonia during amino acid deamination, adapted to the acidic environment and protecting cells from acidic stress (Gänzle, 2015; Vrancken et al., 2009; Tang et al., 2023).

8.8.2 Arginine Deiminase Pathway

At pH values over 3.5, LAB switches to arginine metabolism via the ADI pathway during pH homeostasis and the stationary phase of lactic acid fermentation, producing ornithine and enhancing acid tolerance. The enzyme ornithine transcarbamylase transforms arginine into ornithine and carbamoylase phosphate after ADI transforms it into citrulline and ammonia (Vrancken et al., 2009). Uptake and synthesis are aided by the arginine/ornithine antiporter (Majsnerowska et al., 2018; Tang et al., 2023).

8.8.3 Glutamate Decarboxylase Pathway

Glutamate decarboxylase is an enzyme that converts glutamate into γ-aminobutyric acid, and it is produced by LAB in the presence of L-glutamic acid. Gamma amino butyric acid (GABA) is a nonproteinaceous 4-carbon amino acid, it is an inhibitory neurotransmitter in the brain, promoting relaxation and dampening activity. It plays various physiological functions, including improving sleeplessness, controlling blood pressure, reducing anxiety, and supporting liver function. Microbial fermentation is preferred due to environmental concerns to produce GABA, facilitated by the GAD gene in LAB (Small & Waterman, 1998).

The GAD enzyme is a crucial component of GABA synthesis, encoded by gadA or gadB genes and gadC, facilitating decarboxylation within a compact structure. GadC protein facilitates L-glutamate uptake and GABA release through the same channel. The enzyme GAD catalyzes the transformation of L-glutamine into GABA, resulting in carbon dioxide production and extracellular transport of newly synthesized GABA (Tang et al., 2023).

As a secondary pathway, glycolysis followed by TCA will produce α-ketoglutarate, and it will act as a precursor to the L-glutamate. The enzyme known as L-glutamate dehydrogenase (GDH) will transform it into L-glutamate. The converted L-glutamate will also get decarboxylated to GABA (Figure 8.1).

Lactic acid bacteria like *Lactiplantibacillus plantarum, Levilactobacillus brevis, Lactobacillus delbrueckii subsp. bulgaricus, Lactobacillus helveticus, Lactobacillus paracasei, Lactobacillus fermentum, Lactococcus lactis, Lactobacillus casei, Lactobacillus buchneri,* and *Streptococcus salivarius* are considered as good GABA-producing strains (Gänzle, 2015; Kook & Cho, 2013; Park et al., 2021; Wang et al., 2021). Other than LAB, few other organisms which can produce adequate GABA are *Saccharomyces cerevisiae, Aspergillus oryzae, and Kluyveromyces marxianus.* Among all these, LAB strains produce higher amounts of GABA and they are considered as safe. Amino acids other than Glutamate produce biogenic amines under decarboxylation reactions. Decarboxylation of tryptophan, lysine, histidine, tyrosine, histamine, and ornithine can produce cadaverine, tyramine, tryptamine, putrescine, histamine, etc. (Barbieri et al., 2019). These biogenic amines have negative effects on human health when ingested. The selection of carboxylase-negative lactic acid strains can regulate the production of biogenic amines (Hayat et al., 2015).

8.8.4 Metabolism of Branched-Chain and Aromatic Amino Acids

Lactobacillus metabolism includes the breakdown of some amino acids to produce sulfur-containing volatile compounds, branched chain volatiles, and aromatic chemicals. Aromatic volatiles are produced by amino acids including phenylalanine, tyrosine, and tryptophan. On the other hand, amino acids such as leucine, valine, and isoleucine produce branched-chain volatiles including 3-methylbutan-1-ol, 3-methylpropan-1-ol, and 3-methyl butanoic acid, while methionine and cysteine make sulfur-containing volatiles. These substances are made in minimal amounts (Park & Kim, 2019).

Leucine is also responsible for producing Leucic acid or HICA (2-hydroxyisocaproic acid) It is an organic acid metabolite, that can be used as a muscle-building supplement. It also has fungicidal and antibacterial properties. HICA is a naturally occurring substance that is created by LAB and humans via the leucine degradation pathway (Park et al., 2017).

Phenyllactic acid is a natural metabolite that can be synthesized under the metabolism of phenylalanine (Sun et al., 2019). This natural metabolism can be observed in several microbes including LAB. Phenyllactic acid prevents the growth of harmful bacteria and can be used as a preservative in food, cosmetics, and pharmaceuticals.

A specific enzyme called deoxynucleic acid aldolase will metabolize thymine and produce aromatic compounds like acetaldehyde and glyceraldehyde. Threonine will be metabolized by *streptococcus thermophilus* to glycine by the serine hydroxymethyl transferase (SHMT) along with the threonine aldolase enzyme (Chaves et al., 2002). Nonetheless, LAB also produce flavor compounds by metabolizing aspartic acid. Diacetyl, acetoin, and 1,3-butanediol will be synthesized by *Lacticaseibacillus paracasei* (Thage et al., 2005). *Lacticaseibacillus casei, Lacticaseibacillus rhamnosus, and streptococcus thermophilus* are some of the LAB responsible for producing branched-chain amino acids and aromatic amino acid derivatives (Table 8.5) (Bancalari et al., 2017).

8.8.5 D-Amino Acid Production

Lactic acid bacteria contain distinct genes that code for D-amino acid metabolism. According to the recent research, LAB are capable of producing large amounts of

Table 8.5 Metabolism of Aromatic and Branched-Chain Amino Acids

Amino Acids	Aromatic Compounds/Branched-Chain Amino Acids/Organic Acids	Reference
Leucine	2-Hydroxyisocaproic acid (HICA)	Park et al. (2017)
Phenylalanine	Phenyllactic acid	Sun et al. (2019)
Thymine	Acetaldehyde and glyceraldehyde phosphate	Chaves et al. (2002)
Threonine	Glycine and acetaldehyde	Chaves et al. (2002)
Aspartic acid	Diacetyl, acetoin, and 1,3-butanediol	Thage et al. (2005)

dextrorotatory amino acids, including D-alanine, D-glutamate, and D-aspartate. Enzymes such as AlaR, AspR, and GluR (glutamate racemase) catalyze the reversible conversion of L-amino acids, which produce all of these amino acids. A few of them can be identified in the bacterial cell wall components (Kobayashi, 2019; Marcone et al., 2020).

8.9 Utilization and Metabolism of Malic Acid and Citric Acid

During alcoholic-based fermentations, malic acid undergoes decarboxylation by LAB like *Oenococcus, Lactobacillus, Pediococcus, Lactococcus*, and *Leuconostoc*, releasing lactic acid and CO_2 into the environment (Paramithiotis et al., 2022). When supplemented with citrate, these bacteria produce aromatic compounds like diacetyl, acetoin, and butanediol (Bandell et al., 1997). They also perform oxaloacetate decarboxylation on oxaloacetate, producing pyruvate and CO_2. The pyruvate can also be metabolized into other compounds like lactate, formate, acetate, and ethanol (Sarantinopoulos et al., 2001). *Lactococcus lactis* and *Leuconostoc* species are majorly identified as having the ability to metabolize citrate. If LAB cannot complete citrate metabolism, succinic acid is formed through malic acid and fumaric acid (Table 8.6) (Torino et al., 2005; Y. Wang et al., 2021).

Table 8.6 Utilization and Metabolism of Malic Acid and Citric Acid

Organic Acid	Organism	Metabolite	References
Malic acid	*Lactiplantibacillus plantarum, Oenococcus oeni, Lactococcus lactis, Pediococcus, Leuconostoc,*	L-lactate and CO_2	Paramithiotis et al. (2022); Wang et al. (2021)
Citric acid	*Lactobacilli*	Succinate/lactate or acetylacetone	Gänzle (2015); Wang et al. (2021)
	T. halophilus	Acetate and ethanol	
	Lc. lactis, Leuconostoc spp. and O. oeni	Acetoin or lactate	

8.10 Metabolite Profiles of Lactobacillus Ferments

8.10.1 Organic Acids

Lactic acid bacteria produce various organic acids, including lactic acid, formic acid, acetic acid, succinic acid, and SCFAs like propionic acid and butyric acid, in homo- and heterolactic fermentations (Tang et al., 2023; Tekin & Dincer, 2023). The production of these acids depends on the carbon source and amino acid availability in the substrate. The aldolase enzyme is crucial in glucose metabolism in both homolactic and heterolactic fermentations. Lactic acid, a key metabolite, can be produced in two forms: Dextrorotatory (D-lactic acid) and Levorotatory (L-lactic acid) (Kim et al., 2020; Kuley et al., 2020; Kumar et al., 2013; Shirai et al., 2001). The optical rotation of D- and L-lactic acids depends on the enzymes D-lactate dehydrogenase and L-lactate dehydrogenase in pyruvate conversion (Figure 8.1) (Kim et al., 2020). In aerobic fermentation conditions, α-acetolactate can be obtained through α-acetolactate synthase. Other organic acids like 2-hydroxyisocaproic acid and phenyllactic acid can be produced through amino acid catabolism, such as leucine and phenylalanine (Park et al., 2017; Sun et al., 2019). The biotechnologically developed lactic acid is considered safe with GRAS (Generally Recommended as Safe) status, authorized by the U.S. Food and Drug Administration (FDA) (Table 8.7) (Abedi & Hashemi, 2020; Huang et al., 2023b).

8.10.2 Bacteriocins

Bacteriocins are small extracellular ribosomically synthesized peptide molecules produced by Gram-positive and Gram-negative bacteria, with antimicrobial activity. They inhibit the growth of the organism and are heat and pH-tolerant (Perez et al., 2014). They are odorless, colorless, and tasteless. Bacteriocins are categorized into Class I, II, and III varieties based on physiochemical and structural properties (Zacharof & Lovitt, 2012).

Class I also termed as lantibiotics. They are small heat-stable peptides with 2-5 kDa molecular weight and made of polycyclic thioether amino acids like lanthionine, methyllan-thione, and unsaturated amino acids like dehydroalanine and 2-amino isobutyric acid.

Table 8.7 Metabolism of Carbohydrates and Amino Acids with Subsequent Production of Organic Acids

Carbohydrates	Organic Acids	References
Glucose, lactose, sucrose, starch, glycerol, etc.	Lactic acid, formic acid, succinic acid, 3-hydroxypropionic acid	Kumar et al. (2013), Shirai et al. (2001), Kuley et al. (2020), Kim et al. (2020)
Resistant starch	Acetic acid, propionic acid, butyric acid	Tekin and Dincer (2023)
Amino Acids		
Leucine	2-hydroxyisocaproic acid (HICA)	Park et al. (2017)
Phenylalanine	Phenyllactic acid	Sun et al. (2019)

Lantibiotics are further categorized into two varieties, based on the charge of the molecules (Kaur & Kaur, 2015).

Type A lantibiotics are positively charged peptides with 2–4 kDa. Nisin and lacticin are considered into type A lantibiotics (Kaur & Kaur, 2015).

Type B lantibiotics are neutral or negatively charged peptides with 2–3 kDa. Mersacidin is considered into type B lantibiotics (Sahl & Bierbaum, 1998).

Class II bacteriocins are small heat-stable peptides with less than 10 kDa molecular weight. These peptides do not contain lanthionine (Kaur & Kaur, 2015).

These are further categorized into three subclasses.

subclass IIa comprising Pediocin PA-1 and sakacin A (Kaur & Kaur, 2015).
subclass IIb comprising Lactacin F and lactococcin G (Kaur & Kaur, 2015)
subclass IIc comprising Gassericin A, circularin A, and carnocyclin A (Cotter et al., 2005; Kawai et al., 2004).

Class III: These are heat sensitive protein molecules with greater than 30 kDa molecular weight. These include colicins, helveticin I, and enterolysin (Kaur & Kaur, 2015).

Bacteriocins inhibit the growth of target organisms in two main modes: Class I inhibits peptidoglycan synthesis, and Class II and Class III form pores in cells to kill bacteria (Cotter et al., 2013). The production of bacteriocins is influenced by the use of complex growth media and the selected strain's growth conditions during fermentation. Bacteriocins are produced during the exponential phase of the growth cycle (Lv et al., 2005).

Complex media composition with glucose, nitrogen sources, yeast extract, peptone, and protein hydrolysate-rich compositions will give adequate production of bacteriocins (Altaf et al., 2007). Starch has a negative impact on the production. Surfactants like tween 80 are also essential components (Zendo et al., 2005). Low temperatures, optimal pH, aeration, and oxygen supply are directly connected to bacteriocin production. NaCl and ethanol can be used to downregulate bacteriocin production (Leroy et al., 2003; Verluyten et al., 2004).

8.11 Vitamins

Lactic acid bacteria can synthesize vitamins like folic acid, riboflavin, vitamin C, thiamine, pyridoxal, and cobalamin when fermented with different substrates and added supplements (Figure 8.1) (Capozzi et al., 2012; LeBlanc et al., 2020; Taranto et al., 2003). Riboflavin is produced by riboflavin synthase coding genes on a rib operon, involving seven enzymatic reactions (Chen et al., 2017; Ge et al., 2020; Levit et al., 2021; Perez Visñuk et al., 2022; Spacova et al., 2022). Similarly, folic acid can be made when supplemented with P-aminobenzoic acid (pABA) through the folic acid synthesis pathway following 17 enzymatic reactions (Albano et al., 2020; Levit et al., 2021; Mosso et al., 2018). Other vitamins produced by LAB include Vitamin K2 by *Lactococcus lactis*, Vitamin B2 (riboflavin), B3 (nicotinamide and nicotinic acid), and B6 (pyridoxine) by *Latilactobacillus sakei*, and cobalamin by *Lactobacillus coryniformis* strains. These bacteria produce various vitamins and minerals through their synthesis pathways (Liu et al., 2019; Oguro et al., 2017; Torres et al., 2018; Wang et al., 2021).

8.12 Short-chain Fatty Acids

Short-chain fatty acids, including acetate, butyrate, propionate, isobutyrate, valerate, and isovalerate, are fatty acids produced by microorganisms through the fermentation of RS or dietary fiber. Acetate, propionate, and butyrate are the main SCFAs produced by lactobacillus organisms. These fermented fatty acids can be used for various health benefits, including anti-inflammatory, antiobesity, antidiabetes, anticancer, hepatoprotection, cardiovascular protection, neuroprotection, constipation treatment, and immunoregulation (Xiong et al., 2022). Intake of suitable RS will help in weight loss and rebalance the gut microbiota (Li et al., 2024). Over the past decade, research on the effects of SCFAs on rodents has shown a positive impact on mitigating skin diseases like Eczema, Psoriasis, Atopic dermatitis, Acne, and Chronic spontaneous Utricaria, with different topical applications (Coppola et al., 2022; Xiao et al., 2023).

8.13 Exopolysaccharides

Exopolysaccharides are polysaccharide macromolecular substances produced by the polymerization of multiple monosaccharides or their derivatives. For example, Lasiodiplodan, produced by Lasiodiplodia theobromae, is an exopolysaccharide which has numerous biological properties and can be used in cosmetics, pharmaceutical, and nutraceutical industry (Ascencio et al., 2021). Lactic acid-producing organisms like *Streptococcus thermophilus, Limosilactobacillus reuteri, Lacticaseibacillus casei, and Lactiplantibacillus plantarum* can synthesize EPS when supplemented with different carbon sources like fructose, sucrose, glucose, galactose, and lactose(Sørensen et al., 2022). There are at least one or more clusters of EPS synthesizing genes in the genome of these bacteria (Cerning, 1990). HoPS and HePS are two types of EPS that typically exist as a slimy layer on bacterial cell surfaces. HoPS are polymers of similar monosaccharide units like dextran or fructan, while HePS are polymers of different monosaccharide units including glucose, galactose, or rhamnose in different combinations (Sanalibaba & Cakmak, 2016). EPS, such as glucan, can be synthesized by LAB in the bread-making process or dough fermentation (Wang et al., 2021). Glucan can be used as a gelling agent, thickener, stabilizer, and emulsifier in food and cosmetic products (Nguyen et al., 2020; Sanalibaba & Cakmak, 2016; Silva et al., 2019).

8.14 Antioxidant Substances

Glutathione is one of the major antioxidants produced during lactic acid fermentation. It is a nonprotein thiol and a low-molecular-weight tripeptide molecule. It is made with a combination of glutamate, cysteine, and glycine (Pophaly et al., 2012). During fermentation, lactic acid-producing bacteria require complex media like MRS, Typticase soya broth, M-17, lactobacillus media, etc. to produce glutathione (Wang et al., 2021).

8.15 Production of Polyols

Lactic acid bacteria ferment plant-based materials like vegetables, cabbage, soybean, and wheat, converting free fermentable oligosaccharides and monosaccharides into polyols. These sugar alcohols, known as polyols, can be used as low-calorie sweeteners in the food industry and as humectants in the cosmetic industry due to their moisture-holding properties. LAB species like *Levilactobacillus brevis* and *Leuconostoc* can reduce fructose to mannitol, while other organisms like *Oenococcus, Leuconostoc mesenteroides,* and *Fructi Lactobacillus sanfranciscensis* can produce erythritol. Soybean fermentation by *Leuconostoc, T. halophilus, and E. faecium* species produces mannitol and xylitol (Table 8.8) (Ortiz et al., 2013).

8.16 Metabolomic Profiles of Different Lactic Acid Bacteria in the Rice Fermentation

Rice, a staple food, is rich in nutrients, lipids, protein, crude fiber, dietary fiber, ash, and bioactive compounds, including γ-oryzanol, phytic acid, ferulic acid, γ-aminobutyric acid, and tocopherol (Jin et al., 2023). It is also a rich source of amino acids like glutamine,

Table 8.8 Production of Polyols by Lactic Acid Bacteria

Substrate	Organism	Polyols	References
Mixed vegetables (green tomatoes, carrots, and cauliflowers)	*Lb. plantarum, Lb. brevis, Leuconostoc mesenteroides, Leuconostoc citreum*	Mannitol	Wouters et al. (2013); Ortiz et al. (2013)
Chinese cabbage	*Leuconostoc mesenteroides, Lactobacillus spp.*	Mannitol	Jung et al. (2011, 2012)
Soybean fermented food	*Leuconostoc mesenteroides, Tetragenococcus halophilus, E. faecium*	Mannitol and Xylitol	Kim et al. (2009); Namgung et al. (2010)
Cocoa pulp & Cocoa beans	*Leuconostoc pseudomesenteroides, Lb. fermentum, Fructobacillus tropaeola*	Mannitol	Papalexandratou et al. (2011)
Wheat and spelt sourdoughs	*Lb. plantarum, Lb. fermentum, Lb. rossiae, Lb. brevis, Lb. paraplantarum*	Mannitol and Erythritol	Papalexandratou et al. (2011); Van der Meulen et al. (2007)
Gluten-free sourdough	*L. curvatus, L. reuteri, L. animalis*	Mannitol	Rühmkorf et al. (2012)
Sorghum and wheat sourdoughs	*L. sanfranciscensis*	Mannitol	Galle et al. (2010)

proline, leucine, alanine, tryptophan, methionine, histidine, lysine, isoleucine, glycine, serine, tyrosine, and valine (Jayaprakash et al., 2022). Due to its high glycemic index values (64–93%), it has become a leading dietary cause of high blood sugar levels (Pereira et al., 2021).

Rice fermentation with LAB and advanced cooking methods can reduce rice's glycemic index and blood sugar levels by 50% (Dundar & Gocmen, 2013; Pereira et al., 2021). These fermentations improve nutritional availability by digesting rice proteins, peptides, organic acids, carbohydrates, aldehydes, esters, ketones, hydrocarbons, terpenes, alcohols, furan derivatives, and benzene derivatives. Different lactic acid producers, such as *Lacticaseibacillus paracasei, Latilactobacillus sakei, Lactiplantibacillus pentosus, Pediococcus pentosus, Pediococcus acidilactici, Levilactobacillus brevis, Lentilactobacillus hilgardii, Leuconostoc pseudomesenteroides,* and *Weissella cibaria,* produce various bioactive compounds in rice ferments (Table 8.9) (Lee et al., 2019; Park & Kim, 2019).

8.16.1 Nonvolatile Compounds

i) Amino acids including Alanine, Arginine, Citrulline, Tryptophan, Valine, Leucine, Isoleucine, Proline, Glycine, Serine, Threonine, Methionine, Aspartic acid, Glutamic acid, Phenylalanine, Ornithine, and Arginine and a nonproteinaceous amino acid GABA will be available in the rice ferments developed by LAB (Sugino et al., 2008; Vrancken et al., 2009; Yvon & Rijnen, 2001).
ii) Organic acids like lactic acid, citric acid, propionic acid, oxalic acid, and succinic acid. During obligatory heterofermentation, fumaric acid and malic acid have been detected (Cselovszky et al., 1992; García-Quintáns et al., 2008; Xiao & Lu, 2014).
iii) Different carbohydrates have been detected after the denaturation of starch like ribofuranose, xylose, arabitol, myoinositol, sucrose, maltose, trehalose, sophorose, mannose, fructose, galactose, and glucose (Lee et al., 2019).

8.16.2 Volatile Compounds

iv) Aldehydes like 2-methylbutanal, 3-methylbutanal, hexanal, nonanal, and benzaldehyde have been detected in the fermented liquids. 2-methylbutanal, 3-methylbutanal, and nonanal have not found in the obligatory heterofermentation processed with *Levilactobacillus brevis, Lb. hilgardii, Leuconostoc pseudomesenteroides, and Weissella cibaria* (Filipiak et al., 2012; Kaseleht et al., 2011a, 2011b; Vermeulen et al., 2007).
v) Esters including ethyl acetate, methyl butanoate, betyl acetate, hexyl acetate have been found in all varieties of ferments (Holland et al., 2005; Lee et al., 2018).
vi) Along with them, ketones including propan-2-one, butan-2-one, butane-2,3dione, 3-hydroxybutan-2-one, 6-methylhept-5-en-2-one, and nonan-2-one have been detected in all the ferments (Lee et al., 2016).

8.16.3 Other Volatile Compounds

vii) Other volatile compounds like alcohols including ethanol, butan-1-ol, 3-methylbutan-1-ol, pentan-1-ol, hexan-1-ol, oct-1-en-3-ol, 2-ethylhexan-1-ol, octan-1-ol, nonan-1-ol,

Facultatively Heterofermentative			Obligatory Homofermentative			Obligatory Heterofermentative		
Lacticaseibacillus paracasei *Lacticaseibacillus paracasei* *Latilactobacillus sakei* *Lactiplantibacillus pentosus*			*Pediococcus pentosus* *Pediococcus acidilactici*			*Levilactobacillus brevis* *Levilactobacillus brevis* *Lentilactobacillus hilgardii* *Leuconostoc pseudomesenteroides* *Leuconostoc pseudomesenteroides* *Weissella ciberia*		
Nonvolatile Compounds								
Amino Acids	**Organic Acids**	**Carbohydrates**	**Amino Acids**	**Organic Acids**	**Carbohydrates**	**Amino Acids**	**Organic Acids**	**Carbohydrates**
Alanine	Lactic acid	Ribofuranose	Alanine	Lactic acid	Ribofuranose	Alanine	Lactic acid	Ribofuranose
Arginine	Citric acid	Arabitol	Arginine	Citric acid	Arabitol	Arginine	Citric acid	Arabitol
Citrulline	Propanoic acid	Xylose	Citrulline	Propanoic acid	Xylose	Citrulline	Propanoic acid	Xylose
Gamma Amino Butyric Acid (GABA)	Oxalic acid	Myo-Inositol	Gamma amino butyric acid (GABA)	Oxalic acid	Myo-Inositol	Gamma amino butyric acid (GABA)	Oxalic acid	Myo-Inositol
Tryptophan	Succinic acid	Sucrose	Tryptophan	Succinic acid	Sucrose	Tryptophan	Succinic acid	Sucrose
Valine		Maltose	Valine		Maltose	Valine	Fumaric acid	Maltose
Leucine		Trehalose	Leucine		Trehalose	Leucine	Malic acid	Trehalose
Isoleucine		Sophorose	Isoleucine		Sophorose	Isoleucine		Sophorose
Proline		Mannose	Proline		Mannose	Proline		Mannose
Glycine		Fructose	Glycine		Fructose	Glycine		Fructose
Serine		Galactose	Serine		Galactose	Serine		Galactose
Threonine		Glucose	Threonine		Glucose	Threonine		Glucose
Methionine			Methionine			Methionine		Mannitol
Aspartic acid			Aspartic acid			Aspartic acid		
Glutamic acid			Glutamic acid			Glutamic acid		
Phenylalanine			Phenylalanine			Phenylalanine		
Ornithine			Ornithine			Ornithine		

(*Continued*)

Table 8.9 (Continued)

Volatile Compounds								
Aldehydes	**Esters**	**Ketones**	**Aldehydes**	**Esters**	**Ketones**	**Aldehydes**	**Esters**	**Ketones**
2-Methylbutanal	Ethyl Acetate	Propan-2-One	2-Methylbutanal	Ethyl Acetate	Propan-2-One		Ethyl Acetate	Propan-2-One
3-Methylbutanal	Methyl Butanoate	Butan-2-One	3-Methylbutanal	Methyl Butanoate	Butan-2-One		Methyl Butanoate	Butan-2-One
Hexanal	Butylacetate	Butane-2,3 Dione	Hexanal	Butylacetate	Butane-2,3 Dione	Hexanal	Butylacetate	Butane-2,3 Dione
Nonanal	Hexylacetate	3-Hydroxybutan-2-One	Nonanal	Hexylacetate	3-Hydroxybutan-2-One		Hexylacetate	3-Hydroxybutan-2-One
Benzaldehyde		6-Methyl Hept-5-En-2-One	Benzaldehyde		6-Methyl Hept-5-En-2-One	Benzaldehyde		6-Methyl Hept-5-En-2-One
		Nonan-2-One			Nonan-2-One			Nonan-2-One

Volatile Compounds								
Alcohols	**Furan Derivatives**	**Benzene and Benzene Derivatives**	**Alcohols**	**Furan Derivatives**	**Benzene and Benzene Derivatives**	**Alcohols**	**Furan Derivatives**	**Benzene and Benzene Derivatives**
Ethanol	2-Ethylfuran	Toluene	Ethanol	2-Ethylfuran	Toluene	Ethanol	2-Ethylfuran	Toluene
Butan-1-ol	2-Propylfuran	Ethylbenzene	Butan-1-ol	2-Propylfuran	Ethylbenzene	Butan-1-ol	2-Propylfuran	Ethylbenzene
3-Methylbutan-1-ol	2-Butylfuran	1,4-Xylene	3-Methylbutan-1-ol	2-Butylfuran	1,4-Xylene	3-Methylbutan-1-ol	2-Butylfuran	1,4-Xylene
Pentan-1-ol	2-Pentylfuran		Pentan-1-ol	2-Pentylfuran		Pentan-1-ol	2-Pentylfuran	1,2-Xylene
		Propylbenzene			Propylbenzene	Heptan-2-ol		Propylbenzene

Hexan-1-ol		1-Ethyl-2-methylbenzene	Hexan-1-ol		1-Ethyl-2-methylbenzene	Hexan-1-ol		1-Ethyl-2-methylbenzene
Oct-1-en-3-ol		1-Ethyl-3-methylbenzene	Oct-1-en-3-ol		1-Ethyl-3-methylbenzene	Oct-1-en-3-ol		1-Ethyl-3-methylbenzene
2-Ethylhexan-1-ol		1,3,5-Trimethylbenzene	2-Ethylhexan-1-ol		1,3,5-Trimethylbenzene	2-Ethylhexan-1-ol		1,3,5-Trimethylbenzene
		1,2,4-Trimethylbenzene				(E) hept-2-en-1-ol		1,2,4-Trimethylbenzene
Octan-1-ol		1,2,3-Trimethylbenzene	Octan-1-ol		1,2,3-Trimethylbenzene	Octan-1-ol		1,2,3-Trimethylbenzene
						(E) oct-2-en-1-ol		
Nonan-1-ol			Nonan-1-ol			Nonan-1-ol		
Butane-1,4 diol			Butane-1,4 diol			Butane-1,4 diol		

Other Volatile Compounds

Acids	Hydrocarbons	Terpenes	Acids	Hydrocarbons	Terpenes	Acids	Hydrocarbons	Terpenes
Acetic acid	Octane		Acetic acid	Octane		Acetic acid	Octane	1-Methyl-4-prop-1-en-2-ylcyclohexene (limonene)
Octanoic acid			Octanoic acid			Octanoic acid		
Hexanoic acid			Hexanoic acid			Hexanoic acid		

butane-1,4 diol, (E) hept-2-en-1-ol, (E) oct-2-en-1-ol, and heptan-2-ol will be available as volatiles but heptan-2-ol, (E) hept-2-en-1-ol, and (E) oct-2-en-1-ol are only found in obligatory heteroferments of rice fermentations (Annuk et al., 2003; Assaf et al., 1997; Laëtitia et al., 2014; Matsui et al., 2003; Vermeulen et al., 2007).

viii) Furan derivatives like 2-ethylfuran, 2-propylfuran, 2-butylfuran, and 2-pentylfuran molecules have been detected in all varieties of fermentations (Lampi et al., 2015; Yang et al., 2008).

ix) Benzene and benzene derivatives like toluene, ethylbenzene, 1,4-xylene, propylbenzene, 1-ethyl-2-methylbenzene, 1-ethyl-3-methylbenzene, 1,3,5-trimethylbenzene, 1,2,4-trimethylbenzene, and 1,2,3-trimethylbenzene are available in all the varieties of ferments, but 1,2-xylene is only available in obligatory heterofermented liquids (Lee et al., 2019).

x) Acids like acetic acid, octanoic acid, and hexanoic acid are the three different acidic forms that have been observed in three varieties of fermentations.

xi) Hydrocarbon called octane has been detected in all the fermentations.

xii) Terpenes like 1-methyl-4-prop-1-en-2-ylcyclohexene (limonene) has been detected only in the obligatory heterofermentation.

All these compounds have been detected in the rice-fermented liquids displayed in Table 8.9 based on the metabolomics libraries, and some of the molecules have been analyzed based on standards.

Acknowledgments

This research is supported and funded by the Department of Biotechnology (DBT), the Biotechnology Industry Research Assistance Council (BIRAC), the Government of India, and the Society for Innovation and Entrepreneurship (SINE IIT Bombay). We acknowledge our attribute to "Biorender" for vectors, images, illustrations, and icons.

References

Abdel-Rahman, M. A., Tashiro, Y., & Sonomoto, K. (2011). Lactic acid production from lignocellulose-derived sugars using lactic acid bacteria: Overview and limits. *Journal of Biotechnology*, *156*(4), 286–301. https://doi.org/10.1016/J.JBIOTEC.2011.06.017

Abedi, E., & Hashemi, S. M. B. (2020). Lactic acid production – Producing microorganisms and substrates sources-state of art. *Heliyon*, *6*(10), e04974. https://doi.org/10.1016/j.heliyon.2020.e04974

Albano, C., Silvetti, T., & Brasca, M. (2020). Screening of lactic acid bacteria producing folate and their potential use as adjunct cultures for cheese bio-enrichment. *FEMS Microbiology Letters*, *367*(9). https://doi.org/10.1093/femsle/fnaa059

Altaf, M., Naveena, B. J., & Reddy, G. (2007). Use of inexpensive nitrogen sources and starch for l(+) lactic acid production in anaerobic submerged fermentation. *Bioresource Technology*, *98*(3), 498–503. https://doi.org/10.1016/j.biortech.2006.02.013

Annuk, H., Shchepetova, J., Kullisaar, T., Songisepp, E., Zilmer, M., & Mikelsaar, M. (2003). Characterization of intestinal lactobacilli as putative probiotic candidates. *Journal of Applied Microbiology, 94*(3), 403–412. https://doi.org/10.1046/j.1365-2672.2003.01847.x

Ascencio, J. J., Philippini, R. R., Gomes, F. M., Pereira, F. M., da Silva, S. S., Kumar, V., & Chandel, A. K. (2021). Comparative highly efficient production of β-glucan by *Lasiodiplodia theobromae* CCT 3966 and its multiscale characterization. *Fermentation, 7*, 108. https://doi.org/10.3390/fermentation7030108

Assaf, S., Hadar, Y., & Dosoretz, C. G. (1997). 1-Octen-3-ol and 13-hydroperoxylinoleate are products of distinct pathways in the oxidative breakdown of linoleic acid by Pleurotus pulmonarius. *Enzyme and Microbial Technology, 21*(7), 484–490. https://doi.org/10.1016/S0141-0229(97)00019-7

Bancalari, E., Savo Sardaro, M. L., Levante, A., Marseglia, A., Caligiani, A., Lazzi, C., Neviani, E., & Gatti, M. (2017). An integrated strategy to discover Lactobacillus casei group strains for their potential use as aromatic starters. *Food Research International, 100*, 682–690. https://doi.org/10.1016/j.foodres.2017.07.066

Bandell, M., Ansanay, V., Rachidi, N., Dequin, S., & Lolkema, J. S. (1997). Membrane potential-generating malate (MleP) and citrate (CitP) transporters of lactic acid bacteria are homologous proteins. *Journal of Biological Chemistry, 272*(29), 18140–18146. https://doi.org/10.1074/jbc.272.29.18140

Barbieri, F., Montanari, C., Gardini, F., & Tabanelli, G. (2019). Biogenic amine production by lactic acid bacteria: A review. *Food, 8*(1), 17. https://doi.org/10.3390/foods8010017

Behera, S. S., Ray, R. C., & Zdolec, N. (2018). *Lactobacillus plantarum* with functional properties: An approach to increase safety and shelf-life of fermented foods. *BioMed Research International, 2018*, 1–18. https://doi.org/10.1155/2018/9361614

Bintsis, T. (2018). Lactic acid bacteria as starter cultures: An update in their metabolism and genetics. *AIMS Microbiology, 4*(4), 665–684. https://doi.org/10.3934/microbiol.2018.4.665

Burke, R. C., Leet, T. H., & Buettner-Januscht, V. (1966). Free amino acids and water soluble peptides in stratum corneum and skin surface film in human beings. *The Yale Journal of Biology and Medicine, 38*, 355–373.

Calabia, B. P., Tokiwa, Y., & Aiba, S. (2011). Fermentative production of l-(+)-lactic acid by an alkaliphilic marine microorganism. *Biotechnology Letters, 33*, 1429–1433.

Capozzi, V., Russo, P., Dueñas, M. T., López, P., & Spano, G. (2012). Lactic acid bacteria producing B-group vitamins: A great potential for functional cereals products. *Applied Microbiology and Biotechnology, 96*(6), 1383–1394. https://doi.org/10.1007/s00253-012-4440-2

Cerning, J. (1990). Exocellular polysaccharides produced by lactic acid bacteria. *FEMS Microbiology Letters, 87*(1–2), 113–130. https://doi.org/10.1111/j.1574-6968.1990.tb04883.x

Chaves, A. C. S. D., Fernandez, M., Lerayer, A. L. S., Mierau, I., Kleerebezem, M., & Hugenholtz, J. (2002). Metabolic engineering of acetaldehyde production by *Streptococcus thermophilus*. *Applied and Environmental Microbiology, 68*(11), 5656–5662. https://doi.org/10.1128/AEM.68.11.5656-5662.2002

Chen, J., Vestergaard, M., Jensen, T. G., Shen, J., Dufva, M., Solem, C., & Jensen, P. R. (2017). Finding the needle in the haystack—The use of microfluidic droplet technology to identify vitamin-secreting lactic acid bacteria. *MBio, 8*(3). https://doi.org/10.1128/mBio.00526-17

Chopin, A. (1993). Organization and regulation of genes for amino acid biosynthesis in lactic acid bacteria. *FEMS Microbiology Reviews, 12*, 21–37.

Coppola, S., Avagliano, C., Sacchi, A., Laneri, S., Calignano, A., Voto, L., Luzzetti, A., & Canani, R. B. (2022). Potential clinical applications of the postbiotic butyrate in human skin diseases. *Molecules*, *27*(6). https://doi.org/10.3390/molecules27061849

Cotter, P. D., Hill, C., & Ross, R. P. (2005). Bacteriocins: Developing innate immunity for food. *Nature Reviews Microbiology*, *3*(10), 777–788. https://doi.org/10.1038/nrmicro1273

Cotter, P. D., Ross, R. P., & Hill, C. (2013). Bacteriocins—A viable alternative to antibiotics? *Nature Reviews Microbiology*, *11*(2), 95–105. https://doi.org/10.1038/nrmicro2937

Crost, E. H., Le Gall, G., Laverde-Gomez, J. A., Mukhopadhya, I., Flint, H. J., & Juge, N. (2018). Mechanistic insights into the cross-feeding of Ruminococcus gnavus and Ruminococcus bromii on host and dietary carbohydrates. *Frontiers in Microbiology*, *9*(Nov). https://doi.org/10.3389/fmicb.2018.02558

Cselovszky, J., Wolf, G., & Hammes, W. P. (1992). Production of formate, acetate, and succinate by anaerobic fermentation of Lactobacillus pentosus in the presence of citrate. *Applied Microbiology and Biotechnology*, *37*(1). https://doi.org/10.1007/BF00174210

de Oliveira, P. M., Santos, L. P., Coelho, L. F., Avila Neto, P. M., Sass, D. C., & Contiero, J. (2021). Production of L (+) lactic acid by Lactobacillus casei Ke11: Fed batch fermentation strategies. *Fermentation*, *7*(3), 151. https://doi.org/10.3390/fermentation7030151

Dundar, A. N., & Gocmen, D. (2013). Effects of autoclaving temperature and storing time on resistant starch formation and its functional and physicochemical properties. *Carbohydrate Polymers*, *97*(2), 764–771. https://doi.org/10.1016/j.carbpol.2013.04.083

Fernandes, A., Rodrigues, P. M., Pintado, M., & Tavaria, F. K. (2023). A systematic review of natural products for skin applications: Targeting inflammation, wound healing, and photo-aging. *Phytomedicine*, *115*. https://doi.org/10.1016/j.phymed.2023.154824

Filipiak, W., Sponring, A., Baur, M. M., Ager, C., Filipiak, A., Wiesenhofer, H., Nagl, M., Troppmair, J., & Amann, A. (2012). Characterization of volatile metabolites taken up by or released from Streptococcus pneumoniae and Haemophilus influenzae by using GC-MS. *Microbiology*, *158*(12), 3044–3053. https://doi.org/10.1099/mic.0.062687-0

Galle, S., Schwab, C., Arendt, E., & Gänzle, M. (2010). Exopolysaccharide-forming *Weissella* strains as starter cultures for sorghum and wheat sourdoughs. *Journal of Agricultural and Food Chemistry*, *58*(9), 5834–5841. https://doi.org/10.1021/jf1002683

Gänzle, M. G. (2015). Lactic metabolism revisited: Metabolism of lactic acid bacteria in food fermentations and food spoilage. *Current Opinion in Food Science*, *2*, 106–117. https://doi.org/10.1016/j.cofs.2015.03.001

García-Quintáns, N., Repizo, G., Martín, M., Magni, C., & López, P. (2008). Activation of the diacetyl/acetoin pathway in *Lactococcus lactis* subsp. *lactis* bv. diacetylactis CRL264 by acidic growth. *Applied and Environmental Microbiology*, *74*(7), 1988–1996. https://doi.org/10.1128/AEM.01851-07

Ge, Y.-Y., Zhang, J.-R., Corke, H., & Gan, R.-Y. (2020). Screening and spontaneous mutation of pickle-derived Lactobacillus plantarum with overproduction of riboflavin, related mechanism, and food application. *Food*, *9*(1), 88. https://doi.org/10.3390/foods9010088

Giraffa, G., Chanishvili, N., & Widyastuti, Y. (2010). Importance of lactobacilli in food and feed biotechnology. *Research in Microbiology*, *161*(6), 480–487. https://doi.org/10.1016/j.resmic.2010.03.001

Gobbetti, M., Lavermicocca, P., Minervini, F., De Angelis, M., & Corsetti, A. (2000). Arabinose fermentation by Lactobacillus plantarum in sourdough with added pentosans and alphaalpha-L-arabinofuranosidase: A tool to increase the production of acetic acid. *Journal of Applied Microbiology*, *88*(2), 317–324. https://doi.org/10.1046/j.1365-2672.2000.00962.x

Grossiord, B., Vaughan, E. E., Luesink, E., & de Vos, W. M. (1998). Genetics of galactose utilisation via the Leloir pathway in lactic acid bacteria. *Le Lait*, *78*(1), 77–84. https://doi.org/10.1051/lait:1998110

Hayat, A., Jahangir, T. M., Khuhawar, M. Y., Alamgir, M., Hussain, Z., Haq, F. U., & Musharraf, S. G. (2015). HPLC determination of gamma amino butyric acid (GABA) and some biogenic amines (BAs) in controlled, germinated, and fermented brown rice by pre-column derivatization. *Journal of Cereal Science*, *64*, 56–62. https://doi.org/10.1016/j.jcs.2015.04.014

Holland, R., Liu, S.-Q., Crow, V. L., Delabre, M.-L., Lubbers, M., Bennett, M., & Norris, G. (2005). Esterases of lactic acid bacteria and cheese flavour: Milk fat hydrolysis, alcoholysis and esterification. *International Dairy Journal*, *15*(6–9), 711–718. https://doi.org/10.1016/j.idairyj.2004.09.012

Huang, J., Wang, J., & Liu, S. (2023a). Advanced fermentation techniques for lactic acid production from agricultural waste. *Fermentation*, *9*(8), 765. https://doi.org/10.3390/fermentation9080765

Huang, Y., Wang, Y., Shang, N., & Li, P. (2023b). Microbial fermentation processes of lactic acid: Challenges, solutions, and future prospects. *Food*, *12*(12), 2311. https://doi.org/10.3390/foods12122311

Iskandar, C. F., Cailliez-Grimal, C., Borges, F., & Revol-Junelles, A.-M. (2019). Review of lactose and galactose metabolism in lactic acid bacteria dedicated to expert genomic annotation. *Trends in Food Science & Technology*, *88*, 121–132. https://doi.org/10.1016/j.tifs.2019.03.020

Jayaprakash, G., Bains, A., Chawla, P., Fogarasi, M., & Fogarasi, S. (2022). A narrative review on rice proteins: Current scenario and food industrial application. *Polymers*, *14*(15), 3003. https://doi.org/10.3390/polym14153003

Jin, Z., Peng, S., & Nie, L. (2023). Active compounds: A new direction for rice value addition. *Food Chemistry: X*, *19*, 100781. https://doi.org/10.1016/j.fochx.2023.100781

Jung, D. H., Kim, G. Y., Kim, I. Y., Seo, D. H., Do Nam, Y., Kang, H., Song, Y., & Park, C. S. (2019). Bifidobacterium adolescentis P2P3, a human gut bacterium having strong non-gelatinized resistant starch-degrading activity. *Journal of Microbiology and Biotechnology*, *29*(12), 1904–1915. https://doi.org/10.4014/jmb.1909.09010

Jung, J. Y., Lee, S. H., Kim, J. M., Park, M. S., Bae, J.-W., Hahn, Y., Madsen, E. L., & Jeon, C. O. (2011). Metagenomic analysis of kimchi, a traditional Korean fermented food. *Applied and Environmental Microbiology*, *77*(7), 2264–2274. https://doi.org/10.1128/AEM.02157-10

Jung, J. Y., Lee, S. H., Lee, H. J., Seo, H.-Y., Park, W.-S., & Jeon, C. O. (2012). Effects of Leuconostoc mesenteroides starter cultures on microbial communities and metabolites during kimchi fermentation. *International Journal of Food Microbiology*, *153*(3), 378–387. https://doi.org/10.1016/j.ijfoodmicro.2011.11.030

Kang, T. S., Korber, D. R., & Tanaka, T. (2013). Regulation of dual glycolytic pathways for fructose metabolism in heterofermentative Lactobacillus panis PM1. *Applied and Environmental Microbiology*, *79*(24), 7818–7826. https://doi.org/10.1128/AEM.02377-13

Kaseleht, K., Paalme, T., Mihhalevski, A., & Sarand, I. (2011a). Analysis of volatile compounds produced by different species of lactobacilli in rye sourdough using multiple headspace extraction. *International Journal of Food Science & Technology*, *46*(9), 1940–1946. https://doi.org/10.1111/j.1365-2621.2011.02705.x

Kaseleht, K., Paalme, T., Mihhalevski, A., & Sarand, I. (2011b). Analysis of volatile compounds produced by different species of lactobacilli in rye sourdough using multiple headspace extraction. *International Journal of Food Science & Technology*, *46*(9), 1940–1946. https://doi.org/10.1111/j.1365-2621.2011.02705.x

Kaur, S., & Kaur, S. (2015). Bacteriocins as potential anticancer agents. *Frontiers in Pharmacology*, *6*. https://doi.org/10.3389/fphar.2015.00272

Kawai, M., Tsuchiya, A., Ishida, J., Yoda, N., Yashiki-Yamasaki, S., & Katakura, Y. (2020). Suppression of lactate production in fed-batch culture of some lactic acid bacteria with sucrose as the carbon source. *Journal of Bioscience and Bioengineering*, *129*(5), 535–540. https://doi.org/10.1016/J.JBIOSC.2019.11.009

Kawai, Y., Kemperman, R., Kok, J., & Saito, T. (2004). The circular bacteriocins gassericin A and circularin A. *Current Protein & Peptide Science*, *5*(5), 393–398. https://doi.org/10.2174/1389203043379549

Kieliszek, M., Pobiega, K., Piwowarek, K., & Kot, A. M. (2021). Characteristics of the proteolytic enzymes produced by lactic acid bacteria. *Molecules*, *26*(7), 1858. https://doi.org/10.3390/molecules26071858

Kim, K. H., Chun, B. H., Baek, J. H., Roh, S. W., Lee, S. H., & Jeon, C. O. (2020). Genomic and metabolic features of Lactobacillus sakei as revealed by its pan-genome and the metatranscriptome of kimchi fermentation. *Food Microbiology*, *86*, 103341. https://doi.org/10.1016/j.fm.2019.103341

Kim, T.-W., Lee, J.-H., Kim, S.-E., Park, M.-H., Chang, H. C., & Kim, H.-Y. (2009). Analysis of microbial communities in doenjang, a Korean fermented soybean paste, using nested PCR-denaturing gradient gel electrophoresis. *International Journal of Food Microbiology*, *131*(2–3), 265–271. https://doi.org/10.1016/j.ijfoodmicro.2009.03.001

Kobayashi, J. (2019). D-amino acids and lactic acid bacteria. *Microorganisms*, *7*(12). https://doi.org/10.3390/microorganisms7120690

Kook, M. C., & Cho, S. C. (2013). Production of GABA (gamma amino butyric acid) by lactic acid bacteria. *Korean Journal for Food Science of Animal Resources*, *33*(3), 377–389. https://doi.org/10.5851/kosfa.2013.33.3.377

Kuley, E., Özyurt, G., Özogul, I., Boga, M., Akyol, I., Rocha, J. M., & Özogul, F. (2020). The role of selected lactic acid bacteria on organic acid accumulation during wet and spray-dried fish-based silages. Contributions to the winning combination of microbial food safety and environmental sustainability. *Microorganisms*, *8*(2), 172. https://doi.org/10.3390/microorganisms8020172

Kumar, V., Ashok, S., & Park, S. (2013). Recent advances in biological production of 3-hydroxypropionic acid. *Biotechnology Advances*, *31*(6), 945–961. https://doi.org/10.1016/j.biotechadv.2013.02.008

Laëtitia, G., Pascal, D., & Yann, D. (2014). The citrate metabolism in homo- and heterofermentative LAB: A selective means of becoming dominant over other microorganisms in complex ecosystems. *Food and Nutrition Sciences*, *5*(10), 953–969. https://doi.org/10.4236/fns.2014.510106

Lampi, A.-M., Damerau, A., Li, J., Moisio, T., Partanen, R., Forssell, P., & Piironen, V. (2015). Changes in lipids and volatile compounds of oat flours and extrudates during processing and storage. *Journal of Cereal Science*, *62*, 102–109. https://doi.org/10.1016/j.jcs.2014.12.011

LeBlanc, J. G., Levit, R., Savoy de Giori, G., & de Moreno de LeBlanc, A. (2020). Application of vitamin-producing lactic acid bacteria to treat intestinal inflammatory diseases. *Applied Microbiology and Biotechnology, 104*(8), 3331–3337. https://doi.org/10.1007/s00253-020-10487-1

Lee, S. M., Hwang, Y. R., Kim, M. S., Chung, M. S., & Kim, Y.-S. (2019). Comparison of volatile and nonvolatile compounds in rice fermented by different lactic acid bacteria. *Molecules, 24*(6), 1183. https://doi.org/10.3390/molecules24061183

Lee, S. M., Lim, H. J., Chang, J. W., Hurh, B.-S., & Kim, Y.-S. (2018). Investigation on the formations of volatile compounds, fatty acids, and γ-lactones in white and brown rice during fermentation. *Food Chemistry, 269*, 347–354. https://doi.org/10.1016/j.foodchem.2018.07.037

Lee, S. M., Oh, J., Hurh, B., Jeong, G., Shin, Y., & Kim, Y. (2016). Volatile compounds produced by *Lactobacillus paracasei* during oat fermentation. *Journal of Food Science, 81*(12). https://doi.org/10.1111/1750-3841.13547

Leroy, F., Vankrunkelsven, S., De Greef, J., & De Vuyst, L. (2003). The stimulating effect of a harsh environment on the bacteriocin activity by Enterococcus faecium RZS C5 and dependency on the environmental stress factor used. *International Journal of Food Microbiology, 83*(1), 27–38. https://doi.org/10.1016/S0168-1605(02)00316-1

Levit, R., Savoy de Giori, G., Moreno de LeBlanc, A., & LeBlanc, J. G. (2021). Recent update on lactic acid bacteria producing riboflavin and folates: application for food fortification and treatment of intestinal inflammation. *Journal of Applied Microbiology, 130*(5), 1412–1424. https://doi.org/10.1111/jam.14854

Li, H., Zhang, L., Li, J., Wu, Q., Qian, L., He, J., Ni, Y., Kovatcheva-Datchary, P., Yuan, R., Liu, S., Shen, L., Zhang, M., Sheng, B., Li, P., Kang, K., Wu, L., Fang, Q., Long, X., Wang, X., . . . Jia, W. (2024). Resistant starch intake facilitates weight loss in humans by reshaping the gut microbiota. *Nature Metabolism, 6*(3), 578–597. https://doi.org/10.1038/s42255-024-00988-y

Liu, S. (2003). Practical implications of lactate and pyruvate metabolism by lactic acid bacteria in food and beverage fermentations. *International Journal of Food Microbiology, 83*(2), 115–131. https://doi.org/10.1016/S0168-1605(02)00366-5

Liu, Y., van Bennekom, E. O., Zhang, Y., Abee, T., & Smid, E. J. (2019). Long-chain vitamin K2 production in Lactococcus lactis is influenced by temperature, carbon source, aeration and mode of energy metabolism. *Microbial Cell Factories, 18*(1), 129. https://doi.org/10.1186/s12934-019-1179-9

Lv, W., Zhang, X., & Cong, W. (2005). Modelling the production of nisin by Lactococcus lactis in fed-batch culture. *Applied Microbiology and Biotechnology, 68*(3), 322–326. https://doi.org/10.1007/s00253-005-1892-7

Majsnerowska, M., Noens, E. E. E., & Lolkema, J. S. (2018). Arginine and citrulline catabolic pathways encoded by the *arc* gene cluster of Lactobacillus brevis ATCC 367. *Journal of Bacteriology, 200*(14). https://doi.org/10.1128/JB.00182-18

Marcone, G. L., Rosini, E., Crespi, E., & Pollegioni, L. (2020). D-amino acids in foods. *Applied Microbiology and Biotechnology, 104*(2), 555–574. https://doi.org/10.1007/s00253-019-10264-9

Matsui, K., Sasahara, S., Akakabe, Y., & Kajiwara, T. (2003). Linoleic acid 10-hydroperoxide as an intermediate during formation of 1-Octen-3-ol from linoleic acid in *Lentinus decadetes.*

Bioscience, Biotechnology, and Biochemistry, *67*(10), 2280–2282. https://doi.org/10.1271/bbb.67.2280

Mosso, A. L., Jimenez, M. E., Vignolo, G., LeBlanc, J. G., & Samman, N. C. (2018). Increasing the folate content of tuber based foods using potentially probiotic lactic acid bacteria. *Food Research International*, *109*, 168–174. https://doi.org/10.1016/j.foodres.2018.03.073

Nagarjun, P. A., Rao, R. S., Rajesham, S., & Rao, L. V. (2005). Optimization of lactic acid production in SSF by Lactobacillus amylovorus NRRL B-4542 using Taguchi Methodology. *Journal of Microbiology*, *43*(1).

Namgung, H.-J., Park, H.-J., Cho, I. H., Choi, H.-K., Kwon, D.-Y., Shim, S.-M., & Kim, Y.-S. (2010). Metabolite profiling of doenjang, fermented soybean paste, during fermentation. *Journal of the Science of Food and Agriculture*. https://doi.org/10.1002/jsfa.4036

Nguyen, P.-T., Nguyen, T.-T., Bui, D.-C., Hong, P.-T., Hoang, Q.-K., & Nguyen, H.-T. (2020). Exopolysaccharide production by lactic acid bacteria: The manipulation of environmental stresses for industrial applications. *AIMS Microbiology*, *6*(4), 451–469. https://doi.org/10.3934/microbiol.2020027

Noguchi, A., & Djerassi, D. (2009). Amino acids and peptides: Building blocks for skin proteins. In *Nutritional Cosmetics: Beauty from Within* (pp. 287–317). Elsevier. https://doi.org/10.1016/B978-0-8155-2029-0.50022-3

Oguro, Y., Nishiwaki, T., Shinada, R., Kobayashi, K., & Kurahashi, A. (2017). Metabolite profile of koji amazake and its lactic acid fermentation product by Lactobacillus sakei UONUMA. *Journal of Bioscience and Bioengineering*, *124*(2), 178–183. https://doi.org/10.1016/j.jbiosc.2017.03.011

Ortiz, M. E., Bleckwedel, J., Raya, R. R., & Mozzi, F. (2013). Biotechnological and in situ food production of polyols by lactic acid bacteria. *Applied Microbiology and Biotechnology*, *97*(11), 4713–4726. https://doi.org/10.1007/s00253-013-4884-z

Papalexandratou, Z., Vrancken, G., De Bruyne, K., Vandamme, P., & De Vuyst, L. (2011). Spontaneous organic cocoa bean box fermentations in Brazil are characterized by a restricted species diversity of lactic acid bacteria and acetic acid bacteria. *Food Microbiology*, *28*(7), 1326–1338. https://doi.org/10.1016/j.fm.2011.06.003

Paramithiotis, S., Stasinou, V., Tzamourani, A., Kotseridis, Y., & Dimopoulou, M. (2022). Malolactic fermentation—Theoretical advances and practical considerations. *Fermentation*, *8*(10), 521. https://doi.org/10.3390/fermentation8100521

Park, B., Hwang, H., Chang, J. Y., Hong, S. W., Lee, S. H., Jung, M. Y., Sohn, S.-O., Park, H. W., & Lee, J.-H. (2017). Identification of 2-hydroxyisocaproic acid production in lactic acid bacteria and evaluation of microbial dynamics during kimchi ripening. *Scientific Reports*, *7*(1), 10904. https://doi.org/10.1038/s41598-017-10948-0

Park, J., Sung, J. M., Choi, Y. S., & Park, J. D. (2020). Effect of natural fermentation on milled rice grains: Physicochemical and functional properties of rice flour. *Food Hydrocolloids*, *108*. https://doi.org/10.1016/j.foodhyd.2020.106005

Park, M. K., & Kim, Y.-S. (2019). Distinctive formation of volatile compounds in fermented rice inoculated by different molds, yeasts, and lactic acid bacteria. *Molecules*, *24*(11), 2123. https://doi.org/10.3390/molecules24112123

Park, S. J., Kim, D. H., Kang, H. J., Shin, M., Yang, S. Y., Yang, J., & Jung, Y. H. (2021). Enhanced production of γ-aminobutyric acid (GABA) using Lactobacillus plantarum EJ2014 with simple medium composition. *LWT*, *137*. https://doi.org/10.1016/j.lwt.2020.110443

Pereira, C., Lourenço, V. M., Menezes, R., & Brites, C. (2021). Rice compounds with impact on diabetes control. *Food*, *10*(9), 1992. https://doi.org/10.3390/foods10091992

Perez, R. H., Zendo, T., & Sonomoto, K. (2014). Novel bacteriocins from lactic acid bacteria (LAB): Various structures and applications. *Microbial Cell Factories, 13 Suppl 1(Suppl 1)*, S3. https://doi.org/10.1186/1475-2859-13-S1-S3

Perez Visñuk, D., Teran, M. D. M., Savoy de Giori, G., LeBlanc, J. G., & de Moreno de LeBlanc, A. (2022). Neuroprotective effect of riboflavin Producing lactic acid bacteria in parkinsonian models. *Neurochemical Research*, *47*(5), 1269–1279. https://doi.org/10.1007/s11064-021-03520-w

Petrova, P., & Petrov, K. (2012). Direct starch conversion into L-(+)-lactic acid by a novel amylolytic strain of Lactobacillus paracasei B41. *Starch/Staerke*, *64*(1), 10–17. https://doi.org/10.1002/star.201100074

Petrova, P., Petrov, K., & Stoyancheva, G. (2013). Starch-modifying enzymes of lactic acid bacteria - Structures, properties, and applications. *Starch/Staerke*, *65*(1–2), 34–47. https://doi.org/10.1002/star.201200192

Pophaly, S. D., Singh, R., Pophaly, S. D., Kaushik, J. K., & Tomar, S. K. (2012). Current status and emerging role of glutathione in food grade lactic acid bacteria. *Microbial Cell Factories*, *11*(1), 114. https://doi.org/10.1186/1475-2859-11-114

Qiao, N., Wittouck, S., Mattarelli, P., Zheng, J., Lebeer, S., Felis, G. E., & Gänzle, M. G. (2022). After the storm—Perspectives on the taxonomy of Lactobacillaceae. *JDS Communications*, *3*(3), 222–227. https://doi.org/10.3168/jdsc.2021-0183

Rao, R. S., Kumar, C. G., Prakasham, R. S., & Hobbs, P. J. (2008). The Taguchi methodology as a statistical tool for biotechnological applications: A critical appraisal. *Biotechnology Journal*, *3*(4), 510–523. https://doi.org/10.1002/biot.200700201

Renye, J. A., White, A. K., & Hotchkiss, A. T. (2021). Identification of lactobacillus strains capable of fermenting fructo-oligosaccharides and inulin†. *Microorganisms*, *9*(10). https://doi.org/10.3390/microorganisms9102020

Rühmkorf, C., Jungkunz, S., Wagner, M., & Vogel, R. F. (2012). Optimization of homoexopolysaccharide formation by lactobacilli in gluten-free sourdoughs. *Food Microbiology*, *32*(2), 286–294. https://doi.org/10.1016/j.fm.2012.07.002

Sahl, H.-G., & Bierbaum, G. (1998). Lantibiotics: Biosynthesis and biological activities of uniquely modified peptides from gram-positive bacteria. *Annual Review of Microbiology*, *52*(1), 41–79. https://doi.org/10.1146/annurev.micro.52.1.41

Sanalibaba, P., & Cakmak, G. A. (2016). Exopolysaccharides production by lactic acid bacteria. *Applied Microbiology: Open Access*, *2*(2). https://doi.org/10.4172/2471-9315.1000115

Sanchez, S., Rodríguez-Sanoja, R., Ramos, A., & Demain, A. L. (2018). Our microbes not only produce antibiotics, they also overproduce amino acids. *Journal of Antibiotics*, *71*(1), 26–36. https://doi.org/10.1038/ja.2017.142

Sarantinopoulos, P., Kalantzopoulos, G., & Tsakalidou, E. (2001). Citrate metabolism by *Enterococcus faecalis* FAIR-E 229. *Applied and Environmental Microbiology*, *67*(12), 5482–5487. https://doi.org/10.1128/AEM.67.12.5482-5487.2001

Shirai, K., Guerrero, I., Huerta, S., Saucedo, G., Castillo, A., Obdulia Gonzalez, R., & Hall, G. M. (2001). Effect of initial glucose concentration and inoculation level of lactic acid bacteria in shrimp waste ensilation. *Enzyme and Microbial Technology*, *28*(4–5), 446–452. https://doi.org/10.1016/S0141-0229(00)00338-0

Silva, L. A., Lopes Neto, J. H. P., & Cardarelli, H. R. (2019). Exopolysaccharides produced by Lactobacillus plantarum: Technological properties, biological activity, and potential application in the food industry. *Annals of Microbiology*, *69*(4), 321–328. https://doi.org/10.1007/s13213-019-01456-9

Sionek, B., Szydłowska, A., Küçükgöz, K., & Kołożyn-Krajewska, D. (2023). Traditional and new microorganisms in lactic acid fermentation of food. *Fermentation*, *9*(12). https://doi.org/10.3390/fermentation9121019

Small, P. L. C., & Waterman, S. R. (1998). Acid stress, anaerobiosis and gadCB: Lessons from Lactococcus lactis and Escherichia coli. *Trends in Microbiology*, *6*(6), 214–216. https://doi.org/10.1016/S0966-842X(98)01285-2

Sørensen, H. M., Rochfort, K. D., Maye, S., MacLeod, G., Brabazon, D., Loscher, C., & Freeland, B. (2022). Exopolysaccharides of lactic acid bacteria: Production, purification and health benefits towards functional food. *Nutrients*, *14*(14), 2938. https://doi.org/10.3390/nu14142938

Spacova, I., Ahannach, S., Breynaert, A., Erreygers, I., Wittouck, S., Bron, P. A., Van Beeck, W., Eilers, T., Alloul, A., Blansaer, N., Vlaeminck, S. E., Hermans, N., & Lebeer, S. (2022). Spontaneous riboflavin-overproducing Limosilactobacillus reuteri for biofortification of fermented foods. *Frontiers in Nutrition*, *9*. https://doi.org/10.3389/fnut.2022.916607

Sugino, T., Shirai, T., Kajimoto, Y., & Kajimoto, O. (2008). l-Ornithine supplementation attenuates physical fatigue in healthy volunteers by modulating lipid and amino acid metabolism. *Nutrition Research*, *28*(11), 738–743. https://doi.org/10.1016/j.nutres.2008.08.008

Sun, D., Li, H., Song, D., Zhang, L., Zhao, X., & Xu, X. (2019). Genome, transcriptome and fermentation analyses of Lactobacillus plantarum LY-78 provide new insights into the mechanism of phenyllactate biosynthesis in lactic acid bacteria. *Biochemical and Biophysical Research Communications*, *519*(2), 351–357. https://doi.org/10.1016/j.bbrc.2019.09.011

Tang, H., Huang, W., & Yao, Y.-F. (2023). The metabolites of lactic acid bacteria: Classification, biosynthesis and modulation of gut microbiota. *Microbial Cell*, *10*(3), 49–62. https://doi.org/10.15698/mic2023.03.792

Taranto, M. P., Vera, J. L., Hugenholtz, J., De Valdez, G. F., & Sesma, F. (2003). *Lactobacillus reuteri* CRL1098 produces cobalamin. *Journal of Bacteriology*, *185*(18), 5643–5647. https://doi.org/10.1128/JB.185.18.5643-5647.2003

Tekin, T., & Dincer, E. (2023). Effect of resistant starch types as a prebiotic. *Applied Microbiology and Biotechnology*, *107*(2–3), 491–515. https://doi.org/10.1007/s00253-022-12325-y

Thage, B. V., Broe, M. L., Petersen, M. H., Petersen, M. A., Bennedsen, M., & Ardö, Y. (2005). Aroma development in semi-hard reduced-fat cheese inoculated with Lactobacillus paracasei strains with different aminotransferase profiles. *International Dairy Journal*, *15*(6–9), 795–805. https://doi.org/10.1016/j.idairyj.2004.08.026

Torino, M. I., Taranto, M. P., & Font de Valdez, G. (2005). Citrate catabolism and production of acetate and succinate by Lactobacillus helveticus ATCC 15807. *Applied Microbiology and Biotechnology*, *69*(1), 79–85. https://doi.org/10.1007/s00253-005-1949-7

Torres, A. C., Vannini, V., Font, G., Saavedra, L., & Taranto, M. P. (2018). Novel pathway for corrinoid compounds production in lactobacillus. *Frontiers in Microbiology*, *9*. https://doi.org/10.3389/fmicb.2018.02256

Turner, K. W., & Martley, F. G. (1983). Galactose fermentation and classification of thermophilic lactobacilli. *Applied and Environmental Microbiology*, *45*(6), 1932–1934. https://doi.org/10.1128/aem.45.6.1932-1934.1983

Van der Meulen, R., Scheirlinck, I., Van Schoor, A., Huys, G., Vancanneyt, M., Vandamme, P., & De Vuyst, L. (2007). Population dynamics and metabolite target analysis of lactic acid bacteria during laboratory fermentations of wheat and spelt sourdoughs. *Applied and Environmental Microbiology*, *73*(15), 4741–4750. https://doi.org/10.1128/AEM.00315-07

Verluyten, J., Leroy, F., & de Vuyst, L. (2004). Influence of complex nutrient source on growth of and curvacin A production by sausage isolate *Lactobacillus curvatus* LTH 1174. *Applied and Environmental Microbiology*, *70*(9), 5081–5088. https://doi.org/10.1128/AEM.70.9.5081-5088.2004

Vermeulen, N., Czerny, M., Gänzle, M. G., Schieberle, P., & Vogel, R. F. (2007). Reduction of (E)-2-nonenal and (E,E)-2,4-decadienal during sourdough fermentation. *Journal of Cereal Science*, *45*(1), 78–87. https://doi.org/10.1016/j.jcs.2006.07.002

Vrancken, G., Rimaux, T., Weckx, S., De Vuyst, L., & Leroy, F. (2009). Environmental pH determines citrulline and ornithine release through the arginine deiminase pathway in Lactobacillus fermentum IMDO 130101. *International Journal of Food Microbiology*, *135*(3), 216–222. https://doi.org/10.1016/j.ijfoodmicro.2009.07.035

Wang, M., Chen, X., Zhou, L., Li, Y., Yang, J., Ji, N., Xiong, L., & Sun, Q. (2022). Prebiotic effects of resistant starch nanoparticles on growth and proliferation of the probiotic Lactiplantibacillus plantarum subsp. plantarum. *LWT*, *154*. https://doi.org/10.1016/j.lwt.2021.112572

Wang, Y., Wu, J., Lv, M., Shao, Z., Hungwe, M., Wang, J., Bai, X., Xie, J., Wang, Y., & Geng, W. (2021). Metabolism characteristics of lactic acid bacteria and the expanding applications in food industry. *Frontiers in Bioengineering and Biotechnology*, *9*. https://doi.org/10.3389/fbioe.2021.612285

White, A. G. C., Steele, R. H., & Pierce, W. A. (1955). The effect of ph on the fermentation of glucose and galactose by Streptococcus pyogenes. *Journal of Bacteriology*, *70*(1), 82–86. https://doi.org/10.1128/jb.70.1.82-86.1955

Wouters, D., Grosu-Tudor, S., Zamfir, M., & De Vuyst, L. (2013). Bacterial community dynamics, lactic acid bacteria species diversity and metabolite kinetics of traditional Romanian vegetable fermentations. *Journal of the Science of Food and Agriculture*, *93*(4), 749–760. https://doi.org/10.1002/jsfa.5788

Wu, W., & Zhao, N. (2019). Metabolomics of lactic acid bacteria. In *Lactic Acid Bacteria* (pp. 167–182). Singapore: Springer Nature Singapore. https://doi.org/10.1007/978-981-13-7832-4_6

Xiao, X., Hu, X., Yao, J., Cao, W., Zou, Z., Wang, L., Qin, H., Zhong, D., Li, Y., Xue, P., Jin, R., Li, Y., Shi, Y., & Li, J. (2023). The role of short-chain fatty acids in inflammatory skin diseases. *Frontiers in Microbiology*, *13*. https://doi.org/10.3389/fmicb.2022.1083432

Xiao, Z., & Lu, J. R. (2014). Strategies for enhancing fermentative production of acetoin: A review. *Biotechnology Advances*, *32*(2), 492–503. https://doi.org/10.1016/j.biotechadv.2014.01.002

Xiong, R. G., Zhou, D. D., Wu, S. X., Huang, S. Y., Saimaiti, A., Yang, Z. J., Shang, A., Zhao, C. N., Gan, R. Y., & Li, H. (2022). Health benefits and side effects of short-chain fatty acids. *Food*, *11*(18). https://doi.org/10.3390/foods11182863

Yang, D. S., Shewfelt, R. L., Lee, K.-S., & Kays, S. J. (2008). Comparison of odor-active compounds from six distinctly different rice flavor types. *Journal of Agricultural and Food Chemistry*, *56*(8), 2780–2787. https://doi.org/10.1021/jf072685t

Yvon, M., & Rijnen, L. (2001). Cheese flavour formation by amino acid catabolism. *International Dairy Journal*, *11*(4–7), 185–201. https://doi.org/10.1016/S0958-6946(01)00049-8

Zacharof, M. P., & Lovitt, R. W. (2012). Bacteriocins produced by lactic acid bacteria a review article. *APCBEE Procedia*, *2*, 50–56. https://doi.org/10.1016/j.apcbee.2012.06.010

Ze, X., Duncan, S. H., Louis, P., & Flint, H. J. (2012). Ruminococcus bromii is a keystone species for the degradation of resistant starch in the human colon. *ISME Journal*, *6*(8), 1535–1543. https://doi.org/10.1038/ismej.2012.4

Zendo, T., Eungruttanagorn, N., Fujioka, S., Tashiro, Y., Nomura, K., Sera, Y., Kobayashi, G., Nakayama, J., Ishizaki, A., & Sonomoto, K. (2005). Identification and production of a bacteriocin from Enterococcus mundtii QU 2 isolated from soybean. *Journal of Applied Microbiology*, *99*(5), 1181–1190. https://doi.org/10.1111/j.1365-2672.2005.02704.x

9

Biofoundry in Microbial Protein Production

Process Challenges and Future Scenario

Simab Kanwal[1], *Sher Zaman Safi*[2], *Aphichart Karnchanatat*[3], *and Piroonporn Srimongkol*[1]

[1] *High-Value Food from Mushrooms and Bioactive Plants in the Green Economy Value Chain Research Group, The Institute of Biotechnology and Genetic Engineering, Chulalongkorn University, Pathumwan, Bangkok, Thailand*
[2] *Department of Biochemistry, Faculty of Medicine, MAHSA University, Jenjarom, Selangor, Malaysia*
[3] *Center of Excellence in Bioconversion and Bioseparation for Platform Chemical Production, The Institute of Biotechnology and Genetic Engineering, Chulalongkorn University, Pathumwan, Bangkok, Thailand*

9.1 Introduction

A biofoundry is a platform designed to automate and accelerate the construction, testing, and optimization of biological systems, particularly in the context of synthetic biology and biotechnology. In the realm of microbial protein production, biofoundries play a pivotal role in streamlining the development of microorganisms that can efficiently produce high-value metabolites, such as enzymes, therapeutics, or industrial chemicals, by leveraging cutting-edge technologies in genetic engineering, high-throughput screening, and automation. In recent years, as industries seek to reduce their dependence on fossil fuels, the biofoundries tend to emerge largely as a source of renewable alternatives (Rosch et al., 2024). Due to that, over the past decade, biofoundries have played a significant role in advancing biotechnology and shifting towards more sustainable production methods (Hillson et al., 2019).

The fundamentals of biofoundries in microbial protein production comprise the automated genetic engineering, microbial strain development, optimization of production conditions, and high-throughput screening of microbial strains. Biofoundries integrate automation tools to design and synthesize genetic parts (genes, promoters, ribosome binding sites, etc.), which can be assembled into microbial genomes for optimized protein production (Chao et al., 2017). This high-throughput approach significantly accelerates the creation of genetically modified organisms (GMOs). Once microbial chassis (such as *E. coli*, yeast, or other model organisms) are chosen, biofoundries enable the rapid development of strains with engineered pathways for enhanced protein yield, stability, and quality (Ganjoo et al., 2022; Moffat et al., 2021). Moreover, biofoundries combine sophisticated monitoring systems and experimental workflows to optimize fermentation and cultivation

Biofoundry Techniques for Biotechnology Applications, First Edition. Edited by Anuj Kumar Chandel.

conditions, ensuring that microorganisms are cultivated under the most favorable conditions for protein production. By automating testing procedures, biofoundries can rapidly assess the performance of numerous microbial strains and process conditions to identify those that produce the target protein most efficiently. On the other hand, through the screening and optimization processes, the obtained data is analyzed using machine learning and other computational tools to guide the next steps in design and optimization; that is also considered a key element in biofoundries along with the scale-up that is achieved by identifying the optimal microbial strains and conditions. Biofoundries can facilitate the scale-up of microbial cultures from laboratory-scale to industrial-scale bioreactors, making the production of proteins commercially viable. This iterative feedback loop increases the chances of success in achieving high production levels.

Among the few applications of microbial biofoundries in protein production are enzyme production, therapeutic proteins (including monoclonal antibodies, vaccines, and insulin), sustainable production of food and feed proteins, and bio-based chemicals (to replace petrochemical-derived products). Beyond chemicals and fuels, biofoundries have begun exploring new applications, including the production of sustainable materials, textile fibers, and even alternative food sources like lab-grown meat and algae-based products. The next frontier seems to be expanding into the agricultural and food industries, where biofoundries could help in developing novel, more sustainable food production systems. This chapter will provide a brief survey of the basics of microbial bioprocessing, cultivation systems as well as the strategies in synthetic bioengineering for sustainable protein production in a biofoundry. The chapter also outlines the challenges and future perspectives associated with biofoundries development that should prove useful to biotechnologists, food technologists, and other stakeholders who wish to plan and develop a sustainable optimization of microbial production of valuable compounds other than proteins, ranging from fuels to pharmaceuticals.

9.2 Microorganisms and Protein Production

Bacteria, fungi, and algae are the highly explored microbial resources for the production of high-value metabolites, including proteins. The global population is projected to surpass 9 billion by 2050 and will reach a peak of 9.73 billion by 2064 (Adam, 2021; Vollset et al., 2020). These demographic shifts and the continuous growth of the global population are exerting tremendous pressure on the food and agriculture sectors to provide larger quantities and diverse types of food, with a particular emphasis on protein resources (Daszkiewicz, 2022). Van Dijk et al. (2021) estimate that global food production will need to rise by 70% to accommodate the projected population growth by 2050. However, population growth is not the sole factor driving the increasing demand for food, as factors like economic development, urbanization, changing diets, and environmental challenges also contribute, collectively driving the need for increased and diversified food production.

Due to concerns about food scarcity, alternative proteins are being developed as substitutes for traditional proteins. These alternative sources are designed to require less resource-intensive production methods (Hadi & Brightwell, 2021). In 1966, Professor Carroll Wilson at Massachussetts Institute of Technology (M.I.T.) coined the term "single-cell protein

(SCP)," which refers to proteins derived from microorganisms such as bacteria, fungi, yeast, and microalgae (Goldberg, 1985). In addition to proteins, SCP products may contain free amino acids, carbohydrates, lipids, vitamins, and minerals. SCP is believed to offer higher nutritional value compared to animal or plant proteins (Molfetta et al., 2022).

SCP proteins are primarily utilized as protein-rich food supplements or ingredients for human and animal nutrition. These proteins are also used in various industries, such as paper and leather processing, and as foam stabilizers (Mensah & Twumasi, 2016). Compared to agricultural proteins, SCP production is more environmentally friendly, consuming less water, requiring smaller land areas, and exerting a significantly lower impact on climate change (Sharif et al., 2021). As feed, SCP can replace traditional protein supplements like fishmeal and soymeal. However, competition for fishmeal among poultry, swine, aquaculture, and other animal industries has created challenges in meeting demand without overexploiting forage fish fisheries. To support industry expansion and maintain feed performance while protecting aquaculture health, alternative protein ingredients are urgently needed (Jones et al. 2020).

9.3 Strain Selection for Protein Production

Careful selection of microbial strains plays an important role in meeting the production requirements at various stages of the value chain, which enhances the feasibility of process integration and scale-up (Patil et al., 2020). For instance, the mutant microalgae strain *Chlorella* sp. AT1 thrives in 10% CO_2 while producing high levels of lipids and carbohydrates, whereas the biosorbent yeast *Saccharomyces cerevisiae* assimilates zinc to generate enriched biomass containing 50% protein by weight. The former highlights the potential of utilizing concentrated CO_2 as a feedstock, while the latter underscores the role of metabolic processes in reducing the need for extensive feedstock purification (Patil et al., 2020; Saraei et al., 2022).

A variety of host cells and their corresponding genetic engineering toolkits have been developed for recombinant protein production. Among prokaryotic expression systems, *Escherichia coli* has been widely utilized for protein overexpression owing to its rapid growth, suitability for high-cell-density fermentation, cost-effectiveness, and, most notably, the availability of robust genetic tools. The optimization of recombinant protein expression in *E. coli* has primarily relied on trial-and-error approaches involving adjustments to parameters such as expression vectors, host strains, media composition, and growth temperature (Burgess-Brown et al., 2008). A broad range of commercially available expression vectors is offered, featuring various origins of replication, promoters, translation initiation regions, antibiotic resistance markers, and transcription terminators. In many cases, selecting an appropriate vector, combined with the use of codon-optimized genes, is sufficient to achieve significant accumulation of the target protein (Hatfield & Roth, 2007).

Bacillus subtilis is also an effective host for recombinant protein production. Its Generally Recognized As Safe (GRAS) status and its exceptional natural ability to take up and integrate exogenous DNA into its genome make it an excellent platform for the heterologous expression of bioactive substances. Furthermore, *B. subtilis* has proven to be a robust host and an efficient technological platform for expressing bioactive proteins and high-yield protein production, with excellent scalability for bioreactor applications (Heravi et al., 2015). *Pseudomonas fluorescens*, another bacterium, has demonstrated its potential in protein manufacturing. Through its ATP-binding cassette (ABC) transporter and type I secretion system, it has successfully produced recombinant proteins. However, alongside the target proteins, *P. fluorescens* also secretes extraneous background proteins, which can complicate purification and downstream processing (Fabia et al., 2021).

Yeast is an excellent host for producing recombinant proteins with industrial and medical applications. The advantages, combined with recent advancements, have increased its popularity. Yeast can secrete heterologous proteins, making it ideal for producing complex eukaryotic proteins. Common yeast expression hosts include *Pichia pastoris*, *Yarrowia lipolytica*, *Saccharomyces cerevisiae*, *Kluyveromyces lactis*, *Hansenula polymorpha*, *Schizosaccharomyces pombe*, and *Arxula adeninivorans* (Baghban et al., 2019).

Beyond bacteria, various systems are used for recombinant protein production, each with unique advantages. Insect cell lines like Sf9 and High Five are suitable for complex proteins via the baculovirus expression vector system, commonly used for vaccines and therapeutic proteins. Mammalian cells such as CHO and HEK293 are preferred for high-quality proteins needing post-translational modifications like monoclonal antibodies (Vieira Gomes et al., 2018). Plant systems, like tobacco and rice, offer large-scale, cost-effective production. Algal systems such as *Chlamydomonas reinhardtii* provide sustainable alternatives, while cell-free systems allow rapid, scalable expression. Additionally, moss systems like *Physcomitrella patens* show promise for cost-effective bioactive protein production (Jayakrishnan et al., 2024).

9.4 Protein-Rich Biomass Production

Innovative food and feed products are among the fastest-growing trends in the food industry. Microorganisms, such as bacteria, fungi, and algae, can convert low-cost substrates like agricultural waste, organic residues, and industrial by-products into high-protein biomass. At the same time, ensuring the availability of high-quality food for the rapidly expanding global population presents a significant challenge. Protein-rich products derived from algae, fungi, and bacteria have gained consumer attention as healthier and more sustainable alternatives to red meat, and they also hold promise as valuable ingredients in animal feed (Gharibzahedi et al., 2018).

Bacteria are among the most commonly used microorganisms for protein-rich biomass production. Certain bacterial strains, such as *Corynebacterium glutamicum*, *E. coli*, and *B. subtilis*, are capable of efficiently converting carbon sources like glucose, waste sugars, and agricultural residues into protein (Hatfield & Roth, 2007; Heravi et al., 2015). These bacteria grow rapidly and can produce high yields of protein in short periods of time. In addition to their speed, bacteria can be genetically engineered to enhance protein production or to produce proteins with specific amino acid profiles, making them versatile candidates for various applications. Some bacteria are also capable of synthesizing essential amino acids, which can contribute to the nutritional value of the biomass produced (Hatfield & Roth, 2007; Heravi et al., 2015).

Mycoprotein is classified as GRAS, and its products, such as miso, tofu, oncom, tempeh, red rice, shoyu, and hamanatto, have traditionally been produced using various types of filamentous fungi, including *Monascus purpureus*, *Neurospora intermedia*, *Rhizopus oryzae*, and *Aspergillus oryzae* (Gharibzahedi et al., 2018; Ravindra et al., 2009). These microorganisms possess robust enzymatic machinery that enables them to degrade complex organic materials into simpler compounds, which can then be utilized for protein synthesis. Moreover, these can grow on various media and produce a wide range of products, including enzymes, nutrient-rich fungal biomass, small molecule drugs, organic acids, antibiotics, and several enzymes that can be used in the saccharification of lignocellulosic materials for bioethanol production (Ferreira, 2015).

Algae, such as microalgae (*Chlorella vulgaris*, *Spirulina platensis*) and macroalgae (e.g., kelps), are increasingly recognized for their protein content. Microalgae grow rapidly and accumulate protein, lipids, and bioactive compounds, making them suitable for food, feed, and biofuel industries. They efficiently convert solar energy into biomass and can be cultivated in photobioreactors with minimal land and freshwater (Pipliya et al., 2025). Yeast, such as *Saccharomyces cerevisiae* and *Pichia pastoris*, also produce high-protein biomass by converting carbon sources into protein, with applications in food, feed, and biotechnology. Yeast offers rapid growth, ease of genetic modification, and the ability to produce specific proteins or enzymes, making it a versatile option for protein production (Yang et al., 2024b).

9.5 Microbial Bioprocessing

Microbial processes play a key role in industries such as pharmaceuticals, cosmetics, food, chemicals, and the development of new materials. In the last two decades, new and more efficient industrial processes utilizing microorganisms have been introduced, resulting in purer, more cost-effective products or substances that cannot be produced through traditional methods (Ghosh et al., 2016). Microbial biotechnology provides a key approach for sustainable bioprocesses, where microorganisms and their enzymes are used to convert glycerol, carbohydrates, and lignin into renewable resources like bioenergy (Gaur et al., 2024).

The accumulation of agricultural and industrial waste in fields creates significant environmental challenges. To mitigate this issue, microorganisms are of great economic value in various biotechnological processes, including microbial fermentation methods (Sharma et al., 2022). Additionally, microbes are versatile organisms that contribute to the utilization and bioconversion of biomass, as they are essential sources of enzymes with biotechnological potential.

Although mixed microbial consortia generally outperform single microorganisms, their full potential for cost-efficient industrial bioprocessing applications has yet to be fully realized. To optimize synthetic microbial consortia, numerical modeling has become an essential tool for predicting compatibility and designing (Hanly & Henson, 2013; Höffner & Barton, 2014). While synthetic microbial consortia created using a "bottom-up" approach undoubtedly represent progress, this method still encounters several significant challenges. The primary issue is the inability to co-culture microorganisms that are known to form stable and functional communities in advance. More than 99% of microorganisms have yet to be cultivated or genetically characterized, meaning the "bottom-up" construction of synthetic consortia is inherently constrained by the limited knowledge of characterized strains (Rodríguez Amor & Dal Bello, 2019; Wu et al., 2024).

9.6 Cultivation Systems

Establishment of a sustainable biofoundry is usually considered a challenging task that requires the operational and technical considerations to be addressed first (Holowko et al. 2021). In a microbial biofoundry, the microbial proliferation is controlled in a microbial biocontainment system where microorganisms (like bacteria, yeast, fungi, or algae) are grown in the cultivation systems under controlled environments for various applications in research, industrial production, and biotechnology (Pantoja et al., 2022). The cultivation systems provide the necessary conditions for microbial growth and can be used for processes including fermentation, protein production, or waste treatment. The small-scale microbial cultivation is done in shake flasks for the primary use in research and early-stage development and is suitable for small-scale fermentation, optimizing growth conditions, screening microbial strains, or protein expression studies. It consists of a flask filled with a growth medium and a shaking mechanism that provides aeration and mixing. For large-scale microbial cultivation, bioreactors are used, which are the devices or systems that provide a controlled environment for cultivating microorganisms. These systems are widely used in biotechnology, and pharmaceutical industries for producing therapeutic or recombinant proteins, enzymes, antibodies, vaccines, and other valuable biomolecules in microbial systems (Ruiz et al., 2010; Hanišáková et al., 2022).

9.7 Bioreactors for Protein Production

Some common types of bioreactors used for microbial protein production are (Barboza-Rodríguez et al., 2024; Boruta et al., 2021; Shanmugam et al., 2022):

I) Stirred-Tank Bioreactors: These are the most common type for large-scale protein production. They use mechanical stirring to keep cells suspended and ensure adequate mixing of gases (like oxygen) and nutrients. They can be operated in batch, fed-batch, or continuous modes.
II) Air-Lift Bioreactors: These use the difference in gas flow (usually air) to circulate the medium, offering a gentler mixing action than stirred tanks, which can be helpful for more fragile cells, making it suitable for more delicate or shear-sensitive microorganisms.
III) Packed-Bed Bioreactors: These have a solid support (like beads, fibers, or particles) where cells or microorganisms adhere and grow. They are often used for immobilized cell systems and used for continuous cultivation of microorganisms, especially for processes like enzyme production.

IV) Fed-Batch and Continuous Culture: In a fed-batch system, nutrients are gradually added to the culture to extend the growth phase and increase yields. Continuous systems provide a steady flow of nutrients and medium while removing waste, allowing for more consistent production.

V) Photobioreactors: These are designed to grow phototrophic microorganisms like algae or cyanobacteria, which require light for their metabolic processes (Figure 9.1). Photobioreactors optimize light exposure, CO_2 supply, and nutrient delivery for metabolites production in phototrophic microorganisms.

Bacteria like *E. coli* are a common choice for simpler proteins, while fungi are considered an excellent choice for the production of SCPs for their higher protein content. The growth medium provides essential nutrients like sugars, amino acids, and vitamins, which are required for the growth of cells and the expression of the target protein (Bajić et al., 2023). Bioreactors allow precise control of temperature, pH, dissolved oxygen levels, and agitation. These parameters are critical to optimizing cell growth and protein expression. For the production of the target protein, the engineered microbes are typically cultured in large bioreactors. The proteins can be harvested and purified using various biochemical and chromatography techniques. Advances in fermentation and bioreactor design are key to improving yields and reducing costs.

9.8 Downstream Processing

Downstream processing is a crucial part of industrial biotechnology and comprises a series of steps including separation, release of intracellular products, concentration, purification, and finally the formulation (Rathore & Shirke, 2011). Once the cells have produced the protein, the supernatant (or culture medium) is often collected, and protein purification steps follow, which is referred to as downstream processing. This involves the series of steps used to isolate, purify, and concentrate the proteins of interest from the microbial fermentation broth and separate them from cellular debris, residual culture medium, and other contaminants. For intracellular protein production, the microbial cells are broken open (lysis) using mechanical methods (like high-pressure homogenization) or chemical treatments to release the target protein (Wang et al., 2019b). For some proteins, the protein may be aggregated or insolubilized after lysis, often with denaturants like urea or guanidine hydrochloride. For large-scale filtration and concentration, tangential flow filtration is also used (Agrawal et al., 2023). After purification, the protein may be formulated for storage or use by adjusting the protein's pH, adding stabilizers (like glycerol), or lyophilizing (freeze-drying) for long-term storage. A variety of assays such as SDS-PAGE, Western blotting, and functional activity assays, are performed to ensure the protein is pure and functional. General steps involved in the bacterial protein production are illustrated in Figure 9.2. Endotoxin removal is often a critical step in quality control, particularly for therapeutic proteins, since endotoxins can cause adverse effects. Scalability, yield and purity, and cost of purification are the significant factors that need to be considered in downstream processing. For the production of proteins and high-value bioproducts such as pigments, the

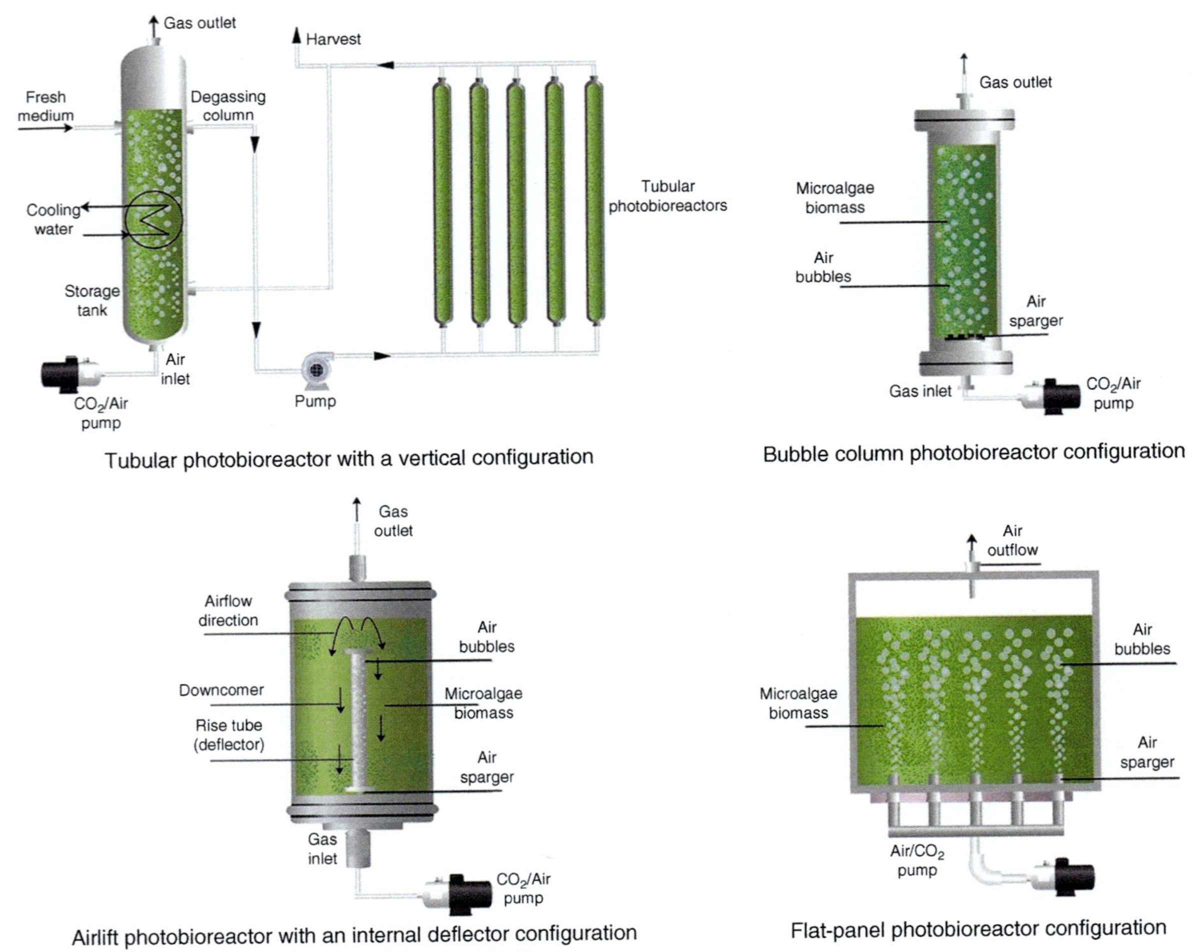

Figure 9.1 Configurations of photobioreactor for cultivation of microalgae biomass for biorefinery. *Source:* Barboza-Rodríguez et al. (2024) / with permission of Elsevier.

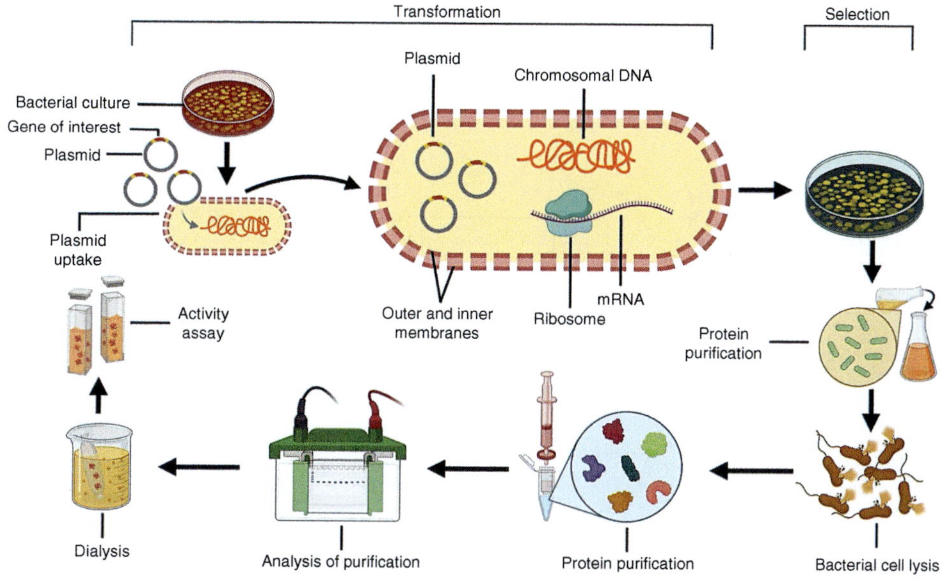

Figure 9.2 The protein production process uses bacterial transformation. It depicts plasmid uptake, gene expression, selection, bacterial lysis, protein purification, and analysis. The workflow includes dialysis and activity assays to verify protein function.

third-generation (3G) biorefineries are used that utilize feedstock from aquatic biomass (algae or cyanobacteria), serving as a part of the circular bioeconomy (Bhatia et al., 2022).

9.9 Strategies in Synthetic Bioengineering

In microbial biofoundries, synthetic bioengineering is a rapidly advancing field that focuses on using engineered microorganisms to produce proteins and other valuable compounds. The approach combines synthetic biology, bioengineering, and industrial microbiology to improve the biofoundry processes or production facilities intended to streamline the design, synthesis, and optimization of biological systems (Lama et al., 2024). This is achieved by optimizing microbial strains and metabolic pathways for the sustainable and renewable chemicals' production. A few strategies employed in synthetic bioengineering to achieve efficient biorefining are discussed as follows.

9.9.1 Microbial Engineering

In microbial biofoundries, microorganisms such as bacteria, yeast, or fungi are genetically engineered to produce target proteins or metabolites. *E. coli*, *Saccharomyces cerevisiae* (yeast), and *B. subtilis* are commonly used due to their well-understood genetics, ease of cultivation, and fast growth. In those microbial strains, genetic modifications can be made to optimize protein expression, enhance metabolic pathways, or improve the stability and yield of the desired product. The genetic modifications in the metabolic pathways are usually done by

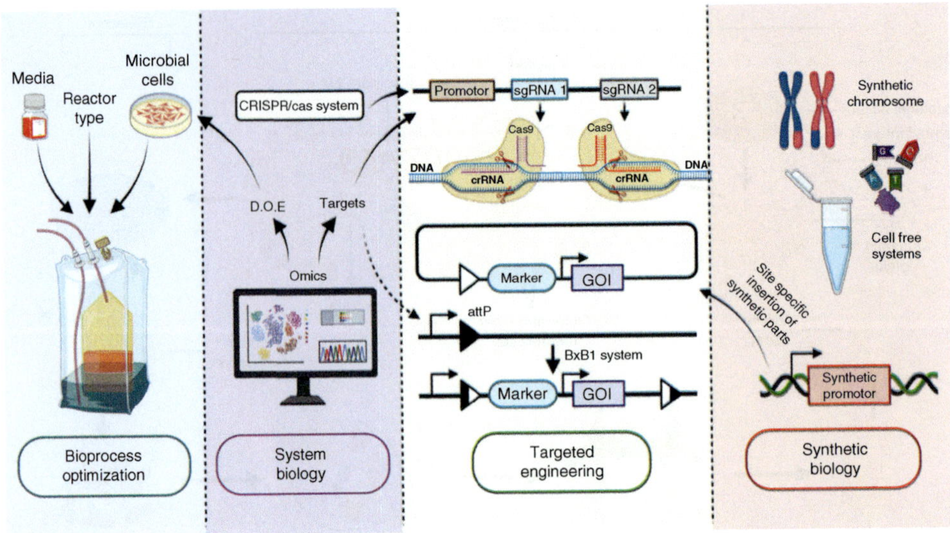

Figure 9.3 The integration of bioprocess optimization, system biology, targeted engineering, and synthetic biology to enhance protein production efficiency. Media composition, omics data, CRISPR/Cas systems, and synthetic parts are employed to achieve precise genetic modifications to collectively advance the development of sustainable and efficient protein production systems.

synthesis or redesigning of promoters, terminators, ribosomal binding sites, site-directed mutagenesis (by deleting or inserting genes) or by altering chromosomal expression. Specific integration plasmids, chassis-independent recombinase-assisted genome engineering (CRAGE), XPORT, and CRISPR/Cas9 are among the few commonly used tools employed for those purposes (Figure 9.3) (Englund et al., 2015; Rusk, 2018; Wang et al., 2019a).

9.9.2 Metabolic Pathway Optimization

The production of specific proteins is enhanced in microbial biofoundries involved in the redesign of metabolic networks. For instance, pathways that comprise enzymes or intermediates crucial for protein biosynthesis can be altered to favor the production of the desired target protein. The use of modular synthetic biology techniques allows for the assembly of new metabolic pathways that might not naturally exist in the organism (Englund et al., 2015). This is made possible due to the development of databases such as MetaCys, Expasy, KEGG, CyanoCys, BRENDA, and antiSMASH, and software tools including NP. searcher, ClusterFinder, RetroPath 2.0, NaPDoS, Primer3web, and Clustal (Kanwal & Incharoensakdi, 2019, 2020; Lama et al., 2024).

9.9.3 High-Throughput Screening

One of the strengths of biofoundries is the use of automated high-throughput screening techniques that allow researchers to test large numbers of genetic variants rapidly. This helps in identifying the best microbial strains for optimal protein production. Techniques

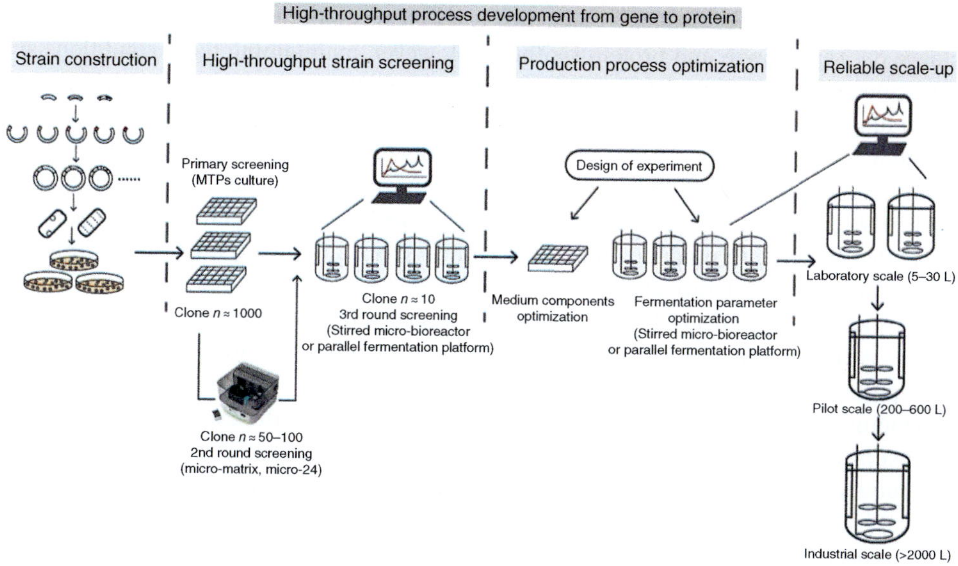

Figure 9.4 A summary of high-throughput process development from gene cloning to protein (recombinant) production in the current bioindustry sector. The process comprises the screening and construction of clone, process optimization for high-throughput production, and reliable scale-up of production process.

like DNA assembly, gene synthesis, and library screening help in testing different genetic modifications and conditions. Computational models, including the genome-scale models and omics techniques, are serving as powerful tools in optimizing the growth of synthetic microbial communities as well as understanding and guiding their metabolic engineering, phenotypic behaviors, and cellular metabolism (Colarusso et al., 2021; Lama et al., 2024). Figure 9.4 represents a scheme of high-throughput process development for recombinant protein production via gene cloning that can be further scaled up for commercial the production utilizing a large-scale bioreactor (Sun et al. 2023).

9.10 Challenges and Future Prospects

As described in Section 9.9, the production of protein from microorganisms has garnered increasing attention as a potential solution to the growing global demand for sustainable and alternative protein sources. However, despite these promising benefits, there are numerous challenges associated with microbial protein production that need to be addressed before they can be scaled up to meet global protein demands (Li et al., 2024).

Optimizing fermentation processes and scaling them up to industrial levels are two of the most significant challenges in microbial protein production. Fermentation is the core method for growing microorganisms to produce protein, and it requires precise control over various factors, such as temperature, pH, oxygen levels, and nutrient availability (Du et al., 2022). Each microorganism has its own set of ideal conditions, which must be met to achieve optimal growth and protein synthesis. However, maintaining these conditions at large scales, such as in fermentation vessels that can hold thousands of liters, is technically demanding. In large-scale systems, challenges like homogeneous mixing, temperature distribution, and oxygen transfer become more pronounced, and even slight fluctuations can result in suboptimal yields and variations in protein quality (Michel et al., 2024; Singh et al., 2017). Furthermore, achieving consistency across large-scale operations is crucial to ensure that the protein produced meets the desired specifications for nutritional content, texture, and flavor (Colgrave et al., 2021).

Cost is another significant barrier to the widespread adoption of microbial protein production. Although the cost has been decreasing over the years, it is still higher than traditional protein sources, such as plant-based or animal-derived proteins (D'Almeida & de Albuquerque, 2025). The main contributors to this cost include the expense of fermentation media, energy requirements, and the need for advanced infrastructure (D'Almeida & de Albuquerque, 2025; Malila et al., 2024). Many microorganisms require expensive nutrients and growth media, such as nitrogen, carbon sources, and vitamins, adding to the production cost (Chai & Adnan, 2018).

Despite advancements in bioreactor design and energy-efficient technologies, large-scale commercial microbial protein production still requires significant infrastructure investment (Asin-Garcia et al., 2024). Furthermore, the regulatory landscape for microbial proteins complicates their adoption, as stringent safety assessments and approval processes are required before they can enter the food supply (Sutay Kocabaş & Grumet, 2019). Concerns about GMOs also pose challenges, as public perception and regulatory bodies may raise concerns regarding safety, environmental impacts, and ethical issues. Overcoming these challenges requires continued research, technological advancements, consumer education, and careful consideration of public perception, but with innovation, microbial proteins have the potential to become a key component of future sustainable food systems. Advances in biotechnology, genetic engineering, and fermentation technologies are expected to drive significant improvements in microbial protein production. In particular, genetic engineering and synthetic biology offer opportunities to design microorganisms with optimized metabolic pathways, enabling them to produce larger quantities of protein more efficiently (Francois et al., 2020; Yang et al., 2024a). Additionally, the development of microbial strains that can use low-cost raw materials like agricultural waste or even carbon dioxide could make production more economically feasible, enhancing the sustainability of microbial protein production (Francois et al., 2020).

In addition to advancements in genetic engineering, significant improvements in fermentation technology will play a crucial role in the future of microbial protein (Abbaspour, 2024). Enhanced bioreactor designs and energy-efficient technologies will help overcome current scaling challenges and improve the efficiency of fermentation processes (Xia et al., 2021). The use of sustainable feedstocks, such as organic waste and carbon dioxide, also promises to reduce costs and environmental impact (Zahid et al., 2024). Ultimately, as these technologies mature, microbial proteins will likely become a key component of the global food system, providing sustainable and scalable solutions to meet future protein demands. Overall, microbial biofoundries are revolutionizing protein production by making the process more efficient, scalable, and versatile. The technology is enabling the development of new bioproducts and providing solutions to challenges in sectors like healthcare, agriculture, and environmental sustainability.

9.11 Conclusions

In conclusion, biofoundries represent a transformative platform in microbial protein production, playing a crucial role in automating and optimizing the development of genetically engineered microorganisms. By integrating automation, high-throughput screening, and machine learning, biofoundries enhance the design and scale-up of microbial processes, driving progress in various sectors, including biotechnology, food, and pharmaceuticals. Establishing a sustainable biofoundry involves addressing both operational and technical challenges, particularly in the management of microbial cultivation systems. Advancements in bioreactor design, fermentation, and downstream processing will continue to be pivotal in optimizing microbial production and supporting industrial biotechnology applications. By genetically engineering microorganisms and redesigning metabolic networks, biofoundries can optimize the efficiency and yield of target products. Tools like CRISPR/Cas9 and modular synthetic biology techniques are essential for these modifications, enabling the creation of novel metabolic pathways. As the global demand for sustainable production methods rises with global population growth, biofoundries hold great promise in addressing challenges related to food security, resource efficiency, and environmental sustainability. The continued development of biofoundry technologies will be essential for advancing the production of alternative proteins and other valuable compounds, ultimately contributing to a more sustainable and resilient future.

References

Abbaspour, N. (2024). Fermentation's pivotal role in shaping the future of plant-based foods: An integrative review of fermentation processes and their impact on sensory and health benefits. *Applied Food Research*, 100468.

Adam, D. (2021). How far will the global population rise? Researchers can't agree. *Nature*, *597*, 462–465.

Agrawal, P., Wilkstein, K., Guinn, E., Mason, M., SerranoMartinez, C. I., & Saylae, J. (2023). A review of tangential flow filtration: Process development and applications in the pharmaceutical industry. *Organic Process Research and Development*, *27*(4), 571–591.

Asin-Garcia, E., Fawcett, J. D., Batianis, C., & Dos Santos, V. A. M. (2024). A snapshot of biomanufacturing and the need for enabling research infrastructure. *Trends in Biotechnology*, *43*(5), 1000–1014.

Baghban, R., Farajnia, S., Rajabibazl, M., Ghasemi, Y., Mafi, A. A., Hoseinpoor, R., Rahbarnia, L., & Aria, M. (2019). Yeast expression systems: Overview and recent advances. *Molecular Biotechnology*, *61*, 365–384.

Bajić, B., Vučurović, D., Vasić, Đ., Jevtić-Mučibabić, R., & Dodić, S. (2023). Biotechnological production of sustainable microbial proteins from agro-industrial residues and by-products. *Food*, *12*, 107. https://doi.org/10.3390/foods12010107

Barboza-Rodríguez, R., Rodríguez-Jasso, R. M., Rosero-Chasoy, G., Rosales Aguado, M. L., & Ruiz, H. A. (2024). Photobioreactor configurations in cultivating microalgae biomass for biorefinery. *Bioresource Technology*, *394*, 130208. https://doi.org/10.1016/j.biortech.2023.130208

Bhatia, L., Bachheti, R. K., Garlapati, V. K., & Chandel, A. K. (2022). Third-generation biorefineries: A sustainable platform for food, clean energy, and nutraceuticals production. *Biomass Conversion and Biorefinery*, *12*, 4215–4230. https://doi.org/10.1007/s13399-020-00843-6

Boruta, T., Ścigaczewska, A., & Bizukojć, M. (2021). "Microbial wars" in a stirred tank bioreactor: Investigating the co-cultures of *Streptomyces rimosus* and *Aspergillus terreus*, filamentous microorganisms equipped with a rich arsenal of secondary metabolites. *Frontiers in Bioengineering and Biotechnology*, *9*, 713639. https://doi.org/10.3389/fbioe.2021.713639

Burgess-Brown, N. A., Sharma, S., Sobott, F., Loenarz, C., Oppermann, U., & Gileadi, O. (2008). Codon optimization can improve expression of human genes in *Escherichia coli*: A multi-gene study. *Protein Expression and Purification*, *59*(1), 94–102.

Chai, J. M., & Adnan, A. (2018). Effect of different nitrogen source combinations on microbial cellulose production by Pseudomonas aeruginosa in batch fermentation. In *IOP Conference Series: Materials Science and Engineering* (Vol. 440, pp. 012044). IOP Publishing.

Chao, R., Mishra, S., Si, T., & Zhao, H. (2017). Engineering biological systems using automated biofoundries. *Metabolic Engineering*, *42*, 98–108. https://doi.org/10.1016/j.ymben.2017.06.003

Colarusso, A. V., Goodchild-Michelman, I., Rayle, M., & Zomorrodi, A. R. (2021). Computational modeling of metabolism in microbial communities on a genome-scale. *Current Opinion in Systems Biology*, *26*, 46–57. https://doi.org/10.1016/j.coisb.2021.04.001

Colgrave, M. L., Dominik, S., Tobin, A. B., Stockmann, R., Simon, C., Howitt, C. A., Belobrajdic, D. P., Paull, C., & Vanhercke, T. (2021). Perspectives on future protein production. *Journal of Agricultural and Food Chemistry*, *69*(50), 15076–15083.

D'Almeida, A. P., & de Albuquerque, T. L. (2025). Is it possible to produce meat without animals? The potential of microorganisms as protein sources. *Fermentation*, *11*(1), 24.

Daszkiewicz, T. (2022). Food production in the context of global developmental challenges. *Agriculture*, *12*(6), 832.

Du, Y. H., Wang, M. Y., Yang, L. H., Tong, L. L., Guo, D. S., & Ji, X. J. (2022). Optimization and scale-up of fermentation processes driven by models. *Bioengineering (Basel)*, *9*(9), 473.

Englund, E., Andersen-Ranberg, J., Miao, R., Hamberger, B., & Lindberg, P. (2015). Metabolic engineering of *Synechocystis* sp. PCC 6803 for production of the plant diterpenoid manoyl oxide. *ACS Synthetic Biology*. https://doi.org/10.1021/acssynbio.5b00070

Fabia, B. U., Bingwa, J., Park, J., Hieu, N. M., & Ahn, J. H. (2021). Utilizing the ABC transporter for growth factor production by fleQ deletion mutant of *Pseudomonas fluorescens*. *Biomedicine*, *9*(6), 679.

Ferreira, J. Integration of Filamentous Fungi in Ethanol Dry-Mill Biorefinery. Ph.D. Thesis, University of Borås, Borås, Sweden, 2015.

Francois, J. M., Alkim, C., & Morin, N. (2020). Engineering microbial pathways for production of bio-based chemicals from lignocellulosic sugars: Current status and perspectives. *Biotechnology for Biofuels*, *13*, 118.

Ganjoo, A., Sharma, N., Shafeeq, H., Bhat, N. A., Dubey, K. K., & Babu, V. (2022). Progress and challenges in the biofoundry of immunosuppressants: From process to practice. *Biotechnology and Bioengineering, 119*, 3339–3369.

Gaur, S., Kaur, M., Kalra, R., Rene, E. R., & Goel, M. (2024). Application of microbial resources in biorefineries: Current trend and future prospects. *Heliyon, 10*, e28615.

Gharibzahedi, S. M. T., Roohinejad, S., George, S., Barba, F. J., Greiner, R., Barbosa-Cánovas, G. V., & Mallikarjunan, K. (2018). Innovative food processing technologies on the transglutaminase functionality in protein-based food products: Trends, opportunities and drawbacks. *Trends in Food Science & Technology, 75*, 194–205.

Ghosh, S., Chowdhury, R., & Bhattacharya, P. (2016). Mixed consortia in bioprocesses: Role of microbial interactions. *Applied Microbiology and Biotechnology, 100*, 4283–4295.

Goldberg, I. (1985). *Single cell protein* (pp. 3). Berlin, Germany: Springer.

Hadi, J., & Brightwell, G. (2021). Safety of alternative proteins: Technological, environmental and regulatory aspects of cultured meat, plant-based meat, insect protein and single-cell protein. *Foods, 10*, 1226.

Hanišáková, N., Vítězová, M., & Rittmann, S. K.-M. R. (2022). The historical development of cultivation techniques for methanogens and other strict anaerobes and their application in modern microbiology. *Microorganisms, 10*, 412. https://doi.org/10.3390/microorganisms10020412

Hanly, T. J., & Henson, M. A. (2013). Unstructured modeling of a synthetic microbial consortium for consolidated production of ethanol. *IFAC Proceedings Volumes, 46*, 157–162.

Hatfield, G. W., & Roth, D. A. (2007). Optimizing scaleup yield for protein production: Computationally optimized DNA assembly (CODA) and translation engineering. *Biotechnology Annual Review, 13*, 27–42.

Heravi, K. M., Watzlawick, H., & Altenbuchner, J. (2015). Development of an anhydrotetracycline-inducible expression system for expression of a neopullulanase in *B. subtilis*. *Plasmid, 82*, 35–42.

Hillson, N., Caddick, M., Cai, Y., Carrasco, J. A., Chang, M. W., Curach, N. C., Bell, D. J., Le Feuvre, R., Friedman, D. C., Fu, X., Gold, N. D., Herrgård, M. J., Holowko, M. B., Johnson, J. R., Johnson, R. A., Keasling, J. D., Kitney, R. I., Kondo, A., Liu, C., Martin, V. J. J., Menolascina, F., Ogino, C., Patron, N. J., Pavan, M., Poh, C. L., Pretorius, I. S., Rosser, S. J., Scrutton, N. S., Storch, M., Tekotte, H., Travnik, E., Vickers, C. E., Yew, W. S., Yuan, Y., Zhao, H., & Freemont, P. S. (2019). Building a global alliance of biofoundries. *Nature Communications, 10*(1), 2040. https://doi.org/10.1038/s41467-019-10079-2 Erratum in: Nature Communications 2019 Jul 11;10(1):3132. doi: 10.1038/s41467-019-10862-1. PMID: 31068573; PMCID: PMC6506534

Höffner, K., & Barton, P. I. (2014). Design of microbial consortia for industrial biotechnology. In M. R. Eden (Ed.), *Computer Aided Chemical Engineering* (pp. 65–74). Elsevier.

Holowko, M. B., Frow, E. K., Reid, J. C., Rourke, M., & Vickers, C. E. (2021). Building a biofoundry. *Synthetic Biology, 6*(1), ysaa026. https://doi.org/10.1093/synbio/ysaa026

Jayakrishnan, A., Wan Rosli, W. R., Tahir, A. R. M., Razak, F. S. A., Kee, P. E., Ng, H. S., Chew, Y.-L., Lee, S.-K., Ramasamy, M., Tan, C. S., et al. (2024). Evolving paradigms of recombinant protein production in pharmaceutical industry: A rigorous review. *The Sciences, 6*(1), 9.

Jones, S. W., Karpol, A., Friedman, S., Maru, B. T., & Tracy, B. P. (2020). Recent advances in single cell protein use as a feed ingredient in aquaculture. *Current Opinion in Biotechnology, 61*, 189–197.

Kanwal, S., & Incharoensakdi, A. (2019). The role of GAD pathway for regulation of GABA accumulation and C/N balance in *Synechocystis* sp. PCC6803. *Journal of Applied Phycology, 31*, 3503–3514.

Kanwal, S., & Incharoensakdi, A. (2020). GABA synthesis mediated by-aminobutanal dehydrogenase in *Synechocystis* sp. PCC6803 with disrupted glutamate and-ketoglutarate decarboxylase genes. *Plant Science, 290*, 110287.

Lama, S., Thapa, L. P., Upadhayaya, S. K., Gauchan, D. P., & Singh, A. (2024). Metabolic engineering in lignocellulose biorefining for high-value chemicals: Recent advances, challenges, and outlook for enabling a bioeconomy. *Frontiers in Industrial Microbiology, 1*, 1319774. https://doi.org/10.3389/finmi.2023.1319774

Li, Y. P., Ahmadi, F., Kariman, K., & Lackner, M. (2024). Recent advances and challenges in single cell protein (SCP) technologies for food and feed production. *npj Science of Food, 8*, 66.

Malila, Y., Owolabi, I. O., Chotanaphuti, T., Sakdibhornssup, N., Elliott, C. T., Visessanguan, W., Karoonuthaisiri, N., & Petchkongkaew, A. (2024). Current challenges of alternative proteins as future foods. *npj Science of Food, 8*, 53.

Mensah, J. K. M., & Twumasi, P. (2016). Use of pineapple waste for single cell protein (SCP) production and the effect of substrate concentration on the yield. *Journal of Food Biochemistry, 40*, e12478.

Michel, M., Eldridge, A. L., Hartmann, C., Klassen, P., Ingram, J., & Meijer, G. W. (2024). Benefits and challenges of food processing in the context of food systems, value chains and sustainable development goals. *Trends in Food Science & Technology, 104703.*

Moffat, A. D., Elliston, A., Patron, N. J., Truman, A. W., & Carrasco Lopez, J. A. (2021). A biofoundry workflow for the identification of genetic determinants of microbial growth inhibition. *Synthetic Biology (Oxford), 6*(1), ysab004. https://doi.org/10.1093/synbio/ysab004

Molfetta, M., Morais, E. G., Barreira, L., Bruno, G. L., Porcelli, F., Dugat-Bony, E., Bonnarme, P., & Minervini, F. (2022). Protein sources alternative to meat: State of the art and involvement of fermentation. *Food, 11*, 2065.

Pantoja Angles, A., Valle-Pérez, A. U., Hauser, C., & Mahfouz, M. M. (2022). Microbial biocontainment systems for clinical, agricultural, and industrial applications. *Frontiers in Bioengineering and Biotechnology, 10*, 830200. https://doi.org/10.3389/fbioe.2022.830200

Patil, R. A., Kausley, S. B., Joshi, S. M., & Pandit, A. B. (2020). Process intensification applied to microalgae-based processes and products. In *Handbook of Microalgae-based Processes and Products* (pp. 737–769).

Pipliya, S., Kumar, S., Gupta, R. K., Das, R. S., Meena, D., Srivastav, P. P., . . . Garcia-Vaquero, M. (2025). The future of algal proteins: Innovations in extraction and modifications, functional properties, and sustainable food applications. *Future Foods*, 100549.

Rathore, A. S., & Shirke, A. (2011). Recent developments in membrane-based separations in biotechnology processes: Review. *Preparative Biochemistry & Biotechnology, 41*, 398–421.

Ravindra, R., Rudravaram, R., Chandel, A. K., Rao, L. V., & Yim, Z. H. (2009). The effect of de-oiled rice bran for single cell protein production using fungal cultures under solid state fermentation. *International Journal of Food Engineering, 5*(2), 2009.

Rodríguez Amor, D., & Dal Bello, M. (2019). Bottom-up approaches to synthetic cooperation in microbial communities. *Life (Basel), 9*(1), 22.

Rosch, T. M., Tenhaef, J., Stoltmann, T., Redeker, T., Kösters, D., Hollmann, N., Krumbach, K., Wiechert, W., Bott, M., Matamouros, S., Marienhagen, J., & Noack, S. (2024). AutoBioTech—A versatile biofoundry for automated strain engineering. *ACS Synthetic Biology*, *13*(7), 2227–2237. https://doi.org/10.1021/acssynbio.4c00298

Ruiz, B., Chávez, A., Forero, A., García-Huante, Y., Romero, A., Sánchez, M., Rocha, D., Sánchez, B., Rodríguez-Sanoja, R., Sánchez, S., & Langley, E. (2010). Production of microbial secondary metabolites: Regulation by the carbon source. *Critical Reviews in Microbiology*, *36*, 146–167. https://doi.org/10.3109/10408410903489576

Rusk, N. (2018). Engineering wild bacteria. *Nature Methods*, *15*, 764. https://doi.org/10.1038/s41592-018-0160-6

Saraei, F., Amini, K., Haddadi, A., & Larypoor, M. (2022). Application of Saccharomyces cerevisiae isolated from industrial effluent for zinc biosorption and zinc-enriched SCP production for human and animal. *Food Science and Technology*, *42*. https://doi.org/10.1590/fst.82021

Shanmugam, M. K., Mandari, V., Devarai, S. K., & Gummadi, S. N. (2022). Chapter 1 - Types of bioreactors and important design considerations. In R. Sirohi, A. Pandey, M. J. Taherzadeh, & C. Larroche (Eds.), *Current developments in biotechnology and bioengineering* (pp. 3–30). Elsevier. https://doi.org/10.1016/B978-0-323-91167-2.00008-3

Sharif, M., Zafar, M. H., Aqib, A. I., Saeed, M., Farag, M. R., & Alagawany, M. (2021). Single cell protein: Sources, mechanism of production, nutritional value and its uses in aquaculture nutrition. *Aquaculture*, *531*, 735885.

Sharma, P., Gaur, V. K., Gupta, S., Varjani, S., Pandey, A., Gnansounou, E., You, S., Ngo, H. H., & Wong, J. W. C. (2022). Trends in mitigation of industrial waste: Global health hazards, environmental implications and waste derived economy for environmental sustainability. *Science of the Total Environment*, *811*, 152357.

Singh, V., Haque, S., Niwas, R., Srivastava, A., Pasupuleti, M., & Tripathi, C. (2017). Strategies for fermentation medium optimization: An in-depth review. *Frontiers in Microbiology*, *7*, 2087.

Sun, M., Gao, A. X., Liu, X., Yang, Y., Ledesma-Amaro, R., & Bai, Z. (2023). High-throughput process development from gene cloning to protein production. *Microbial Cell Factories*, *22*, 182. https://doi.org/10.1186/s12934-023-02184-1

Sutay Kocabaş, D., & Grumet, R. (2019). Evolving regulatory policies regarding food enzymes produced by recombinant microorganisms. *GM Crops & Food*, *10*(4), 191–207.

Van Dijk, M., Morley, T., Rau, M. L., & Saghai, Y. (2021). A meta-analysis of projected global food demand and population at risk of hunger for the period 2010–2050. *Nature Food*, *2*, 494–501.

Vieira Gomes, A. M., Souza Carmo, T., Silva Carvalho, L., Mendonça Bahia, F., & Parachin, N. S. (2018). Comparison of yeasts as hosts for recombinant protein production. *Microorganisms*, *6*(2), 38.

Vollset, S. E., Goren, E., Yuan, C.-W., Cao, J., Smith, A. E., Hsiao, T., Bisignano, C., Azhar, G. S., Castro, E., Chalek, J., Dolgert, A. J., Frank, T., Fukutaki, K., Hay, S. I., Lozano, R., Mokdad, A. H., Nandakumar, V., Pierce, M., Pletcher, M., Robalik, T., Steuben, K. M., Wunrow, H. Y., Zlavog, B. S., & Murray, C. J. L. (2020). Fertility, mortality, migration, and population scenarios for 195 countries and territories from 2017 to 2100: A forecasting analysiss for the Global Burden of Disease Study. *Lancet*, *396*, 1285–1306.

Wang, G., Zhao, Z., Ke, J., Engel, Y., Shi, Y. M., Robinson, D., Bingol, K., Zhang, Z., Bowen, B., Louie, K., Wang, B., Evans, R., Miyamoto, Y., Cheng, K., Kosina, S., de Raad, M., Silva, L., Luhrs, A., Lubbe, A., Hoyt, D. W., Francavilla, C., Otani, H., Deutsch, S., Washton, N. M., Rubin, E. M., Mouncey, N. J., Visel, A., Northen, T., Cheng, J. F., Bode, H. B., & Yoshikuni, Y. (2019a). CRAGE enables rapid activation of biosynthetic gene clusters in undomesticated bacteria. *Nature Microbiology*, *4*, 2498–2510. https://doi.org/10.1038/s41564-019-0573-8

Wang, Y., Ling, C., Chen, Y., Jiang, X., & Chen, G. Q. (2019b). Microbial engineering for easy downstream processing. *Biotechnology Advances*, *37*, 107365.

Wu, S., Zhou, Y., Dai, L., Yang, A., & Qiao, J. (2024). Assembly of functional microbial ecosystems: From molecular circuits to communities. *FEMS Microbiology Reviews*, *48*(6), fuae026.

Xia, J., Wang, G., Fan, M., Chen, M., Wang, Z., & Zhuang, Y. (2021). Understanding the scale-up of fermentation processes from the viewpoint of the flow field in bioreactors and the physiological response of strains. *Chinese Journal of Chemical Engineering*, *30*, 178–184.

Yang, P., Condrich, A., Lu, L., Scranton, S., Hebner, C., Sheykhhasan, M., & Ali, M. A. (2024a). Genetic engineering in bacteria, fungi, and oomycetes, taking advantage of CRISPR. *DNA*, *4*(4), 427–454.

Yang, S., Song, L., Wang, J., Zhao, J., Tang, H., & Bao, X. (2024b). Engineering Saccharomyces cerevisiae for efficient production of recombinant proteins. *Engineering Microbiology*, *4*(1), 100122.

Zahid, I., Nazir, M. H., Chiang, K., Christo, F., & Ameen, M. (2024). Current outlook on sustainable feedstocks and processes for sustainable aviation fuel production. *Current Opinion in Green and Sustainable Chemistry*, 100959.

10

Nanotechnological Interventions in the Advancement of Lignocellulose Bio-Foundry

Current Status and Future Prospects

Carlos Lopez-Ortiz[1], *Alan Chavez-Hita Wong*[2], *Aldo Sosa*[2], *and Nagamani Balagurusamy*[2]

[1] *Gus R. Douglass Institute and Department of Biology, West Virginia State University, Institute, WV, USA*
[2] *Laboratorio de Biorremediación, Facultad de Ciencias Biológicas, Universidad Autónoma de Coahuila, Torreón, Coahuila, México*

10.1 Introduction

The growing need for sustainable and renewable resources has accelerated the transition from fossil-based systems to bio-based economies (Yang et al., 2021). Lignocellulose biomass, derived from plant and agricultural residues, offers immense potential as a feedstock for producing biofuels, biochemicals, and biomaterials (Mujtaba et al., 2023). Global annual production of lignocellulosic biomass is estimated to exceed 181.5 billion tons, yet only 4.5% is currently utilized due to its complex structure. This structure is primarily composed of cellulose (38–50%), hemicellulose (23–32%), and lignin (15–25%) (Mujtaba et al., 2023; Ojo, 2023; Patel and Shah, 2021). This structural complexity and the inherent recalcitrance of lignin to enzymatic degradation present significant challenges for efficient deconstruction, valorization, and industrial scalability (Jeffri et al., 2024).

Recently, nanotechnology has emerged as a transformative tool to address these bottlenecks, offering precise, efficient, and innovative solutions (Wahab et al., 2024). With materials and processes operating at the nanoscale, nanotechnology enhances the properties of lignocellulosic materials and provides innovative approaches for biomass pretreatment, enzymatic hydrolysis, and product synthesis (Sankaran et al., 2021). Nanotechnological interventions and engineered nanomaterials have transformed the landscape of lignocellulose bio-foundry through enhanced enzyme stability, improved biomass pretreatment, and enabling controlled conversion processes (Hussain et al., 2024). For instance, nanoscale catalysts and nanocarriers have improved the efficiency of biofuel production by allowing highly precise, cost-effective, and environmentally friendly processes. In addition, nanoscale instrumentation has provided further insight into lignocellulosic cell wall ultrastructure and enzymatic degradation mechanisms, optimizing biomass conversion processes (Dutta et al., 2023; Pramanik et al., 2023).

Biofoundry Techniques for Biotechnology Applications, First Edition. Edited by Anuj Kumar Chandel.

Moreover, advancements in nanomaterials have facilitated the selective separation of lignocellulose components and the synthesis of high-value nanocellulose derivatives. Integrating nanotechnology into lignocellulose bio-foundries, merging synthetic biology, systems biology, and process engineering, can potentially revolutionize the sustainable production of biofuels and biomaterials (Gupta & Shukla, 2020; Zhu et al., 2021). These biofoundries leverage cutting-edge technologies to transform lignocellulosic biomass into high-value products, contributing to global demand to reduce fossil fuel consumption and mitigate environmental impact (El-Araby, 2024).

This chapter explores nanotechnology's current status and future prospects in lignocellulose biofoundries, focusing on breakthroughs, challenges, and opportunities. It examines nanoscale materials for catalysis, innovations in biomass processing, and the development of biocompatible and biodegradable materials. Additionally, it highlights the integration of nanotechnology with automation and high-throughput screening to enhance lignocellulose valorization. The discussion underscores nanotechnology's pivotal role in advancing renewable energy and sustainable production, shaping the next generation of lignocellulose biofoundries for a sustainable bioeconomy.

10.2 Advancing Lignocellulose Bio-Foundries: Pretreatment Strategies and Nanotechnology Integration

Lignocellulose bio-foundries represent a transformative approach to converting plant-based biomass into valuable biofuels, chemicals, and materials. However, the use of lignocellulosic materials, such as sugarcane bagasse, coconut, soy, corn, sorghum, kenaf, grasses, plantain, and hemp (Maheshwari, 2018; Reshmy et al., 2022) for bioenergy production is constrained by the inherent physical and chemical complexity of these raw materials. Extracting and processing these materials to obtain precursors is challenging due to the high pressure and temperature required to process components like lignin, hemicellulose, and cellulose (Abolore et al., 2024). To overcome these challenges, pretreatment processes are critical for facilitating the recovery and availability of fermentable sugars in cellulose, hemicellulose, and their respective oligomers and monomers (Shukla et al., 2023). Current chemical pretreatment methods—including acidic, alkaline, oxidative, and ionic liquid techniques—are widely adopted for disrupting the structural integrity of biomass components (Chen et al., 2024). Similarly, physical methods, such as thermal treatment, ultrasound, and irradiation, enhance process efficiency by modifying the physical characteristics of the material (Figure 10.1). However, these methods often face limitations, including high energy demands, extensive chemical usage, and the generation of inhibitory by-products such as lignin monomers (Jönsson & Martín, 2016). In addition, recent progress in metabolic engineering and synthetic biology has enabled the design of robust microbial strains capable of converting a broader range of feedstocks into high-value products (Peña-Castro et al., 2023). Despite these advances, low yields and the high sensitivity of these strains to inhibitors present in lignocellulosic hydrolysates remain significant barriers to efficiency and scalability. Over the past decade, significant advancements have been made in developing processes and nanotechnologies to efficiently deconstruct and valorize lignocellulosic biomass (Ning et al., 2021). Nanotechnological breakthroughs have primarily focused on enhancing the pretreatment,

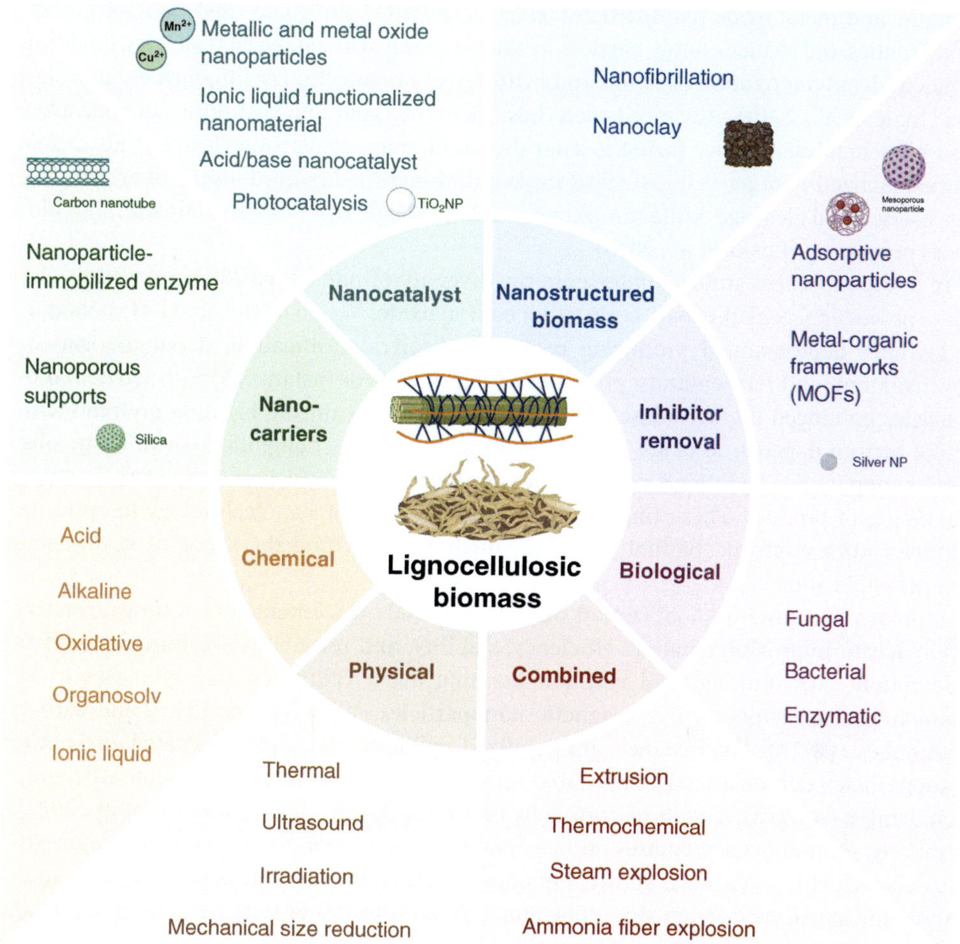

Figure 10.1 Classical and nanotechnology-based strategies for the pretreatment and functionalization of lignocellulosic biomass.

hydrolysis, and fermentation stages of lignocellulose processing (Arora et al., 2020). This has been achieved through the development of nanocatalysts, nanocarriers, adsorptive nanoparticles, enzyme immobilization, and nanostructured materials, each tailored to address specific challenges in biomass conversion, as shown in Figure 10.1 (Khan & Rashid, 2024).

10.3 Catalytic Nanomaterials and Enzyme Immobilization for Lignocellulose Biomass Conversion

Nanotechnology has revolutionized the pretreatment and functionalization of lignocellulosic biomass, addressing critical challenges in biofoundries through its versatility across physical, chemical, and biological processes (Roy et al., 2021). Nanocatalysts, such as metallic and metal oxide nanoparticles (e.g., TiO_2, MnO_2, ZnO, and CuO), enhance reaction kinetics and reduce energy barriers in oxidative and acid-catalyzed reactions, enabling efficient depolymerization of lignin and hydrolysis of hemicellulose into fermentable sugars (Ingle et al., 2019). Advanced materials, including gold and platinum nanoparticles, and photocatalysis further facilitate solar-driven biomass conversion (Katre et al., 2024). Functionalized nanoparticles, such as ionic liquid-modified nanocatalysts, offer selective glycosidic bond cleavage while minimizing undesired side reactions and streamlining biomass processing (Luska et al., 2015).

In addition, recent studies underscore the diversity of nanomaterials employed in biomass processing. Nickel-cobalt, cerium-doped iron oxide, Mg-Zn ferrite, and ZnO nanoparticles have demonstrated promising results in catalytic gasification, depolymerization, pretreatment, and fermentation processes (Table 10.1). For instance, Mg-Zn ferrite nanoparticles enhanced β-glucosidase activity by 18.63–20.94% during biomass pretreatment, while cerium-doped iron oxide nanoparticles achieved 44% delignification in corn cobs. ZnO nanoparticles facilitated rice straw fermentation with *Fusarium oxysporum*, yielding 0.0359 g/g of product. These findings reveal the potential of nanotechnology to optimize biomass utilization and biofuel synthesis, further broadening the scope of sustainable energy applications.

Moreover, enzyme immobilization on nanomaterials has emerged as a transformative approach to improving catalytic efficiency, stability, and reusability. Techniques such as adsorption, cross-linking, and covalent bonding allow cellulases and xylanases to be immobilized on supports like magnetic nanoparticles, silica nanoparticles, and carbon nanotubes (CNTs). For instance, immobilized cellulases on chitosan-coated magnetite nanoparticles demonstrated remarkable reusability and substrate conversion efficiency, achieving 64.44% conversion of corn cobs into fermentable sugars (Zanuso et al., 2022). Similarly, immobilized xylanases on mesoporous silica or iron oxide magnetic nanoparticles coated with silica have shown enhanced thermostability and activity, significantly increasing sugar yields from rice straw and corn cob substrates (Ariaeenejad et al., 2020; Hamid et al., 2022).

Nanostructured materials, including nanofibrils, nanoclays, and nanocellulose, improve the porosity and accessibility of lignocellulosic biomass, facilitating enzymatic or chemical interactions (Kaur et al., 2021). Additionally, adsorptive nanoparticles, such as metal–organic frameworks (MOFs) and chitosan-based materials, effectively remove fermentation inhibitors like furfural, HMF, and lignin-derived phenolics, ensuring cleaner substrates for downstream microbial processes (Gan et al. 2024). In the delignification of biomass, immobilized laccases on nanocomposites such as chitosan-SiO_2–Fe_3O_4 have demonstrated superior performance, achieving an 84.46% delignification rate in sweet sorghum stover (Shanmugam et al., 2020).

Moreover, magnetic nanoparticles, doped with cerium or other metals, have further advanced delignification, enabling efficient reuse and cost-effective processing (Rajak et al., 2021). The immobilization of multienzyme complexes, such as xylanase-lichenase or cellulase-β-glucosidase-pectinase systems on silica or Eudragit polymer matrices, has reduced by-product formation, enhanced enzyme synergy, and improved hydrolytic performance (Lin et al., 2021; Yu et al., 2024).

Table 10.1 Applications of Nanomaterials in Biomass Processing and Biofuel Production

Nanomaterials	Processes	Substrates	Enzymes/Organisms	Results	References
Nickel and cobalt nanoparticles	Nanocatalytic gasification	*Carthamous oxyacantha* *Asphodelus tenuifolius* *Chenopodium album*	–	65.47% Biodiesel	Ali et al. (2020)
Cerium-doped iron oxide nanoparticles	Lignin depolymerization	Corncob	–	44% delignification	Rajak et al. (2021)
Mg–Zn ferrite nanoparticles	Pretreatment	*Chlorella vulgaris* biomass	β-Glucosidase	18.63% increase in enzyme activity	Elshobary et al. (2024)
Mg–Zn ferrite nanoparticles	Pretreatment	*Alkalinema pantanalense*	β-Glucosidase	20.94% increase in enzyme activity	Elshobary et al. (2024)
ZnO nanoparticles	Fermentation	Rice straw	*Fusarium oxysporum*	0.0359 g/g	Gupta and Chundawat (2020)
Chitin—Fe_3O_4 magnetic nanoparticles	Fermentation	Rice straw	β-Glucosidase	31.24% total sugar	Alnadari et al. (2020)
AgO	Enzymatic hydrolysis	Coconut cake	Cellulase	57.804 U/mL/min	Hussain et al. (2023)
FeO	Enzymatic hydrolysis	Coconut cake	Cellulase	32.73 U/mL/min	Hussain et al. (2023)
Calcium hydroxyapatite nanoparticles	Enzymatic hydrolysis	Rice husk	Xylanase	29% of reducing sugars	Dutta et al. (2014)
Calcium hydroxyapatite nanoparticles	Enzymatic hydrolysis	Rice straw	Xylanase	41% of reducing sugars	Dutta et al. (2014)
Magnetite nanoparticles	Enzymatic biodiesel production	*p*-Nitrophenyl palmitate	Lipases	~45% biodiesel per hour	Maroju et al. (2023)
Chitosan-coated magnetic nanoparticles	Lignocellulose hydrolysis	*Agave atrovirens*	Cellulase	~68.14	Sanchez-Ramirez et al. (2017)
Calcium carbonate nanoparticles	Transesterification	Soybean oil	Lipases	$3.319 \pm 0.180\,\mu M\,hr^{-1}\,mg^{-1}$	Choi et al. (2024)
$MgO/MgFe_2O_4$	Transesterification	Sunflower oil	–	91.2% biodiesel production	Alaei et al. (2018)

Beyond pretreatment, nanotechnology also enhances the efficiency of fermentation processes by mitigating the effects of inhibitory compounds. Nanomaterial coatings improve microbial and enzymatic tolerance to toxic by-products, while immobilized yeast systems exhibit increased stability, stress resistance, and substrate utilization (Sharma et al., 2023). Magnetic nanoparticles and carbon-based supports enable efficient enzyme recovery, reducing process costs and ensuring sustainability (Mariño et al., 2021). Thus, by integrating nanotechnological tools into every stage of lignocellulosic biomass conversion—from pretreatment to fermentation—biofoundries can achieve higher yields, reduced process complexity, and enhanced viability (Blasi et al., 2023).

10.4 Underlying the Interactions of Nanotechnology Mechanism in Lignocellulose Bio-Foundry

Nanotechnology has become a powerful tool for improving the efficiency of lignocellulosic pretreatment, hydrolysis, and fermentation processes. Utilizing the unique properties of nanomaterials, innovative methods to effectively deconstruct lignocellulosic biomass have been developed to overcome the challenges associated with its complex structure. For instance, acid-functionalized magnetic nanoparticles (AFMNs) are strong acid nanoparticles that mimic the catalytic action of mineral acids during lignocellulosic biomass decomposition, facilitating the weakening of ester and ether bonds between lignin and hemicellulose (Changmai et al., 2021). The magnetic properties of these nanoparticles enable their efficient recovery using an external magnet, making them reusable and cost-effective (Hosseini, 2022). AFMNs include alkyl-sulfonic acid-functionalized nanoparticles, silica-protected cobalt-bearing spinel ferrite nanoparticles, and silica-propyl sulfonic-functionalized nanoparticles (Chandel et al., 2024). These materials offer the potential for synergistic interactions, enhancing the efficiency of lignocellulosic biomass pretreatment. Additionally, AFMNs can be integrated with other pretreatment methods to improve biomass decomposition further. Unlike conventional acid and alkaline pretreatment methods, which often corrode processing equipment, AFMNs minimize corrosion, reducing maintenance costs (Arya et al., 2021).

Likewise, CNTs have emerged as valuable nanomaterials in bioethanol production due to their unique physicochemical properties, including high surface area, excellent thermal stability, and remarkable mechanical strength (Cosnier et al., 2014). These properties make CNTs highly effective in addressing challenges associated with lignocellulosic biomass conversion, enzymatic hydrolysis, and fermentation. One of the primary roles of CNTs in bioethanol production is enzyme immobilization (Ninawe & Tariq, 2021). This immobilization enhances enzyme stability, protects against denaturation, and allows enzymes to be reused in multiple cycles, significantly reducing production costs. The high surface area of CNTs ensures close contact between enzymes and substrates, improving catalytic efficiency during biomass hydrolysis (Barik et al., 2023).

CNTs also play a critical role in biomass pretreatment. Functionalized CNTs can act as catalysts or co-catalysts in the disruption between lignin and hemicellulose, weakening the structural integrity of lignocellulosic biomass (Arya et al., 2021) (Figure 10.2). This pretreatment enhances the accessibility of cellulose and hemicellulose to enzymatic hydrolysis, increasing the

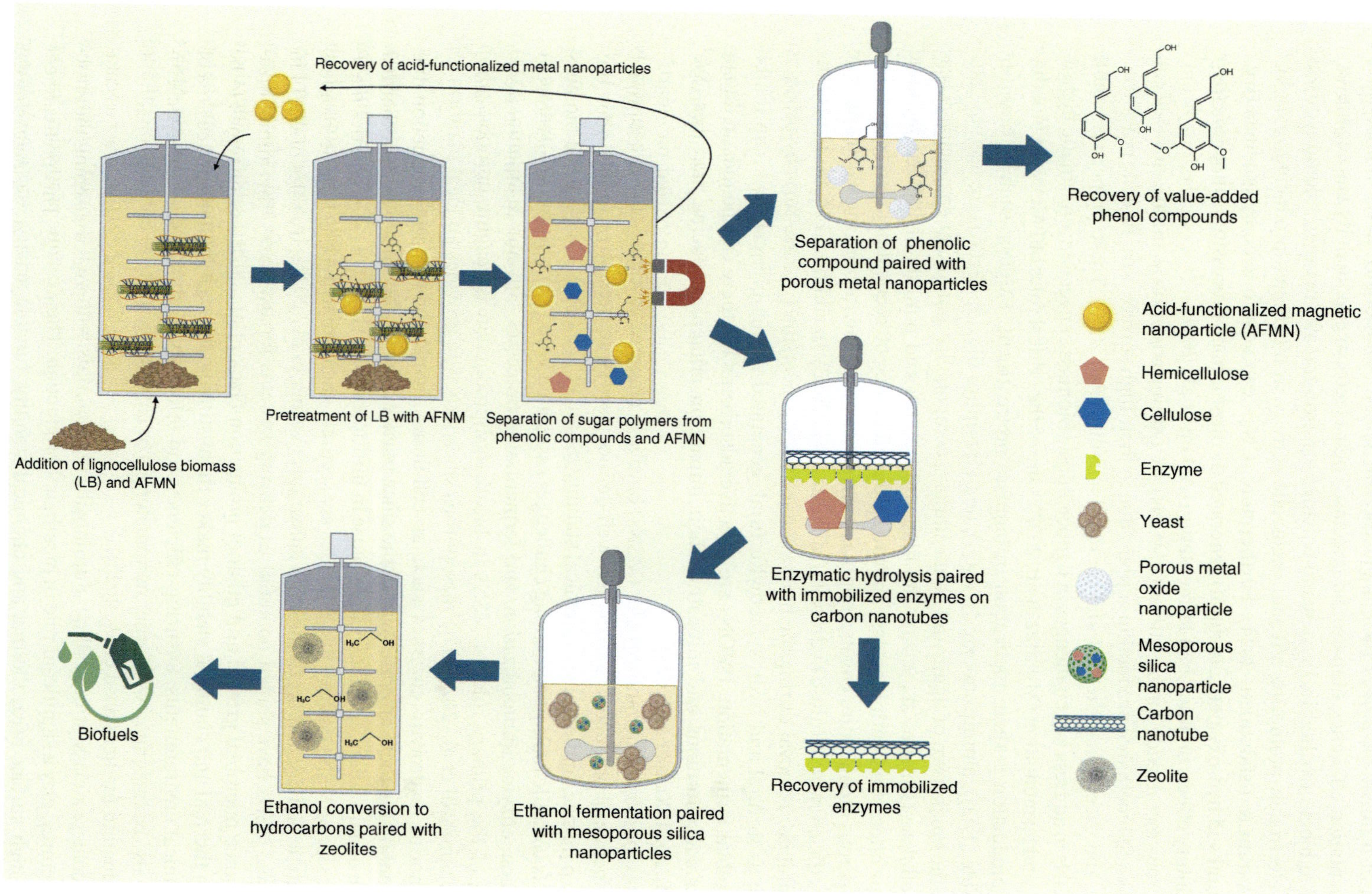

Figure 10.2 Schematic representation of lignocellulosic biomass conversion using nanotechnology.

yield of fermentable sugars such as glucose and xylose. Additionally, CNTs contribute to fermentation efficiency in microbial systems. Their electrical conductivity has been leveraged in bioelectrochemical systems to enhance electron transfer between microorganisms and electrodes, promoting efficient metabolic activity in bioethanol-producing microbes (Mokaram & Esfahanian, 2024). Furthermore, CNTs can adsorb toxic by-products (e.g., furfural and hydroxymethylfurfural) generated during biomass processing, mitigating their inhibitory effects on microbial fermentation (Aslam et al., 2021).

Porous metal oxide nanoparticles (PMONs) are gaining attention as multifunctional catalysts and bioethanol production support due to their high surface area, tunable pore structure, and exceptional thermal and chemical stability (Pal, 2020). These properties enable PMONs to address key challenges in lignocellulosic biomass conversion, enzymatic hydrolysis, and fermentation (Mumtaz et al., 2022). One of the critical roles of PMONs in bioethanol production is their application in biomass pretreatment. PMONs, such as titanium dioxide (TiO_2), aluminum oxide (Al_2O_3), and zirconium dioxide (ZrO_2), are used to catalyze the breakdown of lignin and hemicellulose, components that hinder enzymatic access to cellulose (Deng et al., 2023). The high surface area and porosity of PMONs facilitate better interaction with biomass, improving the efficiency of pretreatment processes. The nonporous architecture ensures optimal enzyme-substrate interaction, increasing the rate and efficiency of cellulose hydrolysis (Mumtaz et al., 2022). In addition, PMONs contribute to inhibitor removal during the biomass-to-ethanol conversion process. Toxic by-products such as furfural and hydroxymethylfurfural, generated during pretreatment, can inhibit microbial fermentation. PMONs are effective adsorbents for these compounds, reducing their concentrations and mitigating their impact on ethanol-producing microbes (Wei et al., 2024; Xiao et al., 2023).

Mesoporous silica nanoparticles (MSNs) have also emerged as highly effective materials in bioethanol production due to their well-defined pore structures, large surface area, tunable pore size, and excellent chemical stability (Shahi et al., 2021). These properties make MSNs versatile tools in addressing challenges associated with lignocellulosic biomass conversion, enzyme immobilization, and fermentation processes (Silviana & Dalanta, 2024). One of the primary roles of MSNs in bioethanol production is enzyme immobilization (Mohammadi et al., 2020). The mesoporous silica structure provides an ideal platform for anchoring hydrolytic enzymes such as cellulases, hemicellulases, and β-glucosidases. Immobilizing these enzymes within the silica pores enhances their stability, protects them from denaturation, and allows their reuse in multiple cycles (Lin et al., 2021). The uniform pore size of MSNs ensures that enzymes are securely encapsulated while maintaining high catalytic activity and accessibility to substrates (Verma et al., 2022). In fermentation processes, MSNs have shown potential as delivery systems for nutrients and cofactors that enhance microbial growth and ethanol production (Jack et al., 2020). Additionally, their high thermal and chemical stability makes them suitable for high-temperature and harsh chemical environments, ensuring efficiency and longevity (Dabirian et al., 2023). MSNs can also be integrated with other nanomaterials or catalytic systems to enhance bioethanol production further (Bosu et al., 2022).

Zeolites, a class of crystalline aluminosilicate materials with well-defined microporous structures, play a significant role in bioethanol production. Their unique properties, including high surface area, thermal and chemical stability, tunable acidity, and ion-exchange capabilities, make them versatile catalysts and supports in various stages of the bioethanol production process (Mardiana et al., 2022). In fermentation enhancement, zeolites can act as carriers for nutrients or cofactors that improve microbial activity. Additionally, their ion-exchange capabilities can help regulate pH levels and remove excess ions that may inhibit microbial growth, creating a more favorable environment for ethanol-producing microorganisms (Nuhma et al., 2021). Zeolites also find application in bioethanol purification. Their ability to selectively adsorb water and other impurities makes them effective in ethanol dehydration and purification processes (Khalil et al., 2020). Molecular sieves, a form of zeolite, are widely used in the final stages of ethanol production to achieve fuel-grade ethanol by removing residual water through selective adsorption (Fereidooni et al., 2021). The use of zeolites can help the catalytic upgrading of bioethanol and by-products. For example, zeolites are employed in processes such as ethanol-to-olefins (ETO) or ethanol-to-hydrocarbons (ETH) for converting ethanol into higher-value chemicals and fuels, adding versatility to bioethanol production chains (Polishchuk et al., 2020). Although nanoparticles offer numerous advantages for the treatment of lignocellulosic biomass, further research is needed to optimize their efficiency and fully understand their underlying mechanisms.

10.5 Factors Affecting Nanotechnology Use and Its Performance in Bio-Foundry Using Lignocellulosic Biomass

Optimizing nanotechnology applications in bio-foundries processing lignocellulosic biomass is influenced by several critical factors. The physicochemical properties of nanomaterials, including size, shape, surface area, and functionalization, significantly impact their catalytic efficiency and interactions with biomass components (Dutta et al., 2023). Additionally, parameters such as temperature, pH, reaction time, nanoparticle concentration, and size significantly influence the performance of nanomaterials in total biomass conversion.

pH plays a significant role in determining the performance of nanotechnology in bio-foundries, as it affects enzymatic activity, fermentation efficiency, and the stability of nanomaterials (Sekoai et al., 2019). Inhibitory compounds such as acetic acid and phenolic compounds, commonly generated during the pretreatment of lignocellulosic biomass, can lower the pH and adversely impact fermentation efficiency (Jönsson & Martín, 2016). Magnetic chitosan nanoparticles, commonly used for enzyme immobilization, exhibit varying stability across different pH levels (Zhang et al., 2024). For cellulolytic enzymes involved in processing lignocellulosic biomass, optimal performance is typically observed within a pH range of 6–8 (Romsaiyud et al., 2009). In contrast, gold nanoparticle-conjugated cellulase enzymes achieve peak functionality at a pH of 5.5. Furthermore, pH influences the interactions between nanomaterials and lignocellulosic substrates, particularly enzyme adsorption and desorption kinetics on nanoparticle surfaces, which are crucial for optimizing hydrolysis efficiency (Sankaran et al., 2021). Likewise, temperature is another important factor in lignocellulose conversion. Increasing temperature enhances the reaction rate to a certain point, but excessive heat can lead to denaturation, reducing or inhibiting enzymatic activity. Some cellulolytic enzymes exhibit optimal operating temperatures ranging from 35 °C to 77 °C (Bala & Singh, 2019). Moreover, enzymatic immobilization methods using nanomaterials also face limitations due to factors such as the solubility of the support matrix, the presence of functional groups, dimensions, porosity/permeability, mechanical strength, reusability, and size. These challenges are compounded by the time and cost invested in nanoparticle synthesis and the immobilization process, along with issues like low surface coverage, susceptibility to contamination, and degradation (Hettiarachchy et al., 2018).

Nanoparticle concentration must be carefully controlled to prevent aggregation. It has been reported that cellulosic nanocomposites' mechanical properties improve with cellulose nanoparticles' addition; however, excessive concentrations can lead to agglomeration, compromising mechanical performance (Qasim et al., 2021). Similarly, excessively high nanoparticle concentrations can cause clustering in enzymatic hydrolysis, reducing the effective surface area available for enzymatic reactions (Peña et al., 2011). Reaction time is critical in enzymatic hydrolysis, significantly impacting sugar yields. Acid-functionalized nanoparticles improve cellobiose hydrolysis; however, extended reaction times can lead to the degradation of sugars into fermentation inhibitors such as furfural and hydroxymethylfurfural (Peña et al., 2011). Similarly, enzymatic hydrolysis of lignocellulose can produce substantial amounts of xylose, but shorter reaction times may result in lower conversion rates due to the inhibitory effects of compounds present in lignocellulose hydrolysate (Tang et al., 2023).

10.6 Challenges and Considerations Using Nanotechnology in Lignocellulose Bio-Foundry

The application of nanotechnology in lignocellulose bio-refineries holds significant potential but also presents unresolved technical, economic, environmental, and safety challenges (Figure 10.3). Among the technical difficulties, the development and implementation of efficient processes remain a critical concern. For example, acid-functionalized metal nanoparticles, commonly called solid acid nanocatalysts, are frequently utilized for lignocellulosic biomass hydrolysis. However, their tendency to aggregate and oxidize rapidly poses a major obstacle, severely limiting their reusability and overall efficiency (Rizvi et al., 2022). Moreover, research on standardizing hydrolysis parameters for these nanocatalysts is still lacking, often leading to suboptimal reaction conditions when applied to hydrolysates (Arora et al., 2020; Ingle et al., 2019). These inconsistencies in process parameters result in performance variability, further complicating their practical application (Hussain et al., 2024).

Using magnetic nanoparticles in conjunction with enzymes for cellulose hydrolysis introduces additional complexities. The design of surface reactivity is critical for ensuring stable immobilization bonds that can withstand reactor conditions (Lee & Au-Duong, 2018). Immobilization can reduce or completely inhibit enzyme activity without careful engineering, rendering the system inefficient (Darwesh et al., 2020). Covalent bonding, a commonly used method for attaching enzymes to nanoparticles due to its strength, may restrict enzyme flexibility and reduce catalytic efficiency (Bidir et al., 2021). On the other hand, weaker bonds, such as Van der Waals forces or ionic interactions, are susceptible to disruption by environmental factors like temperature or pH fluctuations (Arora et al., 2020). Other immobilization techniques, such as entrapment and adsorption, can obstruct

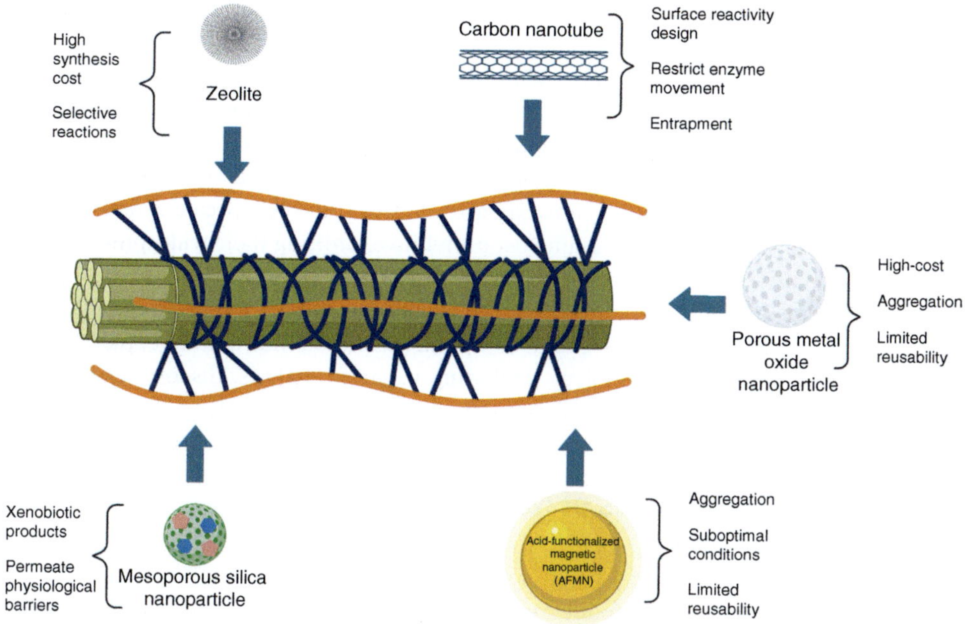

Figure 10.3 Challenges and functional characteristics of nanomaterials in lignocellulosic biomass conversion.

substrate access to the enzyme's active site, leading to prolonged reaction times and reduced yields (Meena et al., 2021). These challenges underscore both the complexities and the promising potential of nanotechnology in advancing bio-refinery processes.

Nanotechnology's potential in bio-refineries goes beyond technical advancements, presenting opportunities to enhance biofuel production by accelerating key process stages. Catalyst recovery and enzyme efficiency improvements can reduce production costs and promote sustainability (Dey et al., 2022). However, despite these benefits, the environmental implications of nanotechnology remain insufficiently understood, as it is still an emerging field. Xenobiotic byproducts, often generated during the synthesis of nanomaterials, may pose environmental and biological risks. For instance, nanoparticles can migrate through soil and water systems, potentially accumulating in ecosystems and creating environmental hazards (Hussain et al., 2024). Bioaccumulation is another significant concern, as nanoparticles may interact with living organisms, enter trophic chains, and lead to unknown ecological consequences (Kuehr et al., 2021). Due to their nanoscale dimensions, nanoparticles can penetrate physiological barriers in living organisms, potentially eliciting harmful biological responses. Studies have identified that commonly used nanoparticles, such as zinc oxide, titanium dioxide, cerium oxide, ferric oxide, and silicon dioxide, can induce adverse effects in humans. These effects include reduced cell viability, oxidative stress, DNA damage, apoptosis, and inflammatory responses (Kumah et al., 2023). These findings highlight the necessity of regulatory oversight. Organizations like the European Research Council and EURO-NanoTox play vital roles in assessing the safety of nanotechnology applications in industrial and biological processes (Zielińska et al., 2020).

Establishing clear guidelines and safety standards is essential to mitigate the risks associated with these advanced materials.

From an economic perspective, the high cost of biofuel production is primarily driven by pretreatment processes, which account for approximately 40% of total processing costs (Ingle et al., 2017). One of the primary goals of incorporating nanotechnology into biorefineries is to reduce these costs by enabling the recovery and reuse of catalysts, which would otherwise represent a significant expense. However, the large-scale synthesis and immobilization of nanoparticles remain cost-prohibitive, limiting their widespread application. Additionally, the need for cost-effective methods for synthesizing nanoparticles and integrating them into bio-refinery processes further escalates expenses (GadelHak et al., 2023). There is also a notable gap in comprehensive economic studies exploring the costs of implementing nanotechnology, including payback analysis, disposal costs, and comparisons with traditional pretreatment methods (Pandit et al., 2023).

10.7 Future Perspectives of Nanotechnology in Biofuel Production

Optimizing and scaling biofuel production remain critical challenges in advancing renewable energy solutions. Synthetic biology offers transformative potential by enabling the engineering of microorganisms to manage every stage of the production process. A key innovation in this field is the integration of synthetic microorganisms with immobilization techniques facilitated by nanotechnology, an essential step in scaling biofuel production. These synthetic organisms are manufactured to have improved pathways related to the process they are in charge of, reducing wastes and toxic by-products (Liu et al., 2022).

Nanotechnology also provides unparalleled opportunities for real-time monitoring of lignocellulose degradation processes. Nanosensors, designed to detect specific chemical intermediates or changes during biomass breakdown, deliver immediate feedback on reaction progress and efficiency. When embedded in bioreactors, these sensors can monitor key parameters such as enzyme activity, substrate concentration, and pH levels with exceptional sensitivity (Gupta & Krasnoslobodtsev, 2024). This real-time monitoring enables adaptive process adjustments, maximizing conversion efficiency and minimizing resource wastage. Additionally, experimental and computational studies incorporating machine learning approaches must be integrated to better understand the mechanisms involved in biofuel production (Bosu et al., 2022). Advances in algorithms and computational power enable more efficient screening applications for identifying high-yield biofuel-producing nanoparticles; when paired with biosensors, machine-learning approaches are able to optimize and regulate the biological process within the reactor.

10.8 Conclusion

Nanotechnology is transforming lignocellulose bio-foundries by addressing key challenges and unlocking new possibilities for biomass conversion. Improving processes like pretreatment, hydrolysis, and fermentation has enhanced efficiency, increased product yields, and reduced costs. Advanced materials like magnetic nanoparticles, CNTs, and mesoporous silica have enabled breakthroughs in enzyme immobilization, inhibitor removal, and real-time monitoring, making biofuel production more practical and scalable. However, challenges such as standardizing reaction conditions, minimizing environmental impacts, and managing costs remain. Despite these obstacles, the combination of nanotechnology with synthetic biology and cutting-edge tools like machine learning offers immense promise for lignocellulose bio-foundries to play a central role in creating sustainable energy and materials, helping us move closer to a greener, more circular economy.

References

Abolore, R. S., Jaiswal, S., & Jaiswal, A. K. (2024). Green and sustainable pretreatment methods for cellulose extraction from lignocellulosic biomass and its applications: A review. *Carbohydrate Polymer Technologies and Applications*, *7*, 100396.

Alaei, S., Haghighi, M., Toghiani, J., & Vahid, B. R. (2018). Magnetic and reusable MgO/MgFe2O4 nanocatalyst for biodiesel production from sunflower oil: Influence of fuel ratio in combustion synthesis on catalytic properties and performance. *Industrial Crops and Products*, *117*, 322–332.

Ali, S., Shafique, O., Mahmood, S., Mahmood, T., Khan, B. A., & Ahmad, I. (2020). Biofuels production from weed biomass using nanocatalyst technology. *Biomass and Bioenergy*, *139*, 105595.

Alnadari, F., Xue, Y., Zhou, L., Hamed, Y. S., Taha, M., & Foda, M. F. (2020). Immobilization of β-glucosidase from Thermatoga maritima on chitin-functionalized magnetic nanoparticle via a novel thermostable chitin-binding domain. *Scientific Reports*, *10*, 1663.

Ariaeenejad, S., Jokar, F., Hadian, P., Ma'mani, L., Gharaghani, S., Fereidoonnezhad, M., & Salekdeh, G. H. (2020). An efficient nano-biocatalyst for lignocellulosic biomass hydrolysis: Xylanase immobilization on organically modified biogenic mesoporous silica nanoparticles. *International Journal of Biological Macromolecules*, *164*, 3462–3473.

Arora, A., Nandal, P., Singh, J., & Verma, M. L. (2020). Nanobiotechnological advancements in lignocellulosic biomass pretreatment. *Materials Science for Energy Technologies*, *3*, 308–318.

Arya, I., Poona, A., Dikshit, P. K., Pandit, S., Kumar, J., Singh, H. N., Jha, N. K., Rudayni, H. A., Chaudhary, A. A., & Kumar, S. (2021). Current trends and future prospects of nanotechnology in biofuel production. *Catalysts*, *11*, 1308.

Aslam, M. M.-A., Kuo, H.-W., Den, W., Usman, M., Sultan, M., & Ashraf, H. (2021). Functionalized carbon nanotubes (CNTs) for water and wastewater treatment: Preparation to application. *Sustainability*, *13*, 5717.

Bala, A., & Singh, B. (2019). Cellulolytic and xylanolytic enzymes of thermophiles for the production of renewable biofuels. *Renewable Energy*, *136*, 1231–1244.

Barik, S., Dash, A. K., & Saharay, M. (2023). Immobilization of cellulase enzymes on single-walled carbon nanotubes for recycling of enzymes and better yield of bioethanol using computer simulations. *Journal of Chemical Information and Modeling*, *63*, 5192–5203.

Bidir, M. G., Millerjothi, N., Adaramola, M. S., & Hagos, F. Y. (2021). The role of nanoparticles on biofuel production and as an additive in ternary blend fuelled diesel engine: A review. *Energy Reports*, *7*, 3614–3627.

Blasi, A., Verardi, A., Lopresto, C. G., Siciliano, S., & Sangiorgio, P. (2023). Lignocellulosic agricultural waste valorization to obtain valuable products: An overview. *Recycling*, *8*, 61.

Bosu, S., Pooja, R., & Rajasimman, M. (2022). Role of nanomaterials in enhanced ethanol production through biological methods–Review on operating factors and machine learning applications. *Fuel*, *320*, 123905.

Chandel, H., Kumar, P., Chandel, A. K., & Verma, M. L. (2024). Biotechnological advances in biomass pretreatment for bio-renewable production through nanotechnological intervention. *Biomass Conversion and Biorefinery*, *14*, 2959–2981.

Changmai, B., Wheatley, A. E., Rano, R., Halder, G., Selvaraj, M., Rashid, U., & Rokhum, S. L. (2021). A magnetically separable acid-functionalized nanocatalyst for biodiesel production. *Fuel*, *305*, 121576.

Chen, J., Ma, X., Liang, M., Guo, Z., Cai, Y., Zhu, C., Wang, Z., Wang, S., Xu, J., & Ying, H. (2024). Physical–chemical–biological pretreatment for biomass degradation and industrial applications: A review. *Waste*, 451–473.

Choi, Y. S., Jeon, H. W., & Hwang, E. T. (2024). In-situ stabilized lipase in calcium carbonate microparticles for activation in solvent-free transesterification for biodiesel production. *Bioresource Technology*, *412*, 131394.

Cosnier, S., Holzinger, M., & Le Goff, A. (2014). Recent advances in carbon nanotube-based enzymatic fuel cells. *Frontiers in Bioengineering and Biotechnology*, *2*, 45.

Dabirian, E., Hajipour, A., Mehrizi, A. A., Karaman, C., Karimi, F., Loke-Show, P., & Karaman, O. (2023). Nanoparticles application on fuel production from biological resources: A review. *Fuel*, *331*, 125682.

Darwesh, O. M., Ali, S. S., Matter, I. A., Elsamahy, T., & Mahmoud, Y. A. (2020). Enzymes immobilization onto magnetic nanoparticles to improve industrial and environmental applications. *Methods in Enzymology*, 481–502.

Deng, W., Feng, Y., Fu, J., Guo, H., Guo, Y., Han, B., Jiang, Z., Kong, L., Li, C., Liu, H., Nguyen, P. T. T., Ren, P., Wang, F., Wang, S., Wang, Y., Wang, Y., Wong, S. S., Yan, K., Yan, N., Yang, X., Zhang, Y., Zhang, Z., Zeng, X., & Zhou, H. (2023). Catalytic conversion of lignocellulosic biomass into chemicals and fuels. *Green Energy & Environment*, *8*, 10–114.

Dey, N., Kumar, G., Vickram, A., Mohan, M., Singhania, R. R., Patel, A. K., Dong, C.-D., Anbarasu, K., Thanigaivel, S., & Ponnusamy, V. K. (2022). Nanotechnology-assisted production of value-added biopotent energy-yielding products from lignocellulosic biomass refinery–A review. *Bioresource Technology*, *344*, 126171.

Dutta, N., Mukhopadhyay, A., Dasgupta, A. K., & Chakrabarti, K. (2014). Improved production of reducing sugars from rice husk and rice straw using bacterial cellulase and xylanase activated with hydroxyapatite nanoparticles. *Bioresource Technology*, *153*, 269–277.

Dutta, S., Saravanabhupathy, S., Anusha, R. R. C., Banerjee, R., Dikshit, P. K., Padigala, C. T., Das, A. K., & Kim, B. S. (2023). Recent developments in lignocellulosic biofuel production with nanotechnological intervention: An emphasis on ethanol. *Catalysts*, *13*, 1439.

El-Araby, R. (2024). Biofuel production: Exploring renewable energy solutions for a greener future. *Biotechnology for Biofuels and Bioproducts*, *17*, 129.

Elshobary, M., Abdullah, E., Abdel-Basset, R., Metwally, M., & El-Sheekh, M. (2024). Maximizing biofuel production from algal biomass: A study on biohydrogen and bioethanol production using MgZn ferrite nanoparticles. *Algal Research*, *81*, 103595.

Fereidooni, L., Enayati, M., & Abbaspourrad, A. (2021). Purification technology for renewable production of fuel from methanolysis of waste sunflower oil in the presence of high silica zeolite beta. *Green Chemistry Letters and Reviews*, *14*, 2–14.

GadelHak, Y., El-Azazy, M., Shibl, M. F., & Mahmoud, R. K. (2023). Cost estimation of synthesis and utilization of nano-adsorbents on the laboratory and industrial scales: A detailed review. *Science of the Total Environment*, *875*, 162629.

Gan, J., Iqbal, H. M., Show, P. L., Rahdar, A., & Bilal, M. (2024). Upgrading recalcitrant lignocellulosic biomass hydrolysis by immobilized cellulolytic enzyme–based nanobiocatalytic systems: A review. *Biomass Conversion and Biorefinery*, *14*, 4485–4509.

Gupta, K., & Chundawat, T. S. (2020). Zinc oxide nanoparticles synthesized using Fusarium oxysporum to enhance bioethanol production from rice-straw. *Biomass and Bioenergy*, *143*, 105840.

Gupta, A. K., & Krasnoslobodtsev, A. V. (2024). Fueling the future: The emergence of self-powered enzymatic biofuel cell biosensors. *Biosensors*, *14*, 316.

Gupta, G. K., & Shukla, P. (2020). Lignocellulosic biomass for the synthesis of nanocellulose and its eco-friendly advanced applications. *Frontiers in Chemistry*, *8*, 601256.

Hamid, A., Zafar, A., Liaqat, I., Afzal, M. S., Peng, L., Rauf, M. K., Ul Haq, I., Ali, S., & Aftab, M. N. (2022). Effective utilization of magnetic nano-coupled cloned β-xylanase in saccharification process. *RSC Advances*, *12*, 6463–6475.

Hettiarachchy, N., Feliz, D., Edwards, J., & Horax, R. (2018). The use of immobilized enzymes to improve functionality. In R. Y. Yada (Ed.), *Proteins in Food Processing* (pp. 569–597). Elsevier.

Hosseini, S. A. (2022). Nanocatalysts for biodiesel production. *Arabian Journal of Chemistry*, *15*, 104152.

Hussain, S., Yasin, M. T., Ahmad, K., Khan, S., Ahmad, R., Khan, J., Ghani, A., Shah, M. M., Ahmed, M., Tariq, H., Rehman, H., Hussain, A., Faheem, M., & Bokhari, S. A. I. (2023). Enhancement effect of AgO nanoparticles on fermentative cellulase activity from thermophilic Bacillus subtilis Ag-PQ. *Journal of Genetic Engineering and Biotechnology*, *21*, 151.

Hussain, A., Parveen, F., Saxena, A., & Ashfaque, M. (2024). A review of nanotechnology in enzyme cascade to address challenges in pre-treating biomass. *International Journal of Biological Macromolecules*, *270*, 132466.

Ingle, A. P., Rathod, J., Pandit, R., da Silva, S. S., & Rai, M. (2017). Comparative evaluation of free and immobilized cellulase for enzymatic hydrolysis of lignocellulosic biomass for sustainable bioethanol production. *Cellulose*, *24*, 5529–5540.

Ingle, A. P., Chandel, A. K., Antunes, F. A., Rai, M., & da Silva, S. S. (2019). New trends in application of nanotechnology for the pretreatment of lignocellulosic biomass. *Biofuels, Bioproducts and Biorefining*, *13*, 776–788.

Jack, J., Lo, J., Donohue, B., Maness, P.-C., & Ren, Z. J. (2020). High rate CO2 valorization to organics via CO mediated silica nanoparticle enhanced fermentation. *Applied Energy*, *279*, 115725.

Jeffri, N. I., Rawi, N. F. M., Kassim, M. H. B. M., & Abdullah, C. K. (2024). Unlocking the potential: Evolving role of technical lignin in diverse applications and overcoming challenges. *International Journal of Biological Macromolecules*, *274*, 133506.

Jönsson, L. J., & Martín, C. (2016). Pretreatment of lignocellulose: Formation of inhibitory by-products and strategies for minimizing their effects. *Bioresource Technology, 199*, 103–112.

Katre, S., Baghmare, P., & Giri, A. S. (2024). Photocatalytic nanomaterials and their implications towards biomass conversion for renewable chemical and fuel production. *Nanoscale Advances, 6*, 5258–5284.

Kaur, P., Sharma, N., Munagala, M., Rajkhowa, R., Aallardyce, B., Shastri, Y., & Agrawal, R. (2021). Nanocellulose: Resources, physio-chemical properties, current uses and future applications. *Frontiers in Nanotechnology, 3*, 747329.

Khalil, I., Jabraoui, H., Lebègue, S., Kim, W. J., Aguilera, L.-J., Thomas, K., Maugé, F., & Badawi, M. (2020). Biofuel purification: Coupling experimental and theoretical investigations for efficient separation of phenol from aromatics by zeolites. *Chemical Engineering Journal, 402*, 126264.

Khan, N. R., & Rashid, A. B. (2024). Carbon-based nanomaterials: A paradigm shift in biofuel synthesis and processing for a sustainable energy future. *Energy Conversion and Management: X, 22*, 100590.

Kuehr, S., Kosfeld, V., & Schlechtriem, C. (2021). Bioaccumulation assessment of nanomaterials using freshwater invertebrate species. *Environmental Sciences Europe, 33*, 1–36.

Kumah, E. A., Fopa, R. D., Harati, S., Boadu, P., Zohoori, F. V., & Pak, T. (2023). Human and environmental impacts of nanoparticles: A scoping review of the current literature. *BMC Public Health, 23*, 1059.

Lee, C. K., & Au-Duong, A. N. (2018). Enzyme immobilization on nanoparticles: Recent applications. *Emerging Areas in Bioengineering, 1*, 67–80.

Lin, Y., Jin, W., Cai, L., Liu, X., Qiu, Y., & Zhang, G. (2021). Green preparation of covalently co-immobilized multienzymes on silica nanoparticles for clean production of reducing sugar from lignocellulosic biomass. *Journal of Cleaner Production, 314*, 127994.

Liu, Z., Wang, J., & Nielsen, J. (2022). Yeast synthetic biology advances biofuel production. *Current Opinion in Microbiology, 65*, 33–39.

Luska, K., Migowski, P., & Leitner, W. (2015). Ionic liquid-stabilized nanoparticles as catalysts for the conversion of biomass. *Green Chemistry, 17*, 3195–3206.

Maheshwari, N. V. (2018). Agro-industrial lignocellulosic waste: An alternative to unravel the future bioenergy. In *Biofuels: Greenhouse gas mitigation and global warming: Next generation biofuels and role of biotechnology* (pp. 291–305).

Mardiana, S., Azhari, N. J., Ilmi, T., & Kadja, G. T. (2022). Hierarchical zeolite for biomass conversion to biofuel: A review. *Fuel, 309*, 122119.

Mariño, M. A., Fulaz, S., & Tasic, L. (2021). Magnetic nanomaterials as biocatalyst carriers for biomass processing: Immobilization strategies, reusability, and applications. *Magnetochemistry, 7*, 133.

Maroju, P. A., Ganesan, R., & Dutta, J. R. (2023). Biofuel generation from food waste through immobilized enzymes on magnetic nanoparticles. *Materials Today: Proceedings, 72*, 62–66.

Meena, J., Gupta, A., Ahuja, R., Singh, M., & Panda, A. K. (2021). Recent advances in nano-engineered approaches used for enzyme immobilization with enhanced activity. *Journal of Molecular Liquids, 338*, 116602.

Mohammadi, N. S., Khiabani, M. S., Ghanbarzadeh, B., & Mokarram, R. R. (2020). Improvement of lipase biochemical properties via a two-step immobilization method: Adsorption onto silicon dioxide nanoparticles and entrapment in a polyvinyl alcohol/ alginate hydrogel. *Journal of Biotechnology*, *323*, 189–202.

Mokaram, M. M., & Esfahanian, M. (2024). Bioethanol fermentation integrated with PDMS membrane modified by carbon nanotube and graphene oxide from sugarcane bagasse. *Biomass Conversion and Biorefinery*, *15*, 1–15.

Mujtaba, M., Fraceto, L. F., Fazeli, M., Mukherjee, S., Savassa, S. M., de Medeiros, G. A., Santo Pereira, A. E., Mancini, S. D., Lipponen, J., & Vilaplana, F. (2023). Lignocellulosic biomass from agricultural waste to the circular economy: A review with focus on biofuels, biocomposites and bioplastics. *Journal of Cleaner Production*, *402*, 136815.

Mumtaz, M., Baqar, Z., Hussain, N., Bilal, M., Azam, H. M. H., & Iqbal, H. M. (2022). Application of nanomaterials for enhanced production of biodiesel, biooil, biogas, bioethanol, and biohydrogen via lignocellulosic biomass transformation. *Fuel*, *315*, 122840.

Ninawe, G., & Tariq, M. (2021). Influence of carbon nanotubes as additive in diesel-biodiesel blends in CI engine-an experimental investigation. *International Journal of Sustainable Engineering*, *14*, 1110–1121.

Ning, P., Yang, G., Hu, L., Sun, J., Shi, L., Zhou, Y., Wang, Z., & Yang, J. (2021). Recent advances in the valorization of plant biomass. *Biotechnology for Biofuels*, *14*, 102.

Nuhma, M. J., Alias, H., Tahir, M., & Jazie, A. A. (2021). Microalgae biomass conversion into biofuel using modified HZSM-5 zeolite catalyst: A review. *Materials Today: Proceedings*, *42*, 2308–2313.

Ojo, A. O. (2023). An overview of lignocellulose and its biotechnological importance in high-value product production. *Fermentation*, *9*, 990.

Pal, N. (2020). Nanoporous metal oxide composite materials: A journey from the past, present to future. *Advances in Colloid and Interface Science*, *280*, 102156.

Pandit, C., Banerjee, S., Pandit, S., Lahiri, D., Kumar, V., Chaubey, K. K., Al-Balushi, R., Al-Bahry, S., & Joshi, S. J. (2023). Recent advances and challenges in the utilization of nanomaterials in transesterification for biodiesel production. *Heliyon*, *9*, e15475.

Patel, A., & Shah, A. R. (2021). Integrated lignocellulosic biorefinery: Gateway for production of second generation ethanol and value added products. *Journal of Bioresources and Bioproducts*, *6*, 108–128.

Peña, L., Ikenberry, M., Ware, B., Hohn, K. L., Boyle, D., Sun, X. S., & Wang, D. (2011). Cellobiose hydrolysis using acid-functionalized nanoparticles. *Biotechnology and Bioprocess Engineering*, *16*, 1214–1222.

Peña-Castro, J. M., Muñoz-Páez, K. M., Robledo-Narvaez, P. N., & Vázquez-Núñez, E. (2023). Engineering the metabolic landscape of microorganisms for lignocellulosic conversion. *Microorganisms*, *11*, 2197.

Polishchuk, V., Tarasenko, S., Antypov, I., Kozak, N., Zhyltsov, A., & Bereziuk, A. (2020). *Investigation of the efficiency of wet biodiesel purification. E3S Web of Conferences* (pp. 02006). EDP Sciences.

Pramanik, A., Chaudhary, A. A., Sinha, A., Chaubey, K. K., Ashraf, M. S., Basher, N. S., Rudayni, H. A., Dayal, D., & Kumar, S. (2023). Nanocatalyst-based biofuel generation: An update, challenges and future possibilities. *Sustainability*, *15*, 6180.

Qasim, U., Osman, A. I., AaH, A.-M., Farrell, C., Al-Abri, M., Ali, M., Vo, D.-V. N., Jamil, F., & Rooney, D. W. (2021). Renewable cellulosic nanocomposites for food packaging to avoid fossil fuel plastic pollution: A review. *Environmental Chemistry Letters*, *19*, 613–641.

Rajak, R. C., Saha, P., Singhvi, M., Kwak, D., Kim, D., Lee, H., Deshmukh, A. R., Bu, Y., & Kim, B. S. (2021). An eco-friendly biomass pretreatment strategy utilizing reusable enzyme mimicking nanoparticles for lignin depolymerization and biofuel production. *Green Chemistry*, *23*, 5584–5599.

Reshmy, R., Philip, E., Madhavan, A., Sirohi, R., Pugazhendhi, A., Binod, P., Awasthi, M. K., Vivek, N., Kumar, V., & Sindhu, R. (2022). Lignocellulose in future biorefineries: Strategies for cost-effective production of biomaterials and bioenergy. *Bioresource Technology*, *344*, 126241.

Rizvi, M., Gerengi, H., & Gupta, P. (2022). *Functionalization of nanomaterials: Synthesis and characterization. Functionalized nanomaterials for corrosion mitigation: Synthesis, characterization, and applications* (pp. 1–26). ACS Publications.

Romsaiyud, A., Songkasiri, W., Nopharatana, A., & Chaiprasert, P. (2009). Combination effect of pH and acetate on enzymatic cellulose hydrolysis. *Journal of Environmental Sciences*, *21*, 965–970.

Roy, S., Dikshit, P. K., Sherpa, K. C., Singh, A., Jacob, S., & Rajak, R. C. (2021). Recent nanobiotechnological advancements in lignocellulosic biomass valorization: A review. *Journal of Environmental Management*, *297*, 113422.

Sanchez-Ramirez, J., Martinez-Hernandez, J. L., Segura-Ceniceros, P., López, G., Saade, H., Medina-Morales, M. A., Ramos-González, R., Aguilar, C. N., & Ilyina, A. (2017). Cellulases immobilization on chitosan-coated magnetic nanoparticles: Application for Agave Atrovirens lignocellulosic biomass hydrolysis. *Bioprocess and Biosystems Engineering*, *40*, 9–22.

Sankaran, R., Markandan, K., Khoo, K. S., Cheng, C. K., Ashokkumar, V., Deepanraj, B., & Show, P. L. (2021). The expansion of lignocellulose biomass conversion into bioenergy via nanobiotechnology. *Frontiers in Nanotechnology*, *3*, 793528.

Sekoai, P. T., Ouma, C. N. M., Du Preez, S. P., Modisha, P., Engelbrecht, N., Bessarabov, D. G., & Ghimire, A. (2019). Application of nanoparticles in biofuels: An overview. *Fuel*, *237*, 380–397.

Shahi, N., Wang, P., Adhikari, S., Min, B., & Rangari, V. K. (2021). Biopolymers fractionation and synthesis of nanocellulose/silica nanoparticles from agricultural byproducts. *ACS Sustainable Chemistry & Engineering*, *9*, 6284–6295.

Shanmugam, S., Krishnaswamy, S., Chandrababu, R., Veerabagu, U., Pugazhendhi, A., & Mathimani, T. (2020). Optimal immobilization of Trichoderma asperellum laccase on polymer coated Fe3O4@ SiO2 nanoparticles for enhanced biohydrogen production from delignified lignocellulosic biomass. *Fuel*, *273*, 117777.

Sharma, Y., Manro, B., Thakur, P., Khera, A., Dey, P., Gupta, D., Khajuria, A., Jaiswal, P. K., Barnwal, R. P., & Singh, G. (2023). Nanotechnology as a vital science in accelerating biofuel production, a boon or bane. *Biofuels, Bioproducts and Biorefining*, *17*, 616–663.

Shukla, A., Kumar, D., Girdhar, M., Kumar, A., Goyal, A., Malik, T., & Mohan, A. (2023). Strategies of pretreatment of feedstocks for optimized bioethanol production: Distinct and integrated approaches. *Biotechnology for Biofuels and Bioproducts*, *16*, 44.

Silviana, S., & Dalanta, F. (2024). *Advances in modified silica nanoparticles utilization for various applications: Now and future. E3S Web of Conferences* (pp. 09004). EDP Sciences.

Tang, H., Chen, Z., Shao, Y., Ju, X., & Li, L. (2023). Development of an enzymatic cascade to systematically utilize lignocellulosic monosaccharide. *Journal of the Science of Food and Agriculture, 103*, 1974–1980.

Verma, D., Paul, J. S., Tiwari, S., & Jadhav, S. (2022). A review on role of nanomaterials in bioconversion of sustainable fuel bioethanol. *Waste and Biomass Valorization, 13*, 4651–4667.

Wahab, A., Muhammad, M., Ullah, S., Abdi, G., Shah, G. M., Zaman, W., & Ayaz, A. (2024). Agriculture and environmental management through nanotechnology: Eco-friendly nanomaterial synthesis for soil-plant systems, food safety, and sustainability. *Science of the Total Environment, 926*, 171862.

Wei, Z., Shao, Q., Yuan, Y., Jin, H., Cao, J., Liu, W., Zhao, G., & Luo, J. (2024). Efficient removal of ammonia nitrogen using biochar derived from the co-fermentation residue of waste activated and orange peel waste: Linking structure properties and reaction kinetics. *Process Safety and Environmental Protection, 189*, 146–153.

Xiao, W., Sun, R., Hu, S., Meng, C., Xie, B., Yi, M., & Wu, Y. (2023). Recent advances and future perspective on lignocellulose-based materials as adsorbents in diverse water treatment applications. *International Journal of Biological Macromolecules, 253*, 126984.

Yang, L., Wang, X.-C., Dai, M., Chen, B., Qiao, Y., Deng, H., Zhang, D., Zhang, Y., de Almeida, C. M. V. B., & Chiu, A. S. (2021). Shifting from fossil-based economy to bio-based economy: Status quo, challenges, and prospects. *Energy, 228*, 120533.

Yu, X., Li, J., Sun, Y., Xie, Y., Su, Y., Tang, S., Bian, S., Liu, L., Huo, F., Huang, Q., & Chen, G. (2024). Co-immobilized multi-enzyme biocatalytic system on reversible and soluble carrier for saccharification of corn straw cellulose. *Bioresource Technology, 395*, 130325.

Zanuso, E., Ruiz, H. A., Domingues, L., & Teixeira, J. A. (2022). Magnetic nanoparticles as support for cellulase immobilization strategy for enzymatic hydrolysis using hydrothermally pretreated corn cob biomass. *Bioenergy Research, 15*, 1946–1957.

Zhang, J., Lovell, J. F., Shi, J., & Zhang, Y. (2024). Nanomaterials for co-immobilization of multiple enzymes. *BMEMat, 3*, e12080.

Zhu, J., Agarwal, U. P., Ciesielski, P. N., Himmel, M. E., Gao, R., Deng, Y., Morits, M., & Österberg, M. (2021). Towards sustainable production and utilization of plant-biomass-based nanomaterials: A review and analysis of recent developments. *Biotechnology for Biofuels, 14*, 114.

Zielińska, A., Costa, B., Ferreira, M. V., Miguéis, D., Louros, J. M., Durazzo, A., Lucarini, M., Eder, P., Chaud, M., & Morsink, M. (2020). Nanotoxicology and nanosafety: Safety-by-design and testing at a glance. *International Journal of Environmental Research and Public Health, 17*, 4657.

11

Synthetic Biology in the Realm of Genome Engineering for Improved Biocatalysts and Production

José Daniel Cano Montoya[1,2], *Diego Hernandez*[1], *and Josman Velasco*[1]

[1] *Departamento de Ciencias Biológicas, Universidad de los Andes, Bogotá, Colombia*
[2] *Departamento de Química, Universidad de los Andes, Bogotá, Colombia*

11.1 Introduction

Although the term synthetic biology (SynBio) was coined in the year 1912 (Lorenzo & Danchin, 2008), it has been in the last decade that its contributions have revolutionized biotechnological applications in the industrial, bioremediation, energy, food and beverage, health, and materials sectors. However, the scope of SynBio is largely due to its integration with recent advances in genome-editing techniques, computational biology, machine learning, and the use of technologies such as artificial intelligence. This chapter describes the main tools used in SynBio to achieve genome engineering and the construction of the main cellular platforms used in bio-based industries.

Before describing the main tools of SynBio, it is necessary to clarify the concepts of SynBio and genetic engineering, their differences, and points of convergence. SynBio is a multidisciplinary field that integrates principles from molecular biology, engineering, and computer science to design and construct new biological parts, devices, and systems or to redesign existing biological systems for useful purposes (Garner, 2021; Plante, 2023). Genetic engineering, by contrast, involves the direct manipulation, modification, and recombination of DNA or other nucleic acid molecules to alter an organism. This process typically includes the transfer of individual genes from one organism to another, enabling the expression of desired characteristics. The primary difference between SynBio and genetic engineering lies in their scope and approach. While genetic engineering focuses on the modification of existing genes within an organism, SynBio envisions the creation of entirely new genetic sequences, metabolic pathways, or systems (Rosenberg, 2017). SynBio often involves the assembly of long DNA sequences, which can include genes found in other organisms or entirely novel sequences, to create new biological functions. In contrast, genetic engineering typically involves more targeted modifications, such as the insertion, deletion, or alteration of specific genes within an organism's genome.

Despite these differences, a close relationship exists between SynBio and genetic engineering. SynBio builds upon the foundational techniques of genetic engineering, utilizing

Biofoundry Techniques for Biotechnology Applications, First Edition. Edited by Anuj Kumar Chandel.

them to achieve more ambitious and complex goals. This relationship highlights the complementary nature of these fields, with SynBio extending the capabilities of genetic engineering to create more sophisticated biological systems. Although SynBio has multiple applications, in this chapter, we will focus on its use in the modification of cellular genomes for the construction of cell factories, understanding that platform organisms are those that have revolutionized production systems, making biologically based industries a great alternative to materialize the much-sought global economic and energy transition.

11.2 The Design–Build–Test–Learn Cycle for Optimizing Biological Systems

The design–build–test–learn (DBTL) methodology is a cyclical process widely used to develop and optimize biological systems. This iterative cycle begins with the *design phase*, where specific biological functions are conceptualized and planned. Next, in the *build phase*, these designs are physically constructed, often involving genetic modifications. The *test phase* follows, where the constructed systems are evaluated for performance and functionality. Finally, the *learn phase* involves analyzing the test results to gain insights and refine the design, thus completing the cycle and preparing for the next iteration (Figure 11.1).

The utility of the DBTL methodology lies in its structured approach to engineering biological systems, which allows for systematic optimization and innovation by iteratively

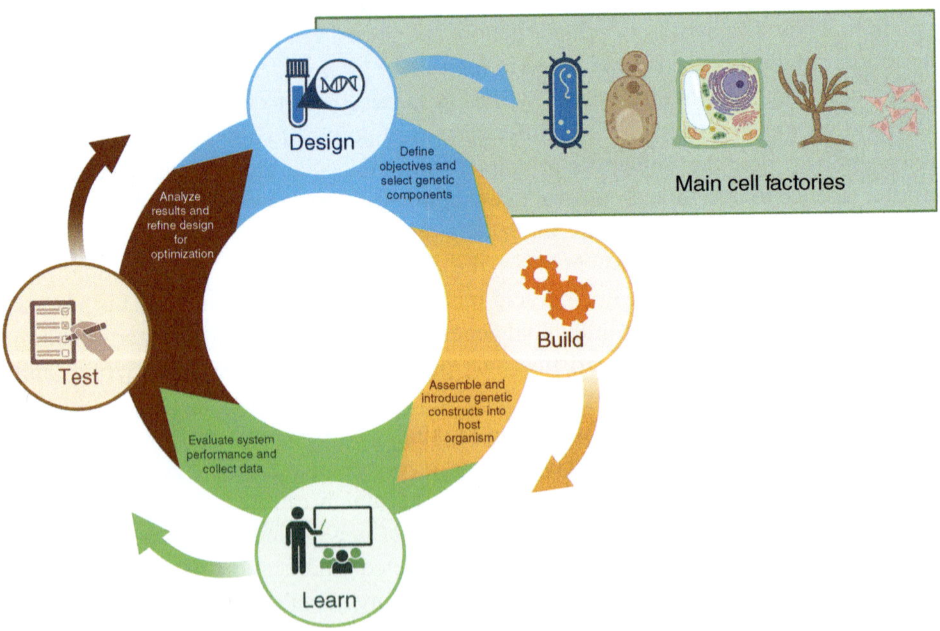

Figure 11.1 Design–build–test–learn cycle. The image illustrates the design–build–test–learn cycle as a process that can be used for optimization of cell factories like bacteria, yeast, filamentous fungi, plants, and mammalian cells.

refining designs based on empirical data, and researchers can achieve more reliable and efficient outcomes (Matzko & Konur, 2024). This methodology is particularly valuable in SynBio, where the complexity of biological systems often requires multiple rounds of optimization to achieve the desired characteristics. The DBTL cycle enhances the efficiency of developing genetically engineered organisms for various applications, including the production of biofuels, pharmaceuticals, and other valuable compounds.

The key tools used in the DBTL methodology include computational modeling and simulation software, high-throughput DNA synthesis, sequencing technologies, and automated laboratory equipment. Computational tools, such as machine learning algorithms and bioinformatics software, are essential for designing and predicting the behavior of genetic constructs. High-throughput technologies enable the rapid construction and testing of numerous genetic variants, while automation reduces human error and increases the reproducibility of experiments (Milessi et al., 2025a). These tools collectively facilitate the efficient execution of the DBTL cycle, allowing for the rapid iteration and optimization of biological systems (Garner, 2021; Matzko & Konur, 2024). The DBTL cycle is crucial for optimizing cell factories, enabling the fine-tuning of metabolic pathways to maximize yield and efficiency.

11.3 The Synthetic Biology Toolkit for Genome Engineering

Genome engineering is a highly complex process that involves the use of various technologies to modify the genetic structure of an organism. Such modifications can range from silencing and/or deletion of native genes, expression or overexpression of native genes, and the incorporation and expression of foreign genes (heterologous expression) to the incorporation of large DNA fragments such as synthetic chromosomes.

Whatever the type of modification desired, it is necessary to comply with some main stages such as the assembly of the genetic constructs and their transfer and maintenance within the target organism. In this section, we will begin with a description of the most relevant tools that are used to assemble pieces of DNA and then mention the main genome-editing techniques.

11.3.1 DNA Fragment Assembly Tools

The assembly of DNA fragments is a key stage in SynBio, enabling the construction of complex genetic circuits and pathways. This methodology facilitates the precise manipulation and combination of genetic elements, allowing the design and engineering of novel biological systems with desired functions.

Assembly methods can be classified as (a) restriction enzyme-dependent methods with iconic examples such as the BioBricks system, one of the first assembly strategies that allow the sequential coupling of fragments and (b) methods based on sequence homology that can occur either *in vivo* or *in vitro*, for example, *in vitro* methods include overlap extension polymerase chain reaction, which enables scarless assembly of DNA parts, and commercial kits such as In-Fusion™ (Clontech) and Gateway™ (Life Technologies) (Chao et al., 2015). Despite the great diversity of techniques that may exist, in this section, we will briefly review the three most commonly used methodologies for assembling DNA fragments based on sequence homology.

11.3.1.1 Ligation-Independent Cloning

The ligation-independent cloning (LIC) methodology provides alternatives to overcome the limitations of conventional DNA fragment assembly techniques, which require specific restriction sites and time-consuming ligation steps that can be cumbersome and inefficient. LIC utilizes the 3′ to 5′ exonuclease activity of T4 DNA polymerase to generate long and complementary overhangs on both the vector and the insert, typically around 10–12 base pairs. This method allows for the stable annealing of DNA fragments without the need for ligation, significantly streamlining the cloning process (Aslanidis & De Jong, 1990). The process begins with the linearization of the vector, either by restriction enzyme digestion or PCR. The linearized vector and the insert are then treated with T4 DNA polymerase in the presence of a single dNTP, typically dGTP. This treatment generates overhangs on both the vector and the insert, which are designed to be complementary, enabling stable annealing of the fragments. The resulting product contains nicks that are repaired by the host cell's machinery during transformation. This scarless cloning method ensures that no extraneous sequences are introduced at the junctions, maintaining the integrity of the assembled DNA (Chao et al., 2015). LIC also has some limitations; it may not be suitable for all types of sequence modifications, and the initial setup can be more complex compared to other methods. Over the years, LIC has been refined and adapted, leading to variations such as sequence and ligation-independent cloning, which further enhances its flexibility and efficiency through recombination of DNA intermediates generated *in vitro* that are subsequently transformed into cells. Endogenous DNA repair machinery is used to finish the repair and generate recombinant DNA molecules. Another method derived from LIC technology is called seamless ligation cloning extract (Zhang et al., 2012) that uses *Escherichia coli* cell extracts to drive homology-mediated DNA assembly, which significantly reduces the cost. However, a disadvantage is that the length of single-stranded overlaps cannot be precisely controlled during the chew-back reaction. Despite these drawbacks, LIC remains a powerful tool in SynBio, enabling the rapid and precise assembly of DNA fragments for various applications, including the construction of genetic circuits.

11.3.1.2 Gibson Assembly

Developed by Daniel G. Gibson in 2009 at the J. Craig Venter Institute, the Gibson assembly (GA) method revolutionized molecular cloning by enabling the seamless joining of multiple DNA fragments in a single, isothermal reaction (Gibson, 2009). This homology-based, *in vitro* DNA assembly method was designed to facilitate the construction of large and complex DNA molecules, including entire genomes. The method's ability to assemble DNA fragments without the need for restriction enzymes has made it a useful tool for SynBio advances.

The key components of GA include three enzymatic activities: a T5 exonuclease, a DNA polymerase, and a DNA ligase. The process begins with the 5′ exonuclease, which chews back the 5′ ends of the DNA fragments to create single-stranded overhangs. These overhangs are designed to be complementary, allowing the fragments to anneal. The DNA polymerase then fills in the gaps, and the DNA ligase seals the nicks, resulting in a continuous double-stranded DNA molecule. This reaction occurs in a single tube at 50 °C, making it a straightforward and efficient method for DNA assembly (Chao et al., 2015; Gibson, 2009). The use of overlapping sequences ensures that the fragments are joined in the correct order, and the seamless nature of the assembly means that no extraneous sequences are introduced at the junctions.

GA offers several advantages over other DNA assembly techniques; it is highly efficient and can join multiple fragments simultaneously, regardless of their length or end compatibility. Additionally, the isothermal nature of the reaction simplifies the process and reduces the risk of errors. However, there are some limitations to consider. The method requires precise design of overlapping sequences, which can be challenging for some applications. Additionally, the reaction can be sensitive to the quality and concentration of the DNA fragments, and the efficiency may decrease with very large or highly repetitive sequences.

11.3.1.3 Yeast-Assisted DNA Assembly

This *in vivo* DNA assembly method utilizes the cellular machinery of yeast, such as *Saccharomyces cerevisiae*, leveraging its highly efficient homologous recombination (HR) system—a feature that has been exploited since the early 2000s to construct large and complex DNA molecules. This yeast species has become a preferred host for DNA assembly due to its ability to recombine multiple DNA fragments with overlapping ends *in vivo*, convenient molecular manipulation methods, and large DNA-carrying capacity (Chao et al., 2015).

This method, often referred to as yeast HR cloning or yeast life cycle assembly, has been instrumental in SynBio and metabolic engineering, enabling the assembly of entire biochemical pathways and even synthetic genomes (Gibson et al., 2008; He et al., 2023). The primary components of this methodology include the yeast cells themselves, which provide the HR machinery, and the DNA fragments to be assembled, which are typically designed with overlapping sequences of 30–50 base pairs. The process begins with the transformation of yeast cells with these DNA fragments; once inside the cell, the HR system recognizes the overlapping ends and facilitates the recombination of the fragments into a continuous DNA molecule. This *in vivo* assembly process is highly efficient and can handle large DNA constructs that would be challenging to assemble using *in vitro* methods (Chao et al., 2015; He et al., 2023). The assembled DNA can then be extracted from the yeast cells for further use or analysis. Additionally, the *in vivo* nature of the process reduces the need for extensive *in vitro* manipulations, which can be time-consuming and prone to errors since *in vitro* manipulation of DNA fragments with a length greater than 100 kb increases the risks of breakage. However, there are also some disadvantages such as the requirement for overlapping sequences can complicate the design of DNA fragments, and the efficiency of assembly can be affected by the length and complexity of the DNA. Furthermore, the extraction and purification of the assembled DNA from yeast cells can be challenging, especially for very large constructs.

11.3.2 Genome-Editing Techniques

11.3.2.1 Clustered Regularly Interspaced Short Palindromic Repeats

Clustered Regularly Interspaced Short Palindromic Repeats (CRISPR) has revolutionized gene editing since its discovery. The journey of CRISPR began in 1987 when scientists first identified unusual repetitive DNA sequences in the genome of *E. coli* (Nidhi et al., 2021). However, it was not until 2005 that these sequences were linked to an adaptive immune system in bacteria, which uses CRISPR-associated (Cas) proteins to cut foreign DNA (Figure 11.2).

In 2012, Jennifer Doudna and Emmanuelle Charpentier published a groundbreaking paper demonstrating that the CRISPR-Cas9 system could be programmed to target specific DNA sequences, enabling precise genome editing (Jinek et al., 2012). The core components of CRISPR technology include the guide RNA (gRNA) and the Cas endonucleases, which work together to target and modify specific DNA sequences. The gRNA is a short synthetic RNA composed of two parts: a scaffold sequence that binds to the Cas protein and a spacer sequence that is complementary to the target DNA. The gRNA directs the Cas endonuclease to the precise location in the genome where the edit is to be made. This specificity is achieved through base pairing between the spacer sequence of the gRNA and the target DNA sequence. Once the gRNA binds to the target DNA, the Cas protein induces a break at that location, initiating the DNA repair process. Cas endonucleases are the molecular scissors of the CRISPR system; to date, Cas endonucleases ranging from 1 to 14 have been described, some of which also act as RNA recognizers, binding to or breaking this nucleic acid (Khan et al., 2024; Nidhi et al., 2021). Each type of Cas protein has unique properties that make it suitable for different applications in genome editing.

The most commonly used Cas protein is Cas9, derived from *Streptococcus pyogenes*; since its discovery, Cas9 from this organism has been the most studied nuclease associated with CRISPR. Cas9 is a versatile enzyme that can be programmed to target nearly any DNA sequence by changing the spacer sequence of the gRNA. Other Cas proteins include Cas12a (Cpf1), which creates staggered cuts in DNA, and Cas13, which targets RNA instead of DNA. Other Cas proteins, such as Cas14, are being explored for their potential in genome editing. The diversity of Cas proteins expands the toolkit available for researchers, allowing for more precise and varied genetic modifications.

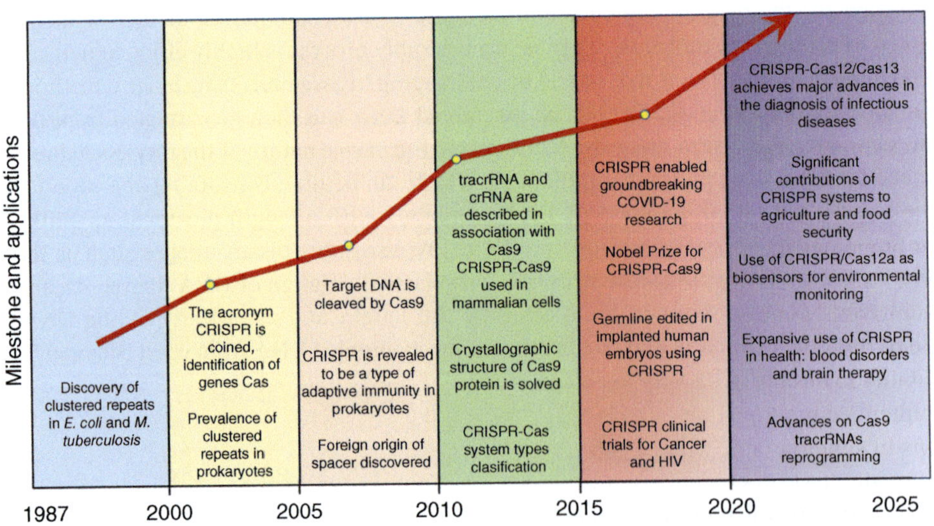

Figure 11.2 Advances, achievements, and applications of Clustered Regularly Interspaced Short Palindromic Repeats-associated technology over time. *Source:* Adapted from Nidhi et al. (2021).

CRISPR technology has rapidly advanced and has been used in mammalian cells, opening the door to its application in human genetics (Nidhi et al., 2021). This led to a surge in research and development, with CRISPR being used to create genetically modified organisms, develop disease models, and explore gene therapy.

CRISPR has also been instrumental in agricultural biotechnology, where scientists have used CRISPR to develop crops with improved traits, such as drought resistance and enhanced nutritional content. This has significant implications for food security and sustainable agriculture (Khan et al., 2024).

More recently, CRISPR has been employed in the fight against infectious diseases. In 2021, researchers developed CRISPR-based diagnostic tools for rapid and accurate detection of pathogens, including the SARS-CoV-2 virus responsible for COVID-19. Additionally, CRISPR is being explored as a potential tool for combating antibiotic resistance by targeting and disabling resistance genes in bacteria. The versatility and precision of CRISPR continue to drive innovation. Newer techniques, such as base editing and prime editing, have been developed to make even more precise genetic modifications without causing double-strand breaks (DSBs) in DNA. These advancements hold great promise for the future of SynBio research.

11.3.2.2 Transcription Activator-Like Effector Nucleases

Transcription activator-like effector nucleases (TALENs) are a class of engineered nucleases used for precise genome editing and are composed of two main components: a DNA-binding domain derived from transcription activator-like effectors (TALEs) and a DNA-cleaving domain from the FokI endonuclease. The TALEs proteins were discovered in *Xanthomonas* bacteria, which naturally bind to specific DNA sequences in plant hosts (Matsumoto & Nomura, 2023). By customizing the TALEs to recognize specific DNA sequences, scientists can target TALENs to precise locations in the genome. The DNA-binding domain of TALENs consists of a series of repeat-variable diresidues (RVDs), each recognizing a single nucleotide. These RVDs are arranged in tandem repeats, allowing for the recognition of specific DNA sequences. The specificity of TALENs is determined by the sequence of these RVDs, which can be engineered to bind to virtually any DNA sequence. This modularity makes TALENs highly versatile tools for genome editing.

The DNA-cleaving domain of TALENs is derived from the FokI endonuclease, which requires dimerization to cleave DNA. When two TALENs bind to adjacent sites on the DNA, the FokI domains dimerize and create a DSB at the target site. This DSB can then be repaired by the cell's natural repair mechanisms, either through nonhomologous end joining or homology-directed repair. These repair processes can introduce mutations or insertions at the target site, enabling precise genetic modifications (Foss & Norris, 2023; Matsumoto & Nomura, 2023).

One of the key advantages of TALENs is their high specificity and low off-target effects. The ability to design TALENs to recognize specific DNA sequences reduces the likelihood of unintended modifications elsewhere in the genome. This specificity is crucial for applications in gene therapy, where precise editing is necessary to avoid potential side effects. Additionally, TALENs can be used in a wide range of organisms, from plants to animals, making them versatile tools for both basic research and applied biotechnology.

In summary, TALENs are powerful genome-editing tools composed of customizable DNA-binding domains and a DNA-cleaving domain. The high specificity and versatility of

TALENs make them valuable for various applications, including gene therapy, functional genomics, and crop improvement.

11.3.2.3 Zinc Finger Nucleases

Zinc finger (ZF) proteins are used to control gene transcription through their recognition and subsequent binding to DNA. In ZF proteins, zinc refers to its histidine and cysteine (Cys) amino acid binding to zinc ions, and each ZF protein module can recognize 64 possible combinations of three DNA bases (Foss & Norris, 2023; Kim et al., 1996). Zinc finger nucleases (ZFNs) are created by fusing ZF DNA-binding domains with a nuclease domain, typically the FokI endonuclease, the same used in TALEN technology that introduces DSBs at the targeted DNA site (see earlier sections).

The mechanism of action of ZFNs involves the binding of the ZF domains to their specific DNA sequences, bringing the FokI nuclease domains into close proximity. The FokI domains then dimerize and create a DSB in the DNA. This targeted break can be exploited to introduce specific genetic modifications (Matsumoto & Nomura, 2023). Compared with other gene-editing technologies like TALENs and CRISPR-Cas9, ZFNs offer high specificity due to the modular nature of ZF domains. However, designing ZFNs for new targets can be complex and time-consuming as each new target sequence requires the creation of a new set of ZF domains. TALENs, which use a similar mechanism but with different DNA-binding domains, are easier to design but are larger and more difficult to deliver into cells. CRISPR-Cas9, on the other hand, is highly versatile and easier to design as it uses a gRNA to direct the Cas9 nuclease to the target DNA sequence. Despite their advantages, the complexity of designing ZFNs remains challenging and needs to be addressed.

11.4 Production and Improvement of Biocatalysts

In the next section, we will explore the crucial role of protein in biotechnology, with a particular emphasis on enzymes. This section will provide a comprehensive analysis of the various species of microorganisms utilized for enzyme production at both research and industrial scales. Each type of microorganism will be presented in a way that highlights its historical importance and contemporary relevance, including the advantages and disadvantages of its use, notable discoveries associated with it, and its limitations in both current and future contexts.

Biocatalysis has evolved significantly over the past few decades into a more mature and widely employed technology. In the early 2000s, its focus was primarily on the synthesis or resolution of optically active intermediates. Today, it has expanded into a comprehensive tool for chemical synthesis and manufacturing. This rapid growth can be attributed to advancements in bioinformatics and computer modeling, which have facilitated the discovery of new enzyme variants and supported enzyme engineering. The evolution of biocatalysis into the versatile processes we see today has been driven by advancements in rational design (RD), directed evolution, high-throughput screening tools, and various design methods (Albayati et al., 2024). The recent acceleration in biocatalyst development has stemmed from the affordable availability of synthetic genes, allowing for rapid screening of diverse enzyme variants. These advancements have made it easier to select the most suitable enzyme and reaction for targeted products. However, despite its recognized importance, biocatalysis also has limitations, with enzyme stability and cost being prominent drawbacks that hinder its full development and industrial application (Wu et al., 2021). In order to mitigate these disadvantages, there is an increase in technologies that seek to (a) obtain more efficient cellular and microbial platforms to obtain high titers of proteins and/or that produce enzymes with better catalytic attributes and (b) carry out improvements in the proteins themselves through protein engineering and other technologies mentioned above. Whatever the case, SynBio is present as a powerful tool in said improvements.

11.4.1 Chassis Organisms for the Production of Biocatalysts

An effective chassis should possess several key features to facilitate the creation of reliable production strains. *Foundational knowledge*: a solid understanding of the organism is essential, informing the design of culture media and bioprocesses. *Nutritional simplicity*: the chassis should have straightforward nutritional requirements, utilizing readily available carbon and nitrogen sources. *Built-in stress resistance*: high resistance to physicochemical stress is crucial for maintaining cellular integrity under varying conditions. *Rapid growth*: efficient growth rates are necessary to enhance productivity and reduce production timelines. *Genetic tool availability and gene stability*: access and ease of development novel tools for targeted genome manipulation (such as CRISPR/Cas9 systems or others described in this chapter) are essential for precise engineering; in addition, the genetic structure of the organism must remain stable, ensuring that its productive attributes are maintained over time. *Efficient secretion systems*: effective secretion mechanisms, whether natural or engineered, are vital for facilitating downstream purification processes. *Extreme condition tolerance*: the ability to endure extreme environmental conditions, such as high temperatures, can expand the range of potential applications.

Despite these ideal characteristics, few microorganisms fulfill all these criteria naturally. Traditional chassis like *E. coli, Bacillus subtilis, Streptomyces* spp., *Pseudomonas putida, Corynebacterium glutamicum, Pichia pastoris, Aspergillus* sp., and *Trichoderma* sp. have been extensively characterized and utilized in protein production due to their well-documented properties. However, there is increasing interest in exploring alternative hosts to diversify the available chassis as plant and mammalian cells. A summary of the main platforms of organisms for protein production is provided in Table 11.1.

11.4.1.1 *Escherichia coli*

E. coli has emerged as the best-studied prokaryotic organism over more than 50 years of intensive research, making it a preferred chassis for the development of cell factories (Calero & Nikel, 2019). Its pivotal role in biotechnology traces back to its use in the 1960s for the production of L-threonine and insulin. *E. coli* exhibits several key characteristics that make it advantageous for bioproduction: rapid growth rates, low nutrient requirements, high-cell-density cultures, versatile metabolism, and genetic manipulation tools available. Despite being a widely used microbial platform for recombinant protein production, *E. coli* has some disadvantages such as the inability to perform post-translational modifications (PTMs), effective protein secretion systems into the extracellular medium, and the formation of inclusion bodies (Schramm et al., 2019). However, the above disadvantages are intended to be overcome using SynBio tools.

Table 11.1 Advantages and Disadvantages of the Main Platforms Organisms Used for Protein Production

Platform	Advantages	Disadvantages	Main Enzymes	Key Species	References
Bacteria	Fast growth rates	Limited stability outside optimal temperature range	Amylases	*Escherichia coli*	Contesini et al. (2018); Fasim et al. (2021); Schramm et al. (2019)
	Well-studied, with established genetic engineering techniques	Low tolerance for extreme pH and high pressures	Proteases	*Bacillus subtilis*	
	Safe and easy to culture	Potential for misfolded proteins and inclusion bodies	Lipases		
Fungi	Capable of producing a wide range of enzymes	Slower growth rates compared to bacteria	Cellulases	*Aspergillus niger*	Corbu et al. (2023); Kango et al. (2019); Liu et al. (2023); Meyer et al. (2016)
	Ability to perform post-translational modifications	Complex cultivation requirements	Laccases	*Trichoderma reesei*	
	Generally recognized as safe	Potential for contamination in large-scale production	Lipases		
Plants	Natural production of complex compounds	Slow growth and longer cultivation times	Secondary metabolites	*Nicotiana benthamiana*	Clemente et al. (2019); Eljounaidi and Lichman (2020); Karki et al. (2021); Schillberg et al. (2019)
	Ability to produce glycosylated proteins	Regulatory hurdles for genetic modifications	Proteinase inhibitors		
		Limited scalability			
Chinese hamster ovary cells (animal cells)	High fidelity in post-translational modifications	Expensive to culture and maintain	Glycosylated enzymes	Chinese hamster ovary	Fischer et al. (2015); S. Gupta et al. (2023); S. K. Gupta et al. (2017); Majumdar et al. (2023)
	Well-established for therapeutic protein production	Risk of contamination	Antibodies		
	Ability to produce complex biologics	Limited growth compared with bacteria			
Thermophiles	High stability at elevated temperatures	Low natural biomass yield (typically <2 g/L)	Thermozymes	*Pyrococcus furiosus*	Sitara et al. (2024); Vavitsas et al. (2022); Vieille and Zeikus (2001); Zeldes et al. (2015)
	Tolerance to denaturing solvents and high pressure	Genetic engineering techniques may be less established	Alkaline proteases	*Bacillus stearothermophilus*	
	Suitable for high-temperature industrial processes	More complex production systems needed	Thermostable amylases	*Thermus thermophilus*	

Recent advances have led to the development of reduced-genome strains that display significant physiological advantages over the K-12 strain MG1655. Examples include the MDS41, MDS42, and MDS43 strains, which have 15% of their genome deleted, exhibiting high electroporation efficiency and increased stability for cloning difficult DNA sequences. The Minimum Genome Factory project also produced strains like MGF-01, which demonstrated increased cell density and higher yields of L-threonine (Calero & Nikel, 2019). A vast toolbox of plasmid vectors, gene expression systems, and regulatory elements has been developed for *E. coli*, enhancing its utility in SynBio. Key developments include CRISPR/Cas9 technologies, allowing for precise genome editing and gene regulation.

11.4.1.2 *Bacillus subtilis*

B. subtilis is an aerobic, rod-shaped, Gram-positive bacterium commonly found in soil and the plant rhizosphere. It is one of the most well-characterized gram-positive microorganisms, noted for its natural competence for DNA transformation and its ability to form spores, which serve as resistant, nonreproductive structures (Danilova & Sharipova, 2020). This organism has been extensively studied, leading to the development of various genetic tools for genome manipulation, including a genome-scale reconstruction of its metabolic network. Recognized for being free of endotoxins, *B. subtilis* has obtained the status of generally recognized as safe (GRAS) for protein production. The availability of numerous genetic tools, including vectors and gene expression systems, makes *B. subtilis* a manageable organism for research and industrial applications. Notably, engineered strains with reduced genomes, such as MGB874, have demonstrated the ability to produce higher concentrations of heterologous protease.

Recombinant proteins, as well as natural enzymes and proteases, constitute the primary industrially relevant products derived from *B. subtilis* due to its efficient secretory machinery. This capability allows for the transport of proteins into the culture medium, achieving concentrations in the range of grams per liter, thereby minimizing purification and recovery costs. Proper signal peptides are essential for the efficient secretion of target proteins, which consist of three domains that facilitate the translocation of proteins from the cytoplasm through the membrane to the external environment. The Sec and Tat pathways are the best-studied protein secretory systems in *B. subtilis* (Contesini et al., 2018).

While *B. subtilis* naturally produces several proteases, which can hinder the efficiency of heterologous protein production, the creation of protease-defective mutants has been crucial for enhancing its capabilities as a production chassis. The most commonly used mutants have had multiple proteases eliminated. Additionally, the development of *B. subtilis* for heterologous protein production has involved designing various gene expression systems utilizing strong constitutive and inducible promoters. Enzymes produced using *B. subtilis* include subtilisin, amylases, and laccases, with numerous studies focused on optimizing the production of these and other enzymes (Contesini et al., 2018; Danilova & Sharipova, 2020). Overall, this bacterium stands out as a robust chassis for both the production of endogenous enzymes and heterologous proteins, although further research is needed to enhance its production efficiencies. This includes genome reduction, signal peptide optimization, and the development of more efficient secretion and expression systems.

11.4.1.3 *Pseudomonas putida*

P. putida, a ubiquitous, saprophytic, and gram-negative rod-shaped bacterium, is increasingly recognized for its potential as a chassis in industrial biotechnology, particularly in harsh operating conditions. The most studied strain, *P. putida* KT2440, has been designated as GRAS for recombinant protein production and is considered a reliable host for heterologous gene expression. This bacterium exhibits several desirable traits for industrial applications, including rapid growth, minimal nutritional requirements, and advanced genetic manipulation tools (Calero & Nikel, 2019). Its versatile metabolism enables the utilization of a wide range of carbon sources, including aromatic compounds derived from lignin degradation, making it particularly suitable for lignin-derived feedstocks (Jiménez et al., 2002). *P. putida* is notable for its remarkable tolerance to chemical stress as high concentrations of aromatic solvents, for example, toluene, xylenes, and styrene, which can be detrimental to most microorganisms. This tolerance is largely attributed to a suite of efflux pumps that extrude toxic compounds from the cell, a process that, while energetically demanding, is crucial for survival in hostile environments. Furthermore, *P. putida* adapts to solvent exposure by modifying its phospholipid composition, which influences membrane fluidity and enhances resistance to solvent-induced disruptions (Calero & Nikel, 2019). These adaptations are complemented by a general stress response, including the expression of molecular chaperones and stress-related enzymes. Significant advancements have been made in genetic and genome engineering tools for *P. putida*, including the development of modular vectors and HR techniques. These tools enable precise genetic modifications, such as the creation of reduced-genome chassis with improved metabolic capabilities.

11.4.1.4 Filamentous Fungi

Filamentous fungi have established themselves as versatile and highly productive cell factories, contributing significantly to various industrial applications for over a century. Their remarkable ability to produce enzymes and small-molecule compounds, including antibiotics and organic acids, has made them invaluable in biotechnology. Recent advancements have also positioned filamentous fungi as potential sources for sustainable materials that could replace conventional plastics (Meyer et al., 2016).

The filamentous growth form of fungi allows for effective colonization of substrates, providing a large surface area that facilitates nutrient uptake. Fungi secrete a diverse range of enzymes that degrade complex polymers in organic waste, such as (holo)cellulose, lipids, and starch, into smaller molecules suitable for consumption. This enzymatic activity is complemented by the secretion of primary and secondary metabolites, which not only serve as energy sources for the fungi but also play crucial roles in nutrient release and interactions with other organisms (Corbu et al., 2023). Filamentous fungi are the organisms best equipped by nature for the production and secretion of high protein titers. They possess a biochemical machinery that allows the realization of PTMs in amino acid sequences. This quality allows fungi to be preferred hosts for the heterologous production of enzymes from eukaryotes. In addition, they can grow in environments with low water activity and using inexpensive carbon sources (Liu et al., 2023; Segato et al., 2014). To optimize the production capabilities of filamentous fungi, several strategies have been employed. Mutagenesis and genetic modification are commonly used to enhance the performance of production strains. Recent advances in single-cell technologies and bulk sample analysis have facilitated the identification of genes that can be targeted for modification (Liu et al., 2023). Techniques such as the introduction of multiple gene copies, utilization of strong promoters, and optimization of codon usage have proven effective in enhancing protein production. Furthermore, mixed culture approaches have emerged as a promising method for discovering novel compounds and improving production efficiency by harnessing the synergistic effects of co-cultured fungi or fungi combined with other microorganisms.

11.4.1.5 *Pichia pastoris*

P. pastoris (renamed as *Komagataella phaffii*) is a methylotrophic yeast that has become a prominent microbial platform for the production of recombinant proteins. Its importance lies in the ability to perform PTMs similar to those in higher eukaryotes, making it an attractive system for producing complex proteins. The use of *P. pastoris* in biotechnology has expanded due to its high growth rate, availability of tools for genetic modification, and capacity for high-density fermentation (Zha et al., 2023).

The production of proteins using *P. pastoris* involves several key procedures. Initially, the gene encoding the target protein is cloned into an expression vector, which is then introduced into the yeast cells. The most commonly used promoter for driving gene expression in *P. pastoris* is the alcohol oxidase 1 (AOX1) promoter, which is induced by methanol. Once the recombinant yeast strain is established, it is cultured in a bioreactor where conditions such as pH, temperature, and nutrient supply are optimized for protein production using methanol to induce the AOX1 promoter, leading to high levels of protein.

P. pastoris has been used to produce a wide range of proteins, including enzymes, antibodies, and vaccines (Duman-Özdamar & Binay, 2021; Zha et al., 2023). For example, it has successfully produced human serum albumin, insulin, and hepatitis B surface antigen. Additional benefits of using this yeast as a protein production platform include its ability to secrete the recombinant protein into the culture medium, simplifying downstream processing. However, there are also some disadvantages, such as the potential hyperglycosylation of proteins, which may affect their functionality, and the need for methanol, which is toxic and requires careful handling.

11.4.1.6 Mammalian Cell Expression Systems

Proteins: Mammalian cell expression systems are crucial for producing recombinant therapeutic proteins (RTPs) due to their ability to closely mimic the molecular structure, physicochemical properties, and biological functions of naturally occurring higher biological proteins. This makes them an important platform for RTP production today. Key components of these systems include expression vectors, serum-free media, and cell lines.

Among the commonly used mammalian cell lines, Chinese hamster ovary (CHO) cells are prominent nonhuman cells utilized for RTP production (S. K. Gupta et al., 2017). Despite their safety and efficacy, proteins produced from nonhuman cells may elicit immune responses in humans due to differences in PTMs, particularly glycosylation patterns. CHO cells have undergone continuous domestication and modification, resulting in various subtypes such as CHO-K1, CHO-S, CHO-DBX11, and CHO-DG44 (S. Gupta et al., 2023). The prevalent expression systems for CHO cells include the methotrexate and the methionine sulfoxide selection systems. These cells are less susceptible to human viruses, can adapt to large-scale serum-free suspension cultures, and can produce proteins with complete PTMs that resemble their natural state. The genes of interest can be stably integrated into the CHO cell genome, allowing for efficient replication and expression. Currently, nearly 70% of approved RTPs are produced using CHO cells, and advances in medium optimization and cell culture technology have significantly increased RTP expression levels. With ongoing progress, monoclonal antibody production in CHO cells has reached titers exceeding 10 g/L in some cases. The market for biopharmaceuticals produced from CHO cells has seen substantial growth, with a majority of new approvals relying on these cell lines as the production platform (S. Gupta et al., 2023; S. K. Gupta et al., 2017; Majumdar et al., 2023).

11.4.1.7 Plant Cells

Plant-Based Production Systems and Plant-Derived Protein Products: Since the 1990s, researchers have actively pursued the production of recombinant proteins in plants, leveraging established species for gene transfer techniques. This approach has led to a diverse array of pharmaceutical products that are stable, easily characterized, and can accumulate high levels (>100 mg/kg fresh plant weight or > 100 mg/L culture medium) (Karki et al., 2021; Schillberg et al., 2019), and their functionality can often be verified through simple binding assays.

Despite these advantages, antibody yields in plants remain significantly lower than those produced in CHO cells, raising concerns about the viability of plants as a routine commercial platform for antibody production. However, plants offer unique opportunities in niche markets that traditional platforms cannot match.

Most studies focus on early-stage objectives like verifying expression and optimizing purification, rather than tackling the commercial development challenges. Financial and organizational hurdles stem from unclear market potential and intellectual property rights, making it difficult to secure funding without a solid business case. This hinders the transition from established microbial and mammalian systems to plant-based ones.

Some plant-derived biopharmaceutical candidates have advanced to clinical trials, leading to the establishment of Good Manufacturing Practice-compliant processes approved by regulatory authorities. A notable example is recombinant glucocerebrosidase (prGCD), marketed as Elelyso, produced in carrot cells by Protalix Biotherapeutics (Karki et al., 2021). To address long timelines and significant investments required for commercial viability, focusing on less-regulated products (such as diagnostic, technical, and cosmetic items) may facilitate quicker market access; the avidin produced in maize and human epidermal growth factor in barley for cosmetics exemplify this approach. Nevertheless, significant challenges remain, including low product yields and high downstream processing costs. Other challenges faced by plant expression systems are (a) enhancing protein stability, strategies include creating fusion proteins, targeting proteins to specific compartments, freeze-drying, and utilizing chloroplast expression systems; and (b) PTMs: differences in glycosylation between plants and other organisms present challenges for protein production (Burnett & Burnett, 2019; Schillberg et al., 2019). Cellular engineering aims to enhance protein production by modulating the plant genome. This field has advanced significantly through modern molecular biology tools, focusing on genes encoding proteases and glycosyltransferases. Inactivating specific genes can lead to new plant cell lines with reduced proteolytic activity and humanized glycosylation patterns. Targeting protease genes for inactivation can mitigate proteolytic degradation. Recent studies have successfully silenced protease genes in BY-2 cells, resulting in higher accumulation of intact antibodies. CRISPR/Cas9 genome editing is emerging as a promising tool to humanize glycosylation. Researchers can create glyco-engineered plant cell lines that lack plant-specific glycosylation enzymes.

Techniques such as RNAi and CRISPR/Cas9 have been employed to achieve this in BY-2 and rice cells, significantly reducing undesirable glycan structures (Karki et al., 2021; Shanmugaraj et al., 2020). While significant progress has been made in modulating plant genomes for protein production, challenges remain, particularly regarding low productivity in plant cell cultures. Achieving substantial increases in productivity is essential for economic viability.

11.4.2 Techniques for the Improvement of Biocatalysts

As already mentioned in this chapter, the improvement of biocatalysts can be carried out by different routes or a combination of them. In the previous section, we described the platform organisms that are used in their production and how SynBio participates in their improvement. In this section, we will briefly describe the main techniques used in improving enzymes, where of course we will see the participation of biological, molecular, and computer tools.

The purpose of improving a biocatalyst can be as specific as the industry that produces it, the characteristics of the process in which it participates, and/or the type of reaction it catalyzes. In some cases, for the same enzyme, its improvement may have divergent purposes. For example, a pectinase can be improved to maintain high catalytic power at low temperatures if what is sought is the degradation of pectins in fruit juices that are stored in refrigeration, or, on the contrary, the objective of the improvement may be an optimal catalytic performance at high temperatures if what is sought is its action on pectins present in animal feed that is subjected to pelleting processes. The improvement techniques described here were chosen for their broad spectrum and versatility to be used, whatever the purpose.

11.4.2.1 Directed Evolution

The method inspired by natural selection has revolutionized protein engineering since its inception in the 1990s. The technique was pioneered by Frances Arnold, who received the Nobel Prize in Chemistry in 2018 for her work in this field. Initially developed to enhance the properties of enzymes, directed evolution has since become a cornerstone in the development of biocatalysts for industrial applications (Albayati et al., 2024; Sellés Vidal et al., 2023). This methodology involves several critical components. (a) Diversity generation—this is achieved through random mutagenesis or recombination techniques to create a library of gene variants; random mutations are introduced into the gene encoding the target enzyme. (b) The gene variants are expressed in a suitable host organism, often bacteria or yeast, to produce the corresponding protein variants. (c) Screening selection, high-throughput screening, or selection methods are employed to identify variants with the desired properties, such as increased activity, stability, or selectivity. This cycle is repeated multiple times to progressively enhance the desired traits of the enzyme (Ferreira et al., 2022; Song et al., 2023). The strategies to induce genetic variability are very varied and include procedures such as error-prone PCR, random insertions and deletions, and

DNA shuffling (Albayati et al., 2024; Fire & Xu, 1995). Directed evolution can be applied to a wide range of enzymes and other proteins; it can rapidly generate enzymes with significantly improved properties compared with RD methods and is capable of producing enzymes with novel functions not found in nature. However, it has the disadvantage of substantial requirements for high-throughput screening and selection. Furthermore, the random nature of mutagenesis can lead to a large number of nonfunctional variants.

11.4.2.2 Rational Design

RD in protein engineering emerged as a method to enhance enzyme properties based on detailed knowledge of protein structure and function. This approach gained prominence with advancements in X-ray crystallography and computational modeling, which provided insights into enzyme mechanisms at the atomic level (Reetz, 2022; Wang et al., 2021). This technology requires (a) detailed knowledge of the enzyme's three-dimensional structure, often obtained through techniques like X-ray crystallography or nuclear magnetic resonance (NMR) spectroscopy. (b) Software for molecular modeling and simulations to predict the effects of mutations on enzyme function and stability. (c) Techniques to introduce specific mutations at targeted sites within the enzyme's gene as site-saturation mutagenesis, combinatorial active-site saturation test, and focused rational iterative site-specific mutagenesis (Albayati et al., 2024; Bao et al., 2024). (d) Systems to express the mutated enzymes in suitable host organisms purify them for testing and methods to evaluate the performance of the engineered enzymes under relevant conditions. RD allows for targeted modifications based on detailed structural knowledge, leading to predictable improvements in enzyme properties, reduces the need for extensive screening by focusing on specific, rationally chosen mutations, can be applied to a wide range of enzymes, and tailored to specific industrial needs. However, this requires high-quality structural information, which may not be available for all enzymes, and the accuracy of predictions depends on the quality of computational models and algorithms.

11.4.2.3 Chemical Modification of Enzymes

Although it is a technique that is not based on SynBio, it has been included here due to its ability to improve almost any enzyme, its wide applications, and the possibility of being used in a complementary way to the techniques mentioned above. Chemical alterations imply the modification of one of the reactive side chains from protein or the formation of a covalent attachment by introducing a novel functional group or co-factor; such modification can induce novel enzyme functions.

Early work focused on simple modifications, such as acetylation and phosphorylation, but the field has evolved to include sophisticated methods like PEGylation and site-specific chemical modifications (Ivens et al., 2015). Covalent modification as cross-linking with glutaraldehyde reagent can facilitate enzyme stabilization, preserving it from denaturation and from the solvent's harmful impacts. On the other hand, enzyme PEGylation involves covalent binding of synthetic polymers of polyethylene glycol (PEG)-specific polymeric functional groups to the free amino acids group on the enzyme surface, resulting in the formation of a PEGylated biocatalyst that is soluble in organic solvents that can be advantageous for producing biopharmaceuticals with reduced antigenicity and enhanced stability.

Chemical modification involves several critical steps: the enzyme to be modified must have identified reactive amino acid residues (e.g., lysine, Cys) that can be subject to modification. The enzyme is exposed to chemical reagents (such as PEG, aldehydes, or carbodiimides) under controlled conditions, leading to the formation of covalent bonds between the reagent and the enzyme (Albayati et al., 2024; Rodriguez-Abetxuko et al., 2020; Yan et al., 2015). The reaction conditions must be optimized to ensure efficient and selective modification. Then, the modified enzyme is purified to isolate it from reaction mixtures and characterized to ensure that the desired modifications have been achieved through analytical methods like mass spectrometry and NMR. The process can be technically challenging, requiring precise control over reaction conditions, extensive optimization, the need for specialized reagents, and analytical tool. Compared with directed evolution, which relies on random mutagenesis and high-throughput screening, chemical modification offers a more targeted approach to enzyme improvement. However, it may not generate the same level of diversity and novel functions as directed evolution. RD, on the other hand, shares the precision of chemical modification but typically involves genetic rather than chemical alterations (Milessi et al., 2025b).

11.5 Conclusions and Final Remarks

On a planet where the transition to circular, environmentally, and socioeconomically sustainable production models is imperative, bio-based industries emerge as a great alternative. In this chapter, we have explored the fundamental role that SynBio plays in the development of new technologies that promote bio-based production systems. Although there are still many challenges to overcome, biotechnological advances in collaboration with computational developments have already allowed the creation of cell factories and enzymes successfully integrated into various economic sectors such as pharmaceutical, energy, health, environmental, agricultural, and food. In this way, the existing capacities to carry out genome engineering and the improvement of biocatalysts have been and will be key pieces in the transition of production systems. This should encourage investment by the public and private sectors in research and promote collaboration and interaction between the government, academia, research centers, and industries where the different disciplines of knowledge can act synergistically to maximize the response capacity to global challenges.

Acknowledgment

We gratefully acknowledge the Facultad de Ciencias of the Universidad de los Andes - Bogotá, Colombia for the support given through the FAPA project PR.3.2024.10867.

Declaration

Figure 11.1 was designed using *BioRender* under a publishing license agreement number JE27LRT942.

References

Albayati, S. H., Nezhad, N. G., Taki, A. G., & Rahman, R. N. Z. R. A. (2024). Efficient and easible biocatalysts: Strategies for enzyme improvement. A review. *International Journal of Biological Macromolecules*, *276*, 133978. https://doi.org/10.1016/j.ijbiomac.2024.133978

Aslanidis, C., & De Jong, P. J. (1990). Ligation-independent cloning of PCR products (LIC-PCR). *Nucleic Acids Research*, *18*(20), 6069–6074.

Bao, Y., Xu, Y., & Huang, X. (2024). Focused rational iterative site-specific mutagenesis (FRISM): A powerful method for enzyme engineering. *Molecular Catalysis*, *553*, 113755. https://doi.org/10.1016/j.mcat.2023.113755

Burnett, M. J. B., & Burnett, A. C. (2020). Therapeutic recombinant protein production in plants: Challenges and opportunities. *Plants, People, Planet*, *2*(2), 121–132. https://doi.org/10.1002/ppp3.10073

Calero, P., & Nikel, P. I. (2019). Chasing bacterial chassis for metabolic engineering: A perspective review from classical to non-traditional microorganisms. *Microbial Biotechnology*, *12*(1), 98–124. https://doi.org/10.1111/1751-7915.13292

Chao, R., Yuan, Y., & Zhao, H. (2014). Recent advances in DNA assembly technologies. *FEMS Yeast Research*, *15*(1), 1–9. https://doi.org/10.1111/1567-1364.12171

Clemente, M., Corigliano, M. G., Pariani, S. A., Sánchez-López, E. F., Sander, V. A., & Ramos-Duarte, V. A. (2019). Plant serine protease inhibitors: Biotechnology application in agriculture and molecular farming. *International Journal of Molecular Sciences*, *20*(6). https://doi.org/10.3390/ijms20061345

Contesini, F. J., de Melo, R. R., & Sato, H. H. (2018). An overview of Bacillus proteases: From production to application. *Critical Reviews in Biotechnology*, *38*(3), 321–334. https://doi.org/10.1080/07388551.2017.1354354

Corbu, V. M., Gheorghe-Barbu, I., Dumbravă, A. Ş., Vrâncianu, C. O., & Şesan, T. E. (2023). Current insights in fungal importance—A comprehensive review. *Microorganisms*, *11*(6). https://doi.org/10.3390/microorganisms11061384

Danilova, I., & Sharipova, M. (2020). The practical potential of bacilli and their enzymes for industrial production. *Frontiers in Microbiology*, *11*. https://doi.org/10.3389/fmicb.2020.01782

Duman-Özdamar, Z. E., & Binay, B. (2021). Production of industrial enzymes via Pichia pastoris as a cell factory in bioreactor: Current status and future aspects. *Protein Journal*, *40*(3), 367–376. https://doi.org/10.1007/s10930-021-09968-7

Eljounaidi, K., & Lichman, B. R. (2020). Nature's chemists: The discovery and engineering of phytochemical biosynthesis. *Frontiers in Chemistry*, *8*. https://doi.org/10.3389/fchem.2020.596479

Fasim, A., More, V. S., & More, S. S. (2021). Large-scale production of enzymes for biotechnology uses. *Current Opinion in Biotechnology*, *69*, 68–76. https://doi.org/10.1016/j.copbio.2020.12.002

Ferreira, P., Fernandes, P. A., & Ramos, M. J. (2022). Modern computational methods for rational enzyme engineering. *Chem Catalysis*, *2*(10), 2481–2498. https://doi.org/10.1016/j.checat.2022.09.036

Fire, A., & Xu, S.-Q. (1995). Rolling replication of short DNA circles. *Proceedings of the National Academy of Sciences of the United States of America*, *92*, 4141–4645.

Fischer, S., Handrick, R., & Otte, K. (2015). The art of CHO cell engineering: A comprehensive retrospect and future perspectives. *Biotechnology Advances*, *33*(8), 1878–1896. https://doi.org/10.1016/j.biotechadv.2015.10.015

Foss, D. V., & Norris, A. L. (2023). Genome editing technologies. In D. F. Dluzen & M. H. Schmidt (Eds.), *Rigor and Reproducibility in Genetics and Genomics*: Peer-reviewed, Published, Cited (pp. 397–423). Elsevier. https://doi.org/10.1016/B978-0-12-817218-6.00011-5

Garner, K. L. (2021). Principles of synthetic biology. *Essays in Biochemistry*, *65*(5), 791–811. https://doi.org/10.1042/EBC20200059

Gibson, D. G. (2009). Synthesis of DNA fragments in yeast by one-step assembly of overlapping oligonucleotides. *Nucleic Acids Research*, *37*(20), 6984–6990. https://doi.org/10.1093/nar/gkp687

Gibson, D. G., Benders, G. A., Axelrod, K. C., Zaveri, J., Algire, M. A., Moodie, M., Montague, M. G., Craig Venter, J., Smith, H. O., & Hutchison Iii, C. A. (2008). One-step assembly in yeast of 25 overlapping DNA fragments to form a complete synthetic Mycoplasma genitalium genome. *Proceedings of the National Academy of Sciences of the United States of America*, *105*(51), 20404–20409.

Gupta, S. K., Srivastava, S. K., Sharma, A., Nalage, V. H. H., Salvi, D., Kushwaha, H., Chitnis, N. B., & Shukla, P. (2017). Metabolic engineering of CHO cells for the development of a robust protein production platform. *PLoS One*, *12*(8), 1–23. https://doi.org/10.1371/journal.pone.0181455

Gupta, S., Shah, B., Fung, C. S., Chan, P. K., Wakefield, D. L., Kuhns, S., Goudar, C. T., & Piret, J. M. (2023). Engineering protein glycosylation in CHO cells to be highly similar to murine host cells. *Frontiers in Bioengineering and Biotechnology*, *11*. https://doi.org/10.3389/fbioe.2023.1113994

He, B., Ma, Y., Tian, F., Zhao, G. R., Wu, Y., & Yuan, Y. J. (2023). YLC-assembly: Large DNA assembly via yeast life cycle. *Nucleic Acids Research*, *51*(15), 8283–8292. https://doi.org/10.1093/nar/gkad599

Ivens, I. A., Achanzar, W., Baumann, A., Brändli-Baiocco, A., Cavagnaro, J., Dempster, M., Depelchin, B. O., Irizarry Rovira, A. R., Dill-Morton, L., Lane, J. H., Reipert, B. M., Salcedo, T., Schweighardt, B., Tsuruda, L. S., Turecek, P. L., & Sims, J. (2015). PEGylated biopharmaceuticals: Current experience and considerations for nonclinical development. *Toxicologic Pathology*, *43*(7), 959–983. https://doi.org/10.1177/0192623315591171

Jiménez, J. I., Miñambres, B., García, J. L., & Díaz, E. (2002). Genomic analysis of the aromatic catabolic pathways from Pseudomonas putida KT2440. *Environmental Microbiology*, *4*(12), 824–841. https://doi.org/10.1046/j.1462-2920.2002.00370.x

Jinek, M., Chylinski, K., Fonfara, I., Hauer, M., Doudna, J. A., & Charpentier, E. (2012). A programmable dual-RNA-guided DNA endonuclease in adaptive bacterial immunity. *Science*, *337*, 816–820. https://www.science.org

Kango, N., Jana, U., & Choukade, R. (2019). Advancing frontiers in mycology & mycotechnology: Basic and applied aspects of fungi. In T. Satyanarayana, S. K. Deshmukh, & D. Mukund (Eds.), *Advancing frontiers in mycology and mycotechnology: Basic and applied aspects of fungi* (pp. 1–675). Springer Singapore. https://doi.org/10.1007/978-981-13-9349-5

Karki, U., Fang, H., Guo, W., Unnold-Cofre, C., & Xu, J. (2021). Cellular engineering of plant cells for improved therapeutic protein production. *Plant Cell Reports*, *40*(7), 1087–1099. https://doi.org/10.1007/s00299-021-02693-6

Khan, S. U., Qian, M., Li, S., Fan, Y., Wang, H., Chang, W., Alam, O., Saeed, S., & Lu, K. (2024). Unveiling CRISPR/Cas in rapeseed: Triumphs, trials, and tomorrow. *New Crops*. https://doi.org/10.1016/j.ncrops.2024.100045

Kim, Y.-G., Cha, J., & Chandrasegaran, S. (1996). Hybrid restriction enzymes: Zinc finger fusions to Fok I cleavage domain. *Proceedings of the National Academy of Sciences of the United States of America*, *93*, 1156–1160.

Liu, D., Garrigues, S., & de Vries, R. P. (2023). Heterologous protein production in filamentous fungi. *Applied Microbiology and Biotechnology*, *107*(16), 5019–5033. https://doi.org/10.1007/s00253-023-12660-8

Lorenzo, V., & Danchin, A. (2008). Synthetic biology: Discovering new worlds and new words. *EMBO Reports*, *9*, 822–827. https://doi.org/10.1038/embor.2008.159

Majumdar, S., Desai, R., Hans, A., Dandekar, P., & Jain, R. (2023). From efficiency to yield: Exploring recent advances in CHO cell line development for monoclonal antibodies. *Molecular Biotechnology*, *67*, 369–392. https://doi.org/10.1007/s12033-024-01060-6

Matsumoto, D., & Nomura, W. (2023). The history of genome editing: Advances from the interface of chemistry & biology. *Chemical Communications*, *59*(50), 7676–7684. https://doi.org/10.1039/d3cc00559c

Matzko, R., & Konur, S. (2024). Technologies for design-build-test-learn automation and computational modelling across the synthetic biology workflow: A review. *Network Modeling Analysis in Health Informatics and Bioinformatics*, *13*(1). https://doi.org/10.1007/s13721-024-00455-4

Meyer, V., Andersen, M. R., Brakhage, A. A., Braus, G. H., Caddick, M. X., Cairns, T. C., de Vries, R. P., Haarmann, T., Hansen, K., Hertz-fowler, C., Krappmann, S., Mortensen, U. H., Peñalva, M. A., Ram, A. F. J., & Head, R. M. (2016). Current challenges of research on filamentous fungi in relation to human welfare and a sustainable bio-economy: A white paper. *Fungal Biology and Biotechnology*, *3*(1), 1–17. https://doi.org/10.1186/s40694-016-0024-8

Milessi, T. S., Reis, C. E., Ramos, M. D. N., Carvalho, A. K. F., Aminabhavi, T., & Chandel, A. K. (2025a). Native and tailor-made microbial cell factories for economic production of biorenewables: A paradigm shift in industrial biotechnology. *Biomass and Bioenergy*, *201*, 108097. https://doi.org/10.1016/j.biombioe.2025.108097

Milessi, T. S., Sandri, J. P., Arruda, P. V., Esteves, T. D., Pinheiro, L. P., Kumar, V., & Chandel, A. K. (2025b). Role of non-genetically modified or native pentose fermenting microorganisms in establishing viable lignocellulosic biorefineries in the Brazilian context. *Critical Reviews in Biotechnology*, *45*, 1287–1305. https://doi.org/10.1080/07388551.2025.2452628

Nidhi, S., Anand, U., Oleksak, P., Tripathi, P., Lal, J. A., Thomas, G., Kuca, K., & Tripathi, V. (2021). Novel CRISPR–cas systems: An updated review of the current achievements, applications, and future research perspectives. *International Journal of Molecular Sciences*, *22*(7), 2–42. https://doi.org/10.3390/ijms22073327

Plante, M. (2023). Epistemology of synthetic biology: A new theoretical framework based on its potential objects and objectives. *Frontiers in Bioengineering and Biotechnology*, *11*. https://doi.org/10.3389/fbioe.2023.1266298

Reetz, M. (2022). Making enzymes suitable for organic chemistry by rational protein design. *ChemBioChem*, *23*(14), e202200049. https://doi.org/10.1002/cbic.202200049

Rodriguez-Abetxuko, A., Sánchez-deAlcázar, D., Muñumer, P., & Beloqui, A. (2020). Tunable polymeric scaffolds for enzyme immobilization. *Frontiers in Bioengineering and Biotechnology*, *8*. https://doi.org/10.3389/fbioe.2020.00830

Rosenberg, E. (2017). Genetic Engineering. In *It's in Your DNA* (pp. 81–93). Elsevier. https://doi.org/10.1016/B978-0-12-812502-1.00010-X

Schillberg, S., Raven, N., Spiegel, H., Rasche, S., & Buntru, M. (2019). Critical analysis of the commercial potential of plants for the production of recombinant proteins. *Frontiers in Plant Science*, *10*. https://doi.org/10.3389/fpls.2019.00720

Schramm, F. D., Schroeder, K., & Jonas, K. (2020). Protein aggregation in bacteria. *FEMS Microbiology Reviews*, *44*(1), 54–72. https://doi.org/10.1093/femsre/fuz026

Segato, F., Damásio, A. R. L., de Lucas, R. C., Squina, F. M., & Prade, R. A. (2014). Genomics review of holocellulose deconstruction by Aspergilli. *Microbiology and Molecular Biology Reviews*, *78*(4), 588–613. https://doi.org/10.1128/MMBR.00019-14

Sellés Vidal, L., Isalan, M., Heap, J. T., & Ledesma-Amaro, R. (2023). A primer to directed evolution: Current methodologies and future directions. *RSC Chemical Biology*, *4*(4), 271–291. https://doi.org/10.1039/d2cb00231k

Shanmugaraj, B., Bulaon, C. J. I., & Phoolcharoen, W. (2020). Plant molecular farming: A viable platform for recombinant biopharmaceutical production. *Plants*, *9*(7), 1–19. https://doi.org/10.3390/plants9070842

Sitara, A., Hocq, R., Horvath, J., & Pflügl, S. (2024). Industrial biotechnology goes thermophilic: Thermoanaerobes as promising hosts in the circular carbon economy. *Bioresource Technology*, *408*, 131164. https://doi.org/10.1016/j.biortech.2024.131164

Song, Z., Zhang, Q., Wu, W., Pu, Z., & Yu, H. (2023). Rational design of enzyme activity and enantioselectivity. *Frontiers in Bioengineering and Biotechnology*, *11*. https://doi.org/10.3389/fbioe.2023.1129149

Vavitsas, K., Glekas, P. D., & Hatzinikolaou, D. G. (2022). Synthetic biology of thermophiles: Taking bioengineering to the extremes? *Applied Microbiology*, *2*(1), 165–174. https://doi.org/10.3390/applmicrobiol2010011

Vieille, C., & Zeikus, G. J. (2001). Hyperthermophilic enzymes: Sources, uses, and molecular mechanisms for thermostability. *Microbiology and Molecular Biology Reviews*, *65*(1), 1–43. https://doi.org/10.1128/mmbr.65.1.1-43.2001

Wang, W., Dasetty, S., Sarupria, S., & Blenner, M. (2021). Rational engineering of low temperature activity in thermoalkalophilic Geobacillus thermocatenulatus lipase. *Biochemical Engineering Journal*, *174*, 108093. https://doi.org/10.1016/j.bej.2021.108093

Wu, S., Snajdrova, R., Moore, J. C., Baldenius, K., & Bornscheuer, U. T. (2021). Biocatalysis: Enzymatic synthesis for industrial applications. *Angewandte Chemie, International Edition*, *60*(1), 88–119. https://doi.org/10.1002/anie.202006648

Yan, E. K., Cao, H. L., Zhang, C. Y., Lu, Q. Q., Ye, Y. J., He, J., Huang, L. J., & Yin, D. C. (2015). Cross-linked protein crystals by glutaraldehyde and their applications. *RSC Advances*, *5*(33), 26163–26174. https://doi.org/10.1039/c5ra01722j

Zeldes, B. M., Keller, M. W., Loder, A. J., Straub, C. T., Adams, M. W. W., & Kelly, R. M. (2015). Extremely thermophilic microorganisms as metabolic engineering platforms for production of fuels and industrial chemicals. *Frontiers in Microbiology*, *6*(Nov). https://doi.org/10.3389/fmicb.2015.01209

Zha, J., Liu, D., Ren, J., Liu, Z., & Wu, X. (2023). Advances in metabolic engineering of Pichia pastoris strains as powerful cell factories. *Journal of Fungi*, *9*(10). https://doi.org/10.3390/jof9101027

Zhang, Y., Werling, U., & Edelmann, W. (2012). SLiCE: A novel bacterial cell extract-based DNA cloning method. *Nucleic Acids Research*, *40*(8), e55. https://doi.org/10.1093/nar/gkr1288

12

Multi-omics Technologies Paving the Way for the Success of Biorefinery

Shruti Ahlawat[1], Somya Gupta[1], Ritika Yadav[1], and Krishna Kant Sharma[2]

[1] *Department of Life Sciences, Faculty of Allied Health Sciences, SGT University, Gurugram, Haryana, India*
[2] *Laboratory of Enzymology and Gut Microbiology, Department of Microbiology, Maharshi Dayanand University, Rohtak, Haryana, India*

12.1 Introduction

Over the years, there has been a rise in the demand of clean and renewable energy sources, such as biofuels due to rising global warming as a result of prolonged use of fossil fuels (Brink et al., 2013). The combustion of fossil fuels releases carbon dioxide (main greenhouse gas) and polyaromatic hydrocarbon aerosols (persistent organic pollutants) in huge amounts (Soltanian et al., 2020), which led to environmental issues, like climate change. Due to these reasons, a switch from fossil-based fuels to clean alternatives, such as biofuels, is required quickly (Aghbashlo et al., 2018). Biofuels include a variety of solid, liquid, and gaseous fuels, such as alcohols (butanol, ethanol, and methanol), biodiesel, biochar, biocrude, bio-oil, Fischer–Tropsch diesel/gasoline, methane, and hydrogen obtained from various bio-originated feedstocks (Farzad et al., 2013). The carbon dioxide released from the combustion of biofuels is inherently biogenic and thus pose no additional negative effects on the environment. From last few decades, the synthesis of biofuels by bioconversion of lignocellulosic (LC) residues has gained huge attention because of its milder operating conditions (Gnansounou & Dauriat, 2011). Overall, biofuels as a substitute for petrochemical-based fuels can effectively solve the problems associated with energy security, environmental degradation, etc. (Soltanian et al., 2020).

Specifically, biofuels derived from starch, vegetable oils, or sugar are known as first-generation biofuels. As these compete with food sources, second-generation or advanced biofuels were developed that are derived from cellulosic products, like wood, wood waste, straw, or long grass. Further, biofuels obtained from microalgae or unicellular algae are known as third-generation biofuels. These unicellular algae are promising feedstock for biofuels due to a rapid growth rate, higher lipid productivity and photosynthetic efficiency, and broader adaptability to various environmental conditions (Ozsoz et al., 2019; Patade et al., 2018). Multi-omics technologies embracing native and tailor made biocatalysts could

Biofoundry Techniques for Biotechnology Applications, First Edition. Edited by Anuj Kumar Chandel.

play a pivotal role in advancing the cost-effective production of biofuel, biochemical, and biomaterials in biorefineries (Milessi et al., 2025a, 2025b).

12.2 Lignocellulosic Biomass

Lignocellulosic biomass includes agricultural residues (bagasse, straws, and stovers), forestry waste (softwood, hardwood, sawdust, etc.), energy crops, like perennial grasses, aquatic plants, like water hyacinth, and municipal solid waste (Patel & Shah, 2021). They are found in great amount in environment; for instance, 181.50 billion tons of LC biomass is generated every year (Dahmen et al., 2019), and it can be effectively converted into various biological value-added chemicals and products. Thus, the accumulation of LC biomass in great quantities in nature not only creates the issue of proper disposal but also causes the loss of these valuable biomaterials (Sánchez, 2009).

Due to varied sources, these feedstocks have variable composition and proportion of basic components. Specifically, LC biomass is constituted up of cellulose (35–55%), hemicellulose (20–40%), and lignin (10–25%) with traces of extractives (extractables, pectin, ash, and proteins) (Varriale & Ulber, 2023), where hydrogen and covalent bonds connect the former two with each other and both link to the lignin, thereby forming an LC matrix that is recalcitrant to depolymerization (Patel & Shah, 2021; Zabed et al., 2016). In LC biomass, cellulose is made up of β-1, 4 linked D-glucopyranose monomers. Hemicellulose, a branched heteropolysaccharide with high crystallinity, consists of different monomers, like pentoses, hexoses, deoxyhexoses, and uronic acids (Bajpai, 2016; Chandra & Madakka, 2019). It assists the chemical bond of cellulose with lignin. Further, lignin is a polycrystalline polymer of phenyl propane linked by C—C/ether bond with irregular repeating units (Limayem & Ricke, 2012; Patel & Shah, 2021). Thus, lignin is constituted of three distinct substituted aromatic phenols (*p*-hydroxyphenyl (H), guaiacyl (G), and syringyl (S) subunits) (Wang et al., 2024) and to convert the biomass into biochemicals and biofuels, they are split into carbohydrates (their building blocks) (Varriale & Ulber, 2023). Overall, lignin is highly-branched cross-linked polymer, which provides protection against abiotic and stresses biotic and imparts mechanical strength (Tiwari et al., 2023).

To generate biofuels and other value-added products (VAPs), LC biomass is first thermochemically pretreated due to highly complex and recalcitrant nature of LC that is followed by the enzymatic hydrolysis to convert polysaccharides into easily fermentable sugars (Sun & Cheng, 2002). For the enzymatic hydrolysis, lignin removal, complete or partial hydrolysis of hemicellulose, reduction in degree of polymerization, and reduction in crystallinity of cellulose are needed, all of which are mediated by pretreatment (Patel & Shah, 2021). In conclusion, LC biomass is among the renewable carbon sources, which could replace the fossil fuels (Isikgor & Becer, 2015).

12.3 Steps in Biorefinery

In LC biomass biorefinery, LC biomass is fractionated using low-cost and sustainable processes and biologically converted into energy and VAPs (Lask et al., 2019).

12.3.1 Step 1-Pretreatment of LC Biomass

Pretreatment decreases the size of biomass, reduces the sugar loss, and enhances lignin removal along with decreased formation of inhibitors. Thus, it makes the process economical (Jamil et al., 2022). A wide array of pretreatment technologies has been used to increase the enzymatic accessibility of LC biomass; among which, steam explosion, dilute acid treatment, and methods based on usage of organic solvents and ionic liquids, etc. have been used majorly at the industrial scale over last few years. However, the choice of pretreatment technology relies on the kind of feedstock and type of desired end products. Specifically, few pretreatment processes enhance plant cell wall accessibility by reorganizing cell wall polymers (ammonium fiber explosion), and others enhance accessibility of cellulose by hydrolytic enzymes via fractionation of biomass by separating hemicellulose (steam explosion) or lignin (alkali) or both (organosolv or ionic liquid pretreatment) from cellulose (Østby et al., 2020).

12.3.1.1 Physical Pretreatment

This pretreatment method involves reduction in the mechanical size of LC biomass by extrusion, milling, microwave irradiation, pyrolysis, and mechanical methods. This is a cost- and energy-intensive process as it generates simultaneous changes in pressure and temperature through high energy consumption, which leads to high production costs (Carrillo-Nieves et al., 2019). Specifically, the extrusion process involves a simultaneous combination of thermal, chemical, and mechanical techniques that leads to structural alteration of LC biomass due to shearing force created by high-power rotation. This leads to high recovery of sugar from biomass (Zhang et al., 2020). Further, milling methods (ball milling, roll milling, wet-disk milling, chipping, and grinding processes) reduces the crystallinity of cellulose in agricultural waste (Yuan et al., 2012). On the other hand, microwave irradiation changes the cellulose structure and degrades lignin and hemicellulose in LC biomass, thereby enhances the vulnerability of biomass toward enzymes (Bak et al., 2009).

12.3.1.2 Chemical Pretreatment

This pretreatment method involves breaking of covalent and hydrogen bonds between hemicellulose, cellulose, and lignin in LC biomass by using various chemicals with a requirement of maintaining the ambient temperature. Thus, this method involves the use of chemicals to transform the crystalline structure of LC biomass into an amorphous form (Shukla et al., 2023).

The alkaline pretreatment method uses chemicals that are non-pollutants and non-corrosive, like ammonia, sodium hydroxide, sulfite, ammonium hydroxide, and lime (calcium hydroxide) (Kim et al., 2016). This method enhances the digestibility of LC biomass by removing lignin from its carbohydrate counterparts. Further, it expands the surface area and porosity of biomass, and reduces the degree of polymerization and crystallinity (Rezania et al., 2020). Among all the chemical pretreatment methods, alkaline pretreatment is thought to be the most efficacious and acceptable. However, the major drawback of this method is high alkali consumption, formation of degradation products in large amounts, and long pretreatment time (Özbek et al., 2020).

Acid pretreatment is carried out by using acid, either concentrated or diluted; however, dilute acid is preferred because of its non-corrosive nature. In addition, concentrated acid shows mechanistic hindrance. During pretreatment with concentrated acid, cellulose from LC biomass releases a catalyst, which breaks the heterocyclic-ether linkages between the chain monomers and polymers of hemicellulose. This shows lesser degradation of lignin and thereby allows less accessibility of cellulose toward enzymatic hydrolysis. Thus, acid pretreatment by using dilute acid is more effective; however, major drawbacks are the generation of inhibitors, need of relatively higher temperature, and acid recovery is costlier (Shukla et al., 2023). Further, acids are toxic, hazardous, and corrosive and need expensive construction materials (Amin et al., 2017).

Recently, the ionic liquid pretreatment method has gained attention due to its ability to break apart the cellulose at ordinary conditions. Moreover, ionic liquid is regarded as non-flammable, non-volatile, and recyclable solvent with high chemical and thermal stability (Sorn et al., 2019). In addition, various reports suggest the effectiveness of addition of aqueous organic solvents, such as acetone, methanol, and acetone, etc. to LC biomass with different pressure and temperature conditions, which occur in the presence of catalysts, acids, and bases (Koo et al., 2011). Among them, glycerol is most preferred due to being non-lethal, easily accessible at more economical rates, and high boiling point (290 °C; Shukla et al., 2023). Another pretreatment method related to organosolvent is deep eutectic solvent, which consists of both hydrogen bond acceptors as well as donors with comparatively lesser toxicity and biocompatibility (Wang et al., 2024).

12.3.1.3 Physio-chemical Pretreatment Processes

This pretreatment method involves both chemical and physical processes together. It includes processes such as wet air oxidation, steam explosion, liquid hot water process, CO_2 explosion, and ammonium fiber explosion that are environment-friendly processes for altering the cellulose structure and for extracting hemicellulose from LC biomass (Shukla et al., 2023). Specifically, the steam explosion pretreatment is the most commonly used method as it breaks the LC structure in environment-friendly manner with high energy efficiency. In this process, LC biomass is exposed to saturated steam at high pressure for a small period, followed by abrupt release of pressure. This causes steam expansion in the feedstock material, thereby resulting in disintegration of individual fibers (Singh et al., 2015). However, main limitations are partial removal of lignin and generation of toxic chemicals, like furfural, acetic acid, and hydroxymethylfurfural (Rezania et al., 2020). Further, liquid hot water pretreatment is like the steam explosion as it needs hot compressed water under high pressure to carry out the pretreatment (Mikulski & Kłosowski, 2020). Overall, it is a simple, eco-friendly, and cost-effective process as it provides substitute to hazardous and expensive chemicals, reduces equipment corrosion, and avoids solvent recycling. However, limitations are high energy requirement and generation of inhibitors (Imman et al., 2017). Moreover, ammonium fiber explosion, same to the steam explosion method, involves the treatment of LC biomass with ammonia (chemical) at high pressure (250–300 psi) and temperature (600–1,000 °C) (physical) for short time (Iram et al., 2019). Low inhibitor toxicity and the absence of a need for water washing are among the advantages of ammonia pretreatment at the industrial level (Shukla et al., 2023). In the CO_2 explosion method, CO_2 is fed into barrel having LC biomass at high pressure and temperature (~200 °C) under agitation. In this process, CO_2 is converted to carbonic acid, which diffuses into biomass to hydrolyze hemicelluloses. After which, pressure is eventually raised that distorts biomass structure, thereby resulting in better accessibility of cellulose (Arumugham et al., 2022). The advantages of this process over others are cost-effective due to requirement of relatively lower temperature and higher reducing sugar yields. However, this method needs costlier reactor to combat high pressure due to CO_2 explosion. Further, LC biomass having low moisture content is the main challenge for carrying out this pretreatment (Shukla et al., 2023).

12.3.1.4 Biological Pretreatment Method

This method involves microorganisms; thus, it is regarded as an environment-friendly and safe process. The microorganisms are chosen based on the substrate. Overall, this method is favored over other methods because of low energy requirement and less generation of pollutants (Kumar & Sharma, 2017). Among microorganisms, white-rot fungi are indicated to have high titers of hydrolytic enzymes, such as laccase and peroxidase (Shankar et al., 2024; Xie et al., 2017). However, enzyme production for pretreatment is a slow process that needs a large amount of carbohydrates for production. Overall, this method needs low capital investment as microorganisms are easily obtained from natural sources, requires less energy, and is efficient, non-hazardous, and environmentally safe. However, due to long pretreatment time, this method is fused with other methods to enhance the overall effect of the process (Shukla et al., 2023). Among microorganisms, fungi produce a wide range of lignocellulolytic enzymes (Reyes-Sosa et al., 2017) and thermophilic fungi have shown to possess more active and stable enzymes for LC biomass disintegration in comparison to enzymes obtained from mesophilic fungi (Berka et al., 2011). Over mesophilic enzymes, thermostable enzymes offer advantages, like elevated thermal stability that decreases enzyme dosage by allowing longer hydrolysis time (Viikari et al., 2007). Additionally, they offer lower viscosity, higher mass-transfer rates, better substrate solubility, and reduced risk of contamination (Turner et al., 2007).

Thus, the LC biomass pretreatment consists of several physical, chemical, physio-chemical, and biological protocols. Individually, these pretreatment techniques for different feedstocks have increased the output toward final biofuel production. However, if these pretreatment methods are integrated, then this can further elevate the output for final product besides reduced chemical usage and time. At present, combinations of two or more pretreatment approaches are applied for effective biomass digestibility and selective biomass recovery (Shukla et al. 2023).

12.3.2 Step 2-Saccharification

Saccharification involves the enzyme-mediated conversion of the complex polysaccharides of hemicellulose and cellulose into fermentable monosaccharides and oligosaccharides, which are main nutrients for subsequent fermentation to produce biofuels and other chemicals (Sharma & Aggarwal, 2020). This process is synergistically catalyzed by multiple enzymes (majorly glycoside hydrolases [GHs], including hemicellulases and cellulases). Cellulases include three different groups of enzymes: β-glucosidases, endoglucanases (EGs), and exoglucanases or cellobiohydrolases (CBHs) (Lynd et al., 2002; Saini et al., 2022). EGs decrease the polymerization degree of cellulose by liberating individual polysaccharide chains by breaking the crystalline structure of cellulose microfibrils (Zhang et al., 2012).

Exoglucanases and CBHs release cellobiose by cutting cellulose from free ends of polysaccharide chains. β-Glucosidases further hydrolyze cellobiose to generate individual molecules of glucose. Together, these enzymes break cellulose into glucose units (Wang et al., 2024). Moreover, besides GHs, auxiliary enzymes, such as lytic polysaccharide monooxygenases (LPMOs), increase the efficiency of the cellulose hydrolysis (Song et al., 2018). Further, complete degradation of hemicellulose requires broader category of enzymes, such as arabinoxylanases, β-xylosidases, esterases, mannanases, and xylanases to release monomers, like mannoses and xyloses (Wang et al., 2024).

Almost all LC-degrading fungi and bacteria secrete an array of hemicellulases and cellulases. Despite various advantages of these enzymes, like mild reaction conditions, specific activity, and environment-friendly, the enzyme cost provides hinderance in economic feasibility of the biorefinery process (Ilić et al., 2023). Further, the saccharification process is suppressed by several factors, such as product inhibition (cello-oligosaccharide and cellobiose inhibit cellulases, and glucose inhibits β-glucosidases) (Konar et al., 2019; Nhim et al., 2022).

12.3.3 Step 3-Fermentation

In LC biorefinery, saccharification is the key step to produce nutrients for successive downstream fermentation; thus, several LC bioconversion approaches have been developed recently, including on-site and off-site saccharification that involves the integration of biocatalyst production methods with downstream and upstream processes (Liu et al., 2020; Saini & Sharma, 2021). Initially, the off-site saccharification approaches are developed that are followed by the on-site saccharification as new approaches, which were developed for low-cost operation (specially cost of enzyme production) by combining the steps of enzyme production and saccharification, and includes consolidated biosaccharification (CBS) and consolidated bioprocessing (CBP) (Wang et al., 2024). Among off-site saccharification approaches, simultaneous saccharification and fermentation (SSF) and separate hydrolysis and fermentation (SHF) strategies are mostly used that uses free cellulases. However, simultaneous saccharification co-fermentation (SSCF) and separate hydrolysis co-fermentation (SHCF) approaches were developed for the joint usage of hexose and pentose downstream (Zhao et al., 2024).

Several species and strains of yeast have been used extensively for industrial bioproduction. Specifically, xylan- and cellulose-degrading yeasts are of great interest for use in SSF and CBP strategies (Šuchová et al., 2022).

12.4 Various Value-Added Products Generated from Lignocellulosic Biomass

Overall, LC biomasses can be utilized to produce various organic chemicals, including ethanol (Buaban et al., 2010), butanol, acetone (Amiri et al., 2014), biomethane (Song et al., 2013), biohydrogen (Lai et al., 2014), and organic acids, citric acid (Kumar & Jain, 2008), acetic acid (Deng et al., 2016), fumaric acid (Wang et al., 2015), lactic acid (Tirpanalan et al., 2015), xylitol (Carvalho et al., 2002), etc. Following are discussed various VAPs obtained from feedstocks (Figure 12.1).

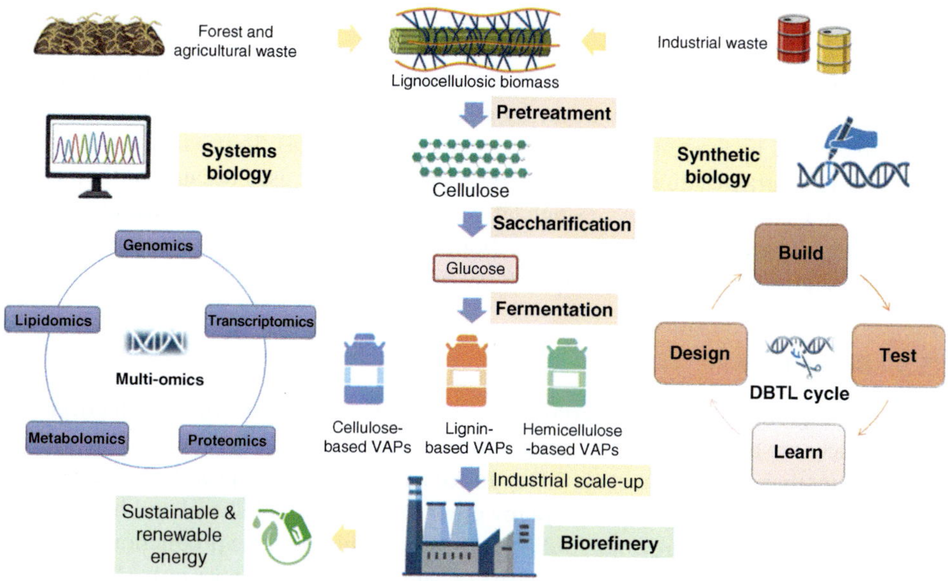

Figure 12.1 Lignocellulosic biomass obtained as agricultural, forestry, and industrial waste can be effectively converted into numerous biological value-added products (VAPs). To generate biofuels and other VAPs, it is first pretreated due to highly complex and recalcitrant nature of lignocellulose that is followed by the enzymatic hydrolysis to convert polysaccharides into easily fermentable sugars, which are subsequently converted to various VAPs. The concept of biorefinery requires the development of economical and sustainable production routes for biofuels and other VAPs from LC biomass using genetically engineered microorganisms for high selectivity and yield. The integration of multi-omics approaches with systems biology, synthetic biology, and genome editing tools helps to identify critical genes and/or pathways, and re-construct and re-design those biological pathways to produce various eco-friendly as well as economical value-added biochemicals and biofuels.

12.5 Cellulose-Based Value-Added Products

12.5.1 Lactic Acid

It is popular to produce various high-value products, including polylactic acid that is a suitable alternate option to petrochemical-based plastics for packaging, fabrication of prosthetic devices, and drug delivery systems due to its biocompatible and biodegradable properties (Cubas-Cano et al., 2019). For microbial fermentation of lactic acid, many high sugar containing crops, like cane molasses, corn cob, rice bran, sugarcane bagasse, wheat bran, etc., and microorganisms, like *Lactobacillus* (*L. brevis* [Guo et al., 2010], *L. coryniformis* [Yáñez et al., 2003], *L. casei* [Xavier & Lonsane, 1994], and *L. buchneri* [Montipo et al., 2018], *Streptococcus thermophiles* [Hofvendahl & Hahn-Hägerdal, 2000], *Bacillus coagulans* [Cubas-Cano et al., 2020], and *Rhizopus oryzae* [Park et al., 2004], have been utilized.

12.5.2 Bioethanol

It is used as a biofuel that assists to lessen the emission of pollutant gases. In the biorefinery, the biomass is converted into sugar by enzymatic hydrolysis, which is further converted into ethanol by fermentation (Romero-Garcia et al., 2014). To produce bioethanol, forest and agricultural residuals, and industrial wastes are most used because of their easy availability and cost-effectiveness (Singh et al., 2014). Ethanol is a promising biofuel because of high energy density and vaporization, and low CO_2 emission; thus, its production is rising enormously as a substitute of fossil fuels (Chandel et al., 2011).

12.5.3 Biomethane

In anaerobic digestion, organic materials such as food waste, manure, or other biodegradable substances are broken down by bacteria in an oxygen-free environment to biologically produce the methane gas (Banu et al., 2021). Indeed, biomethane is a promising fuel due to its higher energy content compared to many other biofuels, like biodiesel and bioethanol (Mayer et al., 2014). Many previous studies suggested the biomethane production from LC biomass, like rice straw, wheat straw, corn stover, sunflower stalks, paper pulp sludge, and fallen leaves (Kumar et al., 2016). This makes them a valuable resource in the transition to more sustainable energy systems. Moreover, the Wood-to-Methane 3+2 (W2M3+2) approach is an innovative industrial process that focuses on converting the wood waste into biomethane using a combination of microbial fermentation and thermophilic anaerobic digestion (Ingersoll, 2023).

12.5.4 Biodiesel

The processes of alkali fatty acid neutralization and transesterification are main steps in the production of biodiesel with a higher content of methyl esters (the primary component of biodiesel) (Banu et al., 2021). Agro-industrial LC biomasses, such as feedstock, are used to produce biodiesel, which offer various advantages, like more seasonal and climate flexibility, faster production, higher yield, reduced labor, and ease of scale-up (Zuccaro et al., 2020).

12.5.5 Biobutanol

However, 1-butanol is considered advantageous over ethanol in various key areas, such as energy density, safety, and engine compatibility because it has higher energy density and is less flammable and corrosive (Al-Shorgani et al., 2015). The generation of biobutanol through acetone/butanol/ethanol (ABE) fermentation is primarily carried out by specific species within the *Clostridium* genus, particularly *Clostridium acetobutylicum* and *Clostridium beijerinckii* (Kumar et al., 2016). However, this faces several significant challenges, like toxicity of butanol that limit its productivity, concentration, and yield. Thus, extensive research has been focused on genetic engineering approaches to improve the performance of strain. These approaches aim to enhance butanol tolerance, optimize solvent production, and increase overall yield through techniques, such as gene knock-out/knock-down, gene insertion/over-expression, and several molecular tools, like TargeTron technology, antisense RNA technology, and recombinant DNA technology (Al-Shorgani et al., 2018).

12.6 Hemicellulose-Based Value-Added Products

Hemicelluloses are indeed one of the most underutilized components during the bioconversion of LC biomass to ethanol and other biofuels. This is due to several challenges related to their chemical structure and sensitivity to degradation during the pretreatment stages of LC biomass processing (like during acid hydrolysis, and organosolv and steam explosion treatment) (Patel & Shah, 2021). Important VAPs obtained from extracted hemicellulose or hemicellulosic hydrolysates are as follows.

12.6.1 Xylitol

It is a sugar alcohol that is commonly used as a low-calorie sweetener in sugar-free chewing gums, candies, and other products. Moreover, it has low-caloric content and anticarcinogenic properties due to which it is widely used as a sugar substitute in the food industry (Irmak et al., 2017). It is manufactured through the catalysis of xylose (Cortez et al., 2016) that is predominantly derived from hardwood sources using bacteria, yeasts, and fungi. Among all, maximum yield of xylitol is achieved using yeast (*Candida* sp.) (Patel & Shah, 2021). Thus, biological routes to produce xylitol, mainly via yeast fermentation, offer a more sustainable and environmental-friendly alternative to the traditional chemical process that uses catalytic hydrogenation. The biological conversion of xylose to xylitol presents various advantages, like lower energy requirements, reduced environmental impact, and the ability to use renewable LC biomass sources (Mussatto, 2012, Cortez et al., 2016).

12.6.2 Xylooligosaccharides (XOS)

These are soluble and non-digestible dietary fibers that are made up of short chains (oligomers) of xylose units, which are linked together via β-1, 4-glycosidic bonds. They have been recognized to have potential prebiotic activity as they have beneficial effects on gut health by promoting the growth of beneficial gut bacteria (e.g., bifidobacteria) (Carvalho et al., 2013). Further, XOS, which contain uronic acids as branch, exhibit antiallergic and antioxidant properties (Jain et al., 2015).

The XOS can be produced from LC biomass using methods, such as autohydrolytic, chemical, enzymatic, or by a union of processes (Carvalho et al., 2013; Kumar & Satyanarayana, 2015). Each method has its benefits and limitations. The autohydrolytic or chemical processes can result in the formation of undesirable by-products, such as furfural and hydroxymethylfurfural (Bian et al., 2014), and produce XOS with a broad range of polymerization, i.e., from DP2 to DP20 (Li et al., 2018). On the other hand, enzymatic processes produce low-DP XOS, i.e., from DP2 to DP4 that is needed for prebiotic application (Carvalho et al., 2015; Sukri & Mimi Sakinah, 2018).

12.6.3 Furfural

It is a heterocyclic compound with an aromatic aldehyde structure. It has a furan ring with an aldehyde group attached to one of the carbons in the ring (Mathew et al., 2018). It is involved in making of products, like adhesives, antacids, fungicides, fertilizers, nematicides, flavoring compounds, plastics, and inks (Raman & Gnansounou, 2015).

Specifically, furfural is a natural dehydration product of xylose (Mathew et al., 2018), which is utilized to produce chemicals, like furan, furfuryl alcohol, levulinic acid (LA), and furonic acid (FA), and tetrahydrofuran (THF) (Xiu & Shahbazi, 2012). As furfural is toxic to fermenting microbes, thus, yeast and bacteria capable of converting furfural into compound with lesser toxicity have been identified (Wierckx et al., 2011). For, e.g., *Gluconobacter oxydans* (ATCC 621H) can oxidize furfural as well as furfuryl alcohol to FA, which is utilized in the synthesis of various agricultural, industrial, and pharmaceutical chemicals (Zhou et al., 2017).

12.7 Lignin-Based Value-Added Products

12.7.1 Biopolymers

Lignin is a versatile polymer with multiple reaction sites or functional groups that make it a promising raw material to produce various biopolymers. Among biopolymers, polyhydroxyalkanoates (PHA) are biodegradable, compostable, and bio-based polyesters that offer a sustainable alternative to conventional petroleum-based polymers (Patel & Shah, 2021). A previous study reported the efficient production of mcl-PHA from p-coumaric acid (a lignin compound) by wild-type and metabolically-engineered strains of *Pseudomonas putida* using varying feeding strategies. Lignin-derived PHA can be used for food packaging and in medical devices. Further, lignin is directly utilized as a polyol in the synthesis of polyurethane, which is used to produce adhesives, foams, and films. Finally, during the synthesis of polyester, lignin monomers are incorporated to limit the polymerization process and facilitate the condensation reaction (Strassberger et al., 2014).

12.7.2 Biochar

It is a carbon-rich material produced from the pyrolysis or gasification of the biomass. It has been well known for its effectiveness in water remediation, particularly in removing the pollutants through adsorption. The comparison between different feedstocks, like palm kernel shells, wheat straw, and pine sawdust, highlights the important differences in biochar yield and reactivity during the gasification process. Specifically, the palm kernel shells displayed higher biochar productivity and lower reactivity during the gasification (Martin-Sampedro et al., 2018; Zhang et al., 2019).

Beside these, lignin-rich feedstock is used to produce phenolic compounds and hydrocarbons to replace the usage of conventional fossil fuels (Wang et al., 2018). Further, lignin can be used as a precursor for the synthesis of carbon fibers (Duval & Lawoko, 2014) and vanillin (a flavoring agent) (Mota et al., 2018).

12.8 CRISPR/Cas9 and -Omics Technologies

Several new tools and strategies have been developed to engineer filamentous fungi for biofuel production. These include next-generation -omics approaches to gain deeper insights into the molecular biology of the strain and novel engineering strategies to genetically modify the strain to produce highly efficient cell factories. In this regard, systems biology tools provide deeper understanding about how these microbes integrate multi-level environmental information to correlate the gene expression and protein production in reaction to varying conditions. Further, synthetic biology allows the genetic modification of the strain, mainly by constructing synthetic promoters, for heterologous expression of genes, thereby either complements their abilities or re-shapes their regulatory networks (Amores et al., 2016; Gurawalia et al., 2024).

Recently, CRISPR/Cas9 technology has been developed as a revolutionary tool for the genetic modification. It utilizes the adaptive immune system of *Streptococcus pyogenes* to modify the genetic sequences or even edit the genome of a specific organism. This is accomplished by a synthetic RNA molecule called single guide RNA (sgRNA), which is specific to the target DNA sequence. sgRNA forms a complex with CRISPR-associated protein 9 (Cas9). In the sgRNA-Cas9 complex, sgRNA-guided Cas9 searches for the matching DNA sequence in the genome. Once this complex finds a target DNA sequence, Cas9 creates a double-strand break in the DNA at that location. This activates the DNA repair mechanisms, which enable the gene editing. More specifically, non-homologous end joining (NHEJ) repair mechanism leads to gene disruption (knockout) and homology-directed repair (HDR) mechanism leads to insertion of a new genetic material (Costa et al., 2017; Gan & Ling, 2022).

The term "omics" refers to a suite of scientific disciplines focused on large-scale measurement and analysis of biological molecules, providing insights into the underlying biology of living organisms at an unprecedented resolution. Genomics, transcriptomics, proteomics, lipidomics, metabolomics, and epigenomics denoted to global analyses genes, RNA, proteins, lipids, and methylated DNA or modified histone proteins in chromosomes, respectively. "Omics" field relies on various high-throughput technologies, including next-generation sequencing (NGS) for transcriptomics and genomics, mass spectrometry (MS) for metabolomics, proteomics, and lipidomics, and microarrays for the gene expression analysis (Omenn et al., 2012).

As proteomics validates predicted genes, correlates stop-codon and start-codon positions, and reveals the novel protein(s), which are missing in genome annotation (Yang et al., 2014). Thus, combining genomics and transcriptomes, and proteomics provides a complete understanding, mainly for the variability between translation and transcription (Lin et al., 2019).

12.9 Utilization of -Omics Technologies Toward Biorefinery Success

Since last few years, enormous amount of -omics data has been created for better understanding of biofuel production, which led to the evolution of fourth-generation biofuels from genetically engineered species for the economical production of renewable energy (Table 12.1) (Patade et al., 2018).

Table 12.1 Multi-omics Technologies Used to Study Pathways for Multi-product Biorefineries

Omic Approach	Microorganism	Gene/Protein/Pathway/Metabolite Identified	Consequence	Reference
Comparative metabolomics	*Crypthecodinium cohnii*	Lower production of xylose and fructose, and increased production of glycerol, glutamate, succinate, and malonate	Glucose tolerance due to decreased osmotic stress and protection of cells against substrate inhibition	Lin et al. (2019)
Multi-omics (proteomics, transcriptomics, and metabolomics), functional annotation, and RB-T DNA sequencing	*Rhodosporidium toruloides*	Protein ID – 9990 (matched D-arabinitol 2-dehydrogenases with high identity), Protein ID – 9837 (matched D-arabinitol 2-dehydrogenase, D-arabinitol 4-dehydrogenase, and D-arabinitol dehydrogenase [NADP+] with low identity), protein ID – 16850 (D-xylulose kinase) had low RNA abundance and weak fitness defects, protein ID – 8905 and protein ID – 9837 had weak fitness defects whereas protein ID – 14368 (D-ribulose kinase) had significant fitness defects	Presence of unique metabolic pathway for D-xylose and L-arabinose utilization rather than traditional xylose pathway	Kim et al. (2021)
Bacterial 16S rRNA gene amplicon sequencing, metagenomic analysis, and molecular fingerprintings	Plant biomass degrading microbial consortia obtained from soil, i.e., efficient in degradation of once-used corn stover (CS-M), switchgrass (SG-M), and wheat straw (WS1-M)	Bacteroidetes members (*Chryseobacterium*, *Flavobacterium*, *Sphingobacterium*, and *Weeksella*) were favored on WS1-M, Proteobacteria members (*Brevundimonas*, *Caulobacter*, *Xanthomonas*, and *Stenotrophomonas*) were selected on CS-M and SG-M; Domination of glycosyl hydrolases (GHs) families, like GH3, GH4, GH10, GH13, GH16, GH28, GH29, GH43, and GH92 and AA6 and AA10 protein families; low abundance of GH5, AA2, AA4, and AA7; Occurrence of arabinoxylan, arabinan, galactomannan, β-glucan, rhamnogalacturonan, and xylan degrading enzymes	Microbial consortia can efficiently carry out the saccharification of LC biomass	Jiménez et al. (2016)

RNA sequencing and proteomics (secretome analysis)	*Parascedosporium putredinis* NO1	Genes encoding 102 putative CAZymes, 47 GHs, 41 AAs, 10 CEs, and 1 PL were up-regulated; GHs (GH1, GH3, GH5, GH7, GH10, GH11, GH31, GH43_1, GH43_11, GH43_21, GH43_22, GH43_26, GH43_36, and GH45,), 3 CE1 family proteins, 3 CE5 family proteins, 69 AAs (AA1_2, AA1_3, AA3, AA6, AA7, AA9, AA10, AA11, AA13, AA14, and AA15) were abundant within the secretome; One unknown protein (c2092_g1_i1) with β etherase activity was also identified that improved saccharification and released tricin from lignin in absence of co-factors	c2092_g1_i1 is suggested to have potential application in the biorefinery industry	Oates et al. (2021)
Genomics and proteomics	*Geobacillus* WSUCF1	ERGO sequencing identified 34 GH enzymes across 19 different families (seven genes from GH13, four genes from GH1 and GH43, three genes from GH18, two genes from GH51, and one gene each from GH2, GH4, GH10, GH23, GH25, GH27, GH32, GH36, GH39, GH42, GH52, GH67, GH127, and GH130); 17 types of hemicellulases (four β-xylosidases, two each of endo-1,2-β-xylanases and α-L-arabinofuranosidases, one each of α-galactosidase, α-amylase, 1,2-α-glucan branching enzyme, endo-1,2-α-L-arabinosidase, exo-α-1,2-glucosidase, TREHALOSE-6-phosphate hydrolase, neopullulanase, pullulanase, and xylan α-1,2-glucuronidase), two types of cellulases (endoglucanase M and 6-phospho-β-glucosidase), 5 lignin-degrading enzymes (2 thioredoxin peroxidases and 1 each of cytochrome c peroxidase, laccase, and glutathione peroxidase); AFC-S (Ammonia fiber expansion [AFEX] pre-treated corn stover) and AVI-S (Avicel) have higher xylanase and cellulase activities, respectively; In comparison to AFEX-CS, extractive ammonia pre-treated corn stover (EA-CS) gave higher glucan conversion (improved saccharification) at low enzyme loadings	CAZymes profiles varied among secretomes depending upon type of substrate and pre-treatment conditions Production of CAZymes in a biorefinery concept by using cheaper substrates, like Avicel and AFEX pre-treated corn stover will significantly reduce the cost of enzymes production and biofuel processing	Bhalla et al. (2022)

(*Continued*)

Table 12.1 (Continued)

Omic Approach	Microorganism	Gene/Protein/Pathway/Metabolite Identified	Consequence	Reference
Transcriptomics and proteomics	*Laetiporus sulphureus* ATCC 52600	On performing transcriptomics, CAZymes content included 133 GHs (GH1, GH3, GH5, GH7, GH10, GH13, GH15, GH16, GH18, GH28, GH37, GH43, GH53, GH55, GH115), 51 AAs (AA1 (AA1_1, AA1_2, AA1_3), AA2, AA3 (AA3_2, AA3_3), AA5_1, AA6, AA7, AA9, AA14), 67 GTs (GT2, GT4, GT8, GT15, GT20, GT21, GT39, GT69), 13 CEs (CE1, CE4, CE8, CE9, CE16), and 4 PLs (PL14 and PL35); via proteomics 112 proteins were identified that included 42 CAZymes (GHs, AAs, and PL35); Most of identified enzyme activities were found at higher levels on sugarcane bagasse (SCB). Replacing 15% of commercial enzyme load with SCB secretome resulted in 7% enhanced glucan conversion	Secretome enrichment can reduce the dependency on commercial enzymes for saccharification, thereby reducing the biorefinery cost	de Figueiredo et al. (2021)
Proteomics	*Thermothielavioides terrestris* LPH172	Higher expression of GH5_7 in $STEX_{180°C/auto}$, $STEX_{210°C/auto}$; GH51 and GH115 in $STEX_{210°C/H2SO4}$; GH43 and GH62 in $STEX_{180°C/auto}$, $STEX_{210°C/auto}$, and untreated steam-pretreated spruce; GH47, GH76, and hemicellulose debranching enzymes in untreated steam-pretreated spruce; supplementation of commercial enzyme mix with $STEX_{180°C/auto}$ secretome resulted in twofold increased release of mannose from steam-pretreated spruce when compared to commercial enzyme mix alone	Saccharification using commercial enzymes can be improved by supplementing with tailored secretomes	Caputo et al. (2024)

The concept of biorefinery requires the development of economical and sustainable production routes for biofuels and commodity chemicals from LC biomass by using genetically engineered microorganisms for high selectivity and yield (Bozell & Petersen, 2010; Nielsen & Keasling, 2016). However, it is successfully applied to a handful of commercial compounds due to cost- and time-intensive nature that is because of complex relation between phenotype and genotype, which involves transcriptional, translational, and metabolic regulation. In this regard, the modern -omics analysis efficiently identifies the genetic, metabolic, and regulatory phenotypes of the engineered microbes but still designing genetic interventions to attain a desirable phenotype is difficult (Liu et al., 2015). Therefore, Design-Build-Test-Learn (DBTL) cycle in the metabolic engineering can be used to generate non-intuitive and effective design predictions (Figure 12.1). It harnesses latest advancements in genetic engineering and high-throughput characterization to effectively screen the wider libraries of strain modifications. The goal is to utilize the characterization of already engineered strains for developing the improved strains (John & Bomble, 2019).

12.10 Role in Efficient Enzyme Production

The utilization of systems biology tools to investigate the deconstruction and depolymerization abilities of fungi provides a universal view and can be used for the economical bioconversion of LC biomass to biofuels. Further, the knowledge of cellulase expressing regulatory networks provides the basis for genetic engineering approaches to construct the recombinant strains for the remarkable performance in an industrial setup. The concept of synthetic biology in respect to optimum production of various industrial products by filamentous fungi is discussed in detail in a chapter published by Amores and co-workers (2016). Overall, in past, genetic engineering of microbes is focused on manipulating individual genes using relatively basic techniques that are rudimentary skills when applied to understand and modify the complex biological systems. However, synthetic biology has expanded the scope and precision of genetic engineering by emphasizing the rational design and construction of both simple as well as complex synthetic gene circuits by using modularized and standardized parts (Amores et al., 2016).

A recent review paper summarizes the consequences of gene editing technique CRISPR-Cas to alter certain genes in extremophilic microbes to enhance the biofuels production efficiency. Overall, the effectiveness of intra-cellular enzymes (such as cellulase, hemicellulase, etc.) has been enhanced by altering the genes related to the thermotolerance and enzymatic activity (Garg et al., 2023). These enzymes provide various advantages against mesothermic enzymes, such as greater thermal stability that decreases enzyme dosage by allowing longer hydrolysis time (Viikari et al., 2007). Furthermore, better substrate solubility is attained by utilizing the higher temperature of 50–60 °C while saccharification. High mass transfer rates, low viscosity, and lesser contamination risks are among other benefits of thermostable enzymes (Turner et al., 2007), which reduces the production costs (Garg et al., 2023). In this regard, recent studies on engineering the extremophiles by using synthetic biology and metabolic engineering strategies for efficient production of biofuels are introduced. Specifically, systems biology approaches identify pathways using integrated -omics approaches to generate lignin conversion synthetic genetic circuits, whereas synthetic biology approaches design schemes for microbes to grow in required conditions and produce various VAPs. Moreover, conventional adaptation and metabolic engineering methods help in the development of microbial strains that can grow at the concentrations of aromatic compounds, which normally hinders the microbial growth. Finally, utilization of extremophiles as chassis cells to produce biofuels and integration of synthetic biology, artificial intelligence, and nanomaterial technology unlocks the next generation of bioenergy approaches (Zhu et al., 2020).

The low titers of cellulase from cellulolytic fungi can be overcome via CRISPR/Cas9-mediated fungal strain engineering, which allows efficient and precise genome editing of cellulase gene by over-expressing cellulase genes, deleting cellulase repressors, and modifying the regulatory elements responsible for the cellulase gene expression (Mondal et al., 2022). For instance, in a study, the authors used CRISPR/Cas-9 to genetically engineer *Myceliophthora thermophila* and *M. heterothallica*. Various cellulase genes, like alp-1, cre-1, gh1–1, and res-1, were selected as the editing targets to generate strains having high lignocellulose activities (Liu et al., 2017). Later, the authors developed an *Acidaminococcus* sp. CRISPR/Cas-12a system for multiplex genome editing. Nine genes of cellulase production pathways in *M. thermophila* were targeted, such that the nonuple mutant M9 displayed 18.5-fold increased activity of lignocellulose (Liu et al., 2019). Further, by performing the CRISPR-Cas9 approach, quintuple and ΔMtalp1 mutants were created that showed two-fold higher cellulase production than wild type (Li et al., 2020). In another report, the authors utilized the CRISPR/Cas9 to precisely disrupt a transcriptional repressor gene (*ACE1*) in hyper-LC *Rasamsonia emersonii* (M36). Their research indicated that this gene negatively controls the expression of xylanase and cellulase genes; thus, it is a potent target in strain engineering to enhance the LC enzymes production ability of this fungus (Singh et al. 2023). These findings highlight the significance of cellulase enzyme, which hydrolyzes the cellulosic fractions of LC biomass into monomeric glucose that can be subsequently converted into biofuels and other VAPs.

Generally, a single enzyme is often not sufficient to efficiently degrade the LC biomass for use in a biorefinery. Thus, a cocktail of enzymes is required. In a similar way, a single enzyme cocktail is not effective for optimal performance across all LC biomasses due to their compositional variability. Based on the above-described bottlenecks, advancements in enzyme formulation have been carried out through growing knowledge regarding the synergism among CAZymes gathered from -omics studies of single microorganisms and metaomics studies of microbiomes. Thus, the mixture of cellulases can be generated by different strategies: by using consortium or microbiome-derived enzyme mixtures having multi-species enzyme preparations, by using one core enzyme from single cellulolytic microorganisms (such as *Trichoderma reesei*, *Penicillium funiculosum*, etc.) with synergistic partners, or by using the synergistic minimal enzyme systems containing specific array of mono-component tailor-made enzymes against specific substrate-of-interest. Specifically, metagenomic techniques identify enzymes from uncultured microbes that are involved in the LC degradation by cataloging an array of genes with their functions. Alterations in genes encoding CAZymes can significantly impact specificity and efficiency of polysaccharide degradation under varying environmental conditions. A detailed review by Champreda and co-authors provides a comprehensive update of studies toward the development of high-performance cellulases for biorefinery (Champreda et al., 2019). In another recent study, a novel cold-adapted xylanase was reported with potential applications in the feed and food industry. The development of an enriched microbial consortium on sugarcane bagasse following integrative -omics analysis is a potent approach to reveal microbial genome, pathways, and biomass related biocatalysts for bioproduct applications (Paixão et al., 2021).

In a study, the authors isolated an unusual LC biomass-deconstructing fungus belonging to the genus *Parascedosporium*, which secretes a diverse variety of the carbohydrate-active enzymes (CAZymes), including an enzyme with a new oxidase activity that works without cofactors. This enzyme splits the beta-ether units in the monocot lignin into flavonoid tricin that increases the digestibility of LC, thereby possess potential for the biorefinery. RNA sequencing of *Parascedosporium putredinis* NO1 growing on wheat straw (as sole carbon source) revealed up-regulation of genes encoding putative CAZy proteins (102), GHs (47), auxiliary activities or AAs (41), carbohydrate esterases or CEs (10), and a polysaccharide lyase (PL). Further, 192 putative carbohydrate-degrading enzymes were identified by carrying out the liquid chromatography with tandem mass spectrometry (LC–MS/MS) analysis (Shankar et al., 2023). These include CAZy families GH6s and GH7s (EGs or CBHs) as cellulases, GH1 and GH3 (glucosidases), GH5 and GH45 (EGs), and GH10 and GH11 (endo β-1, 4-xylanases). Further, five proteins were associated with putative β-1, 4-xylosidase activity, whereas GH43 subfamilies with putative alpha-l-arabinofuranosidase, β-d-xylosidases, and β-1, 3-galactosidase activities were also identified among the secretome. Three proteins of the CE1 family that shows homology to feruloyl esterases or FEs and proteins of putative acetyl xylan esterases were also identified. Further, putative proteins among various AA classes (AA9, AA10, AA11, AA13, AA14, and AA15) that act as LPMOs and five putative multi-copper oxidase (MCO) proteins were also recognized. In the exosecretome, an unknown protein was identified that was subsequently reported to possess β-etherase activity (Oates et al., 2021). Overall, proteomics can be applied to recognize proteins-of-interest, which helps in the identification and development of engineered microbial pathways and monitoring strategies for improving the industrial processes, optimizing the microbial production strains, and enabling more efficient and precise biosynthesis of VAPs (Vélez-Mercado et al., 2021).

The integration of genetic engineering techniques, like CRISPR-Cas, -omics approaches, and protein engineering led to the discovery of novel CAZymes and the improvement of existing enzymes, thereby opening the new opportunities for the industrial biotechnology. Specifically, many microbial (fungal and bacterial) species have been explored for developing CAZymes by applying advance strain enhancement techniques, like CRISPR-Cas and by deploying specific expression systems. Further, specific omics-based approaches and protein engineering reveal new CAZymes or improve the usability of the already existing enzymes. Furthermore, *in-silico* computational research and functional annotation of new CAZymes are promising techniques to produce optimized enzyme cocktails for the biorefinery applications (Chettri et al., 2020).

12.11 Role in Microalgae-Based Biorefinery

Since, India has the longest coastal line, this can be utilized to produce algae-based biodiesel (Sarwan et al., 2024). Further, compared to plants, cyanobacteria and microalgae generate higher productivity due to easier cultivation and higher photosynthetic efficiencies and growth rates. Microalgal biorefinery is based on the concept of CO_2 assimilation and conversion into various VAPs, like fatty acids, nucleic acids, proteins, carotenoids, and vitamins that can be utilized in health foods, aquaculture feed, bioenergy, and medical and pharmaceutical fields. Thus, microalgae are regarded as the third-generation feedstock in biorefinery and bioenergy (Lin et al., 2019). Apart from providing the genome information, the comparative transcriptomics identifies the mutant strains with exceptional qualities. In this regard, in a study, comparative transcriptomics identified up-regulation of photosynthetic genes of a mutant *Chlamydomonas reinhardtii*, which led to 5.2-fold increased H_2 production compared to the wild type (Ban et al., 2019). In another study, a comprehensive transcriptome approach identified "inactivation of carbon concentration machinery" as the process of high CO_2 tolerance in carbonic anhydrase-deficient *Nannochloropsis* sp. that has increased biomass at 5% CO_2 due to which it can be utilized for flue gas-based oil production (Wei et al., 2019). Similarly, comparative metabolomics approach attributed higher production of glycerol, glutamate, succinate, and malonate and lower production of xylose and fructose toward glucose tolerance of a mutant *Crypthecodinium cohnii* (glucose-tolerant green algae) (Li et al., 2017). As stated earlier, combining genomics and transcriptomes approaches with proteomics offers a better understanding; thus, proteomics, metabolomics, and genomics approaches together offered a systematic understanding of cellular behavior of *Synechococcus* sp. PCC7002 producing ethanol (Kopka et al., 2017) and *Synechocystis* sp. PCC6803 producing 3-hydroxypropionate (Wang et al., 2016). Furthermore, based on the multi-omics data (metabolomics, proteomics, transcriptomics, and genomics), highly productive *Synechocystis* sp. PCC6803 was constructed for ethanol production (Nishiguchi et al., 2020).

Another alga, *Chlorella,* behaves as a microbial cell factory for producing lipids, biodiesel, and other VAPs. In a recent study, the authors performed the Algomics (integrated proteomics and metabolomics) to study the stress-induced oleaginous mechanisms in *Chlorella* sp. FC2 IITG under nitrogen deprivation. This approach offers insight to improve the strain use in the production of biodiesel and other high-value products. Specifically, 10 high-value products from 4 biological classes (polyunsaturated fatty acids [PUFA], terpenoids, triacylglycerol [TAG], and phytosterols) were annotated. In the study, two PUFAs, dihomo-gamma-linolenic acid and 9(S)HpOTrE accumulation increased in the alga by 6- and 4-times, respectively, in the nitrogen-deprivation conditions. Further, highest accumulation of TAG (59:10) and TAG (54:4) was observed at 40 hr and 88 hr, respectively, thereby suggesting the induction of various TAG species by regulating the severity of nitrogen limitation. The co-extraction of PUFA and TAG could lower the production cost of biodiesel for commercialization. Additionally, proteomics identified elevated glycolysis, tricarboxylic acid (TCA) cycle, and oxidation, which provides acetyl CoA, pyruvate, and phosphoenolpyruvate (PEP) for accumulation of phytochemicals in response to nitrogen deprivation. Overall, this study highlights a complex metabolic trade-off between biomass production and storage lipid accumulation (Rai et al., 2021).

In a study, the authors have performed a multi-omics analysis to investigate genes and metabolic pathways involved in the carbon (glucose, D-xylose, glucose + D-xylose, L-arabinose, and *p*-coumarate) utilization by the yeast *Rhodotorula toruloides* (Kim et al., 2021), which is well known to accumulate lipids and produce carotenoids that make it a potential host organism to produce biofuels and other VAPs from LC biomass-derived carbon sources (xylose, *p*-coumaric acid, and glucose). Furthermore, it can tolerate the inhibitory compounds found in the LC biomass hydrolysate (Yaegashi et al., 2017). Earlier, many multi-omics studies have been done till date but still it is not clear how the yeast utilizes various LC biomass-derived carbon sources. Thus, a metabolic network model is required to study the metabolism of yeast. Therefore, the authors re-constructed a genome-scale metabolic network of the yeast by utilizing high-quality published models and carried out the manual curation using functional annotation and multi-omics data (Random Bar Code Transposon-Site Sequencing [RB-TDNA] sequencing, transcriptomics, metabolomics, and proteomics) that will be utilized to study and engineer the yeast for LC biomass degradation in future (Kim et al., 2021).

In another paper, the authors highlighted the significance of an oleaginous yeast, i.e., *Yarrowia lipolytica* as a superior host to reform the low-value carbons into high-value fatty acid-based chemicals and secondary metabolites, which is attributed to unique malonyl-CoA/acetyl-CoA supply mode as well as adaptable carbon-utilization pathways. This review paper summarizes the importance of systems biology, synthetic biology, and metabolic engineering in modifying the yeast for the development of a sustainable biorefinery, where specifically multi-omics data and development of genome-scale metabolic (GEM) models will help in attaining the highest yields of a particular pathway. Computational models help in anticipating metabolic responses and optimizing engineering techniques. In addition, -omics approaches help in interpreting the complex phenotypes and understanding the cell metabolism at the systems level. In detail, these are discussed in a review by Ma and co-workers (2020). Another recent review by Kumar and co-authors (2023) provides deep insights in multi-omics strategies, optobiological approaches, genome engineering methods, and artificial illumination-based roadmap for sustainable algal biorefineries.

Overall, in the context of microalgae-based biorefinery, various microalgal species, including *Aureococcus anophagefferens*, *Botryococcus braunii* UTEX 572, *Chlamydomonas reinhardtii*, *Chlorella* sp. NC64A, *Coccomyxa* sp. C-169, *Dunaliella salina*, *Micromonas pusilla*, *Nannochloropsis oceanica*, *Ostreococcus tauri*, *Phaeodactylum tricornutum*, and *Volvox carteri*, have been successfully screened and sequenced by the scientists. Thus, the generated database containing genes provide valuable tools for gene editing and metabolic engineering to increase the lipid production for biofuel generation (Ozsoz et al., 2019).

12.12 Conclusion

Latest approaches of synthetic and systems biology have seen significant growth in the last decade. The omics-based approaches have provided a large-scale information related to gene expression, protein synthesis, metabolic shifts, and cellular adaptations. Based on which, new engineering attempts have been done to construct and characterize novel strains with enhanced performance. Alongside the traditional microbes, like filamentous fungi, yeasts and microalgae have gained significant attention in the fields of biofuels and biorefineries. It is worth noting that although much advancement has been attained in the synthetic biology of common model microbes such as *Escherichia coli,* but similar work on oleaginous yeast, filamentous fungi, and microalgae have only started to appear. Thus, the complete understanding or utilization of these organisms for biotechnological purposes still need further studies. In conclusion, based on the significant advancements in the field of synthetic biology and genetic engineering, most likely within another decade we would be able to develop the consolidated technology to develop engineered microbial strains that will be ideal for large-scale economical utilization in cellulose-based biofuel industries, thereby paving the way for the success of sustainable biorefinery.

Conflict of Interest

The authors have no competing interests to declare.

Author Contributions

All authors read and approved the manuscript.

Funding

The authors acknowledge the infrastructural support from DST-FIST grant (Grant No. 1196SR/FST/LS-I/2017/4) & DBT-BUILDER grant (Grant No. BT/INF/22/SP43043/2021).

References

Aghbashlo, M., Tabatabaei, M., Hosseini, S.S. et al. (2018). Performance assessment of a wind power plant using standard exergy and extended exergy accounting (EEA) approaches. *Journal of Cleaner Production 171*: 127–136.

Al-Shorgani, N.K.N., Shukor, H., Abdeshahian, P. et al. (2015). Process optimization of butanol production by Clostridium saccharoperbutylacetonicum N1-4 (ATCC 13564) using palm oil mill effluent in acetone–butanol–ethanol fermentation. *Biocatalysis and Agricultural Biotechnology 4*(2): 244–249.

Al-Shorgani, N.K.N., Shukor, H., Abdeshahian, P. et al. (2018). Enhanced butanol production by optimization of medium parameters using clostridium acetobutylicum YM1. *Australian Journal of Biological Sciences 25*: 1308–1321.

Amin, F.R., Khalid, H., Zhang, H. et al. (2017). Pretreatment methods of lignocellulosic biomass for anaerobic digestion. *AMB Express 7*: 1–12.

Amiri, H., Karimi, K., and Zilouei, H. (2014). Organosolv pretreatment of rice straw for efficient acetone, butanol, and ethanol production. *Bioresource Technology 152*: 450–456.

Amores, G.R., Guazzaroni, M.E., and Magalhães, L. (2016). Systems and synthetic biology approaches for fungal engineering. *Fungal Biotechnology for Biofuel Production 1*: 395.

Arumugham, T., AlYammahi, J., Rambabu, K. et al. (2022). Supercritical CO2 pretreatment of date fruit biomass for enhanced recovery of fruit sugars. *Sustainable Energy Technologies and Assessments 52*: 102231.

Bajpai, P. (2016). Pretreatment of lignocellulosic biomass. In: *Pretreatment of lignocellulosic biomass for biofuel production* (ed. P. Bajpai), 17–70. Singapore: Springer Singapore.

Bak, J.S., Ko, J.K., Choi, I.G. et al. (2009). Fungal pretreatment of lignocellulose by Phanerochaete chrysosporium to produce ethanol from rice straw. *Biotechnology and Bioengineering 104*(3): 471–482.

Ban, S., Lin, W., Luo, Z., and Luo, J. (2019). Improving hydrogen production of Chlamydomonas reinhardtii by reducing chlorophyll content via atmospheric and room temperature plasma. *Bioresource Technology 275*: 425–429.

Banu, J.R., Kavitha, S., Tyagi, V.K. et al. (2021). Lignocellulosic biomass based biorefinery: A successful platform towards circular bioeconomy. *Fuel 302*: 121086.

Berka, R.M., Grigoriev, I.V., Otillar, R. et al. (2011). Comparative genomic analysis of the thermophilic biomass-degrading fungi *Myceliophthora thermophila* and *Thielavia terrestris*. *Nature Biotechnology 29*(10): 922–927.

Bhalla, A., Arce, J., Ubanwa, B. et al. (2022). *Thermophilic Geobacillus* WSUCF1 Secretome for saccharification of ammonia fiber expansion and extractive ammonia pretreated corn stover. *Frontiers in Microbiology 13*: 844287.

Bian, J., Peng, P., Peng, F. et al. (2014). Microwave-assisted acid hydrolysis to produce xylooligosaccharides from sugarcane bagasse hemicelluloses. *Food Chemistry 156*: 7–13.

Bozell, J.J. and Petersen, G.R. (2010). Technology development for the production of biobased products from biorefinery carbohydrates—the us department of energy's "top 10" revisited. *Green Chemistry 12*: 539.

van den Brink, J., van Muiswinkel, G.C., Theelen, B. et al. (2013). Efficient plant biomass degradation by thermophilic fungus *Myceliophthora heterothallica*. *Applied and Environmental Microbiology 79*(4): 1316–1324.

Buaban, B., Inoue, H., Yano, S. et al. (2010). Bioethanol production from ball milled bagasse using an on-site produced fungal enzyme cocktail and xylose-fermenting Pichia stipitis. *Journal of Bioscience and Bioengineering 110*(1): 18–25.

Caputo, F., Siaperas, R., Dias, C. et al. (2024). Elucidating *Thermothielavioides terrestris* secretome changes for improved saccharification of mild steam-pretreated spruce. *Biotechnology for Biofuels and Bioproducts 17*(1): 127.

Carrillo-Nieves, D., Alanís, M.J.R., de la Cruz Quiroz, R. et al. (2019). Current status and future trends of bioethanol production from agro-industrial wastes in Mexico. *Renewable and Sustainable Energy Reviews 102*: 63–74.

Carvalho, W., Silva, S.S., Vitolo, M. et al. (2002). Improvement in xylitol production from sugarcane bagasse hydrolysate achieved by the use of a repeated-batch immobilized cell system. *Zeitschrift Für Naturforschung. C, A Journal of Biosciences 57*(1–2, 112): 109.

Carvalho, A.F.A., de Oliva Neto, P., Da Silva, D.F., and Pastore, G.M. (2013). Xylo-oligosaccharides from lignocellulosic materials: Chemical structure, health benefits and production by chemical and enzymatic hydrolysis. *Food Research International 51*(1): 75–85.

Carvalho, A.V., da Costa Lopes, A.M., and Bogel-Łukasik, R. (2015). Relevance of the acidic 1-butyl-3-methylimidazolium hydrogen sulphate ionic liquid in the selective catalysis of the biomass hemicellulose fraction. *RSC Advances 5*(58): 47153–47164.

Champreda, V., Mhuantong, W., Lekakarn, H. et al. (2019). Designing cellulolytic enzyme systems for biorefinery: From nature to application. *Journal of Bioscience and Bioengineering 128*(6): 637–654.

Chandel, A.K., Chandrasekhar, G., Radhika, K. et al. (2011). Bioconversion of pentose sugars into ethanol: A review and future directions. *Biotechnology and Molecular Biology Reviews 6*(1): 008–020.

Chandra, M.R.G.S. and Madakka, M. (2019). Comparative biochemistry and kinetics of microbial lignocellulolytic enzymes. In: *Recent developments in applied microbiology and biochemistry* (ed. V. Buddolla), 147–159. Academic Press.

Chettri, D., Verma, A.K., and Verma, A.K. (2020). Innovations in CAZyme gene diversity and its modification for biorefinery applications. *Biotechnology Reports 28*: e00525.

Cortez, D.V., Mussatto, S.I., and Roberto, I.C. (2016). Improvement on d-xylose to xylitol biotransformation by Candida guilliermondii using cells permeabilized with triton X-100 and selected process conditions. *Applied Biochemistry and Biotechnology 180*: 969–979.

Costa, J. R., Bejcek, B. E., McGee, J. E., Fogel, A. I., Brimacombe, K. R., & Ketteler, R. (2017). Genome editing using engineered nucleases and their use in genomic screening. *Assay Guidance Manual [Internet]*. https://www.ncbi.nlm.nih.gov/books/NBK53196/

Cubas-Cano, E., González-Fernández, C., Ballesteros, M., and Tomás-Pejó, E. (2019). Lactobacillus pentosus CECT 4023 T co-utilizes glucose and xylose to produce lactic acid from wheat straw hydrolysate: Anaerobiosis as a key factor. *Biotechnology Progress 35*: e2739.

Cubas-Cano, E., Lopez-Gomez, J.P., Gonzalez-Fernandez, C. et al. (2020). Towards sequential bioethanol and l-lactic acid co-generation: Improving xylose conversion to l-lactic acid in presence of lignocellulosic ethanol with an evolved *Bacillus coagulans*. *Renewable Energy 153*: 759–765. https://doi.org/10.1016/j.renene.2020.02.066.

Dahmen, N., Lewandowski, I., Zibek, S., and Weidtmann, A. (2019). Integrated lignocellulosic value chains in a growing bioeconomy: Status quo and perspectives. *GCB Bioenergy 11*(1): 107–117.

De Figueiredo, F.L., De Oliveira, A.C.P., Terrasan, C.R.F. et al. (2021). Multi-omics analysis provides insights into lignocellulosic biomass degradation by *Laetiporus sulphureus* ATCC 52600. *Biotechnology for Biofuels 14*: 1–17.

Deng, W., Wang, Y., and Yan, N. (2016). Production of organic acids from biomass resources. *Curr Opin Green Sustain Chem 2*: 54–58.

Duval, A. and Lawoko, M. (2014). A review on lignin-based polymeric, micro- and nanostructured materials. *Reactive and Functional Polymers 2014* (85): 78–96.

Farzad, S., Haghtalab, A., and Rashidi, A. (2013). Comprehensive study of nanostructured supports with high surface area for Fischer-Tropsch synthesis. *Journal of Energy Chemistry 22*(4): 573–581.

Gan, W.C. and Ling, A.P.K. (2022). CRISPR/Cas9 in plant biotechnology: Applications and challenges. *Biotechnologia 103*: 81–93.

Garg, D., Samota, M.K., Kontis, N. et al. (2023). Revolutionizing biofuel generation: unleashing the power of CRISPR-Cas mediated gene editing of extremophiles. *Microbiological Research 274*: 127443.

Gnansounou, E. and Dauriat, A. (2011). Technoeconomic analysis of lignocellulosic ethanol. *Biofuels* 123–148.

Guo, W., Jia, W., Li, Y., and Chen, S. (2010). Performances of Lactobacillus brevis for producing lactic acid from hydrolysate of lignocellulosics. *Applied Biochemistry and Biotechnology 161*: 124–136.

Gurawalia, H., Shinde, P.B., and Sharma, K.K. (2024). Embracing Industry 4.0 ingredients in synthetic biology for the development of carbon–neutral economy. In: *Biorefinery and Industry 4.0: Empowering sustainability. Green energy and technology* (ed. A.K. Chandel). Nature Switzerland: Springer ISBN: 978-3-031-51600-9. https://doi.org/10.1007/978-3-031-51601-6_8.

Hofvendahl, K. and Hahn–Hägerdal, B. (2000). Factors affecting the fermentative lactic acid production from renewable resources. *Enzyme and Microbial Technology 26*(2–4): 87–107.

Ilić, N., Milić, M., Beluhan, S., and Dimitrijević-Branković, S. (2023). Cellulases: From lignocellulosic biomass to improved production. *Energies 16*: 3598.

Imman, S., Laosiripojana, N., and Champreda, V. (2017). Effects of liquid hot water pretreatment on enzymatic hydrolysis and physicochemical changes of corncobs. *Applied Biochemistry and Biotechnology 184*: 432–443.

Ingersoll, J.G. (2023). Thermophilic fungi as the microbial agents of choice for the industrial co-fermentation of wood wastes and nitrogen-rich organic wastes to bio-methane. *Microorganisms 11*(10): 2600.

Iram, A., Cekmecelioglu, D., and Demirci, A. (2019). Optimization of dilute sulfuric acid, aqueous ammonia, and steam explosion as the pretreatments steps for distillers' dried grains with solubles as a potential fermentation feedstock. *Bioresource Technology 282*: 475–481.

Irmak, S., Canisag, H., Vokoun, C., and Meryemoglu, B. (2017). Xylitol production from lignocellulosics: Are corn biomass residues good candidates? *Biocatalysis and Agricultural Biotechnology 11*: https://doi.org/10.1016/j.bcab.2017.07.010.

Isikgor, F.H. and Becer, C.R. (2015). Lignocellulosic biomass: A sustainable platform for the production of bio-based chemicals and polymers. *Polymer Chemistry 6*(25): 4497–4559.

Jain, I., Kumar, V., and Satyanarayana, T. (2015). Xylooligosaccharides: An economical prebiotic from agroresidues and their health benefits. *Indian Journal of Experimental Biology 53*: 131–142.

Jamil, F., Aslam, M., Al-Muhtaseb, A.H. et al. (2022). Greener and sustainable production of bioethylene from bioethanol: Current status, opportunities and perspectives. *Reviews in Chemical Engineering 38*: 185–207.

Jiménez, D.J., de Lima Brossi, M.J., Schückel, J. et al. (2016). Characterization of three plant biomass-degrading microbial consortia by metagenomics-and metasecretomics-based approaches. *Applied Microbiology and Biotechnology 100*: 10463–10477.

Kim, J.S., Lee, Y.Y., and Kim, T.H. (2016). A review on alkaline pretreatment technology for bioconversion of lignocellulosic biomass. *Bioresource Technology 199*: 42–48.

Kim, J., Coradetti, S.T., Kim, Y.M. et al. (2021). Multi-omics driven metabolic network reconstruction and analysis of lignocellulosic carbon utilization in *Rhodosporidium toruloides. Frontiers in Bioengineering and Biotechnology 8*: 612832.

Konar, S., Sinha, S.K., Datta, S., and Ghorai, P.K. (2019). Probing the effect of glucose on the activity and stability of β-glucosidase: An all-atom molecular dynamics simulation investigation. *ACS Omega 4*: 11189–11196.

Koo, B.W., Kim, H.Y., Park, N. et al. (2011). Organosolv pretreatment of Liriodendron tulipifera and simultaneous saccharification and fermentation for bioethanol production. *Biomass and Bioenergy 35*(5): 1833–1840.

Kopka, J., Schmidt, S., Dethloff, F. et al. (2017). Systems analysis of ethanol production in the genetically engineered cyanobacterium Synechococcus sp. PCC 7002. *Biotechnology for Biofuels 10*: 1–21.

Kumar, A. and Jain, V.K. (2008). Solid state fermentation studies of citric acid production. *African Journal of Biotechnology 7*(5).

Kumar, V. and Satyanarayana, T. (2015). Generation of xylooligosaccharides from microwave irradiated agroresidues using recombinant thermo-alkali-stable endoxylanase of the polyextremophilic bacterium Bacillus halodurans expressed in Pichia pastoris. *Bioresource Technology 179*: 382–389.

Kumar, A.K. and Sharma, S. (2017). Recent updates on different methods of pretreatment of lignocellulosic feedstocks: A review. *Bioresources and Bioprocessing 4*: 1–19.

Kumar, A., Gautam, A., and Dutt, D. (2016). Biotechnological transformation of lignocellulosic biomass in to industrial products: An overview. *Advances in Bioscience and Biotechnology 7*(3): 149–168.

Kumar, A., Baldia, A., Rajput, D. et al. (2023). Multiomics and optobiotechnological approaches for the development of microalgal strain for production of aviation biofuel and biorefinery. *Bioresource Technology 369*: 128457.

Lai, Z., Zhu, M., Yang, X. et al. (2014). Optimization of key factors affecting hydrogen production from sugarcane bagasse by a thermophilic anaerobic pure culture. *Biotechnology for Biofuels 7*: 1–11.

Lask, J., Wagner, M., Trindade, L., and Lewandowski, I. (2019). Life cycle assessment of ethanol production from miscanthus: A comparison of production pathways at two European sites. Special Issue 'Biobased value chains for a growing bioeconomy' of GCB-Bioenergy. *Global Change Biology Bioenergy 11*(1): 269–288. https://doi.org/10.1111/gcbb.12551.

Li, X., Pei, G., Liu, L. et al. (2017). Metabolomic analysis and lipid accumulation in a glucose tolerant Crypthecodinium cohnii strain obtained by adaptive laboratory evolution. *Bioresource Technology 235*: 87–95.

Li, H.Y., Wang, B., Wen, J.L. et al. (2018). Availability of four energy crops assessing by the enzymatic hydrolysis and structural features of lignin before and after hydrothermal treatment. *Energy Conversion and Management 155*: 58–67.

Li, X., Liu, Q., Sun, W. et al. (2020). Improving cellulases production by Myceliophthora thermophila through disruption of protease genes. *Biotechnology Letters 42*: 219–229.

Limayem, A. and Ricke, S.C. (2012). Lignocellulosic biomass for bioethanol production: Current perspectives, potential issues and future prospects. *Progress in Energy and Combustion Science 38*(4): 449–467.

Lin, W.R., Tan, S.I., Hsiang, C.C. et al. (2019). Challenges and opportunity of recent genome editing and multi-omics in cyanobacteria and microalgae for biorefinery. *Bioresource Technology 291*: 121932.

Liu, R., Bassalo, M.C., Zeitoun, R.I., and Gill, R.T. (2015). Genome scale engineering techniques for metabolic engineering. *Metabolic Engineering 32*: 143–154.

Liu, Q., Gao, R., Li, J. et al. (2017). Development of a genome-editing CRISPR/Cas9 system in thermophilic fungal *Myceliophthora* species and its application to hyper-cellulase production strain engineering. *Biotechnology for Biofuels 10*: 1–14.

Liu, Q., Zhang, Y., Li, F. et al. (2019). Upgrading of efficient and scalable CRISPR–Cas-mediated technology for genetic engineering in thermophilic fungus Myceliophthora thermophila. *Biotechnology for Biofuels 12*: 1–19.

Liu, G., Zhao, X., Chen, C. et al. (2020). Robust production of pigment-free pullulan from lignocellulosic hydrolysate by a new fungus co-utilizing glucose and xylose. *Carbohydrate Polymers 241*: 116400.

Lynd, L.R., Weimer, P.J., Van Zyl, W.H., and Pretorius, I.S. (2002). Microbial cellulose utilization: Fundamentals and biotechnology. *Microbiology and Molecular Biology Reviews 66*(3): 506–577.

Ma, J., Gu, Y., Marsafari, M., and Xu, P. (2020). Synthetic biology, systems biology, and metabolic engineering of *Yarrowia lipolytica* toward a sustainable biorefinery platform. *Journal of Industrial Microbiology & Biotechnology: Official Journal of the Society for Industrial Microbiology and Biotechnology 47*(9–10): 845–862.

Martin-Sampedro, R., Eugenio, M.E., Fillat, U. et al. (2018). Biorefinery of lignocellulosic biomass from an elm clone: Production of fermentable sugars and lignin-derived biochar for energy and environmental applications. *Energy Technology. 7*: 277–287. https://doi.org/10.1002/ente. 201800685.

Mathew, A.K., Abraham, A., Mallapureddy, K.K., and Sukumaran, R.K. (2018). Lignocellulosic biorefinery wastes, or resources? In: *Waste biorefinery*, 267–297. Elsevier.

Mayer, F., Gerin, P.A., Noo, A. et al. (2014). Assessment of energy crops alternative to maize for biogas production in the Greater Region. *Bioresource Technology 166*: 358–367.

Mikulski, D. and Kłosowski, G. (2020). Microwave-assisted dilute acid pretreatment in bioethanol production from wheat and rye stillages. *Biomass and Bioenergy 136*: 105528.

Milessi, T.S., Reis, C.E., Ramos, M.D.N. et al. (2025a). Native and tailor-made microbial cell factories for economic production of biorenewables: A paradigm shift in industrial biotechnology. *Biomass and Bioenergy 201* n.d.: 108097. https://doi.org/10.1016/j.biombioe.2025.108097.

Milessi TS, Sandri JP, Arruda PV, Esteves TD, Pinheiro LP, Kumar V, and Chandel AK (2025b). Role of non-genetically modified or native pentose fermenting microorganisms in establishing viable lignocellulosic biorefineries in the Brazilian context. *Critical Reviews in Biotechnology, 45*. https://doi.org/10.1080/07388551.2025.2452628, 1287, 1305.

Mondal, S., Halder, S.K., and Mondal, K.C. (2022). Tailoring in fungi for next generation cellulase production with special reference to CRISPR/CAS system. *Systems Microbiology and Biomanufacturing 2*: 1–17.

Montipo, S., Ballesteros, I., Fontana, R.C. et al. (2018). Integrated production of second generation ethanol and lactic acid from steam- exploded elephant grass. *Bioresource Technology 249*: 1017–1024. https://doi.org/10.1016/j.biortech.2017.11.001.

Mota, I.F., Pinto, P.R., Ribeiro, A.M. et al. (2018). Downstream processing of an oxidized industrial kraft liquor by membrane fractionation for vanillin and syringaldehyde recovery. *Separation and Purification Technology 197*: 360–371.

Mussatto, S.I. (2012). Application of xylitol in food formulations and benefits for health. In: *D-Xylitol: Fermentative Production, Application and Commercialization*, 309–323. Berlin Heidelberg: Springer-Verlag.

Nhim, S., Waeonukul, R., Uke, A. et al. (2022). Biological cellulose saccharification using a coculture of *Clostridium thermocellum* and *Thermobrachium celere* strain A9. *Applied Microbiology and Biotechnology 106*: 2133–2145.

Nielsen, J. and Keasling, J.D. (2016). Engineering cellular metabolism. *Cell 164*: 1185–1197. https://doi.org/10.1016/j.cell.2016.02.004.

Nishiguchi, H., Liao, J., Shimizu, H., and Matsuda, F. (2020). Novel allosteric inhibition of phosphoribulokinase identified by ensemble kinetic modeling of Synechocystis sp. PCC 6803 metabolism. *Metabolic Engineering Communications 11*: e00153.

Oates, N.C., Abood, A., Schirmacher, A.M. et al. (2021). A multi-omics approach to lignocellulolytic enzyme discovery reveals a new ligninase activity from Parascedosporium putredinis NO1. *Proceedings of the National Academy of Sciences 118*(18): e2008888118.

Omenn, G.S., Nass, S.J., and Micheel, C.M. (ed.) (2012). *Evolution of translational omics: Lessons learned and the path forward*. Washington, DC: National Academies Press.

Østby, H., Hansen, L.D., Horn, S.J. et al. (2020). Enzymatic processing of lignocellulosic biomass: Principles, recent advances and perspectives. *Journal of Industrial Microbiology & Biotechnology: Official Journal of the Society for Industrial Microbiology and Biotechnology 47*(9–10): 623–657.

Özbek, H.N., Yanık, D.K., Fadıloğlu, S., and Göğüş, F. (2020). Ultrasound-assisted alkaline pre-treatment and its sequential combination with microwave for fractionation of pistachio shell. *Renewable Energy 157*: 637–646.

Ozsoz, M., Ibrahim, A.U., and Coston, P.P. (2019). Application of Crispr technology for the generation of biofuels: A review. *J Fundam Renewable Energy Appl 9*: 278.

Paixão, D.A.A., Tomazetto, G., Sodre, V.R. et al. (2021). Microbial enrichment and meta-omics analysis identify CAZymes from mangrove sediments with unique properties. *Enzyme and Microbial Technology 148*: 109820.

Park, E.Y., Anh, P.N., and Okuda, N. (2004). Bioconversion of waste office paper to L (+)-lactic acid by the filamentous fungus Rhizopus oryzae. *Bioresource Technology 93*(1): 77–83.

Patade, V.Y., Meher, L.C., Grover, A. et al. (2018). Omics approaches in biofuel technologies: Toward cost effective, eco-friendly, and renewable energy. In: *Omics technologies and bio-engineering*, 337–351. Elsevier.

Patel, A. and Shah, A.R. (2021). Integrated lignocellulosic biorefinery: Gateway for production of second generation ethanol and value added products. *Journal of Bioresources and Bioproducts 6*(2): 108–128.

Rai, V., Patel, S.K., Muthuraj, M. et al. (2021). Systematic metabolome profiling and multi-omics analysis of the nitrogen-limited non-model oleaginous algae for biorefining. *Biofuel Research Journal 8*(1): 1330–1341.

Raman, J.K. and Gnansounou, E. (2015). Furfural production from empty fruit bunch–A biorefinery approach. *Industrial Crops and Products 69*: 371–377.

Reyes-Sosa, F.M., López Morales, M., Platero Gómez, A.I. et al. (2017). Management of enzyme diversity in high-performance cellulolytic cocktails. *Biotechnology for Biofuels 10*: 1–10.

Rezania, S., Oryani, B., Cho, J. et al. (2020). Different pretreatment technologies of lignocellulosic biomass for bioethanol production: An overview. *Energy 199*: 117457.

Romero-García, J.M., Niño, L., Martínez-Patiño, C. et al. (2014). Biorefinery based on olive biomass. State of the art and future trends. *Bioresource Technology 159*: 421–432.

Saini, S. and Sharma, K.K. (2021). Fungal lignocellulolytic enzymes and lignocellulose: A critical review on their contribution to multiproduct biorefinery and global biofuel research. *Int. J. Biol. Macromol 193*: 2304–2319.

Saini, S., Kumar, A., Singhal, B. et al. (2022). Fungal oxidoreductases and CAZymes effectively degrade lignocellulosic component of switchgrass for bioethanol production. *Fuel 328*: 125341.

Sánchez, C. (2009). Lignocellulosic residues: Biodegradation and bioconversion by fungi. *Biotechnology Advances 27*(2): 185–194.

Sarwan, J., Nair, M.V., Sharma, S. et al. (2024). Omics technology approaches for the generation of biofuels. In: *Biomass energy for sustainable development*, 165–191. CRC Press.

Shankar, A., Jain, K.K., Kuhad, R.C., and Sharma, K.K. (2023). Comparison of lignocellulosic enzymes and CAZymes between ascomycetes (*Trichoderma*) and basidiomycetes (*Ganoderma*) species: A proteomic approach. *Z Naturforsch C J Biosci.* https://doi.org/10.1515/znc-2023-0125.

Shankar, A., Saini, S., and Sharma, K.K. (2024). Fungal-integrated second-generation lignocellulosic biorefinery: Utilization of agricultural biomass for co-production of lignocellulolytic enzymes, mushroom, fungal polysaccharides, and bioethanol. *Biomass Conv. Bioref. 14*: 1117–1131.

Sharma, A. and Aggarwal, N.K. (2020). Strategies for saccharification of lignocellulosic substrate. In: *Water hyacinth: A potential lignocellulosic biomass for bioethanol*. Cham: Springer https://doi.org/10.1007/978-3-030-35632-3_6.

Shukla, A., Kumar, D., Girdhar, M. et al. (2023). Strategies of pretreatment of feedstocks for optimized bioethanol production: Distinct and integrated approaches. *Biotechnology for Biofuels and Bioproducts 16*(1): 44.

Singh, R., Shukla, A., Tiwari, S., and Srivastava, M. (2014). A review on delignification of lignocellulosic biomass for enhancement of ethanol production potential. *Renewable and Sustainable Energy Reviews 32*: 713–728.

Singh, J., Suhag, M., and Dhaka, A. (2015). Augmented digestion of lignocellulose by steam explosion, acid and alkaline pretreatment methods: A review. *Carbohydrate Polymers 117*: 624–631.

Singh, V., Raheja, Y., Basotra, N. et al. (2023). CRISPR/Cas9 mediated gene editing of transcription factor ACE1 for enhanced cellulase production in thermophilic fungus *Rasamsonia emersonii*. *Fungal Biology and Biotechnology 10*(1): 18.

Soltanian, S., Aghbashlo, M., Almasi, F. et al. (2020). A critical review of the effects of pretreatment methods on the exergetic aspects of lignocellulosic biofuels. *Energy Conversion and Management 212*: 112792.

Song, Z., Yang, G., Han, X. et al. (2013). Optimization of the alkaline pretreatment of rice straw for enhanced methane yield. *BioMed Research International 2013* (1): 968692.

Song, B., Li, B., Wang, X. et al. (2018). Real-time imaging reveals that lytic polysaccharide monooxygenase promotes cellulase activity by increasing cellulose accessibility. *Biotechnology for Biofuels 11*: 1–11.

Sorn, V., Chang, K.L., Phitsuwan, P. et al. (2019). Effect of microwave-assisted ionic liquid/acidic ionic liquid pretreatment on the morphology, structure, and enhanced delignification of rice straw. *Bioresource Technology 293*: 121929.

St. John, P.C. and Bomble, Y.J. (2019). Approaches to computational strain design in the multiomics era. *Frontiers in Microbiology 10*: 597.

Strassberger, Z., Tanase, S., and Rothenberg, G. (2014). The pros and cons of lignin valorisation in an integrated biorefinery. *RSC Advances 4*(48): 25310–25318.

Šuchová, K., Fehér, C., Ravn, J.L. et al. (2022). Cellulose-and xylan-degrading yeasts: Enzymes, applications and biotechnological potential. *Biotechnology Advances 59*: 107981.

Sukri, S.S.M. and Mimi Sakinah, A.M. (2018). Production of high commercial value xylooligosaccharides from meranti wood sawdust using immobilised xylanase. *Applied Biochemistry and Biotechnology 184*: 278–290.

Sun, Y. and Cheng, J. (2002). Hydrolysis of lignocellulosic materials for ethanol production: A review. *Bioresource Technology 83*(1): 1–11.

Tirpanalan, Ö., Reisinger, M., Smerilli, M. et al. (2015). Wheat bran biorefinery–an insight into the process chain for the production of lactic acid. *Bioresource Technology 180*: 242–249.

Tiwari, A., Chen, C.W., Haldar, D. et al. (2023). Laccase in biorefinery of lignocellulosic biomass. *Applied Sciences 13*(8): 4673.

Turner, P., Mamo, G., and Karlsson, E.N. (2007). Potential and utilization of thermophiles and thermostable enzymes in biorefining. *Microbial Cell Factories 6*: 1–23.

Varriale, L. and Ulber, R. (2023). Fungal-based biorefinery: From renewable resources to organic acids. *ChemBioEng Reviews 10*(3): 272–292.

Vélez-Mercado, M.I., Talavera-Caro, A.G., Escobedo-Uribe, K.M. et al. (2021). Bioconversion of lignocellulosic biomass into value added products under anaerobic conditions: Insight into proteomic studies. *International Journal of Molecular Sciences 22*(22): 12249.

Viikari, L., Alapuranen, M., Puranen, T. et al. (2007). Thermostable enzymes in lignocellulose hydrolysis. *Biofuels* 121–145.

Wang, G., Huang, D., Li, Y. et al. (2015). A metabolic-based approach to improve xylose utilization for fumaric acid production from acid pretreated wheat bran by Rhizopus oryzae. *Bioresource Technology 180*: 119–127.

Wang, Y., Sun, T., Gao, X. et al. (2016). Biosynthesis of platform chemical 3-hydroxypropionic acid (3-HP) directly from CO2 in cyanobacterium Synechocystis sp. PCC 6803. *Metabolic Engineering 34*: 60–70.

Wang, H., Duan, Y., Zhang, Q., and Yang, B. (2018). Effects of sugars, furans, and their derivatives on hydrodeoxygenation of biorefinery lignin-rich wastes to hydrocarbon. *ChemSusChem 11*(15): 2562–2568. https://doi.org/10.1002/cssc.201801401.

Wang, Y., Zhang, Y., Cui, Q. et al. (2024). Composition of lignocellulose hydrolysate in different biorefinery strategies: Nutrients and inhibitors. *Molecules 29*(10): 2275.

Wei, L., Shen, C., El Hajjami, M. et al. (2019). Knockdown of carbonate anhydrase elevates Nannochloropsis productivity at high CO2 level. *Metabolic Engineering 54*: 96–108.

Wierckx, N., Koopman, F., Ruijssenaars, H.J., and de Winde, J.H. (2011). Microbial degradation of furanic compounds: Biochemistry, genetics, and impact. *Applied Microbiology and Biotechnology 92*: 1095–1105.

Xavier, S. and Lonsane, B.K. (1994). Sugar-cane pressmud as a novel and inexpensive substrate for production of lactic acid in a solid-state fermentation system. *Applied Microbiology and Biotechnology 41*: 291–295.

Xie, C., Gong, W., Yang, Q. et al. (2017). White-rot fungi pretreatment combined with alkaline/oxidative pretreatment to improve enzymatic saccharification of industrial hemp. *Bioresource Technology 243*: 188–195.

Xiu, S. and Shahbazi, A. (2012). Bio-oil production and upgrading research: A review. *Renewable and Sustainable Energy Reviews 16*(7): 4406–4414.

Yaegashi, J., Kirby, J., Ito, M. et al. (2017). *Rhodosporidium toruloides*: A new platform organism for conversion of lignocellulose into terpene biofuels and bioproducts. *Biotechnology for Biofuels 10*: 241. https://doi.org/10.1186/s13068-017-0927-5.

Yáñez, R., Belén Moldes, A., Alonso, J.L., and Parajó, J.C. (2003). Production of D (−)-lactic acid from cellulose by simultaneous saccharification and fermentation using Lactobacillus coryniformis subsp. torquens. *Biotechnology Letters 25*: 1161–1164.

Yang, M.K., Yang, Y.H., Chen, Z. et al. (2014). Proteogenomic analysis and global discovery of posttranslational modifications in prokaryotes. *Proceedings of the National Academy of Sciences 111*(52): E5633–E5642.

Yuan, L., Chen, Z., Zhu, Y. et al. (2012). One step conversion of wheat straw to sugars by simultaneous ball milling, mild acid, and fungus *Penicillium simplicissimum* treatment. *Applied Biochemistry and Biotechnology 167*: 39–51.

Zabed, H., Sahu, J.N., Boyce, A.N., and Faruq, G. (2016). Fuel ethanol production from lignocellulosic biomass: An overview on feedstocks and technological approaches. *Renewable and Sustainable Energy Reviews 66*: 751–774.

Zhang, Z., Donaldson, A.A., and Ma, X. (2012). Advancements and future directions in enzyme technology for biomass conversion. *Biotechnology Advances 30*: 913–919.

Zhang, K., Xu, R., Abomohra, A.E.F. et al. (2019). A sustainable approach for efficient conversion of lignin into biodiesel accompanied by biological pretreatment of corn straw. *Energy Convers Manage 199*: 111928. https://doi.org/10.1016/j.enconman.2019.111928.

Zhang, Y., Li, T., Shen, Y. et al. (2020). Extrusion followed by ultrasound as a chemical-free pretreatment method to enhance enzymatic hydrolysis of rice hull for fermentable sugars production. *Industrial Crops and Products 149*: 112356.

Zhao, X.-Q., Liu, C.-G., and Bai, F.-W. (2024). Making the biochemical conversion of lignocellulose more robust. *Trends in Biotechnology 42*: 418–430.

Zhou, X., Zhou, X., Xu, Y., and Chen, R.R. (2017). Gluconobacter oxydans (ATCC 621H) catalyzed oxidation of furfural for detoxification of furfural and bioproduction of furoic acid. *Journal of Chemical Technology & Biotechnology 92*(6): 1285–1289.

Zhu, D., Adebisi, W.A., Ahmad, F. et al. (2020). Recent development of extremophilic bacteria and their application in biorefinery. *Frontiers in Bioengineering and Biotechnology 8*: 483.

Zuccaro, G., Pirozzi, D., and Yousuf, A. (2020). Lignocellulosic biomass to biodiesel. In: *Lignocellulosic biomass to liquid biofuels*, 127–167. Elsevier.

13

Sustainability Assessment of Genetically Engineered Biocatalysts Producing Biofuels and Biochemicals

Andreza A. Longati[1,2], *Christian de Oliveira Martins*[3], *Gabriel Baioni*[3], *Adilson José da Silva*[1,3], *Thais Suzane Milessi*[1,3], *and Felipe Fernando Furlan*[1,3]

[1] *Department of Chemical Engineering, Federal University of São Carlos, São Carlos, São Paulo, Brazil*
[2] *Institute of Science and Technology, Federal University of Alfenas, Poços de Caldas, Minas Gerais, Brazil*
[3] *Graduate Program of Chemical Engineering, Federal University of São Carlos, São Carlos, São Paulo, Brazil*

13.1 Introduction

The current scenario of global population growth, climate change, and depletion of natural resources has put considerable pressure on the world's finite resources, leading to shortage risks in many sectors. Herein, the global push for sustainability and reduction of carbon emissions is putting pressure on the development of new technologies to diversify the energy and chemical feedstock portfolio (Wang & Azam, 2024). Microorganisms such as bacteria, yeast, and algae can be improved to efficiently biomass and convert it into biofuels and valuable biochemicals through advances in biotechnology tools and techniques based on recombinant DNA technology to create genetically modified organisms (GMOs) (Harish et al., 2015; Wei et al., 2024). The production of bio-based compounds through GMOs offers an innovative approach to address these global challenges, promoting the transition to a renewable economy (Lee et al., 2008).

As the world seeks alternatives to fossil fuels and looks for greener methods for the production of chemicals, GMOs hold several advantages, including greater efficiency, reduced environmental impact, flexibility, and the ability to use nonfood biomass (Kovak et al., 2022). GMOs can be genetically manipulated to have optimized metabolic pathways, allowing them to convert biomass into biofuels or biochemicals more efficiently than their natural version by exploiting their metabolic diversity and flexibility. This can be used to increase the yield of the product, reducing the amount of raw material needed to produce the same amount of biofuel or biochemical (Okoro et al., 2022). For example, GMOs can be engineered to more effectively convert a wider range of feedstocks, including lignocellulosic biomass (nonfood plant material), into biofuels and biochemicals. This increases the process efficiency by fully utilizing the available biomass, decreasing the biofuels environmental footprint (Elias et al., 2021; Longati et al., 2018). The ability to produce biofuels and

Biofoundry Techniques for Biotechnology Applications, First Edition. Edited by Anuj Kumar Chandel.

biochemicals from nonfood biomass sources with lower environmental impact is a major advantage because it does not compete with food production (which addresses concerns about food security and the ethical implications of using food crops for fuel), allows the use of abundant and low-cost biomass, and contributes to the reduction of the environmental burden. Additionally, GMOs can be customized to produce a wide variety of biochemicals that serve as precursors for plastics, pharmaceuticals, and other industrial chemicals, which can replace petroleum-based products, providing renewable alternatives (Kumar et al., 2020; Meyer, 2016). These benefits make GMO technology a promising pathway toward a sustainable bioeconomy that reduces dependence on fossil fuels and contributes to a cleaner and more sustainable future.

Despite the many potential benefits, the use of GMOs is also controversial, with significant criticisms arising regarding their broader sustainability (Akin et al., 2019; Keatley, 2000). Concerns focus mainly around the unknown ecological impacts (if they escape into the environment, they have the potential for gene transfer to nontarget species, leading to a reduction in biodiversity), and the uncertain net energy balance of the entire process (Bawa & Anilakumar, 2013). Additionally, the economic and social dimensions of scaling these technologies raise questions about accessibility, equity, and ethical considerations. This is especially true in developing regions (Garcia-Yi et al., 2014). From an economic perspective, the GMOs' economic feasibility is currently hampered by high research and development (R&D) costs, scale-up challenges, feedstock availability, competition with fossil fuels, infrastructure requirements, and intellectual property issues. Overcoming these economic challenges is decisive for the widespread adoption of GMOs-based biotechnologies in sustainable industries. Thus, while GMOs represent an innovative route to renewable bio-based products, the overall sustainability of the process remains controversial. It is crucial to ensure that their development and use do not compromise the long-term environmental, economic, and social dimensions of sustainability (Gunukula et al., 2017).

Focusing on the parameters that influence the sustainability of their operations, this chapter provides an overview of the sustainability assessment of GMOs producing biofuels and biochemicals. A comprehensive assessment of these processes is essential to assess their broader impacts. By examining environmental, economic, and social dimensions, we can assess whether the use of GMOs in the production of biofuels and biochemicals contributes to the advancement of a greener and more sustainable bioeconomy. This evaluation helps ensure alignment with development goals across these dimensions, showing the bottlenecks that still need to be overcome for these technologies to be broadly implemented.

13.2 The Role of Genetically Modified Organisms in Biorefineries

As previously mentioned, GMOs can be specially engineered to efficiently produce valuable chemicals from renewable resources like lignocellulosic biomass. Their use offers a practical alternative for chemical production as it typically requires mild temperature and pressure conditions (Cho et al., 2022). The expansion of the biorefinery products portfolio is seen as an important strategy to achieve second-generation (2G) industrial plants feasibility by co-producing high value-added products, together with biofuels and energy (Manicardi et al., 2023). In this sense, GMOs offer a promising way to synthesize a variety of compounds in modern biorefineries by the incorporation of synthetic pathways for the production of the desired molecules (Paredes-Barrada et al., 2024; Seo et al., 2013).

To achieve biorefineries feasibility in a circular economy, all biomass components must be utilized: cellulose, hemicellulose, and lignin. Cellulose and hemicellulose are mainly composed of glucose and xylose, respectively, and together account for approximately 70% of biomass (Aguiar et al., 2021). Through a pretreatment step followed by an enzymatic hydrolysis, the fermentable sugars from biomass components are released and can be used as a substrate. The yeast *Saccharomyces cerevisiae* is the most commonly used industrial microorganism, particularly in the ethanol and beverage sectors, due to its efficient fermentation of glucose, allowing for practical applications of cellulose after glucose release. However, it lacks the ability to metabolize xylose (Milessi et al., 2020), which limits its versatility in biomass valorization. Furthermore, native *S. cerevisiae* use is limited to ethanol production, and, in the context of a biorefinery, these biomass sugars could be transformed into other high-value bioproducts, enhancing the economic performance of the process. For example, xylose can be consumed by native microorganisms such as *Scheffersomyces stipitis*, *Spathaspora passalidarum,* and *Kluyveromyces marxianus* (Cadete et al., 2013; Du et al., 2019; Hernández-Pérez et al., 2020). However, these microorganisms generally have low tolerance to ethanol, low substrate consumption rate, high sensitivity to the inhibitors in the biomass hydrolysates, and need a strict control of dissolved oxygen concentrations, which increases operational costs (Saha, 2003). In addition, besides the biodiversity of the naturally C5-consuming microorganism, the variety of products is limited to the natural metabolic routes.

Considering the exposed, strain engineering is an important tool to xylose-efficient consumption and diversify the bioproducts portfolio of biorefineries by integrating newly high value-added products using biomass-derived sugars as a carbon source since more than 23,000 known secondary metabolites have been identified as potential bioproducts to be obtained from metabolic pathways of microorganisms (Salwan & Sharma, 2020). There is a wide variety of GMOs explored in the literature to produce high value-added products from lignocellulosic biomass. Between the most used host microorganisms, *Escherichia coli* and *S. cerevisiae* stand out as the most extensively investigated for biomass valorization due to the large accumulated genomic knowledge (Zhang et al., 2015). Bioethanol is the most widely reported chemical, especially using xylose as a substrate (Demeke et al., 2013; Li et al., 2024; Westman et al., 2014). However, efforts have also been made to produce different biomolecules, such as organic acids, butanol, and 2,3-butanediol, for instance. Huo et al. (2022) developed two industrial strains of *S. cerevisiae* for the synthesis of 2,3-butanediol from glucose, HGS50 and HGS37 strains, achieving more than 120 g/L of 2,3-butanediol (~1.5 g/L/h) in glucose-based media. Kong et al. (2019) developed a strain of *K. marxianus* YKX071 for acid lactic production from corncob and observed 103 g/L of L-lactic acid with 99.5% purity. Rieder et al. (2021) successfully expressed four lytic polysaccharide monooxyhenases (LPMOs) in *Pichia pastoris* platform reaching 42 mg of pure LPMO per liter of cultivation medium. This enzyme is an important catalyst for biomass hydrolysis once it catalyzes the cleavage of the glycoside bounds of polymeric sugars through an oxidative mechanism using oxygen (Wang et al., 2019).

In addition to a wide range of bioproducts to integrate biorefineries portfolio, the improvement of strains' tolerance to inhibitory compounds generated after biomass pretreatment (such as acetic acid, furfural, and hydrohymethylfurfural) is also a focus of GMO development. Meijnen et al. (2016) improved acetic acid tolerance under industrially relevant conditions by inserting a superior HAA1 allele containing a unique single-point mutation into an industrial *S. cerevisiae* strain for 2G bioethanol production.

Finally, considering the high costs associated with the enzyme cocktails involved in biomass degradation, the development of GMOs that produce hydrolytic enzymes has gained attention in the last decade, especially with the advances in strain engineering. GMOs that produce hydrolytic enzymes *in situ* can be applied in the consolidated bioprocessing (CBP) of biomass, in which hydrolysis of cellulose and hemicellulose and the fermentation of released C5 and C6 sugars occur simultaneously in a single reactor (Periyasamy et al., 2023). Taking this into account, the main challenge for the development of GMOs for CBP is the expression and efficient production of all the enzymes necessary for biomass hydrolysis once cellulose degradation is catalyzed by cellulases (endoglucanases, cellobiohydrolases, and β-glucosidases) and hemicellulose degradation is catalyzed by endo-β-1,4-xylanase, β-xylosidase, and accessory enzymes, including α-L-arabinofuranosidase, α-glucuronidase, α-galactosidase, and acetylxylan esterase (Bhardwaj et al., 2019; Ramos et al., 2022). Between the reported GMOs that secrete hydrolytic enzymes, the yeast *S. cerevisiae* AC14 stands out once it is the only strain capable of producing seven different hydrolytic enzymes and consuming C5 and C6 sugars (Claes et al., 2020). Perez et al. (2023) applied the AC14 strain in the CBP and obtained remarkable results: 26.8 g/L of ethanol with 4.46 g/L/h of productivity in synthetic medium containing xylose, glucose, cellobiose, and corncob xylan; and 13.0 g/L of ethanol and 1.86 g/L/h of productivity in hydrothermally pretreated sugarcane bagasse.

13.3 Metabolic Modeling in the Development of Genetically Modified Organisms

A crucial step in enhancing GMOs as biocatalysts for biorefineries involves targeted genetic modifications, such as gene knockdowns, overexpression, or heterologous expression, to optimize specific metabolic pathways. This process is greatly supported by *in silico* metabolic modeling, which enables detailed analysis of how these genetic changes influence metabolic fluxes, addressing key challenges in biocatalyst optimization. Metabolic modeling is a computational method used to simulate metabolic networks within cells, involving biochemical reactions catalyzed by enzymes that convert nutrients into energy and essential biomolecules. Under simplifying assumptions, for example, pseudo steady-state operation, it allows the prediction of cell behavior under various environmental or genetic conditions (Nielsen et al., 2021).

In biofuel and bioproducts production, metabolic modeling is a powerful tool for optimizing GMOs like bacteria and yeasts to enhance renewable fuel production, and alternatives to petroleum-based chemicals, reducing dependence on fossil fuels and promoting sustainability (Patané et al., 2019). By identifying genetic targets, it predicts how a gene or an enzyme manipulation can alter metabolic flows, improving the production efficiency.

13.3.1 Metabolic Modeling

Metabolic modeling begins with the creation of a genome-scale metabolic model (GSMM) by extracting metabolic functions from genomic data, which can be accelerated by using computational tools (Tarzi et al., 2024). Metabolic databases like KEGG and MetaCyc map genes to biochemical reactions, forming the basis for simulations of metabolic pathways (Tibocha-Bonilla et al., 2018). These models can be refined with additional data, such as gene and protein expression (transcriptomics/proteomics) and metabolite concentrations (metabolomics), to improve accuracy. Techniques such as ^{13}C metabolic flux analysis provide accurate mapping of carbon flows, refining real-time flux estimates (Volk et al., 2023). The final model is a stoichiometric matrix that captures metabolic fluxes, showing how metabolites are consumed and produced in the network.

Once the model is complete, the system is assumed to be in a metabolic steady state, forming a linear system of equations representing reaction rates. Since this system is often underdetermined, constraints (based on biological knowledge like flux bounds or metabolite measurements) are applied to narrow the solution space. Despite these, multiple solutions may still exist. To address this, an optimization problem with an objective function, commonly cell growth, is used to find a flux vector that satisfies both the constraints and optimization criteria. This process is known as flux balance analysis (FBA) (Figure 13.1) (Nielsen et al., 2021).

To enhance FBA, kinetic models can be integrated into constraints to link intracellular compounds with environmental conditions, predicting metabolic shifts under varying factors. For example, Mahadevan et al. (2002) used dynamic FBA to construct and solve a nonlinear programming problem to optimize production rates while considering kinetic

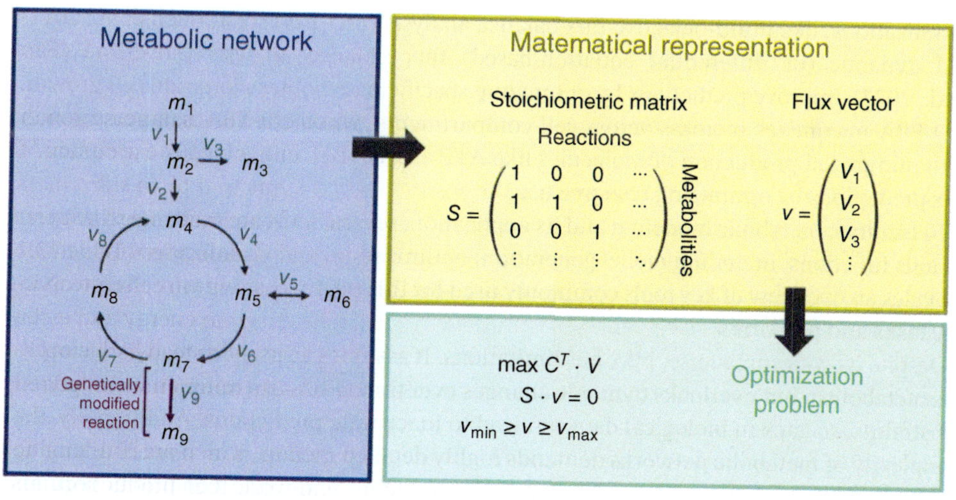

Figure 13.1 Workflow of flux balance analysis.

Table 13.1 Example of Key Tools for Metabolic Modeling and Its Applications

Theme	Tool	Task	Platform	Reference
Genomic databases	NCBI	Genomic data repository	Web-based	NCBI, (2025)
	EMBL	Genomic data repository	Web-based	EMBL, (2025)
Metabolic databases	MetaCyc	Metabolic pathways and models	Web-based	Caspi et al., (2020)
	KEGG	Genomic data annotation	Proprietary software	Kanehisa & Goto, (2000)
Metabolic models database	BIGG	Repository of metabolic models	Web-based	King et al., (2016)
Generate metabolic models	ModelSEED	Automatic generation of genome-scale models (GSMs)	Web-based, Python	Henry et al., (2010)
	CarveMe	Generation of genome-scale models	Python	Machado et al., (2018)
Flux balance analysis and other approaches	COBRA Toolbox	Perform flux balance analysis and related analyses	MATLAB, Python	Heirendt et al., (2019)
	OptFlux	Flux balance analysis-based metabolic engineering	Proprietary software	Rocha et al., (2010)
Visualization of metabolic networks	Escher	Visualize and analyze metabolic maps	JavaScript, Python	King et al., (2015)

constraints. Another method, parsimonious FBA (Lewis et al., 2010), further improves results by maximizing the objective function while minimizing internal fluxes, providing a metabolic solution that closely correlates to the observed data. Additionally, multiobjective optimizations, like multi-objective flux balance analysis (MO-FBA) (Budinich et al., 2017) and dynamic (or differential equation-based) flux balance analysis (DE-FBA) (Fan et al., 2022), improve predictions by optimizing specific metabolites alongside cell growth. MO-FBA maximizes biomass across cell compartments, while DE-FBA balances growth with minimized production of molecules like ATP and NADH, enhancing the accuracy of flux predictions by optimizing resource use.

To facilitate metabolic modeling and its application, several tools are available to support various functions, including model generation, optimization, and visualization. Table 13.1 provides an overview of key tools commonly used for this purpose, along with their respective tasks and platforms.

Despite its widespread use, FBA has limitations. It assumes a steady state, which simplifies metabolism but overlooks dynamic changes over time. FBA also relies on well-defined constraints, so gaps in biological data can lead to inaccurate predictions. Additionally, the complexity of metabolic networks demands highly detailed models, which are challenging to build without extensive data. Accurately reproducing real-world conditions requires deep expertise as small inaccuracies can greatly affect prediction reliability.

13.3.2 Metabolic Modeling for Genetically Modified Organisms

Metabolic models are key tools for predicting the effects of genetic modifications on cell metabolism. By adjusting fluxes or adding exogenous reactions, researchers can simulate genetic modifications like gene knockdowns, overexpression, or heterologous expression, enabling rapid predictions of metabolic outcomes without the need for experiments. This accelerates gene function understanding, guides more efficient metabolic engineering strategies, reduces costs, and broadens potential genetic modifications. Additionally, metabolic models help define and optimize the "optimal phenotype" by integrating stoichiometric ratios, constraints, and algorithms, allowing researchers to identify efficient and biologically feasible pathways for specific outcomes.

For example, Caivano et al. (2023) applied a constraint-based approach for *Clostridium ljungdahlii* to explore how enzymatic constraints, including enzyme turnover rates and molecular masses, could improve the predictive accuracy of growth rates and fermentation profiles. Their model helped identify gene knockouts that increased the production of valuable metabolites like acetate and ethanol under different conditions. This demonstrates how theoretical models can be applied to predict and optimize the effects of genetic perturbations, reducing the need for trial-and-error experimentation.

Qin et al. (2024) used a GSMM to optimize 3-hydroxypropionic acid (3-HP) production using *S. cerevisiae*, focusing on improving CO_2 fixation. The model simulations identified bicarbonate, derived from CO_2, as a limiting factor in achieving high yields. To overcome this, strategies were developed, including the identification of transporters, enzyme domain swapping, and gene deletions to enhance bicarbonate uptake and prevent 3-HP degradation. These metabolic engineering efforts increased 3-HP production, reaching the yield of 0.5625 g/g, which corresponds to 89.3% of the theoretical maximum yield while supporting biomass formation.

One of the specific challenges in applying metabolic models to GMOs is accounting for the complexities introduced by genetic modifications. GMOs often involve intricate interactions between altered genes, regulatory networks, and environmental conditions that are not easily captured by standard models. Modeling these modifications while preserving realistic metabolic behavior demands advanced computational frameworks and thorough biological understanding. The integration of omics data further complicates the process, requiring sophisticated algorithms to accurately reflect the changes brought by genetic alterations and maintain reliable predictions.

13.4 Parameters to Evaluate the Sustainability of Genetically Modified Organisms

In order to make a comprehensive assessment of the sustainability of GMOs and determine the overall benefits and risks of these technologies, it is important to evaluate parameters that span the environmental, economic, and social dimensions (Figure 13.2). This assessment helps ensure that GMOs contribute to environmental and ecological integrity, provide economic benefits without unnecessary costs, and are aligned with societal values of equity, health, and safety. This assessment provides a balanced view of the role of GMOs in

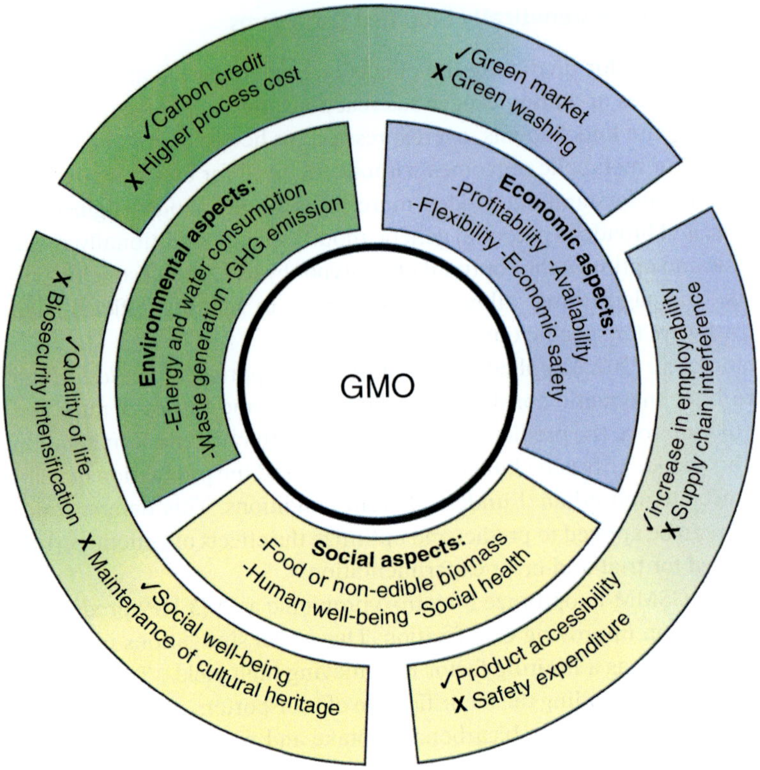

Figure 13.2 Sustainability parameters to assess genetically modified organisms that incorporate environmental, economic, and social dimensions within a broader context of sustainability.

the broader context of sustainable development (Gunukula et al., 2017; Sharma et al., 2022) and provides a comprehensive framework for assessing the potential of processes based on GMO (Sharma et al., 2022).

13.4.1 Environmental Perspective

The environmental perspective of the sustainability assessment of GMOs producing biofuels and biochemicals focuses on understanding the potential benefits, broader ecological impacts, resource efficiency, and potential risks associated with these advanced biotechnologies. Therefore, evaluating the environmental dimension of the sustainability of GMOs involves assessing their long-term environmental impacts considering both positive and negative perspectives (Haris et al., 2023). This happens because while these biocatalysts hold significant promise, their large-scale use also raises a number of environmental challenges that must be thoroughly assessed (Sharma et al., 2022).

A robust assessment of the environmental aspect of the sustainability involves assessing multiple environmental parameters, which can use several methodologies that evaluate the full life cycle impact of GMO-based biofuel and biochemical production across the entire life cycle of these processes, from feedstock sourcing to end-product utilization and disposal (Wei et al., 2024). The goal is to determine how GMOs affect several important parameters such as biodiversity, ecosystem functions, soil health, greenhouse gas (GHG) emissions, energy use, water consumption, and waste generation. Table 13.2 connects the environmental parameters, the methodologies to evaluate these parameters, and relevant references to guide the sustainability assessment of genetically engineered biocatalysts in biofuel and biochemical production. This table highlights the connection between key environmental parameters (such as GHG emissions, energy use, and biodiversity impact) and the methodologies [such as life cycle assessment (LCA), material flow analysis, exergy analysis, and others) that can be applied to measure and evaluate these parameters in the context of GMOs for biofuels and biochemicals.

LCA is one approach that can give a full perspective of the environmental impacts involved in obtaining a product (Table 13.2) (Guinee, 2002). LCA enables the assessment of the environmental impacts associated with the entire production process of the biofuels and biochemicals produced with these biocatalysts, including crucial parameters such as resource consumption, resource efficiency, GHG emissions, waste generation, water consumption, and energy use (Dias et al., 2019; Hauschild & Huijbregts, 2015). Therefore, this methodology provides a holistic approach, ensuring a comprehensive understanding of its effects from "cradle to grave", avoiding problem-shifting, where solutions that reduce impacts at one stage can aggravate the same or a different impact at another stage (Curran & Klöpffer, 2014). An additional advantage of LCA is its quantitative and comparative nature, making it a valuable tool for comparing the environmental performance of different products or processes, an important decision support tool (Guinee, 2002).

The LCA also helps in the identification of the stages of a product's life cycle with the most significant environmental impacts, allowing targeted improvements (Curran & Klöpffer, 2014). However, the effectiveness of this approach depends on data quality (collecting comprehensive and high-quality data can be challenging, time-consuming, and resource-intensive), standardization, uncertainty (the methods used to translate inventory data into environmental impacts may involve scientific uncertainties, especially for long-term impacts like biodiversity or climate change), and its ability to adapt to emerging sustainability challenges (emerging environmental concerns—such as biodiversity loss or water use—standardized methods within LCA are still being developed, which can limit the tool's applicability in these areas) (Björklund, 2002). To overcome these limitations, LCA can be complemented with other methodologies, field studies, ecological assessments, and long-term monitoring programs to provide a more comprehensive evaluation of environmental sustainability (Curran & Klöpffer, 2014).

In the case of sustainability parameters of GMOs for biochemical and biofuel production, key areas of concern include the risks associated with the release or unintended escape of GMOs into natural ecosystems. This fact cannot be overlooked since if GMOs are accidentally released, they may interact with local microbial communities or disrupt ecosystems through horizontal gene transfer, posing potential ecological risks, such as biodiversity impacts (Limayem, 2015; Rutz & Janssen, 2014). Biodiversity assessment often requires more specialized, site-specific studies that examine ecological interactions, species diversity, and long-term effects on ecosystems, which go beyond the standard scope of LCA. To comprehensively evaluate the biodiversity impacts of GMOs, LCA should be combined with ecological risk assessments, field studies, and biodiversity-specific indicators.

Table 13.2 Proposed Parameters and Subparameters for Evaluating the Sustainability of Genetically Modified Organisms (GMOs) According to Environmental Considerations

Environmental Parameter	Methodologies for Evaluation	Description	References
Greenhouse gas (GHG) emissions	Life cycle assessment (LCA)	Assesses total GHG emissions over the entire life cycle, from raw material extraction to disposal, of the biofuel/biochemical produced using GMOs.	Guinee, (2002); ISO 14044, (2006)
	Carbon footprint (CF) analysis	Total amount of carbon emissions caused by an activity directly and indirectly or accumulated of a product over its entire life stages.	ISO 14067: (2018, 2018); Pandey et al., (2011); Wiedmann & Minx, (2008)
Impact on carbon sequestration	LCA	Evaluate the effectiveness of carbon sequestration technologies by including emission sources and carbon sinks across the entire life cycle of a product.	Guinee, (2002); ISO 14044, (2006)
	Ecological footprint analysis	Estimate the resource consumption and waste assimilation of a defined human population or economy in terms of a corresponding productive land area.	Borucke et al., (2013); Global Footprint Network, 2015; Kitzes et al., (2007); Rees et al., (1996)
Carbon sequestration potential	LCA	Estimates the net GHG emissions throughout a product's or system's life cycle.	Guinee, (2002); ISO 14044, (2006); Cherubini et al., (2011); Vogtländer et al., (2014)
	CF	Measures the total GHG emissions, particularly carbon dioxide (CO_2), produced directly or indirectly by human activities.	ISO 14067: (2018, 2018); Pandey et al., (2011); Wiedmann & Minx, (2008)
	Ecological footprint analysis	Quantifies the demand humans place on natural ecosystems by measuring the amount of biologically productive land and water required to produce the resources consumed and to absorb the waste generated, particularly CO_2 emissions.	Borucke et al., (2013); Global Footprint Network, (2015); Kitzes et al., (2007); Rees et al., (1996)
Energy use	LCA	Evaluates energy consumption across the entire life cycle, from raw material extraction to disposal, of the biofuel or bioproduct produced using GMOs.	Guinee, (2002); ISO 14044, (2006)
	Exergy analysis	Focuses on evaluating how efficiently a system uses energy by distinguishing between the usable portion of the energy (exergy) and the nonusable portion (energy or waste energy).	Rosen & Dincer, (2001); Szargut et al., (1987)

Table 13.2 (Continued)

Environmental Parameter	Methodologies for Evaluation	Description	References
Water use and quality	LCA	Assess the environmental impacts related to water throughout the entire life cycle of the biochemicals and biofuels produced from GMOs, including both the quantity of water consumed (water use) and the effects on water quality (water quality).	Guinee, (2002); ISO 14044, (2006)
	Water footprint assessment	Assesses direct and indirect water consumption and location of the water footprint across the life cycle of a product and its effect on water resources and quality. The goal is to analyze how human activities or specific products relate to issues of water scarcity and pollution, and to see how activities and products can become more sustainable from a water perspective.	Hoekstra et al., (2012); ISO 14046, (2014); Mekonnen & Hoekstra, (2010)
Land use and biodiversity	LCA	Evaluates land-use (transformation and occupation) change impacts throughout the life cycle, including on biodiversity. The method considers the impact of land use on ecosystems and biodiversity. It examines how changes in land use (deforestation, agriculture, urbanization) affect habitats, species richness, and ecosystem services.	Guinee, (2002); ISO 14044, (2006)
	Biodiversity impact assessment	Systematic process to evaluate the potential effects of a project, activity, or policy on biological diversity aiming to understand how different species, ecosystems, and genetic resources may be affected, either positively or negatively.	BBOP, (2018); IAIA, (2005)
	Ecological footprint analysis	Estimation of resource consumption and waste assimilation requirements of a defined human population or economy in terms of corresponding productive land use.	Borucke et al., (2013); Global Footprint Network, (2015); Kitzes et al., (2007); Rees et al., (1996)
Resource efficiency	LCA	Measures how resources (e.g., materials, water, energy) are utilized throughout the life cycle.	Guinee, (2002); ISO 14044, (2006)
	Exergy analysis	Refers to how effectively the energy and material inputs are converted into useful outputs (products, services) with minimal waste.	Rosen & Dincer, (2001); Szargut et al., (1987)

(Continued)

Table 13.2 (Continued)

Environmental Parameter	Methodologies for Evaluation	Description	References
Waste generation and by-product management	LCA	Tracks waste generation at each stage of the life cycle, from production waste to end-of-life disposal. It also evaluates the management of by-products, including recycling, reuse, and proper disposal.	Guinee, (2002); ISO 14044, (2006)
	Material flow analysis (MFA)	Analyzes the flow of materials to determine the amount of waste generated during production, consumption, and disposal processes.	Brunner & Rechberger, (2004)
Soil health	LCA	Evaluates the impact of activities on soil quality throughout a product's life cycle.	Guinee, (2002); ISO 14044, (2006)
	MFA	Assess the impact of material flows on soil by tracking inputs like fertilizers, pesticides, or industrial residues that are deposited into the soil.	Brunner & Rechberger, (2004)
Water Pollution	LCA	Measures the potential for water contamination from pollutants, such as chemicals, nutrients (nitrogen and phosphorus), and heavy metals released during the life cycle stages.	Guinee, (2002); ISO 14044, (2006)
	MFA	Tracks the movement of substances that may lead to water pollution, such as nutrient runoff (nitrogen and phosphorus), chemicals, and industrial effluents.	Brunner & Rechberger, (2004)
Resource consumption	LCA	Estimate the total quantity of natural resources used during the life cycle of a product, including raw materials (e.g., minerals, fossil fuels) and renewable resources (e.g., water, biomass).	Guinee, (2002); ISO 14044, (2006)
	MFA	Evaluates the consumption of natural resources (e.g., minerals, water, fossil fuels) throughout a system.	Brunner & Rechberger, (2004)
	Exergy analysis	Quantifies the useful portion of consumed resources and highlights where waste occurs.	Rosen & Dincer, (2001); Szargut et al., (1987)

These approaches complement LCA and provide a fuller picture of how GMOs influence biodiversity (Curran & Klöpffer, 2014; Björklund, 2002; de Baan et al., 2013). Waste generation and the environmental impact of GMO by-products are also important issues that require critical environmental assessment. While GMOs are designed to produce useful biofuels and biochemicals, the potential to produce toxic by-products or pollutants under

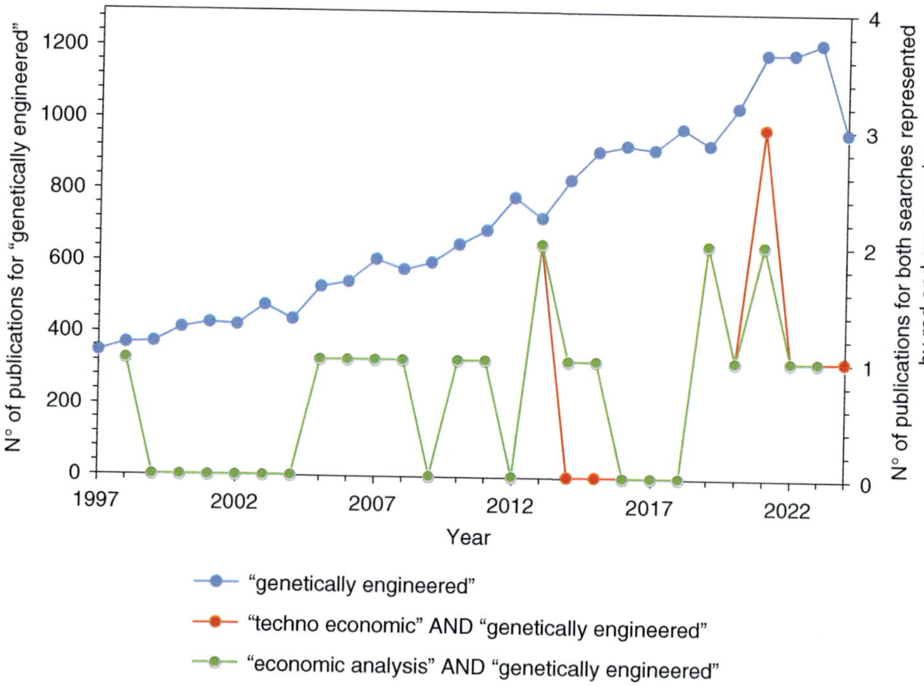

Figure 13.3 Number of publications in the generic engineering area through the period from 1997 to 2024.

Among the studies collected, there is a clear trend in the use of *E. coli* as a platform; 44% of the papers choose to modify this bacterium for the products chosen. In fact, *E. coli* is a popular platform for genetic modification (Yang et al., 2021) since its genome is the most studied one. Therefore, its application in the biochemistry industry is consolidated and the techniques for genetic design of these bacteria are well established. As expected, aerobic conditions are mainly used for *E. coli* cultivation, while anaerobic conditions are used for yeast fermentation, especially for ethanol production.

Beyond the use of genetic engineering in biocatalyst for modification in developed organisms, there are uses that are not involved an independent organism, as the case of the work of Alkanaimsh et al. (2024), where Chinese hamster ovary was used as a cell factory to produce a wealth-related product; in such cases, the cell itself is not an living being, but it is still a genetically engineered catalyst.

In spite of the cultivation conditions, cell separation methods follow a clear tendency for centrifugation and filtration methods, traditional solid–liquid separation methods. Ferreira et al. (2018) argue that the decision between these methods depends on each case; in their work, the separation chosen was the centrifuge. However, they explain that microfiltration can fit better in larger scales. On the other hand, the purification process involves several different unit operations, with distillation and chromatography arising as the main methods. While distillation is one of the most common purifications steps of chemical engineering processes, chromatography is usually regarded as an expensive operation for special

Table 13.4 Set of Case Studies of Genetically Modified Biocatalysts with Economic Analysis Explaining What Is the Catalyst Separation Method, the Main Purification Step, and if the Work Has an Environmental Analysis

Product	Aeration Condition	Separation	Purification	Genetically Modified Organisms	Main Economic Metric	Economic Metric Value	Traditional Value	Is There an Environmental Analysis?	References
1,4-Butanediol	Aerobic	Filtration	Distillation	*Escherichia coli*	Minimum selling price (US$/kg)	1.82	2.50	No	Satam et al., (2019)
β-Glucosidase	Aerobic	Centrifugation	Dead-end filtration/ diafiltration	*E. coli*	Production cost (US$/kg)	37	40–70[a]	No	Ferreira et al., (2018)
Alpha-1 antitrypsin	–	Centrifugation/ filtration	Chromatography/ distillation/ adsorption	Chinese hamster ovary	Cost of goods (US$/g)	378.3	420.5[b]	No	Alkanaimsh et al., (2024)
Ethanol	Anaerobic	Evaporation	Distillation/ adsorption	*Saccharomyces cerevisiae* *Trichoderma reesei*	Minimum selling price (US$/L)	41.22	56.80	No	Dempfle et al., (2021)
Spider silk	Aerobic	Centrifugation	Chromatography	*E. coli*	Minimum selling price (US$/kg)	23.00	NR	Yes	Edlund et al., (2018)
Ethanol	Anaerobic	Centrifugation	Distillation	*S. cerevisiae*	Minimum selling price (US$/L)	0.52	0.52[c]	No	Meyer, (2016)
Acrylic acid	Aerobic	Filtration/ centrifugation	Liquid–liquid extraction	*E. coli*	Minimum selling price (US$/kg)	2.06	1.86	Yes	Brobbey et al., (2023)
Succinic acid	Anaerobic	Centrifugation	Adsorption and evaporation	*S. cerevisiae*	Minimum selling price (US$/kg)	2.29	1.96–2.94	Yes	Baioni e Silva et al., (2024); Furlan et al., (2016)
Adipic acid	Aerobic	Filtration	Crystallization	*Pseudomonas putida*	Multicriteria decision matrix	3.8	2.2	Yes	Unlu et al., (2020)

[a] Other values found in other published studies given by the author.
[b] Base study of the author.
[c] Louw et al. (2023).

high-purity products as, for example, the study of Alkanaimsh et al. (2024), which shows the use of two chromatography steps in the downstream of the conventional alpha-1 antitrypsin process.

As we can see, the main economic metric used in the studies listed in Table 13.4 is the MSP. This metric is especially interesting for innovative processes for which the product market may still not be fully established. Therefore, the metric can be used to compare the biochemical product with others with similar functions in already established markets. This is an easy way to show the profitability, and at some point, the competitiveness of the process. Between the studies listed in Table 13.4, there is no clear trend on the processes' economic feasibility, with many showing MSPs higher than the traditional products. However, it must be stressed that in most cases these processes are still not fully mature and have not been optimized yet. Therefore, there is still room for further improvements. In this sense, many process systems engineering tools can play an important role, such as optimization, process synthesis, and global sensitivity analysis (Meyer, 2016; Dempfle et al., 2021). Another good option is the retro-techno-economic analysis (RTEA) developed by Furlan et al. (2016). This technique shows the main process bottlenecks and the best path to solve them. This is especially interesting when dealing with innovative processes, for which there are still not much data available, and these must be, in the best case, obtained in industrially relevant conditions, which can be obtained through RTEA.

Following a trend similar to the one observed for economic analysis, environmental analysis is still not commonly associated with GMO-based process analysis. Table 13.4 shows that only four of the papers selected united both economic and environmental metrics. Edlund et al. (2018), for example, analyzed the GHG emission of spider silk production process using *E. coli* as the biocatalyst through LCA and found that the product had a GHG emission of 572 kg CO_{2eq}/kg. Nevertheless, since the authors did not compare the biochemical process with the traditional one, it is not possible to know which has the lowest impact. Brobbey et al. (2023) did not focus on a comprehensive environmental analysis of acrylic acid production, doing only a preliminary analysis on GHG emission. Unfortunately, the authors did not compare with other studies or the traditional process. Baioni e Silva et al. (2024), on the other hand, analyzed the succinic acid production using the ReCiPe methodology through the lens of 10 environmental impact metrics involving global warming, toxicity, and biodegradability. The analysis led to the conclusion that the proposed process has the potential to reduce the impacts between 70% and 99%, depending on the metric, compared with the traditional process. Finally, Unlu et al. (2020) continued previous work on adipic acid production using genetically modified *Pseudomonas putida*, integrating performance criteria taking into account social, environmental, and economic analysis. The multicriteria decision matrix considers GHG emissions, toxicity, security, impact on employees and customers in addition to NPV, the payback period, and the rate of return, and it balances all of them to a single value. The authors concluded that the biochemical route had a net score result 73% greater than the conventional process, showing the sustainability potential of such biochemical processes.

13.6 Conclusions

GMOs can have their metabolic pathways optimized to allow conversion of a wider range of feedstocks, including biomass, into biofuels or biochemicals more efficiently than their natural versions. Despite these potential benefits, the use of GMOs has received many criticisms regarding their sustainability and security. Therefore, studies on genetically modified microorganisms would benefit from several tools available to assess both environmental and social dimensions of the process sustainability. In the same sense, the economic perspective is of high importance. Bioprocesses based on GMOs may need more expensive equipment or have additional costs due to regulatory compliances regarding decontamination and safety issues. Therefore, GMO processes will have to show advantages related to higher reaction yields and productivity, simplified downstream stages, or the use of low-cost raw materials in order to attain economic feasibility. Nonetheless, the presence of these advantages in specific processes does not exclude the need for comprehensive economic studies. Although there is an expressive production of knowledge on the application of GMOs for several different processes, most papers do not perform techno-economic and/or environmental analysis. This is a clear bottleneck in the area that must be overcome since it can help guide the experimental effort toward industrially feasible conditions.

Acknowledgments

The authors thank the National Council for Scientific and Technological Development (CNPq) [#152289/2022-4, #310753/2025-2 and #407716/2021-1] and São Paulo Research Foundation (FAPESP) [#2024/02416-4, #2023/09789-8, #2025/12259-6]. This chapter was also supported by the Coordination for the Improvement of Higher Education Personnel—Brazil (CAPES)—Funding Code 001.

References

Abdullah, B., Syed Muhammad, S. A. F., Shokravi, Z., Ismail, S., Kassim, K. A., Mahmood, A. N., & Aziz, M. M. A. (2019). Fourth generation biofuel: A review on risks and mitigation strategies. *Renewable and Sustainable Energy Reviews*, *107*, 37–50.

Aguiar, A., Milessi, T. S., Mulinari, D. R., Lopes, M. S., da Costa, S. M., & Candido, R. G. (2021). Sugarcane straw as a potential second generation feedstock for biorefinery and white biotechnology applications. *Biomass and Bioenergy*, *144*, 105896.

Akin, H., Yeo, S. K., Wirz, C. D., Scheufele, D. A., Brossard, D., Xenos, M. A., & Corley, E. A. (2019). Are attitudes toward labeling nano products linked to attitudes toward GMO? Exploring a potential 'spillover' effect for attitudes toward controversial technologies. *Journal of Responsible Innovation*, *6*(1), 50–74.

Alkanaimsh, S., Alsalal, A. M., & El-Touney, H. (2024). Technoeconomic analysis of intensified PEGylated biopharmaceutical recombinant protein production: Alpha antitrypsin as a model case. *Processes*, *12*(5), 979.

Amyris. Amyris. ©2024. Avaliable from: https://amyris.com/.

Amyris, Inc. AMYRIS, INC. REPORTS FOURTH QUARTER AND FULL YEAR 2022 FINANCIAL RESULTS. 2025. Available from: https://www.prnewswire.com/news-releases/amyris-inc-reports-fourth-quarter-and-full-year-2022-financial-results-301773428.html.

de Baan, L., Alkemade, R., & Koellner, T. (2013). Land use impacts on biodiversity in LCA: A global approach. *International Journal of Life Cycle Assessment*, *18*(6), 1216–1230.

Baioni e Silva, G., Longati, A. A., Sargo, C. R., Furlan, F. F., Capaz, R. S., Lora, E. E. S., & Milessi, T. S. (2024). Life cycle analysis of succinic acid production in the Brazilian biorefinery context. *Sustainability*, *16*(3), 1234.

Bawa, A. S., & Anilakumar, K. R. (2013). Genetically modified foods: Safety, risks and public concerns—A review. *Journal of Food Science and Technology*, *50*(6), 1035–1046.

BBOP. (2018). *Biodiversity offsets*. Forest Trends.

Bell, E. L., Finnigan, W., France, S. P., Green, A. P., Hayes, M. A., Hepworth, L. J., Lovelock, S. L., Niikura, H., Osuna, S., Romero, E., Ryan, K. S., Turner, N. J., & Flitsch, S. L. (2021). Biocatalysis. *Nature Reviews Methods Primers*, *1*(1), 1–21.

Bhardwaj, N., Kumar, B., & Verma, P. (2019). A detailed overview of xylanases: An emerging biomolecule for current and future prospective. *Bioresources and Bioprocessing*, *6*(1), 40.

Björklund, A. E. (2002). Survey of approaches to improve reliability in LCA. *International Journal of Life Cycle Assessment*, *7*(2), 64–72.

Blaine, K., Kamaldeen, S., & Powell, D. (2002). Public perceptions of biotechnology. *Journal of Food Science*, *67*(9), 3200–3208.

Borucke, M., Moore, D., Cranston, G., Gracey, K., Iha, K., Larson, J., Lazarus, E., Morales, J. C., Wackernagel, M., & Galli, A. (2013). Accounting for demand and supply of the biosphere's regenerative capacity: The National Footprint Accounts' underlying methodology and framework. *Ecological Indicators*, *24*, 518–533.

Brobbey, M. S., Louw, J., & Görgens, J. F. (2023). Biobased acrylic acid production in a sugarcane biorefinery: A techno-economic assessment using lactic acid, 3-hydroxypropionic acid and glycerol as intermediates. *Chemical Engineering Research and Design*, *193*, 367–382.

Brunner, P. H., & Rechberger, H. (2004). Practical handbook of material flow analysis. *The International Journal of Life Cycle Assessment*, *9*(5), 337–338.

Budinich, M., Bourdon, J., Larhlimi, A., & Eveillard, D. (2017). A multi-objective constraint-based approach for modeling genome-scale microbial ecosystems. *PLoS One*, *12*(2), e0171744.

Cadete, R. M., Melo, M. A., Zilli, J. E., Vital, M. J. S., Mouro, A., Prompt, A. H., Gomes, F. C. O., Stambuk, B. U., Lachance, M.-A., & Rosa, C. A. (2013). Spathaspora brasiliensis sp. nov., Spathaspora suhii sp. nov., Spathaspora roraimanensis sp. nov. and Spathasporaxylofermentans sp. nov., four novel d-xylose-fermenting yeast species from Brazilian Amazonian forest. *Antonie Van Leeuwenhoek*, *103*(2), 421–431.

Caivano, A., Van Winden, W., Dragone, G., & Mussatto, S. I. (2023). Enzyme-constrained metabolic model and in silico metabolic engineering of Clostridium ljungdahlii for the development of sustainable production processes. *Computational and Structural Biotechnology Journal*, *21*, 4634–4646.

Caspi, R., Billington, R., Keseler, I. M., Kothari, A., Krummenacker, M., Midford, P. E., Ong, W. K., Paley, S., Subhraveti, P., & Karp, P. D. (2020). The MetaCyc database of metabolic pathways and enzymes—A 2019 update. *Nucleic Acids Research*, *48*(D1), D445–D453.

Cherubini, F., Peters, G. P., Berntsen, T., Strømman, A. H., & Hertwich, E. (2011). CO2 emissions from biomass combustion for bioenergy: Atmospheric decay and contribution to global warming. *GCB Bioenergy*, *3*(5), 413–426.

Cho, J. S., Kim, G. B., Eun, H., Moon, C. W., & Lee, S. Y. (2022). Designing microbial cell factories for the production of chemicals. *JACS Au*, *2*(8), 1781–1799.

Claes, A., Deparis, Q., Foulquié-Moreno, M. R., & Thevelein, J. M. (2020). Simultaneous secretion of seven lignocellulolytic enzymes by an industrial second-generation yeast strain enables efficient ethanol production from multiple polymeric substrates. *Metabolic Engineering*, *59*, 131–141.

Cocoon Bioscience. Our Recombinant Protein. ©2025. Available from: https://cocoonbio.com/technology/.

Curran, M. A., & Klöpffer, W. (2014). Strengths and limitations of life cycle assessment. In W. Klöpffer (Ed.), *Background and future prospects in life cycle assessment* (pp. 189–206). Dordrecht: Springer Netherlands.

Damodaran, A. (2010). *Applied corporate finance*. Hoboken, NJ: Wiley.

Demeke, M. M., Dietz, H., Li, Y., Foulquié-Moreno, M. R., Mutturi, S., Deprez, S., Den Abt, T., Bonini, B. M., Liden, G., Dumortier, F., Verplaetse, A., Boles, E., & Thevelein, J. M. (2013). Development of a D-xylose fermenting and inhibitor tolerant industrial *Saccharomyces cerevisiae* strain with high performance in lignocellulose hydrolysates using metabolic and evolutionary engineering. *Biotechnology for Biofuels*, *6*(1), 89.

Dempfle, D., Kröcher, O., & Studer, M. H.-P. (2021). Techno-economic assessment of bioethanol production from lignocellulose by consortium-based consolidated bioprocessing at industrial scale. *New Biotechnology*, *65*, 53–60.

Denby, C. M., Li, R. A., Vu, V. T., Costello, Z., Lin, W., Chan, L. J. G., Williams, J., Donaldson, B., Bamforth, C. W., Petzold, C. J., Scheller, H. V., Martin, H. G., & Keasling, J. D. (2018). Industrial brewing yeast engineered for the production of primary flavor determinants in hopped beer. *Nature Communications*, *9*(1), 965.

Dias, L. C., Freire, F., & Geldermann, J. (2019). Perspectives on multi-criteria decision analysis and life-cycle assessment. In M. Doumpos, J. R. Figueira, S. Greco, & C. Zopounidis (Eds.), *New perspectives in multiple criteria decision making: Innovative applications and case studies* (pp. 315–329). Cham: Springer International Publishing.

Du, C., Li, Y., Zhao, X., Pei, X., Yuan, W., Bai, F., & Jiang, Y. (2019). The production of ethanol from lignocellulosic biomass by Kluyveromyces marxianus CICC 1727-5 and Spathaspora passalidarum ATCC MYA-4345. *Applied Microbiology and Biotechnology*, *103*(6), 2845–2855.

Edlund, A. M., Jones, J., Lewis, R., & Quinn, J. C. (2018). Economic feasibility and environmental impact of synthetic spider silk production from Escherichia coli. *New Biotechnology*, *42*, 12–18.

Elias, A. M., Longati, A. A., de Campos Giordano, R., & Furlan, F. F. (2021). Retro-techno-economic-environmental analysis improves the operation efficiency of 1G-2G bioethanol and bioelectricity facilities. *Applied Energy*, *282*, 116133.

EMBL. (2025). *European Molecular Biology Laboratory*. [online]. European Molecular Biology Laboratory. Available at: https://embl.org/

EU Startups. (2024) 10 fast-growing Spanish startups to watch out for in 2024 and beyond!EU-Startups.

Fan, X., Zhou, J., Xia, J., & Yan, X. (2022). Genome-scale metabolic model's multi-objective solving algorithm based on the inflexion point of Pareto front including maximum energy utilization and its application in *Aspergillus niger* DS03043. *Biotechnology and Bioengineering*, *119*(6), 1539–1555.

Ferreira, R. D. G., Azzoni, A. R., & Freitas, S. (2018). Techno-economic analysis of the industrial production of a low-cost enzyme using E. coli: The case of recombinant β-glucosidase. *Biotechnology for Biofuels*, *11*(1), 81.

Furlan, F. F., Costa, C. B. B., Secchi, A. R., Woodley, J. M., & Giordano, R. C. (2016). Retro-techno-economic analysis: Using (bio)process systems engineering tools to attain process target values. *Industrial and Engineering Chemistry Research*, *55*(37), 9865–9872.

Garcia-Yi, J., Lapikanonth, T., Vionita, H., Vu, H., Yang, S., Zhong, Y., Li, Y., Nagelschneider, V., Schlindwein, B., & Wesseler, J. (2014). What are the socio-economic impacts of genetically modified crops worldwide?A systematic map protocol. *Environmental Evidence*, *3*(1), 24.

Global Footprint Network. (2015) Ecological footprint standards.

Guinee, J. B. (2002). Handbook on life cycle assessment operational guide to the ISO standards. *International Journal of Life Cycle Assessment*, *7*(5), 311–313.

Gunukula, S., Runge, T., & Anex, R. (2017). Assessment of biocatalytic production parameters to determine economic and environmental viability. *ACS Sustainable Chemistry & Engineering*, *5*(9), 8119–8126.

Haris, M., Ajmal, H. M. N., & Bashir, H. (2023). Impacts of genetically modified organisms (GMOs) on environment and agriculture: A comprehensive review. *Trends in Animal and Plant Sciences*, *2*, 9–16.

Harish, B. S., Janaki Ramaiah, M., & Babu Uppuluri, K. (2015). Bioengineering strategies on catalysis for the effective production of renewable and sustainable energy. *Renewable and Sustainable Energy Reviews*, *51*, 533–547.

Hauschild, M. Z., & Huijbregts, M. A. (2015). Introducing life cycle impact assessment. In M. Z. Hauschild & M. A. J. Huijbregts (Eds.), *Life cycle impact assessment* (pp. 1–16). Springer Science+Business Media.

Heirendt, L., Arreckx, S., Pfau, T., Mendoza, S. N., Richelle, A., Heinken, A., Haraldsdóttir, H. S., Wachowiak, J., Keating, S. M., Vlasov, V., Magnusdóttir, S., Ng, C. Y., Preciat, G., Žagare, A., Chan, S. H. J., Aurich, M. K., Clancy, C. M., Modamio, J., Sauls, J. T., Noronha, A., Bordbar, A., Cousins, B., El Assal, D. C., Valcarcel, L. V., Apaolaza, I., Ghaderi, S., Ahookhosh, M., Ben Guebila, M., Kostromins, A., Sompairac, N., Le, H. M., Ma, D., Sun, Y., Wang, L., Yurkovich, J. T., Oliveira, M. A. P., Vuong, P. T., El Assal, L. P., Kuperstein, I., Zinovyev, A., Hinton, H. S., Bryant, W. A., Aragón Artacho, F. J., Planes, F. J., Stalidzans, E., Maass, A., Vempala, S., Hucka, M., Saunders, M. A., Maranas, C. D., Lewis, N. E., Sauter, T., Palsson, B. Ø., Thiele, I., & Fleming, R. M. T. (2019). Creation and analysis of biochemical constraint-based models using the COBRA Toolbox v.3.0. *Nature Protocols*, *14*(3), 639–702.

Henry, C. S., DeJongh, M., Best, A. A., Frybarger, P. M., Linsay, B., & Stevens, R. L. (2010). High-throughput generation, optimization and analysis of genome-scale metabolic models. *Nature Biotechnology*, *28*(9), 977–982.

Hernández-Pérez, A. F., Chaves-Villamil, A. C., de Arruda, P. V., dos Santos, J. C., Felipe, M., & das G. de A. (2020). Sugarcane syrup improves xylitol bioproduction from sugarcane bagasse and straw hemicellulosic hydrolysate. *Waste and Biomass Valorization*, *11*(8), 4215–4224.

Hoekstra, A. Y., Chapagain, A. K., Aldaya, M. M., & Mekonnen, M. M. (2012). *The water footprint assessment manual*. Routledge.

Huo, G., Foulquié-Moreno, M. R., & Thevelein, J. M. (2022). Development of an industrial yeast strain for efficient production of 2,3-butanediol. *Microbial Cell Factories*, *21*(1), 199.

IAIA. (2005) Biodiversity in impact assessment.

ISO 14044. (2006). *Environmental management—Life cycle assessment—Requirements and guidelines*. ISO.

ISO 14046. (2014). *Environmental management—Water footprint—Principles, requirements and guidelines*. ISO.

ISO 14067:2018. (2018). *Greenhouse gases—Carbon footprint of products—Requirements and guidelines for quantification and communication*. ISO.

Kanehisa, M., & Goto, S. (2000). KEGG: Kyoto encyclopedia of genes and genomes. *Nucleic Acids Research*, *28*(1), 27–30.

Keatley, K. L. (2000). Controversy over genetically modified organisms: The governing laws and regulations. *Quality Assurance*, *8*(1), 33–36.

King, Z. A., Dräger, A., Ebrahim, A., Sonnenschein, N., Lewis, N. E., & Palsson, B. O. (2015). Escher: A web application for building, sharing, and embedding data-rich visualizations of biological pathways. *PLoS Computational Biology*, *11*(8), e1004321.

King, Z. A., Lu, J., Dräger, A., Miller, P., Federowicz, S., Lerman, J. A., Ebrahim, A., Palsson, B. O., & Lewis, N. E. (2016). BiGG models: A platform for integrating, standardizing and sharing genome-scale models. *Nucleic Acids Research*, *44*(D1), D515–D522.

Kitzes, J., Peller, A., Goldfinger, S., & Wackernagel, M. (2007). Current methods for calculating national ecological footprint accounts. *Science for Environment & Sustainable Society*, *4*.

Kong, X., Zhang, B., Hua, Y., Zhu, Y., Li, W., Wang, D., & Hong, J. (2019). Efficient l-lactic acid production from corncob residue using metabolically engineered thermo-tolerant yeast. *Bioresource Technology*, *273*, 220–230.

Kovak, E., Blaustein-Rejto, D., & Qaim, M. (2022). Genetically modified crops support climate change mitigation. *Trends in Plant Science*, *27*(7), 627–629.

Kumar, R., Basak, B., & Jeon, B.-H. (2020). Sustainable production and purification of succinic acid: A review of membrane-integrated green approach. *Journal of Cleaner Production*, *277*, 123954.

Lee, S. K., Chou, H., Ham, T. S., Lee, T. S., & Keasling, J. D. (2008). Metabolic engineering of microorganisms for biofuels production: From bugs to synthetic biology to fuels. *Current Opinion in Biotechnology*, *19*(6), 556–563.

Lehmann, A., Zschieschang, E., Traverso, M., Finkbeiner, M., & Schebek, L. (2013). Social aspects for sustainability assessment of technologies—Challenges for social life cycle assessment (SLCA). *International Journal of Life Cycle Assessment*, *18*(8), 1581–1592.

Lewis, N. E., Hixson, K. K., Conrad, T. M., Lerman, J. A., Charusanti, P., Polpitiya, A. D., Adkins, J. N., Schramm, G., Purvine, S. O., Lopez-Ferrer, D., Weitz, K. K., Eils, R., König, R., Smith, R. D., & Palsson, B. Ø. (2010). Omic data from evolved *E. coli* are consistent with computed optimal growth from genome-scale models. *Molecular Systems Biology*, *6*(1), 390.

Li, F., Bai, W., Zhang, Y., Zhang, Z., Zhang, D., Shen, N., Yuan, J., Zhao, G., & Wang, X. (2024). Construction of an economical xylose-utilizing *Saccharomyces cerevisiae* and its ethanol fermentation. *FEMS Yeast Research*, *24*, foae001.

Limayem, A. (2015). Risk assessment on the use of Genetically Modified Organisms (GMOs) for biofuel production. *Agriculture, Food and Analytical Bacteriology*, *5*.

Longati, A. A., Lino, A. R. A., Giordano, R. C., Furlan, F. F., & Cruz, A. J. G. (2018). Defining research & development process targets through retro-techno-economic analysis: The sugarcane biorefinery case. *Bioresource Technology*, *263*, 1–9.

Louw, J., Dogbe, E. S., Yang, B., & Görgens, J. F. (2023). Prioritisation of biomass-derived products for biorefineries based on economic feasibility: A review on the comparability of techno-economic assessment results. *Renewable and Sustainable Energy Reviews*, *188*, 113840.

Machado, D., Andrejev, S., Tramontano, M., & Patil, K. R. (2018). Fast automated reconstruction of genome-scale metabolic models for microbial species and communities. *Nucleic Acids Research*, *46*(15), 7542–7553.

Mahadevan, R., Edwards, J. S., & Doyle, F. J. (2002). Dynamic flux balance analysis of diauxic growth in *Escherichia coli*. *Biophysical Journal*, *83*(3), 1331–1340.

Manicardi, T., Baioni e Silva, G., Longati, A. A., Paiva, T. D., Souza, J. P. M., Pádua, T. F., Furlan, F. F., Giordano, R. L. C., Giordano, R. C., & Milessi, T. S. (2023). Xylooligosaccharides: A bibliometric analysis and current advances of this bioactive food chemical as a potential product in biorefineries' portfolios. *Food*, *12*(16), 3007.

Matches. Reactor cost - autoclave, fermenter, kettle, mixer settler. ©2014. Available from: https://www.matche.com/equipcost/Reactor.html.

Meijnen, J.-P., Randazzo, P., Foulquié-Moreno, M. R., van den Brink, J., Vandecruys, P., Stojiljkovic, M., Dumortier, F., Zalar, P., Boekhout, T., Gunde-Cimerman, N., Kokošar, J., Štajdohar, M., Curk, T., Petrovič, U., & Thevelein, J. M. (2016). Polygenic analysis and targeted improvement of the complex trait of high acetic acid tolerance in the yeast Saccharomyces cerevisiae. *Biotechnology for Biofuels*, *9*, 5.

Mekonnen, M. M., & Hoekstra, A. Y. (2010). A global and high-resolution assessment of the green, blue and grey water footprint of wheat. *Hydrology and Earth System Sciences*, *14*(7), 1259–1276.

Meyer, H.-P. (2016). Biotechnology for the production of chemicals, intermediates, and pharmaceutical ingredients. In R. N. Patel (Ed.), *Green Biocatalysis* (pp. 643–674). John Wiley & Sons, Ltd.

Milessi, T. S., Perez, C. L., Zangirolami, T. C., Corradini, F. A. S., Sandri, J. P., Foulquié-Moreno, M. R., Giordano, R. C., Thevelein, J. M., & Giordano, R. L. C. (2020). Repeated batches as a strategy for high 2G ethanol production from undetoxified hemicellulose hydrolysate using immobilized cells of recombinant Saccharomyces cerevisiae in a fixed-bed reactor. *Biotechnology for Biofuels*, *13*(1), 85.

Mir-Artigues, P., Twyman, R. M., Alvarez, D., Cerda Bennasser, P., Balcells, M., Christou, P., & Capell, T. (2019). A simplified techno-economic model for the molecular pharming of antibodies. *Biotechnology and Bioengineering*, *116*(10), 2526–2539.

NCBI. (2025). *National Center for Biotechnology Information*. [online]. National Library of Medicine. Available at: https://www.ncbi.nlm.nih.gov/

Nielsen, J., Stephanopoulos, G., & Lee, S. Y. (2021). *Metabolic engineering: Concepts and applications*. Wiley.

Nielsen, J., Tillegreen, C. B., & Petranovic, D. (2022). Innovation trends in industrial biotechnology. *Trends in Biotechnology*, *40*(10), 1160–1172.

Okoro, V., Azimov, U., & Munoz, J. (2022). Recent advances in production of bioenergy carrying molecules, microbial fuels, and fuel design—A review. *Fuel*, *316*, 123330.

Pandey, D., Agrawal, M., & Pandey, J. S. (2011). Carbon footprint: Current methods of estimation. *Environmental Monitoring and Assessment*, *178*(1), 135–160.

Paredes-Barrada, M., Kopsiaftis, P., Claassens, N. J., & van Kranenburg, R. (2024). *Parageobacillus thermoglucosidasius* as an emerging thermophilic cell factory. *Metabolic Engineering*, *83*, 39–51.

Patané, A., Jansen, G., Conca, P., Carapezza, G., Costanza, J., & Nicosia, G. (2019). Multi-objective optimization of genome-scale metabolic models: The case of ethanol production. *Annals of Operations Research*, *276*(1–2), 211–227.

Perez, C. L., Milessi, T. S., Sandri, J. P., Ramos, M. D. N., Carvalho, B. T., Claes, A., Demeke, M. M., Thevelein, J. M., & Zangirolami, T. C. (2023). Evaluation of consolidated bioprocessing of sugarcane biomass by a multiple hydrolytic enzyme producer saccharomyces yeast. *Bioenergy Research*, *16*(4), 1973–1989.

Periyasamy, S., Beula Isabel, J., Kavitha, S., Karthik, V., Mohamed, B. A., Gizaw, D. G., Sivashanmugam, P., & Aminabhavi, T. M. (2023). Recent advances in consolidated bioprocessing for conversion of lignocellulosic biomass into bioethanol–A review. *Chemical Engineering Journal*, *453*, 139783.

Qin, N., Li, L., Wan, X., Ji, X., Chen, Y., Li, C., Liu, P., Zhang, Y., Yang, W., Jiang, J., Xia, J., Shi, S., Tan, T., Nielsen, J., Chen, Y., & Liu, Z. (2024). Increased CO2 fixation enables high carbon-yield production of 3-hydroxypropionic acid in yeast. *Nature Communications*, *15*(1), 1591.

Ramos, M. D. N., Milessi, T. S., Candido, R. G., Mendes, A. A., & Aguiar, A. (2022). Enzymatic catalysis as a tool in biofuels production in Brazil: Current status and perspectives. *Energy for Sustainable Development*, *68*, 103–119.

Rees, W. E., Wackernagel, M., & Testemale, P. (1996). *Our ecological footprint: Reducing human impact on the earth*. Gabriola Island, BC: New Catalyst Books.

Ren, J., & Toniolo, S. (2019). *Life cycle sustainability assessment for decision-making*. Elsevier.

Rieder, L., Ebner, K., Glieder, A., & Sørlie, M. (2021). Novel molecular biological tools for the efficient expression of fungal lytic polysaccharide monooxygenases in Pichia pastoris. *Biotechnology for Biofuels*, *14*(1), 122.

Rocha, I., Maia, P., Evangelista, P., Vilaça, P., Soares, S., Pinto, J. P., Nielsen, J., Patil, K. R., Ferreira, E. C., & Rocha, M. (2010). OptFlux: An open-source software platform for in silico metabolic engineering. *BMC Systems Biology*, *4*(1), 45.

Rosen, M. A., & Dincer, I. (2001). Exergy as the confluence of energy, environment and sustainable development. *Exergy, An International Journal*, *1*(1), 3–13.

Rutz, D., & Janssen, R. (2014). *Socio-economic impacts of bioenergy production*. Cham: Springer International Publishing.

Saha, B. C. (2003). Hemicellulose bioconversion. *Journal of Industrial Microbiology and Biotechnology*, *30*(5), 279–291.

Sala, S., Vasta, A., Mancini, L., Dewulf, J., & Rosenbaum, E. (2016). *Social life cycle assessment: State of the art and challenges for supporting product policies*. JRC Publications Repository.

Salwan, R., & Sharma, V. (2020). Molecular and biotechnological aspects of secondary metabolites in actinobacteria. *Microbiological Research*, *231*, 126374.

Satam, C. C., Daub, M., & Realff, M. J. (2019). Techno-economic analysis of 1,4-butanediol production by a single-step bioconversion process. *Biofuels, Bioproducts and Biorefining*, *13*(5), 1261–1273.

Seider, W. D., Lewin, D. R., Seader, J. D., Widagdo, S., Gani, R., & Ng, K. M. (2016). *Product and process design principles: Synthesis, analysis and evaluation*. John Wiley & Sons.

Seo, S. W., Yang, J., Min, B. E., Jang, S., Lim, J. H., Lim, H. G., Kim, S. C., Kim, S. Y., Jeong, J. H., & Jung, G. Y. (2013). Synthetic biology: Tools to design microbes for the production of chemicals and fuels. *Biotechnology Advances*, *31*(6), 811–817.

Sharma, P., Singh, S. P., Iqbal, H. M. N., Parra-Saldivar, R., Varjani, S., & Tong, Y. W. (2022). Genetic modifications associated with sustainability aspects for sustainable developments. *Bioengineered*, *13*(4), 9509–9521.

Szargut, J., Morris, D. R., & Steward, F. R. (1987). *Exergy analysis of thermal, chemical, and metallurgical processes*. Hemisphere Publishing Corporation.

Tarzi, C., Zampieri, G., Sullivan, N., & Angione, C. (2024). Emerging methods for genome-scale metabolic modeling of microbial communities. *Trends in Endocrinology and Metabolism, 35*(6), 533–548.

Tibocha-Bonilla, J. D., Zuñiga, C., Godoy-Silva, R. D., & Zengler, K. (2018). Advances in metabolic modeling of oleaginous microalgae. *Biotechnology for Biofuels, 11*(1), 241.

Turton, R. (2012). *Analysis, synthesis, and design of chemical processes*. Upper Saddle River, NJ: Prentice Hall.

UNEP. (2020) Guidelines for social life cycle assessment of products and organisations 2020—Life cycle initiative.

Unlu, S., Niu, W., & Demirel, Y. A. (2020). Bio-based adipic acid production: Feasibility analysis using a multi-criteria decision matrix. *Biofuels, Bioproducts and Biorefining, 14*(4), 794–807.

Vogtländer, J. G., van der Velden, N. M., & van der Lugt, P. (2014). Carbon sequestration in LCA, a proposal for a new approach based on the global carbon cycle; cases on wood and on bamboo. *International Journal of Life Cycle Assessment, 19*(1), 13–23.

Volk, M. J., Tran, V. G., Tan, S.-I., Mishra, S., Fatma, Z., Boob, A., Li, H., Xue, P., Martin, T. A., & Zhao, H. (2023). Metabolic engineering: Methodologies and applications. *Chemical Reviews, 123*(9), 5521–5570.

Walter, A., & Machado, P. G. (2014). Socio-economic impacts of bioethanol from sugarcane in Brazil. In D. Rutz & R. Janssen (Eds.), *Socio-economic impacts of bioenergy production* (pp. 193–213). Cham: Springer International Publishing.

Wang, J., & Azam, W. (2024). Natural resource scarcity, fossil fuel energy consumption, and total greenhouse gas emissions in top emitting countries. *Geoscience Frontiers, 15*(2), 101757.

Wang, B., Walton, P. H., & Rovira, C. (2019). Molecular mechanisms of oxygen activation and hydrogen peroxide formation in lytic polysaccharide monooxygenases. *ACS Catalysis, 9*(6), 4958–4969.

Wei, W., Grieger, K., Cummings, C. L., Loschin, N., & Kuzma, J. (2024). Identifying sustainability assessment parameters for genetically engineered agrifoods. *Plants, People, Planet, 6*(1), 29–43.

Westman, J. O., Bonander, N., Taherzadeh, M. J., & Franzén, C. J. (2014). Improved sugar co-utilisation by encapsulation of a recombinant *Saccharomyces cerevisiae* strain in alginate-chitosan capsules. *Biotechnology for Biofuels, 7*(1), 102.

Wiedmann, T., & Minx, J. (2008). A definition of carbon footprint. *CC Pertsova, Ecological Economics Research Trends, 2*, 55–65.

Yang, D., Prabowo, C. P. S., Eun, H., Park, S. Y., Cho, I. J., Jiao, S., & Lee, S. Y. (2021). *Escherichia coli* as a platform microbial host for systems metabolic engineering. *Essays in Biochemistry, 65*(2), 225–246.

Zhang, G.-C., Liu, J.-J., Kong, I. I., Kwak, S., & Jin, Y.-S. (2015). Combining C6 and C5 sugar metabolism for enhancing microbial bioconversion. *Current Opinion in Chemical Biology, 29*, 49–57.

14

Lean Manufacturing Toward Minimum Waste Discharge and Potential Gains in the Biorefinery and Biotechnology Industries

Fabricio M. Gomes[1], Messias Borges Silva[1], Giovani Maltempi-Mendes[2], and Anuj Kumar Chandel[2]

[1] *Department of Chemical Engineering, Department of Biotechnology, Engineering School of Lorena, University of São Paulo (EEL-USP), Lorena, São Paulo, Brazil*
[2] *Renewable Carbon and Biology Systems Laboratory-ReCABS, Department of Biotechnology, Engineering School of Lorena, University of São Paulo (EEL-USP), Lorena, São Paulo, Brazil*

14.1 Introduction

Lean manufacturing, a methodology derived from the Toyota Production System (TPS), has gained considerable traction across industries to reduce inefficiencies and improve productivity. The biotechnology industry, marked by complexity, high regulatory standards, and the need to manage biohazardous materials, is particularly suited for adopting lean principles. Minimizing waste discharge, optimizing resource usage, and improving process efficiencies are essential for biotechnology firms to maintain economic competitiveness and align with global sustainability goals.

This chapter examines the implementation of lean manufacturing principles within the biotechnology industry, focusing on minimizing waste discharge. It reviews lean methodologies, provides an in-depth look at specific tools that can be adapted to biotechnology, and explores potential economic, environmental, and operational gains. Case studies illustrate the successful integration of lean practices within biotechnology firms, underscoring the transformative impact of these strategies on overall industry performance.

14.2 The Fundamentals of Lean Manufacturing

Lean manufacturing revolves around eliminating waste, or "muda," which Toyota identified as the primary source of inefficiency in production processes (Liker, 2020). The seven types of waste are overproduction, waiting, transportation, extra processing, inventory, motion, and defects. These inefficiencies can significantly increase production costs, lengthen production cycles, and lose value for both the producer and the customer.

Biofoundry Techniques for Biotechnology Applications, First Edition. Edited by Anuj Kumar Chandel.
 Published 2026 by John Wiley & Sons Ltd.

14.3 The Five Principles of Lean

At its core, lean manufacturing is driven by five fundamental principles (Womack & Jones, 2017):

1) Define value: The first step in lean is identifying value from the customer's perspective. In biotechnology, value may be defined in terms of product quality, safety, and compliance with regulatory standards.
2) Map the value stream: This involves identifying all the steps in the production process that add value to the final product and those that do not. Mapping the value stream in biotechnology can reveal bottlenecks or redundant steps that increase waste and decrease efficiency.
3) Create flow: Once waste has been identified, efforts are made to ensure that value-adding processes flow smoothly and without interruption. In biotechnology, this may mean optimizing equipment maintenance schedules to avoid downtime or automating routine laboratory processes.
4) Establish pull: Rather than pushing products through the production process based on forecasts, lean manufacturing implements a pull system where production is based on actual demand. For biotechnology firms, this may involve using real-time data to align production schedules with market demand or patient needs.
5) Pursue perfection: The final principle of lean is continuous improvement, or "kaizen," where processes are consistently evaluated and refined to reduce waste and improve efficiency. Biotechnology may involve updating standard operating procedures (SOPs) based on new technological developments or regulatory changes.

These principles are not applied in isolation but are interrelated and iterative, providing a framework for ongoing process optimization. The ultimate goal is to create a streamlined, efficient system that delivers high-quality products with minimal waste and at the lowest possible cost.

14.4 Waste Reduction in Biotechnology: Unique Challenges

The biotechnology industry presents unique challenges for the implementation of lean manufacturing. Unlike traditional manufacturing sectors, biotechnology processes often involve sensitive biological materials that require strict environmental and quality controls. The nature of the products—such as vaccines, biologics, and cell therapies—demands a high degree of precision, and the margin for error is small. Any contamination or deviation from regulatory standards can result in significant financial losses and impact patient safety.

14.5 Types of Waste in Biotechnology

The seven types of waste identified in lean manufacturing can be adapted to the biotechnology context as follows:

1) Overproduction: Producing biological products in quantities that exceed current demand can lead to excessive inventory, increasing the risk of expiration and contamination. Overproduction also ties up capital that could be used for research and development.
2) Waiting: Delays between process steps are common in biotechnology due to the time required for quality control, testing, and regulatory approvals. Such delays increase lead times and reduce overall productivity.
3) Transportation: The unnecessary movement of biological materials between production sites or even within a facility can increase the risk of contamination and damage, especially when dealing with temperature-sensitive products.
4) Extra processing: This waste occurs when more work is done on a product than is necessary. In biotechnology, this can result from redundant quality checks or inefficient production methods.
5) Inventory: Holding an excessive stock of raw materials, intermediates, or finished products can be particularly wasteful in biotechnology, where many materials have limited shelf lives and must be stored under specific conditions to maintain their efficacy.
6) Motion: Excessive movement of personnel within a lab or production facility can lead to inefficiencies and increase the risk of contamination in sterile environments.
7) Defects: The occurrence of defects in biotechnology is critical, as any deviation in product quality can lead to the discarding of entire batches, resulting in significant financial and material losses.

14.6 Managing Biohazardous Waste

One of the most significant challenges in biotechnology waste management is handling biohazardous waste, which includes used culture media, contaminated equipment, and biological byproducts that pose a risk to human health and the environment. These materials must be decontaminated and disposed of according to strict regulations, which can be costly and resource-intensive.

Despite these challenges, lean manufacturing offers a pathway for reducing biohazardous waste. For example, single-use technologies (SUTs) such as disposable bioreactors and tubing have become popular in the biotechnology industry due to their ability to reduce the need for cleaning and sterilization between batches, thus minimizing water, chemical, and energy use (Anderson et al., 2021). Using SUTs also reduces the generation of biohazardous waste by eliminating the need to decontaminate reusable equipment.

14.7 Lean Tools for Biotechnology

Several lean tools and techniques can be adapted to the biotechnology context to minimize waste and improve process efficiency. These tools provide practical methods for identifying and eliminating waste in biotechnology processes.

14.7.1 Kaizen

Kaizen, or continuous improvement, is a fundamental lean tool that encourages all employees, from floor workers to top management, to participate in waste reduction efforts. In biotechnology, kaizen events can optimize laboratory workflows, reduce process variability, and improve specialized equipment utilization (Imai, 2019). The collaborative nature of kaizen allows cross-functional teams to identify inefficiencies that may not be visible to individual workers, leading to more holistic process improvements.

14.7.2 Value Stream Mapping (VSM)

Value stream mapping (VSM) is a visual tool for mapping all the steps in a production process, from raw material procurement to product delivery. It helps identify bottlenecks, sources of waste, and opportunities for improvement. In biotechnology, VSM can be particularly useful in tracking the flow of materials through production, including the movement of biological materials and the timing of quality control tests (Rother & Shook, 2020).

An example of VSM can be seen in Figure 14.1.

For instance, VSM could reveal that lengthy quality control procedures are a bottleneck in producing a monoclonal antibody. By analyzing the flow of materials and information, companies can identify opportunities to streamline these procedures, such as automating specific tests or integrating real-time data analytics to speed up decision-making.

14.7.3 5S

The 5S methodology (Sort, Set in Order, Shine, Standardize, and Sustain) is widely used to maintain clean, organized, and efficient workspaces. In biotechnology, 5S principles are especially relevant in laboratory and cleanroom environments, where cleanliness and organization are critical to preventing contamination and ensuring product quality (Hirano, 2018).

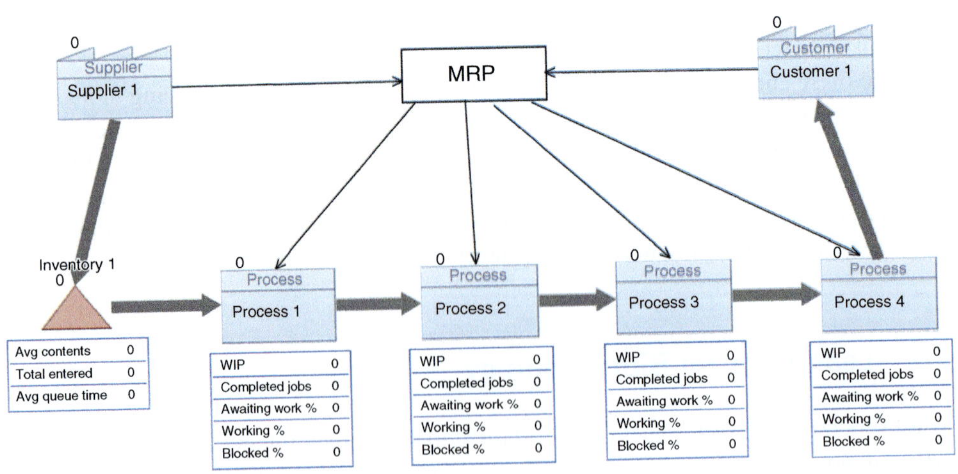

Figure 14.1 Generic representation of a value stream map.

Implementing 5S in biotechnology facilities can lead to immediate productivity improvements by reducing the time employees spend searching for tools and equipment. It can also minimize the risk of contamination by ensuring that all materials and instruments are stored in designated locations and regularly inspected for cleanliness and functionality.

14.7.4 Kanban

Kanban is a lean tool to manage inventory and production flow by visually signaling when new materials or products are needed. In biotechnology, where materials such as cell cultures and reagents have limited shelf life, Kanban can help reduce overproduction and excess inventory by ensuring that materials are only replenished when needed (Gross & McInnis, 2017).

For example, a Kanban system could manage the inventory of critical reagents used in a vaccine production process. By aligning material replenishment with production schedules, companies can reduce the risk of stockouts while avoiding the buildup of expired materials.

14.8 Total Productive Maintenance

Total Productive Maintenance (TPM) focuses on maintaining equipment in optimal condition to prevent breakdowns and minimize downtime. In biotechnology, equipment failures can result in significant losses, as even brief interruptions can lead to the loss of sensitive biological products or the need to revalidate processes. TPM encourages regular maintenance and employee involvement in equipment care, reducing the risk of unexpected failures and extending the lifespan of costly biomanufacturing equipment (Miyake & Enkawa, 2019).

14.9 Lean Manufacturing and Digitalization in Biotechnology

The convergence of lean manufacturing and Industry 4.0 technologies offers new opportunities for waste reduction in biotechnology. Industry 4.0, which includes the use of digital tools such as the Internet of Things (IoT), real-time data analytics, and artificial intelligence (AI), can enhance lean processes by providing more precise control over production parameters and improving decision-making (Rauch et al., 2020).

14.10 Real-Time Data Analytics

Real-time data analytics is a powerful tool for identifying inefficiencies and tracking key performance indicators (KPIs) in biotechnology production. By collecting and analyzing data from equipment, sensors, and quality control systems, biotechnology firms can quickly detect deviations from optimal conditions and take corrective action before defects occur (Sader et al., 2021).

For example, real-time monitoring of bioreactor conditions during fermentation can help identify early signs of contamination or equipment malfunction. By responding to these signals in real-time, companies can prevent the loss of entire batches of biological products and reduce the need for costly rework.

14.11 Digital Twins

Digital twins are virtual models of physical production processes that allow companies to simulate changes and predict their impact before implementing them in the real world. In biotechnology, digital twins can be used to simulate changes to production parameters, such as adjusting the temperature or pH in a bioreactor, and predict how these changes will affect product yield and quality (Rauch et al., 2020).

Digital twins can help biotechnology firms optimize their processes while minimizing the risk of waste. By simulating different scenarios, companies can identify the most efficient production parameters and implement them confidently, reducing the likelihood of defects and improving overall process efficiency.

14.12 Potential Gains from Lean Implementation in Biotechnology

The benefits of lean manufacturing in biotechnology are wide-ranging, encompassing cost savings, improved process efficiency, and enhanced environmental sustainability (Pramanik et al., 2023). Companies that successfully implement lean principles can expect to see significant improvements in both their bottom line and their environmental performance.

14.13 Cost Savings

Reducing waste in biotechnology processes leads to immediate cost savings by lowering the consumption of raw materials, energy, and labor. For example, using SUTs can eliminate the need for costly cleaning and sterilization processes, reducing water and chemical usage (Anderson et al., 2021). Similarly, implementing just-in-time (JIT) production systems can reduce the need for extensive inventories of raw materials and finished products, freeing up capital for other investments (Shah & Ward, 2020).

Lean practices also reduce the occurrence of defects, which can be particularly costly in biotechnology. The loss of a single batch of biologics due to contamination or a quality issue can result in millions of dollars in wasted materials and labor. By standardizing processes and reducing variability, lean manufacturing helps ensure that products are produced consistently and meet stringent regulatory requirements.

14.13.1 Improved Process Efficiency

Lean manufacturing enhances process efficiency by streamlining workflows, reducing lead times, and improving equipment utilization. Biotechnology firms implementing lean principles often see faster production cycles and greater flexibility in meeting market demands (Martin & Osterling, 2021).

For example, continuous processing—a lean technique that replaces traditional batch processing with a constant flow of materials through the production process—can significantly reduce production times and minimize downtime between batches (Lim et al., 2020). Companies can produce biologics more efficiently by adopting continuous processing methods while maintaining consistent product quality.

14.13.2 Environmental Sustainability

In addition to the economic benefits, lean manufacturing contributes to environmental sustainability by reducing energy consumption, water, and raw materials. Biotechnology firms that adopt lean practices often report lower greenhouse gas emissions, reduced wastewater generation, and less hazardous waste production (Khan et al., 2021).

For instance, lean practices emphasizing waste minimization and resource efficiency can help biotechnology companies comply with environmental regulations while reducing their ecological footprint. This is particularly important for firms that handle biohazardous materials, as lowering the volume of waste generated minimizes the need for costly and resource-intensive decontamination and disposal processes.

14.14 Lean Manufacturing's Role in Addressing Sustainability Goals

Sustainability has become a critical concern in the biotechnology industry, driven by regulatory pressures and societal expectations. Lean manufacturing plays a significant role in helping biotechnology companies meet sustainability goals by reducing waste, conserving resources, and minimizing environmental impact. Lean principles align closely with the objectives of the circular economy, which seeks to eliminate waste and maximize resource efficiency by closing the loop on material use (Kirchherr et al., 2018).

One area where biotechnology firms can see significant environmental benefits from lean practices is reducing water and energy consumption. Biopharmaceutical processes, especially those involving cell cultures, often require large amounts of water for cleaning and maintaining sterile conditions. By transitioning to single-use systems, companies can significantly reduce their water consumption and the energy needed for heating, cleaning, and sterilization (Pollock et al., 2021).

Additionally, lean energy management practices focus on optimizing energy use across facilities. Biotechnology companies can implement energy-efficient equipment, monitor energy use in real time, and schedule production to coincide with periods of lower energy demand, thereby reducing their carbon footprint. For example, Amgen, a leading biotechnology firm, implemented lean practices at its Puerto Rico facility, resulting in a 35%

reduction in energy use and a significant reduction in greenhouse gas emissions (Amgen, 2022).

14.15 Regulatory Compliance and Lean in Biotechnology

Biotechnology is a heavily regulated industry with strict requirements for Good Manufacturing Practices (GMP), quality control, and environmental impact assessments. Lean manufacturing supports regulatory compliance by streamlining production processes, reducing the risk of deviations, and ensuring that products are manufactured consistently and safely. Regulatory agencies such as the U.S. Food and Drug Administration (FDA) and the European Medicines Agency (EMA) emphasize the importance of process control, quality by design (QbD), and continuous improvement—all of which are aligned with lean principles (Mahajan et al., 2020).

Implementing Standard Work and Total Quality Management (TQM) within lean systems ensures that processes are efficient and compliant with regulatory requirements. Standard work involves documenting the best practices for each step of the production process and ensuring that all employees follow these procedures. In biotechnology, this can be critical for maintaining the integrity of sensitive biological materials and ensuring that products meet stringent quality standards (Liker & Convis, 2021).

Lean's focus on minimizing defects and rework also helps biotechnology firms reduce the risk of regulatory non-compliance. For example, companies can produce more consistent products by eliminating variability in production processes and reducing the likelihood of batch rejections due to quality control failures. Furthermore, lean tools such as Failure Mode and Effects Analysis (FMEA) can identify potential risks in the production process and implement preventive measures, ensuring that any issues are addressed before they result in non-compliance (Lee et al., 2019).

14.16 Commercial Aspects of Lean Implementation in Biorefineries

Biorefining industries have the potential to support the production of a wide range of valuable bio-based chemicals. Namely, the top chemicals derived from biorefinery carbohydrates were initially reported by the U.S. Department of Energy in 2004 and later revised by Bozell and Petersen (2010). This updated list includes ethanol, furans, glycerol and derivatives, biohydrocarbons, lactic acid, succinic acid, hydroxypropionic acid/aldehyde, levulinic acid, sorbitol, and xylitol (Bozell & Petersen, 2010).

However, the current scenario on the commercialization aspects of biorefineries that foment a circular economy is ruled with challenges, imposing substantial limitations on an otherwise promising foundation for the valorization of bio-based products. In this context, the employment of lean manufacturing principles is presented as a possible strategy to overcome hindrances for establishing commercially sustainable biorefineries (Chandel et al., 2018, 2021).

For instance, the monitoring of data in these units (e.g., input inflow, process efficiency, equipment malfunctioning indicators and tracking of process parameters) is proven to be a cost-heavy and difficult task, suppressing potential scale-ups and optimization efforts. Implementation of data-mapping techniques based on digitalization, such as digital twins, are being developed and proposed for biorefinery designs, with the aim of real-time monitoring, streamlining, and optimizing processes within the biotechnology industry (Gamero et al., 2024).

Real-time equipment-monitoring companies such as Tractian, Vaisala, and ELPRO can offer IoT-based parameter tracking for a wide range of equipment utilized in biochemical processes. The collected data can be of use for AI-driven predictive maintenance, potentially lowering operating costs, reducing downtimes, and enhancing performance and efficiency of the operating equipment. A list of companies that can provide such services is listed in Table 14.1.

Moreover, the integration of other lean manufacturing principles (e.g., Kaizen, Kanban, and Muda) can further aid the categorization, simplification, prioritization, and interpretation of the recovered data (Gamero et al., 2024). The streamlining of data to be analyzed by these units leads to significantly lower management costs and creates opportunities for the development of financially feasible biorefineries.

Similarly, implementing lean principles aimed at increasing resource efficiency, such as value stream mapping (VSM), can highlight current and potential waste sources in biorefineries, while elucidating alternative routes for the recycling of valuable materials and reducing manufacturing costs. Furthermore, adopting and managing accordingly with VSM could bring value to otherwise overlooked byproducts generated in situ, further expanding the product range of an intelligent biorefinery (Schoeman et al., 2021).

In the present scenario, few studies have been conducted in operational biorefineries that assess the economic benefits of individual lean principles being applied in their manufacturing frameworks. However, the numerous advantages of implementing lean manufacturing within adjacent biotechnology industries (e.g., Pfizer and Novartis) have been highlighted in recent reports, evidencing meaningful decreases in energy and water usage, in addition to reduced operational costs.

14.17 Case Study: Lean Implementation at Pfizer

A leading example of lean manufacturing in biotechnology is Pfizer's implementation of lean principles at its biologics production facility in Andover, Massachusetts. The company integrated lean with digital technologies to optimize production processes, reduce waste, and improve efficiency.

Using continuous processing and real-time data monitoring, Pfizer significantly improved resource efficiency. The facility reduced water usage by 30% and energy consumption by 20% while improving production efficiency by 40% (Pfizer, 2022). These gains reduced operational costs and helped Pfizer meet its environmental sustainability goals.

The case of Pfizer demonstrates how lean manufacturing, when combined with digital tools, can drive significant improvements in operational performance and environmental sustainability.

Table 14.1 Companies that Provide Real-Time Data Monitoring Services

Company	Real-Time Monitoring Services Offered	Measured Parameters	Potential Economic Benefits	Reference
Tractian	AI-assisted condition monitoring of equipment Detection and diagnosis of failures Machine health indicators	Temperature (−40 °C to 120 °C) Frequency Acceleration Velocity (RMS) Sample configuration	Cut off operational and maintenance costs Reduction in downtime Revealing of new optimization routes	Tractian: https://tractian.com
ELPRO	Cold chain monitoring Monitoring of rooms and equipment Temperature monitoring in diverse settings	Temperature (−200 °C to 400 °C) Humidity CO_2 Pressure Position and tilt Number of particles Light	Assurance of the quality of products Preservation of materials sensitive to the effects of temperature Revealing of new optimization routes for bioreactors and incubators	ELPRO: https://www.elpro.com
XiltriX	Real-time monitoring of physical parameters with a built-in adjustable alarm system Sensors capable of measuring parameters several times per minute	Temperature (−200 °C to 250 °C) Humidity CO_2 O_2 Pressure Number of particles Airflow Conductivity Volatile organic carbon Shaking H_2O_2 Light	Less disruptive maintenance Alternatives for quality control tools	XiltriX: https://xiltrix.com

Vaisala	Indoor and outdoor monitoring systems for industrial, commercial, and life science environments Wide range of data loggers for diverse settings	Temperature (−40 °C to 60 °C) Humidity CO_2 Pressure Dew point Moisture Dissolved gas in oil Hydrogen Liquid concentration Methane Solar irradiation Vaporized H_2O_2	Potential improvement on the quality of biogas (methane monitoring) Effective monitoring of ecological bio-decontamination (H_2O_2 monitoring)	Vaisala: https://www.vaisala.com
Rees Scientific	Real-time monitoring of environments such as storages, gas tanks, incubators, and clean rooms	Temperature (−196 °C to 1300 °C) Humidity CO_2 O_2 Pressure Light	Meet compliance standards Assurance of the quality of products	Rees Scientific: https://reesscientific.com

14.18 Case Study: Novartis and Lean Implementation in Biopharma

Novartis, a global leader in pharmaceuticals and biotechnology, has successfully implemented lean manufacturing principles to optimize its biopharmaceutical production processes. The company focused on improving production efficiency, reducing waste, and enhancing flexibility in response to market demands.

Through adopting lean practices, Novartis implemented SUTs across its biologics manufacturing facilities, reducing the need for water-intensive cleaning procedures and improving production turnaround times. This lean initiative also included real-time data monitoring systems to track key performance indicators and identify areas for continuous improvement. As a result, the company reduced production cycle times by 25%, decreased operational costs, and improved its environmental footprint by cutting water usage by 40% (Novartis, 2021).

Moreover, Novartis implemented visual management systems to ensure that operators had real-time visibility into production metrics, enabling them to respond quickly to deviations or inefficiencies. This approach empowered employees to take ownership of process improvements and fostered a culture of continuous improvement, aligning with Lean's focus on kaizen (Novartis, 2021).

14.19 Conclusion

Lean manufacturing offers biotechnology firms a robust framework for minimizing waste, improving process efficiency, and enhancing environmental sustainability. As the biotechnology industry continues to evolve, lean principles will play an increasingly important role in helping companies meet the demands of a rapidly changing market while maintaining high standards of product quality and regulatory compliance.

Integrating lean manufacturing with digital technologies, such as real-time data analytics and digital twins, further enhances the potential for waste reduction and process optimization. By leveraging these tools, biotechnology firms can stay competitive, reduce their environmental impact, and drive continuous improvement in their operations.

Acknowledgments

A.K. Chandel gratefully acknowledges the Brazilian National Council for Scientific and Technological Development (CNPq), Brazil for scientific productivity program (Process number: 304451/2025-8).

References

Amgen. (2022). Sustainability report: Lean manufacturing at Amgen's Puerto Rico facility. Retrieved from https://www.amgen.com

Anderson, A., Hofmann, J., & Kanani, D. (2021). The role of single-use technologies in biopharmaceutical manufacturing. *Biotechnology Progress*, *37*(5), e3134.

Bozell, J. J., & Petersen, G. R. (2010). Technology development for the production of biobased products from biorefinery carbohydrates—The US Department of Energy's "Top 10" revisited. *Green Chemistry*, *12*, 539–554.

Chandel, A. K., Garlapati, V. K., Singh, A. K., Antunes, F. A. F., & da Silva, S. S. (2018). The path forward for lignocellulose biorefineries: Bottlenecks, solutions, and perspective on commercialization. *Bioresource Technology*, *264*, 370–381.

Chandel, A. K., Forte, M. B. S., Gonçalves, I. S., Milessi, T. S., Arruda, P. V., Carvalho, W., & Mussatto, S. I. (2021). Brazilian biorefineries from second generation biomass: Critical insights from industry and future perspectives. *Biofuels, Bioproducts and Biorefining*. https://doi.org/10.1002/bbb.2234

ELPRO-BUCHS AG. [Internet]. (Cited 2025, May 17). Retrieved from https://www.elpro.com.

Gamero, E., Shoshi, A., Full, J., Sauer, A., & Miehe, R. (2024). Data management in biorefineries: Conceptual thoughts on lean digital twinning. *Procedia CIRP*, *125*, 48–53.

Gross, J., & McInnis, K. (2017). *Kanban made simple: Demystifying and applying Toyota's legendary manufacturing process*. AMACOM.

Hirano, H. (2018). *5S for operators: 5 pillars of the visual workplace*. Productivity Press.

Imai, M. (2019). *Kaizen: The key to Japan's competitive success*. McGraw-Hill.

Khan, S. A., Yang, W., & Yousaf, M. (2021). Sustainability and lean manufacturing: An overview in the biotechnology industry. *Journal of Cleaner Production*, *298*, 126805.

Kirchherr, J., Reike, D., & Hekkert, M. (2018). Conceptualizing the circular economy: An analysis of 114 definitions. *Resources, Conservation and Recycling*, *127*, 221–232.

Lee, J., Bagheri, B., & Kao, H. A. (2019). A cyber-physical systems architecture for Industry 4.0-based manufacturing systems. *Manufacturing Letters*, *3*(5), 18–23.

Liker, J. K. (2020). *The Toyota way: 14 management principles from the world's greatest manufacturer*. McGraw-Hill.

Liker, J. K., & Convis, G. L. (2021). *The Toyota way to lean leadership: Achieving and sustaining excellence through leadership development*. McGraw-Hill Education.

Lim, A. C., Wong, P. P., & Tan, Z. H. (2020). Advances in continuous bioprocessing for the production of biopharmaceuticals. *Current Opinion in Biotechnology*, *65*, 178–188.

Mahajan, S. S., Pillai, S., & Krishnan, A. (2020). Regulatory frameworks and lean compliance strategies in biopharmaceutical production. *Journal of Regulatory Science*, *8*(1), 15–25.

Martin, K., & Osterling, M. (2021). *Value stream mapping: How to visualize work and align leadership for organizational transformation*. McGraw-Hill Education.

Miyake, D., & Enkawa, T. (2019). Total productive maintenance: A critical review. *Production and Operations Management*, *28*(12), 3152–3168.

Novartis. (2021). Lean manufacturing and operational excellence at Novartis. Retrieved from https://www.novartis.com

Pfizer. (2022). Sustainability report: Continuous biomanufacturing at Andover facility. Retrieved from https://www.pfizer.com.

Pollock, J., Farid, S. S., & Ho, S. V. (2021). The role of single-use technologies in biomanufacturing: Economies of scale, innovation, and sustainability. *Biotechnology Journal*, *16*(4), 3147.

Pramanik, A., Sinha, A., Chaubey, K. K., Hariharan, S., Dayal, D., Bachheti, R. K., Bachheti, A., & Chandel, A. K. (2023). Second-generation bio-fuels: Strategies for employing degraded land for climate change mitigation meeting United Nation-sustainable development goals. *Sustainability*, *15*, 7578. https://doi.org/10.3390/su15097578

Rauch, E., Dallasega, P., & Matt, D. T. (2020). The role of digitalization in lean manufacturing: Lean 4.0. *Procedia CIRP*, *88*, 49–54.

Rees Scientific. [Internet]. (Cited 2025, May 17). Retrieved from https://reesscientific.com.

Rother, M., & Shook, J. (2020). *Learning to see: Value stream mapping to create value and eliminate waste*. Lean Enterprise Institute.

Sader, C., Edens, J., & McKay, L. (2021). Real-time data analytics for lean biomanufacturing. *Biotechnology Advances*, *49*, 107–118.

Schoeman, Y., Oberholster, P., & Somerset, V. (2021). Value stream mapping as a supporting management tool to identify the flow of industrial waste: A case study. *Sustainability*, *13*(1), 91.

Shah, R., & Ward, P. T. (2020). Defining and developing measures of lean production in biopharmaceuticals. *Journal of Operations Management*, *58*, 69–85.

Tractian Technologies Inc. [Internet]. (Cited 2025, May 16). Retrieved from https://tractian.com.

Vaisala. [Internet]. (Cited 2025, May 18). Retrieved from https://www.vaisala.com.

Womack, J. P., & Jones, D. T. (2017). *Lean thinking: Banish waste and create wealth in your corporation*. Simon & Schuster.

XiltriX International. [Internet]. (Cited 2025, May 17). Retrieved from https://xiltrix.com.

Index

Biofoundry Techniques for Biotechnology Applications, First Edition. Edited by Anuj Kumar Chandel.

Printed and bound by CPI Group (UK) Ltd, Croydon, CR0 4YY

17/09/2026

14964297-0006